ANALYTICAL GEOMETRY
of
TWO AND THREE DIMENSIONS

For the students of Mathematics Honours of all Indian universities

ANALYTICAL GEOMETRY *of* TWO AND THREE DIMENSIONS

[with Objective Questions]

Dr A N Das MSc (Gold Medallist), PhD

Reader in Mathematics, Alipurduar College
Ex-Lecturer in Mathematics, St Joseph's College, Darjeeling

New Central Book Agency (P) Ltd

LONDON

HYDERABAD ERNAKULAM BHUBANESWAR

NEW DELHI KOLKATA PUNE GUWAHATI

NCBA

RECD OFFICE
8/1 Chintamoni Das Lane, Kolkata 700 009, India
email: ncbapvtltd@eth.net

OVERSEAS
NCBA (UK) Ltd, 149 Park Avenue North, Northampton, NN3 2HY, UK
email: ncbauk@yahoo.com

EXPORT
NCBA Exports Pvt. Ltd, 212 Shahpur Jat, New Delhi 110 049
email: ncbaexports@ncbapvtltd.com

BRANCHES

2/27 Ansari Road, 1st Floor
Daryaganj, **New Delhi** 110 002
email: ncbadel@ncbapvtltd.com

Shop Nos. 3 & 4, Vinayak Towers
681/B Budhwar Peth, Appa Balwant Chowk
Pune 411 002
email: ncbapune@ncbapvtltd.com

Shop No. 15, Bharati Towers, Ground Floor
Forest Park
Bhubaneswar 751 001
email: ncbabub@ncbapvtltd.com

House No. 3-1-315, 1st Floor
Nimboliadda, Kachiguda, **Hyderabad** 500 027
email: ncbahydb@ncbapvtltd.com

GSS Shopping Complex, 1st Floor
Opposite College Post Office, Convent Road
Ernakulam 682 035
0484-4021054/53

Radhamayee Complex (Second Floor)
Opp. SBI, Panbazar, Jaswanta Road
Guwahati 781 001
email: ncbaguwahati@ncbapvtltd.com

ANALYTICAL GEOMETRY OF TWO AND THREE DIMENSIONS • A N Das

First Published: July 2009
Reprinted: 2013, January 2016

PUBLISHER AND TYPESETTER
New Central Book Agency (P) Ltd
8/1 Chintamoni Das Lane, Kolkata 700 009

PRINTER
New Central Book Agency (P) Ltd
Web-Offset Division, Dhulagarh, Sankrail, Howrah

COVER DESIGNER
Soumen Paul

COVER PRINTER
Biswanath Studio, Kolkata

PROOF CHECKING
Pradip Kumar Chowdhury, Koushik Samanta

PROJECT TEAM
Paritosh Majumder, Swapan Sarkar, Sutapa Kar, and Prabhat Jas

ISBN 978 81 7381 541 6

Price ₹ 350.00

Contents

ANALYTICAL GEOMETRY OF THREE DIMENSIONS

Preface

The main purpose of writing this book is **to provide all the basic topics of analytical coordinate geometry of two and three dimensions to the students of all Indian universities. The syllabi of the Honours courses of mathematics of all Indian universities and different technical courses have also been covered in this book.**

To make it self-explanatory, several problems of coordinate geometry have been solved with diagrams. A large number of problems—taken from different university question papers—have been added at the end of each chapter for practising. Many different and difficult concepts have been explained with diagrams. A few short questions on coordinate geometry have been added at the end of the book.

In spite of my best effort, errors may remain in the book. Any constructive criticism indicating errors, mistakes and suggestions for improvement of the book from helpful teachers and inquisitive students will be gracefully accepted.

I am thankful to my umpteen students who requested me a lot to write this book. Finally, I express my thanks to the publisher for his keen interest in publishing the book in a short span of time.

Dr A N Das

Alipurduar
11.02.2009

ANALYTICAL GEOMETRY
OF
TWO DIMENSIONS

Chapter 1

Transformation of Coordinates and Invariants

1.1 Transformation of Coordinates

A point in a two-dimensional coordinate system is determined by a pair of numbers which depends on the position of axes. If the coordinate axes are changed, by shifting the origin or by rotating the axes or by both, then the coordinates of the point, i.e. the equation of a locus will be changed. These processes are known as *transformation of coordinates*.

1.2 Change of Origin

If the origin $(0, 0)$ of the x-y coordinate system is shifted to the point $A(h, k)$ which is the origin of x'-y' coordinate system such that Ax' and Ay' are parallel to the lines Ox and Oy respectively and if the coordinates of a point P in old and new, i.e. transformed system be (x, y) and (x', y') respectively, then we have

$$x = OM = OB + BM = OB + AN = h + x'$$

and $$y = PM = PN + MN = AB + PN = k + y'$$

i.e., $$x' = x - h \quad \text{and} \quad y' = y - k \tag{1}$$

which are known as *transformation formulae* for translation of axes.

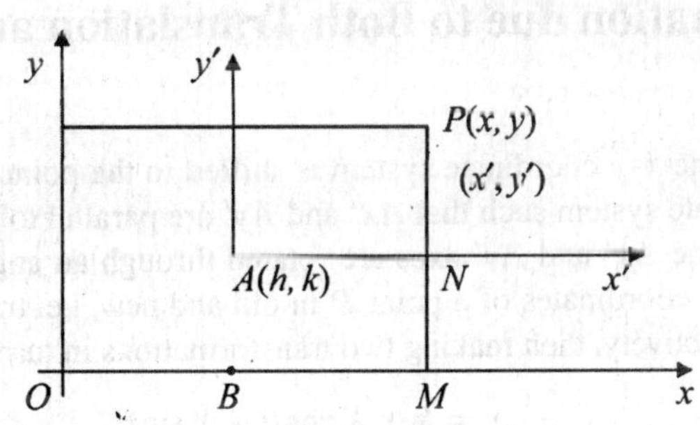

1.3 Rotation of Axes

If the Ox and Oy axes are rotated through an angle θ to the positions Ox' and Oy', and if the coordinates of a point P in old and new, i.e. transformed system be (x, y) and (x', y') respectively, then we have

$$\begin{aligned} x &= OA = OB - AB = OR\cos\theta - QR \\ &= OR\cos\theta - PR\sin\theta \\ &= x'\cos\theta - y'\sin\theta \\ \text{and} \quad y &= PA = PQ + AQ = PQ + BR \\ &= OR\sin\theta + PR\cos\theta \\ &= x'\sin\theta + y'\cos\theta \end{aligned}$$

which are known as *transformation formulae* for rotation of axes.

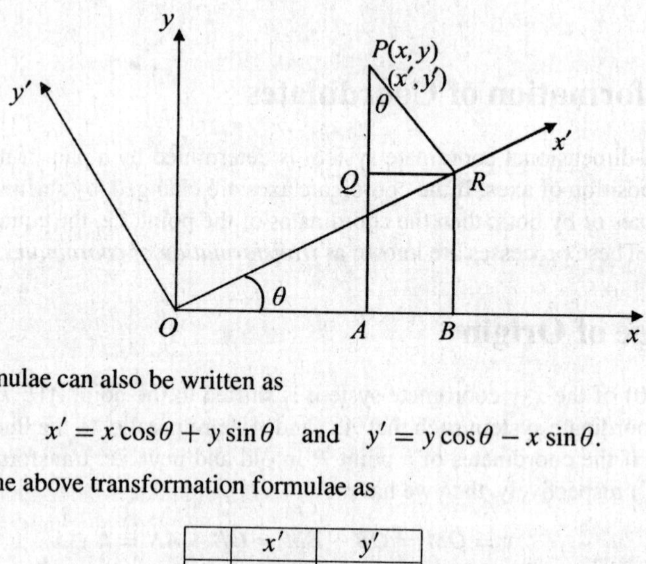

The above formulae can also be written as

$$x' = x\cos\theta + y\sin\theta \quad \text{and} \quad y' = y\cos\theta - x\sin\theta. \tag{2}$$

We can write the above transformation formulae as

	x'	y'
x	$\cos\theta$	$-\sin\theta$
y	$\sin\theta$	$\cos\theta$

1.4 Transformation due to Both Translation and Rotation of Axes

If the origin $(0, 0)$ of the x-y coordinate system is shifted to the point $A(h, k)$ which is the origin of x'-y' coordinate system such that Ax' and Ay' are parallel to the lines Ox and Oy respectively and then the Ax' and Ay' axes are rotated through an angle θ to the positions AX and AY, and if the coordinates of a point P in old and new, i.e. transformed system be (x, y) and (X, Y) respectively, then making two transformations in turn we obtain

$$\begin{aligned} x &= h + X\cos\theta - Y\sin\theta \\ \text{and} \quad y &= k + X\sin\theta + Y\cos\theta. \end{aligned} \tag{3}$$

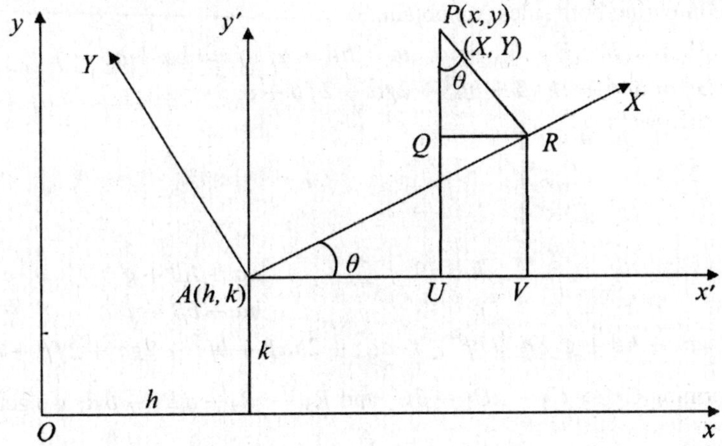

The transformation formulae given by Eqn. (3) can also be written as

$$x = h + X\alpha - Y\beta$$

and $$y = k + X\beta + Y\alpha \tag{4}$$

where $\alpha^2 + \beta^2 = 1$.

1.5 Invariants under Orthogonal Transformation

The transformations of coordinate axes so far discussed are called *orthogonal transformation* when systems of axes are rectangular. Some expressions, especially some relations connecting the coefficients of an expressions remain unchanged under orthogonal transformation, these are called *invariants of orthogonal transformation.*

Here we present few examples of invariants of orthogonal transformation:

(1) The degree of an equation $f(x, y) = 0$ is invariant under orthogonal transformation.
(2) The distance between any two points is invariant under orthogonal transformation.
(3) If under translation, the expression $ax^2 + 2hxy + by^2 + 2gx + 2fy + c$ transforms to $a'x'^2 + 2h'x'y' + b'y'^2 + 2g'x' + 2f'y' + c'$, then

(a) $a = a'$,
(b) $b = b'$,
(c) $h = h'$,
(d) $\Delta = \Delta'$.

Putting $x = x'+\alpha$ and $y = y'+\beta$ in the expression $ax^2+2hxy+by^2+2gx+2fy+c$, we get

$$\begin{aligned}
&ax^2 + 2hxy + by^2 + 2gx + 2fy + c \\
&= a(x' + \alpha)^2 + 2h(x' + \alpha)(y' + \beta) + b(y' + \beta)^2 \\
&\quad + 2g(x' + \alpha) + 2f(y' + \beta) + c \\
&= ax'^2 + 2hx'y' + by'^2 + 2x'(a\alpha + h\beta + g) + 2y'(h\alpha + b\beta + f) \\
&\quad + a\alpha^2 + 2h\alpha\beta + b\beta^2 + 2g\alpha + 2f\beta + c \\
&= a'x'^2 + 2h'x'y' + b'y'^2 + 2g'x' + 2f'y' + c' \quad \text{(given)}.
\end{aligned}$$

So, comparing both sides, we obtain

$a = a'$; $b = b'$; $h = h'$; $g' = a\alpha + h\beta + g$; $f' = h\alpha + b\beta + f$

and $c' = a\alpha^2 + 2h\alpha\beta + b\beta^2 + 2g\alpha + 2f\beta + c$.

Now, $\Delta' = \begin{vmatrix} a' & h' & g' \\ h' & b' & f' \\ g' & f' & c' \end{vmatrix}$

$$= \begin{vmatrix} a & h & a\alpha + h\beta + g \\ h & b & h\alpha + b\beta + f \\ a\alpha + h\beta + g & h\alpha + b\beta + f & a\alpha^2 + 2h\alpha\beta + b\beta^2 + 2g\alpha + 2f\beta + c \end{vmatrix}.$$

Performing $C_3 \to C_3 - \alpha C_1 - \beta C_2$ and $R_3 \to R_3 - \alpha R_1 - \beta R_2$ we get

$$\begin{vmatrix} a & h & g \\ h & b & f \\ g & f & c \end{vmatrix} = \Delta.$$

Another method

$$\begin{aligned}
\Delta' &= a'f'^2 + b'g'^2 + c'h'^2 - 2f'g'h' - a'b'c' \\
&= a(h\alpha + b\beta + f)^2 + b(a\alpha + h\beta + g)^2 \\
&\quad + (a\alpha^2 + 2h\alpha\beta + b\beta^2 + 2g\alpha + 2f\beta + c)h^2 \\
&\quad - 2h(h\alpha + b\beta + f)(a\alpha + h\beta + g) \\
&\quad - ab(a\alpha^2 + 2h\alpha\beta + b\beta^2 + 2g\alpha + 2f\beta + c) \\
&= a(h^2\alpha^2 + b^2\beta^2 + f^2 + 2\alpha\beta hb + 2hf\alpha + 2bf\beta) \\
&\quad + b(a^2\alpha^2 + h^2\beta^2 + g^2 + 2\alpha\beta ha + 2ag\alpha + 2hg\beta) \\
&\quad + (a\alpha^2 + 2h\alpha\beta + b\beta^2 + 2g\alpha + 2f\beta + c)h^2 \\
&\quad - 2h[ah\alpha^2 + bh\beta^2 + fg + (ab + h^2)\alpha\beta + (af + gh)\alpha \\
&\quad + (bg + fh)\beta] - ab(a\alpha^2 + 2h\alpha\beta + b\beta^2 + 2g\alpha + 2f\beta + c) \\
&= af^2 + bg^2 + ch^2 - 2fgh - abc \\
&= \Delta.
\end{aligned}$$

(4) If under rotation, the expression $ax^2 + 2hxy + by^2 + 2gx + 2fy + c$ transforms to $a'x'^2 + 2h'x'y' + b'y'^2 + 2g'x' + 2f'y' + c'$, then

(a) $a + b = a' + b'$,

(b) $ab - h^2 = a'b' - h'^2$,

(c) $f^2 + g^2 = f'^2 + g'^2$,

(d) $bc + ab + ac - f^2 - g^2 - h^2 = b'c' + a'b' + a'c' - f'^2 - g'^2 - h'^2$.

Putting $x = x'\cos\theta - y'\sin\theta$ and $y = x'\sin\theta + y'\cos\theta$ in the expression $ax^2 + 2hxy + by^2 + 2gx + 2fy + c$, we get

$$\begin{aligned}
&a(x'\cos\theta - y'\sin\theta)^2 + 2h(x'\cos\theta - y'\sin\theta)(x'\sin\theta + y'\cos\theta) \\
&+ b(x'\sin\theta + y'\cos\theta)^2 + 2g(x'\cos\theta - y'\sin\theta) \\
&+ 2f(x'\sin\theta + y'\cos\theta) + c
\end{aligned}$$

$$= (a\cos^2\theta + h\sin 2\theta + b\sin^2\theta)x'^2 + [2h\cos 2\theta - (a-b)\sin 2\theta]x'y'$$
$$+ (a\sin^2\theta - h\sin 2\theta + b\cos^2\theta)y'^2$$
$$+ 2(g\cos\theta + f\sin\theta)x' + 2(f\cos\theta - g\sin\theta)y' + c$$
$$= a'x'^2 + 2h'x'y' + b'y'^2 + 2g'x' + 2f'y' + c'.$$

Comparing both sides, we get

$$a' = a\cos^2\theta + h\sin 2\theta + b\sin^2\theta;\ b' = a\sin^2\theta - h\sin 2\theta + b\cos^2\theta;$$
$$h' = h\cos 2\theta - (a-b)\sin\theta\cos\theta;\ g' = g\cos\theta + f\sin\theta;$$
$$f' = f\cos\theta - g\sin\theta \quad \text{and} \quad c' = c.$$

Now, it is easy to see that $a + b = a' + b'$ and $f^2 + g^2 = f'^2 + g'^2$.
Again, from $a' = a\cos^2\theta + h\sin 2\theta + b\sin^2\theta$, we get

$$2a' = a + b + [2h\sin 2\theta + (a-b)\cos 2\theta].$$

Similarly,

$$2b' = a + b - [2h\sin 2\theta + (a-b)\cos 2\theta]$$
$$\text{and} \quad 2h' = 2h\cos 2\theta - (a-b)\sin 2\theta.$$

So,

$$4(a'b' - h'^2) = (a+b)^2 - [2h\sin 2\theta + (a-b)\cos 2\theta]^2$$
$$- [2h\cos 2\theta - (a-b)\sin 2\theta]^2$$
$$= (a+b)^2 - [(a-b)^2 + 4h^2]$$
$$= 4(ab - h^2).$$

Hence,

$$ab - h^2 = a'b' - h'^2.$$

Now,

$$b'c' + a'b' + a'c' - f'^2 - g'^2 - h'^2$$
$$= c'(a' + b') + (a'b' - h'^2) - (f'^2 + g'^2)$$
$$= c(a+b) + (ab - h^2) - (f^2 + g^2)$$
$$= bc + ab + ac - f^2 - g^2 - h^2.$$

Examples

▶▷▶ 1. *Find the equation to the curve* $9x^2 + 4y^2 + 18x - 16y - 11 = 0$ *referred to the parallel axes through the point* $(-1, 2)$.

Solution: Putting $x = X - 1$ and $y = Y + 2$ in the equation $9x^2 + 4y^2 + 18x - 16y - 11 = 0$, we get

$$9(X-1)^2 + 4(Y+2)^2 + 18(X-1) - 16(Y+2) - 11 = 0$$
$$\text{or,} \quad 9X^2 - 18X + 9 + 4Y^2 + 16Y + 16 + 18X - 18 - 16Y - 32 - 11 = 0$$
$$\text{or,} \quad 9X^2 + 4Y^2 = 36.$$

▶▶ **2.** *Transform the equation $x^2 - y^2 = a^2$ if the axes are rotated through an angle 45°.*

Solution: Equations of transformation due to rotation are

$$x' = x\cos 45° + y\sin 45° \quad \text{and} \quad y' = y\cos 45° - x\sin 45°$$
$$\text{or,} \quad x = x'\cos 45° - y'\sin 45° \quad \text{and} \quad y = x'\sin 45° + y'\cos 45°$$
$$\text{i.e.,} \quad x = \frac{x'-y'}{\sqrt{2}} \quad \text{and} \quad y = \frac{x'+y'}{\sqrt{2}}.$$

Putting these values in $x^2 - y^2 = a^2$, we get

$$-2x'y' = a^2$$
$$\text{or,} \quad 2x'y' + a^2 = 0.$$

▶▶ **3.** *What does the equation $3x^2 - 4xy + 25y^2 = 0$ become when the axes are turned through an angle $\tan^{-1} 2$?*

Solution: Equations of transformation due to rotation through an angle θ are

$$x' = x\cos\theta + y\sin\theta \quad \text{and} \quad y' = y\cos\theta - x\sin\theta$$
$$\text{or,} \quad x = x'\cos\theta - y'\sin\theta \quad \text{and} \quad y = x'\sin\theta + y'\cos\theta,$$
$$\text{where} \quad \tan\theta = 2$$
$$\text{i.e.,} \quad x = \frac{x'-2y'}{\sqrt{5}} \quad \text{and} \quad y = \frac{2x'+y'}{\sqrt{5}}.$$

Putting these values in $3x^2 - 4xy + 25y^2 = 0$, we get

$$\frac{3}{5}(x'-2y')^2 - \frac{4}{5}(x'-2y')(2x'+y') + \frac{25}{5}(2x'+y')^2 = 0$$
$$\text{or,} \quad 3[x'^2 - 4x'y' + 4y'^2] - 4[2x'^2 - 3x'y' - 2y'^2] + 25[4x'^2 + 4x'y' + y'^2] = 0$$
$$\text{or,} \quad 95x'^2 + 100x'y' + 45y'^2 = 0$$
$$\text{or,} \quad 19x'^2 + 20x'y' + 9y'^2 = 0.$$

▶▶ **4.** *The origin is shifted to the point $(3, -1)$ and the axes are rotated through an angle $\tan^{-1}(3/4)$. If the coordinates of a point are $(5, 10)$ in the new system, find the coordinates in the old system.*

Solution: In this case the equations of transformation are

$$x = h + X\cos\theta - Y\sin\theta$$
$$\text{and} \quad y = k + X\sin\theta + Y\cos\theta$$

where $X = 5$, $Y = 10$, $h = 3$, $k = -1$ and $\tan\theta = \frac{3}{4}$, so the coordinates of the point in old system are

$$x = 3 + 5(4/5) - 10(3/5) = 1 \quad \text{and} \quad y = -1 + 5(3/5) + 10(4/5) = 10.$$

▶▷▶ 5. *Show that the equation* $3x^2 + 2xy + 3y^2 - 1 = 0$ *is reduced to* $4x'^2 + 2y'^2 = 1$ *when rectangular axes are rotated through an angle* $45°$.

Solution: Equations of transformation due to rotation are

$$x' = x\cos 45° + y\sin 45° \quad \text{and} \quad y' = y\cos 45° - x\sin 45°$$

or, $$x = x'\cos 45° - y'\sin 45° \quad \text{and} \quad y = x'\sin 45° + y'\cos 45°$$

i.e., $$x = \frac{x' - y'}{\sqrt{2}} \quad \text{and} \quad y = \frac{x' + y'}{\sqrt{2}}.$$

Using these in the given equation, we get

$$3(x' - y')^2 + 2(x' - y')(x' + y') + 3(x' + y')^2 - 2 = 0$$

or, $$8x'^2 + 4y'^2 - 2 = 0$$

or, $$4x'^2 + 2y'^2 = 1.$$

▶▷▶ 6. *If* $ax + by$ *transforms to* $a'x' + b'y'$ *due to rotation of axes, then show that*

$$a^2 + b^2 = a'^2 + b'^2.$$

Solution: Putting $x = x'\cos\theta - y'\sin\theta$ and $y = x'\sin\theta + y'\cos\theta$ in $ax + by$, we get

$$\begin{aligned} ax + by &= a(x'\cos\theta - y'\sin\theta) + b(x'\sin\theta + y'\cos\theta) \\ &= (a\cos\theta + b\sin\theta)x' + (-a\sin\theta + b\cos\theta)y'. \end{aligned}$$

So, $a' = a\cos\theta + b\sin\theta$ and $b' = -a\sin\theta + b\cos\theta$.

Now,

$$a'^2 + b'^2 = (a\cos\theta + b\sin\theta)^2 + (-a\sin\theta + b\cos\theta)^2 = a^2 + b^2.$$

▶▷▶ 7. *If by an orthogonal transformation of coordinate axes the coordinates of three non-collinear points* $P(x_1, y_1)$, $Q(x_2, y_2)$, $R(x_3, y_3)$ *become* (x_1', y_1'), (x_2', y_2'), (x_3', y_3') *respectively, show that*

$$\begin{vmatrix} x_1 & y_1 & 1 \\ x_2 & y_2 & 1 \\ x_3 & y_3 & 1 \end{vmatrix} = \begin{vmatrix} x_1' & y_1' & 1 \\ x_2' & y_2' & 1 \\ x_3' & y_3' & 1 \end{vmatrix}$$

Solution: The transformation formulae are given by

$$x = h + X\alpha - Y\beta$$

and $$y = k + X\beta + Y\alpha$$

where $\alpha^2 + \beta^2 = 1$.

So,

$$\begin{vmatrix} x_1' & y_1' & 1 \\ x_2' & y_2' & 1 \\ x_3' & y_3' & 1 \end{vmatrix} = \begin{vmatrix} h + x_1\alpha - y_1\beta & k + x_1\beta + y_1\alpha & 1 \\ h + x_2\alpha - y_2\beta & k + x_2\beta + y_2\alpha & 1 \\ h + x_3\alpha - y_3\beta & k + x_3\beta + y_3\alpha & 1 \end{vmatrix}$$

$$= \begin{vmatrix} x_1 & y_1 & 1 \\ x_2 & y_2 & 1 \\ x_3 & y_3 & 1 \end{vmatrix} \times \begin{vmatrix} \alpha & -\beta & h \\ \beta & \alpha & k \\ 0 & 0 & 1 \end{vmatrix}$$

$$= (\alpha^2 + \beta^2) \begin{vmatrix} x_1 & y_1 & 1 \\ x_2 & y_2 & 1 \\ x_3 & y_3 & 1 \end{vmatrix}$$

$$= \begin{vmatrix} x_1 & y_1 & 1 \\ x_2 & y_2 & 1 \\ x_3 & y_3 & 1 \end{vmatrix}$$

EXERCISE

1. The axes are rotated through an angle 60° without changing the origin. If the coordinates of a point are $(1/2, 5\sqrt{3}/2)$ in old system, what would be its coordinates in new system?
2. The coordinates of A and B are $(5, -1)$ and $(3, 1)$. The origin is shifted to the point A and the axes are rotated in such a way that the new x-axis coincides with the direction of AB. If the rotation is made in positive sense, find the formula for transformation.

   [*Hints*: Slope of the line AB is -1, so the given problem is equivalent to shifting origin to the point $(5, -1)$ and then rotation of axes through an angle 135°. Hence, the equations of transformation are

   $$x = 5 + x' \cos 135° - y' \sin 135° = 5 - \frac{x' + y'}{\sqrt{2}}$$

   and $$y = -1 + x' \sin 135° + y' \cos 135° = -1 + \frac{x' - y'}{\sqrt{2}}.]$$
3. Find the transformed equation of straight line $\frac{x}{a} + \frac{y}{b} = 2$ when the origin is transferred to (a, b).
4. Find the transformed equation of $2x^2 + 4xy + 3y^2 - 2x - 4y + 7 = 0$ when the origin is transferred to $(2, -3)$.
5. Find the angle through which the axes are to be rotated so that the equation $17x^2 - 18xy - 7y^2 = 1$ may be reduced to the form $Ax^2 + By^2 = 1$, $A > 0$. Find also A, B.
6. What does the equation $x^2 - 2xy + 4y^2 + 7x - 18y + 32 = 0$ become when the origin is moved to the point $(2, 3)$ and the axes are rotated through an angle 45°?
7. If $ax + by$ and $cx + dy$ are changed to $a'x' + b'y'$ and $c'x' + d'y'$ respectively for rotation of axes, show that $ad - bc = a'd' - b'c'$.
8. If $ax^2 + 2hxy + by^2$ is changed to $a'x'^2 + 2hx'y' + b'y'^2$ respectively for rotation of axes, show that $(a - b)^2 + 4h^2 = (a' - b')^2 + 4h'^2$.
9. Show that there exists only one point whose coordinates do not alter due to transformation of axes.
10. If under translation, the expression $ax^2 + 2hxy + by^2 + 2gx + 2fy + c$ transforms to $a'x'^2 + 2h'x'y' + b'y'^2 + 2g'x' + 2f'y' + c'$, then show that
    (a) $a = a'$,
    (b) $b = b'$,
    (c) $h = h'$,
    (d) $\Delta = \Delta'$.

11. If under rotation, the expression $ax^2 + 2hxy + by^2 + 2gx + 2fy + c$ transforms to $a'x'^2 + 2h'x'y' + b'y'^2 + 2g'x' + 2f'y' + c'$, then show that
    (a) $a + b = a' + b'$,
    (b) $ab - h^2 = a'b' - h'^2$,
    (c) $f^2 + g^2 = f'^2 + g'^2$,
    (d) $bc + ab + ac - f^2 - g^2 - h^2 = b'c' + a'b' + a'c' - f'^2 - g'^2 - h'^2$.
12. If by rotation of rectangular axes about the origin, $ax^2+2hxy+by^2$ and $cx^2+2gxy+dy^2$ be changed to $a'x'^2+2h'x'y'+b'y'^2$ and $c'x'^2+2g'x'y'+d'y'^2$ respectively, then show that $ad + bc - 2hg = a'd' + b'c' - 2h'g'$ and $ac + bd + 2hg = a'c' + b'c' + 2h'g'$.
13. Transform the equation $2x^2 - xy + y^2 + 2x - 3y + 5 = 0$ to new axes of x and y given by the straight lines $4x + 3y + 1 = 0$ and $3x - 4y + 2 = 0$ respectively.

    [*Hints:* Solving $4x + 3y + 1 = 0$ and $3x - 4y + 2 = 0$, we get $x = -2/5$ and $y = 1/5$. So, shifting origin to the point $(-2/5, 1/5)$, we get the transformed equation as

$$\begin{aligned} & 2(x' - 2/5)^2 - (x' - 2/5)(y' + 1/5) \\ & \quad + (y' + 1/5)^2 + 2(x' - 2/5) - 3(y' + 1/5) + 5 = 0 \\ \text{or,} \quad & 50x'^2 - 25x'y' + 25y'^2 + 5x' - 55y' + 101 = 0. \end{aligned}$$

    Now, to take the line $4x + 3y + 1 = 0$ as x-axis we have to rotate the axes through an angle $\tan^{-1}(-4/3)$. In this case the equations of transformation are

$$x' = \frac{3X + 4Y}{5} \quad \text{and} \quad y' = \frac{3Y - 4X}{5}.$$

    Using these results in the above equation, we get

$$46X^2 + 31XY + 29Y^2 + 47X - 29Y + 101 = 0.]$$

14. If (x, y) and (x', y') be the coordinates of the same point referred to two sets of rectangular axes with same origin and if $ux + vy$ becomes $u'x' + v'y'$, where u and v are independent of x, y, then show that $u^2 + v^2 = u'^2 + v'^2$.

Chapter 2

Pair of Straight Lines

2.1 Equation of a Pair of Straight Lines

The general equation of a straight line is given by $Ax + By + C = 0$, where at least one of A and B is non-zero. Consider two similar equations of straight lines

$$ax + by + c = 0 \tag{1}$$

$$a'x + b'y + c' = 0. \tag{2}$$

Next consider the equation

$$(ax + by + c)(a'x + b'y + c') = 0. \tag{3}$$

Obviously, the points on the lines (1) and (2) will satisfy the Eqn. (3) which is second degree equation in x, y of the form

$$Ax^2 + 2Hxy + By^2 + 2Gx + 2Fy + C = 0. \tag{4}$$

Hence, equation of the form (4) may represent a pair of straight lines, where it is assumed that at least one of A, B, H is non-zero.

2.2 A Homogeneous Equation of nth Degree Represents n Straight Lines through Origin

Let the homogeneous equation of nth degree be

$$a_0y^n + a_1y^{n-1}x + a_2y^{n-2}x^2 + \cdots + a_nx^n = 0 \tag{5}$$

which can be written as

$$a_0z^n + a_1z^{n-1} + a_2z^{n-2} + \cdots + a_n = 0 \tag{6}$$

where $z = y/x$. If the above algebraic equation has n roots (real or imaginary and all of them may not be distinct) $m_1, m_2, \ldots, m_n$ then Eqn. (6) can be written as

$$a_0(z - m_1)(z - m_2)\ldots(z - m_n) = 0$$

or, $$a_0(y - m_1x)(y - m_2x)\ldots(y - m_nx) = 0.$$

Hence, Eqn. (5) represents n straight lines given by

$$y - m_1 x = 0;\ y - m_2 x = 0;\ \ldots;\ y - m_n x = 0$$

all of which clearly pass through origin.

For $n = 2$, let us consider the equation

$$ax^2 + 2hxy + by^2 = 0. \tag{7}$$

If $y - m_1 x = 0$ and $y - m_2 x = 0$ be the individual equations of the line represented by Eqn. (7), then comparing both sides of

$$ax^2 + 2hxy + by^2 = b(y - m_1 x)(y - m_2 x)$$

we obtain

$$m_1 + m_2 = -\frac{2h}{b}; \quad m_1 m_2 = \frac{a}{b}. \tag{8}$$

The straight lines will be real, imaginary or coincident according as m_1, m_2 are real, imaginary or equal. Considering Eqn. (7) as quadratic equation in y we get

$$y = \frac{-h \pm \sqrt{h^2 - ab}}{b} x.$$

Hence, the straight lines will be real, imaginary or coincident according as $h^2 - ab >, <, = 0$. If θ be the angle between the straight lines given by Eqn. (7), then we have

$$\tan\theta = \frac{m_1 - m_2}{1 + m_1 m_2} = \frac{\sqrt{(m_1 + m_2)^2 - 4m_1 m_2}}{1 + m_1 m_2}$$

$$\text{or,} \quad \tan\theta = \frac{\sqrt{\left(\frac{-2h}{b}\right)^2 - \frac{4a}{b}}}{1 + \frac{a}{b}} = \frac{2\sqrt{h^2 - ab}}{a + b}. \tag{9}$$

Now, if the straight lines are parallel/coincident, then

$$h^2 - ab = 0 \tag{10}$$

and if the lines are perpendicular, then

$$a + b = 0. \tag{11}$$

2.3 Bisectors of the Angles between the Pair of Straight Lines $ax^2 + 2hxy + by^2 = 0$

If $y = \tan\theta_1 x$ and $y = \tan\theta_2 x$ be the individual equations of the lines given by $ax^2 + 2hxy + by^2 = 0$, then we have

$$\tan\theta_1 + \tan\theta_2 = -\frac{2h}{b}; \quad \tan\theta_1 \tan\theta_2 = \frac{a}{b}. \tag{12}$$

If θ be the angle that either bisector makes with x-axis, then $\theta = \frac{\theta_1 + \theta_2}{2}$ or $\frac{\pi}{2} + \frac{\theta_1 + \theta_2}{2}$.

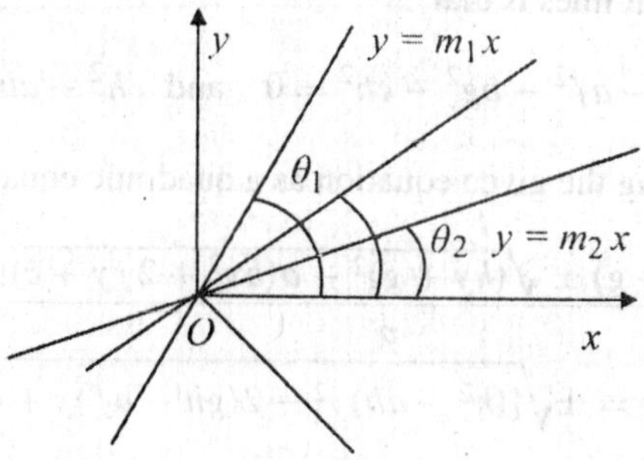

In either case, $\tan 2\theta = \tan(\theta_1 + \theta_2) = \dfrac{\tan\theta_1 + \tan\theta_2}{1 - \tan\theta_1 \tan\theta_2}$

$$\text{or,} \quad \frac{2\tan\theta}{1 - \tan^2\theta} = \frac{-2h/b}{1 - a/b}$$

$$\text{or,} \quad \frac{2y/x}{1 - y^2/x^2} = \frac{-2h/b}{1 - a/b}$$

$$\text{or,} \quad \frac{x^2 - y^2}{a - b} = \frac{xy}{h}.$$

This is the equation of the bisectors of the angles between $ax^2 + 2hxy + by^2 = 0$.

Alternative Method

Let

$$ax^2 + 2hxy + by^2 = (l_1x + m_1y)(l_2x + m_2y).$$

Then we have $l_1l_2 = a;\ l_1m_2 + l_2m_1 = 2h;\ m_1m_2 = b$.

Now the equations of the bisectors of the angles between the lines $l_1x + m_1y = 0$ and $l_2x + m_2y = 0$ are

$$\frac{l_1x + m_1y}{\sqrt{l_1^2 + m_1^2}} = \pm\frac{l_2x + m_2y}{\sqrt{l_2^2 + m_2^2}}$$

$$\text{or,} \quad (l_2^2 + m_2^2)(l_1x + m_1y)^2 = (l_1^2 + m_1^2)(l_2x + m_2y)^2$$

$$\text{or,} \quad (l_1^2m_2^2 - l_2^2m_1^2)(x^2 - y^2) = 2(l_1m_2 - l_2m_1)(l_1l_2 - m_2m_1)xy$$

$$\text{or,} \quad (l_1m_2 + l_2m_1)(x^2 - y^2) = 2(l_1l_2 - m_2m_1)xy$$

$$\text{or,} \quad 2h(x^2 - y^2) = 2(a - b)xy$$

$$\text{or,} \quad \frac{x^2 - y^2}{a - b} = \frac{xy}{h}.$$

2.4 Condition that the General Second Degree Equation May Represent a Pair of Straight Lines

The necessary and sufficient condition that the general second degree equation

$$ax^2 + 2hxy + by^2 + 2gx + 2fy + c = 0$$

may represent a pair of straight lines is that

$$abc + 2fgh - af^2 - bg^2 - ch^2 = 0 \quad \text{and} \quad h^2 - ab \geq 0.$$

Necessary condition: Rewriting the given equation as a quadratic equation in x, we get

$$x = \frac{-(hy+g) \pm \sqrt{(hy+g)^2 - a(by^2 + 2fy + c)}}{a}$$

$$\text{or,} \quad ax + hy + g = \pm\sqrt{[(h^2 - ab)y^2 + 2(gh - af)y + (g^2 - ac)]}. \tag{13}$$

As the given equation represents a pair of straight lines, it can be resolved into two linear factors which imply that the term under radical sign in Eqn. (13) must be a perfect square which gives that

$$4(gh - af)^2 = 4(h^2 - ab)(g^2 - ac) \quad \text{and} \quad h^2 - ab \geq 0$$

$$\text{i.e.,} \quad a(abc + 2fgh - af^2 - bg^2 - ch^2) = 0 \quad \text{and} \quad h^2 - ab \geq 0$$

$$\text{i.e.,} \quad abc + 2fgh - af^2 - bg^2 - ch^2 = 0 \quad \text{and} \quad h^2 - ab \geq 0.$$

Alternative Method

If $l_1x + m_1y + c_1 = 0$ and $l_2x + m_2y + c_2 = 0$ be the individual equations of the lines given by

$$ax^2 + 2hxy + by^2 + 2gx + 2fy + c = 0$$

then we have

$$l_1l_2 = a; \quad m_1m_2 = b; \quad c_1c_2 = c;$$

$$l_2m_1 + l_1m_2 = 2h; \quad l_2c_1 + l_1c_2 = 2g; \quad c_2m_1 + c_1m_2 = 2f. \tag{14}$$

Now consider the following product of determinants:

$$\begin{vmatrix} l_1 & l_2 & 0 \\ m_1 & m_2 & 0 \\ c_1 & c_2 & 0 \end{vmatrix} \times \begin{vmatrix} l_2 & l_1 & 1 \\ m_2 & m_1 & 1 \\ c_2 & c_1 & 1 \end{vmatrix}$$

$$= \begin{vmatrix} 2l_1l_2 & l_1m_2 + l_2m_1 & l_1c_2 + l_2c_1 \\ l_1m_2 + l_2m_1 & 2m_1m_2 & m_1c_2 + m_2c_1 \\ l_1c_2 + l_2c_1 & m_1c_2 + m_2c_1 & 2c_1c_2 \end{vmatrix} = 8\begin{vmatrix} a & h & g \\ h & b & f \\ g & f & c \end{vmatrix}$$

$$\text{or,} \quad \begin{vmatrix} a & h & g \\ h & b & f \\ g & f & c \end{vmatrix} = 0 \quad \text{or,} \quad abc + 2fgh - af^2 - bg^2 - ch^2 = 0. \tag{15}$$

Again,

$$4(h^2 - ab) = (l_1m_2 + l_2m_1)^2 - 4l_1l_2m_1m_2 = (l_1m_2 - l_2m_1)^2 \geq 0$$

$$\text{i.e.,} \quad h^2 - ab \geq 0.$$

Sufficient condition: Let the condition (15) is satisfied, which implies that the following system

$$aX + hY + gZ = 0 \tag{16}$$
$$hX + bY + fZ = 0 \tag{17}$$
$$gX + fY + cZ = 0 \tag{18}$$

has a non-trivial solution. Putting $x' = X/Z$ and $y' = Y/Z$ and multiplying (16) by x', (17) by y', (18) by 1 and then adding, we get

$$ax'^2 + 2hx'y' + by'^2 + 2gx' + 2fy' + c = 0. \tag{19}$$

Now the Eqn. (19) can be transformed to $ax^2 + 2hxy + by^2 = 0$ just by shifting origin and being second degree homogeneous equation in x, y, it represents a pair of straight real lines passing through the origin of x-y coordinate system provided $h^2 - ab \geq 0$.

Thus the necessary and sufficient condition that the general second degree equation $ax^2 + 2hxy + by^2 + 2gx + 2fy + c = 0$ may represent a pair of straight lines is that

$$abc + 2fgh - af^2 - bg^2 - ch^2 = 0 \quad \text{and} \quad h^2 - ab \geq 0.$$

2.5 Point of Intersection of the Lines Given by $ax^2 + 2hxy + by^2 + 2gx + 2fy + c = 0$

Let the lines given by

$$ax^2 + 2hxy + by^2 + 2gx + 2fy + c = 0 \tag{20}$$

intersect at (p, q). On shifting origin to the point (p, q) the Eqn. (20) becomes

$$a(X + p)^2 + 2h(X + p)(Y + q) + b(Y + q)^2 + 2g(X + p) + 2f(Y + q) + c = 0$$

or, $$aX^2 + 2hXY + bY^2 + 2X(pa + hq + g) + 2Y(hp + bq + f) + ap^2 + 2hpq + bq^2 + 2gp + 2fq + c = 0. \tag{21}$$

Since Eqn. (21) represents a pair of straight lines passing through origin, it must be a second degree homogeneous equation in X, Y which leads that

$$pa + hq + g = 0 \tag{22}$$
$$hp + bq + f = 0 \tag{23}$$
$$ap^2 + 2hpq + bq^2 + 2gp + 2fq + c = 0. \tag{24}$$

using Eqn. (22) and Eqn. (23) in Eqn. (24), we get

$$gp + fq + c = 0. \tag{25}$$

Solving Eqn. (22) and Eqn. (23),

$$\frac{p}{bg - hf} = \frac{q}{af - gh} = \frac{1}{h^2 - ab}, \quad \text{i.e.} \quad p = \frac{bg - hf}{h^2 - ab}; \quad q = \frac{af - gh}{h^2 - ab}.$$

Now, if $h^2 - ab = 0$, then the lines meet at infinity, i.e. the lines will be parallel.

The Eqn. (20) can also be written as

$$a(x-p)^2 + 2h(x-p)(y-q) + b(y-q)^2 = 0. \tag{26}$$

Hence, equation of the bisectors of the angles between the lines (20) is

$$\frac{(x-p)^2 - (y-q)^2}{a-b} = \frac{(x-p)(y-q)}{h}. \tag{27}$$

Again, if $l_1x + m_1y + c_1 = 0$ and $l_2x + m_2y + c_2 = 0$ be the individual equations of the lines given by

$$ax^2 + 2hxy + by^2 + 2gx + 2fy + c = 0$$

then we have

$$\begin{aligned} &l_1l_2 = a; \quad m_1m_2 = b; \quad c_1c_2 = c; \\ &l_2m_1 + l_1m_2 = 2h; \quad l_2c_1 + l_1c_2 = 2g; \quad c_2m_1 + c_1m_2 = 2f. \end{aligned} \tag{28}$$

Now, solving $l_1x + m_1y + c_1 = 0$ and $l_2x + m_2y + c_2 = 0$ for their point of intersection, we get

$$\frac{x}{m_1c_2 - m_2c_1} = \frac{-y}{c_2l_1 - c_1l_2} = \frac{1}{l_1m_2 - l_2m_1}$$

$$\text{or,} \quad x = \frac{m_1c_2 - m_2c_1}{l_1m_2 - l_2m_1}; \quad y = \frac{c_1l_2 - c_2l_1}{l_1m_2 - l_2m_1}$$

$$\text{i.e.,} \quad x = \frac{\sqrt{[(m_1c_2 + m_2c_1)^2 - 4m_1c_2m_2c_1]}}{\sqrt{[(m_1l_2 + m_2l_1)^2 - 4m_1l_2m_2l_1]}} = \frac{\sqrt{f^2 - bc}}{\sqrt{h^2 - ab}}$$

$$y = \frac{\sqrt{[(l_1c_2 + l_2c_1)^2 - 4l_1c_2l_2c_1]}}{\sqrt{[(m_1l_2 + m_2l_1)^2 - 4m_1l_2m_2l_1]}} = \frac{\sqrt{g^2 - ac}}{\sqrt{h^2 - ab}}.$$

2.6 To Find the Equation of the Pair of Straight Lines Joining Origin to the Point of Intersection of $lx + my + n = 0$ and $ax^2 + 2hxy + by^2 + 2gx + 2fy + c = 0$

Let the given line meets the curve in P and Q. We have to find out the equation of the pair of straight lines OP and OQ which is a second degree homogeneous equation in x, y. Making homogeneous the equation

$$ax^2 + 2hxy + by^2 + 2gx + 2fy + c = 0 \tag{29}$$

with the help of

$$lx + my + n = 0 \tag{30}$$

as follows:

From the equation of line we get

$$-\frac{lx + my}{n} = 1$$

and using this in the equation $ax^2 + 2hxy + by^2 + 2gx + 2fy + c = 0$ we obtain

$$ax^2 + 2hxy + by^2 - 2(gx + fy)\left(\frac{lx + my}{n}\right) + c\left(\frac{lx + my}{n}\right)^2 = 0$$

$$\text{or,}\quad n^2(ax^2 + 2hxy + by^2) - 2n\{glx^2 + (gm + fl)xy + fmy^2\} + c(l^2x^2 + 2lmxy + m^2y^2) = 0$$

$$\text{or,}\quad (an^2 - 2ngl + cl^2)x^2 + 2(hn^2 - gmn - fln + clm)xy + (bn^2 - 2nfm + cm^2)y^2 = 0 \quad (31)$$

which is a second degree homogeneous equation in x, y and represents a pair of straight lines passing through origin. Now the coordinates of P, Q satisfy both Eqn. (29) and Eqn: (30) and therefore Eqn. (31) too. So Eqn. (31) represents the pair of straight lines OP and OQ.

Examples

▶▶ 1. *Show that equation to the pair of straight lines through origin and perpendicular to $ax^2 + 2hxy + by^2 = 0$ is $bx^2 - 2hxy + ay^2 = 0$.*

Solution: If $y = m_1x$, $y = m_2x$ be the individual equation of the lines given by $ax^2 + 2hxy + by^2 = 0$. Then we have

$$m_2 + m_1 = -\frac{2h}{b}; \quad m_2m_1 = \frac{a}{b}. \quad (1)$$

Now, equation to the pair of straight lines through origin and perpendicular to $ax^2 + 2hxy + by^2 = 0$ is

$$(x + m_1y)(x + m_2y) = 0$$

$$\text{or,}\quad x^2 + (m_2 + m_1)xy + m_2m_1y^2 = 0$$

$$\text{or,}\quad x^2 - \frac{2h}{b}xy + \frac{a}{b}y^2 = 0$$

$$\text{or,}\quad bx^2 - 2hxy + ay^2 = 0.$$

▶▶ 2. *Find the equation of the straight lines which pass through the origin and whose distances from (h, k) are equal to d.*

Solution: Let $y = mx$ be one of the lines. Given that the distance from (h, k) to the line $y = mx$ is 'd'. Then

$$d = \frac{mh - k}{\sqrt{1 + m^2}}$$

$$\text{or,}\quad d^2(1 + m^2) = (mh - k)^2$$

$$\text{or,}\quad d^2\left(1 + \frac{y^2}{x^2}\right) = \left(\frac{yh}{x} - k\right)^2$$

$$\text{or,}\quad d^2(x^2 + y^2) = (yh - xk)^2.$$

▶▶ 3. *Find the area of the triangle formed by $ax^2 + 2hxy + by^2 + 2gx + 2fy + c = 0$ and x-axis.*

Solution: Putting $y = 0$ in the given equation we get $ax^2 + 2gx + c = 0$ and if the given pair of straight lines intersect the x-axis at $(x_1, 0)$ and $(x_2, 0)$, then

$$x_1 + x_2 = -\frac{2g}{a} \quad \text{and} \quad x_1x_2 = \frac{c}{a}. \tag{1}$$

Now,

$$x_1 - x_2 = \sqrt{(x_1 + x_2)^2 - 4x_1x_2} = \sqrt{\frac{4g^2}{a^2} - \frac{4c}{a}} = \frac{2}{a}\sqrt{g^2 - ac}.$$

If the point of intersection of the lines $ax^2 + 2hxy + by^2 + 2gx + 2fy + c = 0$ be (α, β), then we have

$$\alpha = \sqrt{\frac{f^2 - bc}{h^2 - ab}}; \quad \beta = \sqrt{\frac{g^2 - ac}{h^2 - ab}}. \tag{2}$$

Hence, area of the triangle $= \dfrac{1}{2}(x_1 - x_2)\beta = \dfrac{g^2 - ac}{a\sqrt{h^2 - ab}}$.

▶▷ 4. *Find the equation of pair of straight lines obtained by joining origin to the points of intersection of the straight lines $y = mx + c$ and the circle $x^2 + y^2 = a^2$, and prove that they are at right angles, if $2c^2 = a^2(1 + m^2)$.*

Solution: Making homogeneous $x^2 + y^2 = a^2$ with aid of $y = mx + c$ as follows we get

$$x^2 + y^2 = a^2\left(\frac{y - mx}{c}\right)^2$$

$$\text{or,} \quad c^2(x^2 + y^2) = a^2(y - mx)^2$$

$$\text{or,} \quad (c^2 - a^2m^2)x^2 + 2a^2mxy + (c^2 - a^2)y^2 = 0. \tag{1}$$

Now, if the lines represented by Eqn. (1) are perpendicular, then we have

$$\text{coeff of } x^2 + \text{coeff of } y^2 = 0$$

$$\text{or,} \quad (c^2 - a^2m^2) + (c^2 - a^2) = 0$$

$$\text{or,} \quad 2c^2 = a^2(1 + m^2).$$

▶▷ 5. *Show that the lines joining the origin to the points common to $x^2 + hxy - y^2 + gx + fy = 0$ and $fx - gy = \lambda$ are at right angles whatever λ may be.*

Solution: Making homogeneous $x^2 + hxy - y^2 + gx + fy = 0$ with the aid of $fx - gy = \lambda$ as follows we obtain the equation of required pair of straight lines as

$$x^2 + hxy - y^2 + (gx + fy)\left(\frac{fx - gy}{\lambda}\right) = 0$$

$$\text{or,} \quad \left(1 + \frac{fg}{\lambda}\right)x^2 + \left(h + \frac{f^2 - g^2}{\lambda}\right)xy - \left(1 + \frac{fg}{\lambda}\right)y^2 = 0.$$

Now, we see that coeff of x^2 + coeff of $y^2 = 0$ for all values of λ. Hence the lines joining the origin to the points common to $x^2 + hxy - y^2 + gx + fy = 0$ and $fx - gy = \lambda$ are at right angles whatever λ may be.

▶▷▶ **6.** *If the angle between the lines joining origin to the point of intersection of $lx + my = 1$ and $x^2 + y^2 = a^2$ be 45°, show that $4\{a^2(l^2 + m^2) - 1\} = [a^2(l^2 + m^2) - 2]^2$.*

Solution: Making homogeneous $x^2 + y^2 = a^2$ with aid of $lx + my = 1$ as follows we get

$$x^2 + y^2 = a^2(my + lx)^2$$

$$\text{or,} \quad (1 - a^2l^2)x^2 - 2a^2lmxy + (1 - a^2m^2)y^2 = 0. \tag{1}$$

If the angle between the above lines be 45°, then we have

$$[(1 - a^2l^2) + (1 - a^2m^2)]^2 = 4[a^4l^2m^2 - (1 - a^2l^2)(1 - a^2m^2)]$$

$$\text{or,} \quad 4\{a^2(l^2 + m^2) - 1\} = [a^2(l^2 + m^2) - 2]^2.$$

▶▷▶ **7.** *If the two pairs of straight lines $ax^2 + 2hxy + by^2 = 0$ and $a'x^2 + 2h'xy + b'y^2 = 0$ have one line in common and if θ be the angle between the other two, show that*

$$2\cot\theta = \frac{aa'}{ha' - h'a} + \frac{bb'}{h'b - hb'}.$$

Solution: Let $y = mx$ be the equation of common line and $y = m_1x$, $y = m_2x$ respectively be the other lines of the given two pairs of straight lines. Then we have

$$m + m_1 = -\frac{2h}{b}; \quad mm_1 = \frac{a}{b} \tag{1}$$

$$\text{and} \quad m + m_2 = -\frac{2h'}{b'}; \quad mm_2 = \frac{a'}{b'}. \tag{2}$$

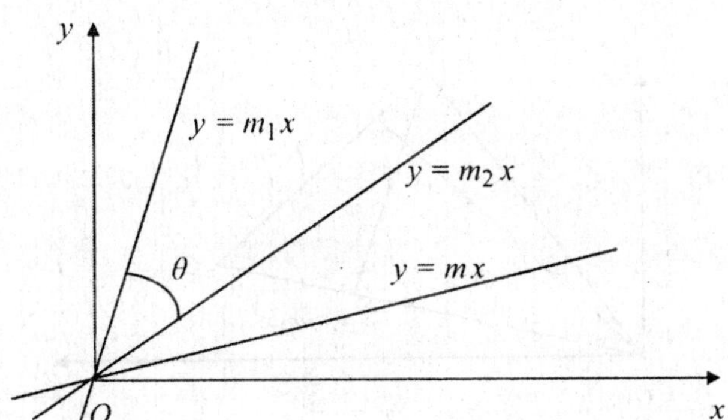

Given that

$$\tan\theta = \frac{m_1 - m_2}{1 + m_1m_2} = \frac{m^2(m_1 - m_2)}{m^2 + mm_1mm_2} = \frac{m^2(2h/b - 2h'/b')}{m^2 + aa'/bb'}$$

$$\Rightarrow \quad 2\cot\theta = \frac{bb'm^2 + aa'}{m^2(h'b - hb')} \Rightarrow 2\cot\theta = \frac{bb'}{h'b - hb'} + \frac{aa'}{m^2(h'b - hb')}. \tag{3}$$

Putting $y = mx$ in the equations $ax^2 + 2hxy + by^2 = 0$ and $a'x^2 + 2h'xy + b'y^2 = 0$, we get

$$a + 2hm + bm^2 = 0;$$

$$a' + 2h'm + b'm^2 = 0.$$

Eliminating m we get

$$m^2 = \frac{a'h - ah'}{h'b - hb'}. \tag{4}$$

Using Eqn. (4) in Eqn. (3) we get

$$2\cot\theta = \frac{aa'}{ha' - h'a} + \frac{bb'}{h'b - hb'}.$$

▶▷ **8.** *Show that the orthocentre of the triangle formed by the lines $ax^2 + 2hxy + by^2 = 0$ and $lx + my = 1$ is given by*

$$\frac{x}{l} = \frac{y}{m} = \frac{a + b}{am^2 - 2hlm + bl^2}.$$

Solution: Let $y = m_1x$ and $y = m_2x$ be the individual equations of the lines given by

$$ax^2 + 2hxy + by^2 = 0.$$

Then we have

$$m_1 + m_2 = -\frac{2h}{b} \quad \text{and} \quad m_1m_2 = \frac{a}{b}. \tag{1}$$

Now, solving $y = m_1x$ and $lx + my = 1$ we get the coordinates of C: $\left(\dfrac{1}{l + mm_1}, \dfrac{m_1}{l + mm_1}\right)$.

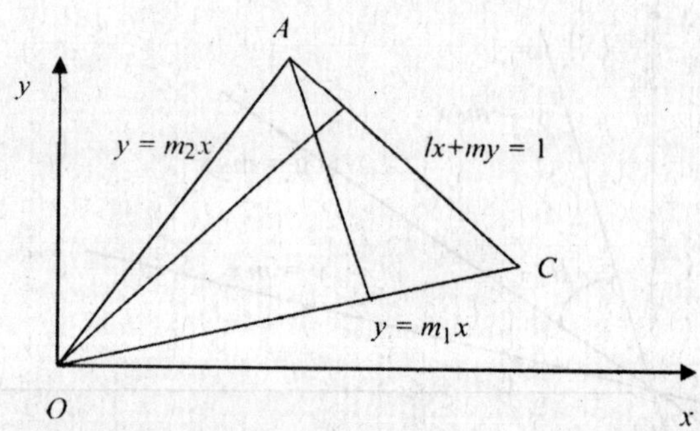

Similarly, solving $y = m_2x$ and $lx + my = 1$ we get the coordinates of A: $\left(\dfrac{1}{l + mm_2}, \dfrac{m_2}{l + mm_2}\right)$. Equation of the perpendicular drawn from origin (0, 0) to the line $lx + my = 1$ is

$$\frac{x}{l} = \frac{y}{m} = k. \tag{2}$$

Now, equation of the perpendicular from A to the line $y = m_1x$ is

$$x + m_1y = M.$$

As it passes through A, we have

$$M = \frac{1}{l + mm_2} + \frac{m_1m_2}{l + mm_2} = \frac{1 + m_1m_2}{l + mm_2}.$$

From (2) we obtain

$$\frac{x}{l} = \frac{y}{m} = \frac{x + m_1 y}{l + mm_1} = \frac{1 + m_1 m_2}{(l + mm_2)(l + mm_1)}$$

$$\Rightarrow \quad \frac{x}{l} = \frac{y}{m} = \frac{1 + a/b}{l^2 + ml(m_2 + m_1) + m^2 m_1 m_2}$$

$$\Rightarrow \quad \frac{x}{l} = \frac{y}{m} = \frac{a + b}{am^2 - 2hlm + bl^2}$$

▶▷ **9.** *Show that the distance from the origin to the orthocentre of the triangle formed by the lines* $\frac{x}{\alpha} + \frac{y}{\beta} = 1$ *and* $ax^2 + 2hxy + by^2 = 0$ *is* $\frac{(a + b)\alpha\beta\sqrt{\alpha^2 + \beta^2}}{a\alpha^2 - 2h\alpha\beta + b\beta^2}$.

Solution: Putting $l = \frac{1}{\alpha}$ and $m = \frac{1}{\beta}$ in the result obtained in problem 8 we get the coordinate of the orthocentre as

$$x\alpha = y\beta = \frac{a + b}{\frac{b}{\alpha^2} - \frac{2h}{\alpha\beta} + \frac{a}{\beta^2}}$$

$$\Rightarrow \quad x = \frac{(a + b)\alpha\beta^2}{a\alpha^2 - 2h\alpha\beta + b\beta^2} \quad \text{and} \quad y = \frac{(a + b)\alpha^2\beta}{a\alpha^2 - 2h\alpha\beta + b\beta^2}.$$

Hence, the distance of orthocentre of the triangle from origin is

$$\sqrt{(x^2 + y^2)} = \frac{(a + b)\alpha\beta\sqrt{\alpha^2 + \beta^2}}{a\alpha^2 - 2h\alpha\beta + b\beta^2}.$$

▶▷ **10.** *Show that the lines* $(A^2 - 3B^2)x^2 + 8ABxy + (B^2 - 3A^2)y^2 = 0$ *form with the line* $Ax + By + C = 0$ *an equilateral triangle. Find also the area of the triangle.*

Solution: Let $y = mx$ be one of the lines given by $(A^2 - 3B^2)x^2 + 8ABxy + (B^2 - 3A^2)y^2 = 0$. If OAB be the equilateral triangle formed by the above lines and the line $Ax + By + C = 0$, then both OA and OB make an angle of magnitude 60° with $Ax + By + C = 0$.

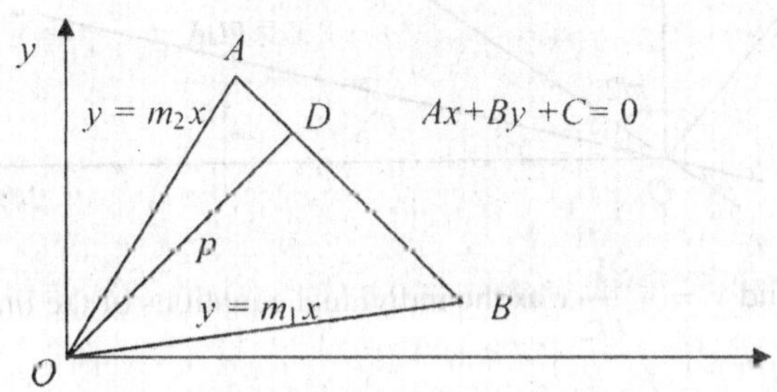

Then we have

$$\tan(\pm 60°) = \frac{m + A/B}{1 - mA/B}$$

$$\Rightarrow \quad \pm\sqrt{3} = \frac{mB + A}{B - mA}$$

$$\Rightarrow \quad 3(B - Am)^2 = (A + Bm)^2.$$

Hence, the equation of the pair of straight lines OA and OB is

$$3\left(B - A\frac{y}{x}\right)^2 = \left(A + B\frac{y}{x}\right)^2$$

$$\Rightarrow \quad (A^2 - 3B^2)x^2 + 8ABxy + (B^2 - 3A^2)y^2 = 0. \tag{1}$$

Now the length of the perpendicular from O on AB is $p = \dfrac{C}{\sqrt{A^2 + B^2}}$.

The area of the triangle $= (1/2)pAB = pAD = p \cdot p \tan 30°$

$$= \frac{C^2}{\sqrt{3}(A^2 + B^2)}.$$

▶▷ **11.** *If one of the lines given by $ax^2 + 2hxy + by^2 = 0$ coincides with one of the lines of $a'x^2 + 2h'xy + b'y^2 = 0$ and the other lines are perpendicular, then show that,*

$$\frac{2h'ab}{b - a} = \frac{2ha'b'}{b' - a'} = \sqrt{-aa'bb'}.$$

Solution: If $y = m_1x$ and $y = m_2x$ be the individual equations of the lines given by $ax^2 + 2hxy + by^2 = 0$, then

$$m_1 + m_2 = -\frac{2h}{b} \quad \text{and} \quad m_1m_2 = \frac{a}{b}. \tag{1}$$

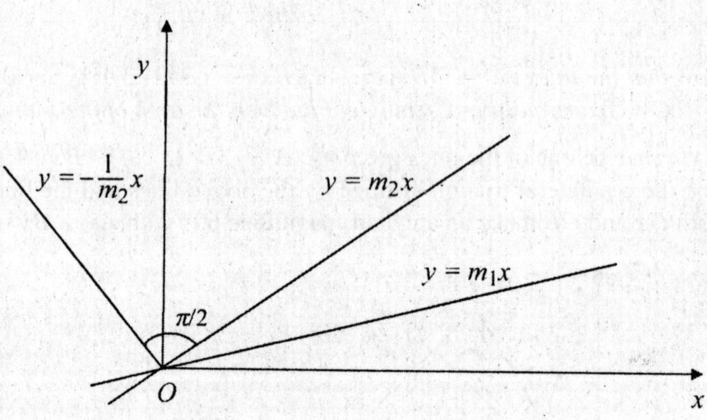

If we take $y = m_1x$ and $y = -\dfrac{1}{m_2}x$ as the individual equations of the lines given by

$$a'x^2 + 2h'xy + b'y^2 = 0$$

then

$$m_1 - \frac{1}{m_2} = -\frac{2h'}{b'} \tag{2}$$

and

$$-\frac{m_1}{m_2} = \frac{a'}{b'}. \tag{3}$$

From Eqn. (2) and Eqn. (3) we get

$$-m_1^2 = \frac{aa'}{bb'} \tag{4}$$

$$\Rightarrow \quad -m_1^2 = \frac{aa'bb'}{b^2b'^2} \tag{5}$$

$$\Rightarrow \quad m_1 = \frac{\sqrt{-aa'bb'}}{bb'}. \tag{6}$$

Using Eqn. (6) in Eqn. (1) we get

$$m_2 = \frac{ab'}{\sqrt{-aa'bb'}}. \tag{7}$$

Now, using Eqn. (6) and Eqn. (7) in Eqn. (2) we get

$$\frac{\sqrt{-aa'bb'}}{b'}\left(\frac{1}{b} - \frac{1}{a}\right) = -\frac{2h'}{b'}. \tag{8}$$

Again, using Eqn. (6) and Eqn. (7) in Eqn. (1) we get

$$\frac{\sqrt{-aa'bb'}}{b}\left(\frac{1}{b'} - \frac{1}{a'}\right) = -\frac{2h}{b}. \tag{9}$$

From Eqn. (8) and (9) we get the result

$$\frac{2h'ab}{b-a} = \frac{2ha'b'}{b'-a'} = \sqrt{-aa'bb'}.$$

▶▶ **12**. *If the lines $ax^2 + 2hxy + by^2 = 0$ be the sides of a parallelogram and the line $lx + my = 1$ be one of the diagonals, show that the equation of other diagonal is $y(bl - hm) = x(am - hl)$.*

Solution: Let $OABC$ be the parallelogram formed by the lines $ax^2 + 2hxy + by^2 = 0$ and the lines parallel to them. Let

$$l_1x + m_1y = 0 \tag{1}$$

$$\text{and} \quad l_2x + m_2y = 0 \tag{2}$$

be the individual equations of the lines given by $ax^2 + 2hxy + by^2 = 0$, then we have

$$l_1l_2 = a; \quad l_1m_2 + l_2m_1 = 2h; \quad m_1m_2 = b.$$

Obviously, the diagonal given by the equation

$$lx + my = 1 \tag{3}$$

is not passing through the origin. So the other diagonal will pass through the origin.

If $A(x_1, y_1)$ be the point of intersection of Eqns. (1) and (3), then

$$x_1 = \frac{m_1}{lm_1 - ml_1}; \quad y_1 = \frac{-l_1}{lm_1 - ml_1}.$$

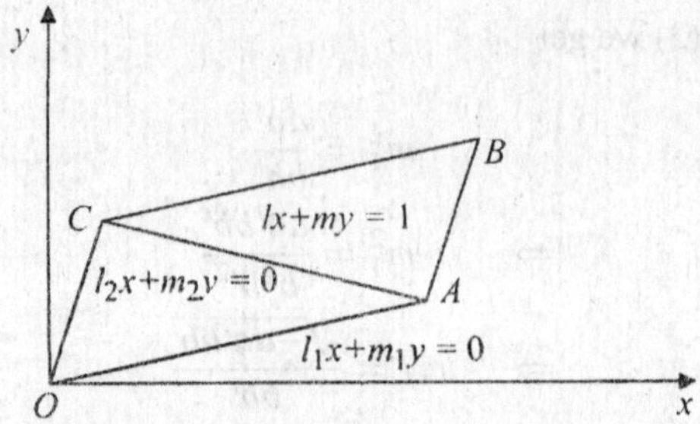

Similarly, the point of intersection $C(x_2, y_2)$ of Eqns. (2) and (3) is given by

$$\left(\frac{m_2}{lm_2 - ml_2}, \frac{-l_2}{lm_2 - ml_2}\right).$$

If D be the middle point of AC, then coordinate of D is

$$\left[\frac{1}{2}\left(\frac{m_2}{lm_2 - ml_2} + \frac{m_1}{lm_1 - ml_1}\right), \frac{1}{2}\left(\frac{-l_2}{lm_2 - ml_2} + \frac{-l_1}{lm_1 - ml_1}\right)\right].$$

On simplification we get the coordinate of D as

$$\left(\frac{bl - hm}{bl^2 - 2hlm + am^2}, \frac{am - hl}{bl^2 - 2hlm + am^2}\right).$$

Hence, equation of the other diagonal is

$$\frac{x-0}{y-0} = \frac{\dfrac{bl - hm}{bl^2 - 2hlm + am^2} - 0}{\dfrac{am - hl}{bl^2 - 2hlm + am^2} - 0}$$

which gives the equation of other diagonal as $y(bl - hm) = x(am - hl)$.

▶▷▶ **13**. *Find the product of the perpendiculars from (p, q) to the lines represented by $ax^2 + 2hxy + by^2 = 0$.*

Solution: If $y = m_1x$ and $y = m_2x$ be the individual equations of the lines given by

$$ax^2 + 2hxy + by^2 = 0$$

then

$$m_1 + m_2 = -\frac{2h}{b} \tag{1}$$

and

$$m_1m_2 = \frac{a}{b}. \tag{2}$$

If p_1, p_2 are the perpendiculars drawn from (p, q) on $y = m_1x$ and $y = m_2x$, then

$$p_1p_2 = \frac{(q - pm_1)(q - pm_2)}{\sqrt{(1 + m_1^2)(1 + m_2^2)}} = \frac{q^2 - (m_1 + m_2)pq + m_1m_2p^2}{\sqrt{1 + m_1^2 + m_2^2 + m_1^2m_2^2}}$$

$$= \frac{ap^2 + 2hpq + bq^2}{\sqrt{(a-b)^2 + 4h^2}}.$$

▶▶ **14.** *Find the length of the line joining the feet of perpendiculars from (p, q) on the lines $ax^2 + 2hxy + by^2 = 0$.*

Solution: Let OA and OB be the lines represented by $ax^2 + 2hxy + by^2 = 0$ and $P(\alpha, \beta)$ be the moving point. PA, PB are the perpendiculars drawn from P on OA and OB; then O, A, P, B lie on a circle having OP as diameter. Let $\angle AOB = \theta$ and $\angle OBA = \phi$. Then we have

$$\tan\theta = \frac{2\sqrt{h^2 - ab}}{a + b}. \quad (1)$$

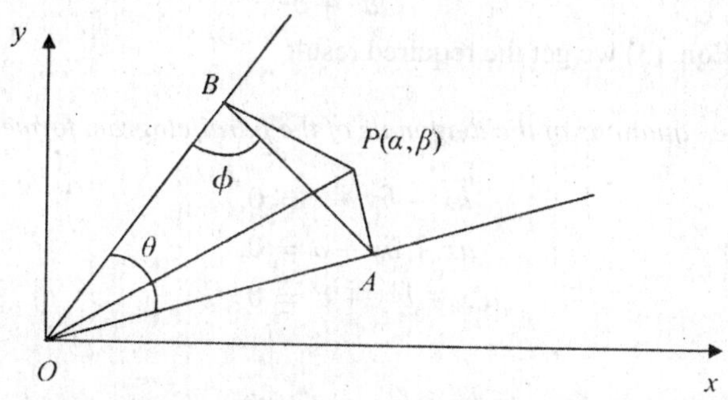

From the triangle ΔAOB we get

$$\frac{AB}{\sin\theta} = \frac{OA}{\sin\phi}. \quad (2)$$

Again, from triangle ΔAOP we get

$$\frac{OP}{\sin 90^\circ} = \frac{OA}{\sin\phi}. \quad (3)$$

From Eqns. (2) and (3)

$$AB = OP\sin\theta = \frac{2\sqrt{(h^2 - ab)(\alpha^2 + \beta^2)}}{\sqrt{(a - b)^2 + 4h^2}}.$$

▶▶ **15.** *If the straight line $ax + by + c = 0$ bisects an angle between a pair of straight lines of which one is $lx + my + n = 0$, show that the other line of the pair is*

$$(lx + my + n)(a^2 + b^2) - 2(al + bm)(ax + by + c) = 0.$$

Solution: Let the equation of AB be

$$lx + my + n = 0 \quad (1)$$

and equation of AC be

$$ax + by + c = 0. \quad (2)$$

Then equation of the other line, i.e. AD can be written as

$$lx + my + n + \lambda(ax + by + c) = 0. \quad (3)$$

Now, slopes of the lines (1), (2) and (3) are

$$m_1 = -\frac{l}{m}; \quad m_2 = -\frac{a}{b}; \quad m_3 = -\frac{l + a\lambda}{m + b\lambda}.$$

Now as angle between Eqns. (1) and (2) is same as angle between Eqns. (2) and (3), we have

$$\frac{\frac{a}{b}-\frac{l+a\lambda}{m+b\lambda}}{1+\frac{a(l+a\lambda)}{b(m+b\lambda)}}=\frac{\frac{l}{m}-\frac{a}{b}}{1+\frac{al}{bm}}.$$

On simplification which yields

$$\lambda=-\frac{2(al+bm)}{a^2+b^2}. \tag{4}$$

Using Eqn. (4) in Eqn. (3) we get the required result.

▶▶ **16**. *Find the equations of the diagonals of the parallelogram formed by the lines*

$$ax+by+c=0, \tag{1}$$
$$ax+by+d=0, \tag{2}$$
$$a'x+b'y+c'=0 \tag{3}$$

and

$$a'x+b'y+d'=0. \tag{4}$$

Show that the parallelogram will be a rhombus if $(a^2+b^2)(c'-d')^2=(a'^2+b'^2)(c-d)^2$; *and area of the parallelogram is* $\dfrac{(c-d)(c'-d')}{ab'-a'b}$.

Solution: Equation of the straight line through the intersection of Eqns. (1) and (4) is

$$ax+by+c+\lambda(a'x+b'y+d')=0$$

and equation of the straight line through the intersection of Eqns. (2) and (3) is

$$ax+by+d+\mu(a'x+b'y+c')=0.$$

If the above lines are identical, then that line will be a diagonal; so $\lambda=\mu$ and $c+\lambda d'=d+\mu c'$. Hence equation of the diagonal is

$$(c'-d')(ax+by+c)+(c-d)(a'x+b'y+d')=0.$$

Proceeding in the same way we get the equation of the other diagonal is

$$(c'-d')(ax+by+d)-(c-d)(a'x+b'y+c')=0.$$

The parallelogram will be rhombus if the distances between the pairs of parallel lines are same. Now distance between the 1st pair of parallel lines is

$$p=\frac{c-d}{\sqrt{a^2+b^2}}$$

and that between the 2nd pair of parallel lines is

$$p'=\frac{c'-d'}{\sqrt{a'^2+b'^2}}.$$

So the required condition is

$$\frac{c-d}{\sqrt{a^2+b^2}} = \frac{c'-d'}{\sqrt{a'^2+b'^2}}$$

i.e., $$(a^2+b^2)(c'-d')^2 = (a'^2+b'^2)(c-d)^2.$$

The required condition can also be obtained using the condition that the diagonals of a rhombus are perpendicular to each other.

If θ be the angle between the straight lines $ax+by+c=0$ and $a'x+b'y+c'=0$, then

$$\tan\theta = \frac{ab'-a'b}{aa'+bb'}.$$

Hence the area of the parallelogram

$$= pp'\operatorname{cosec}\theta = \frac{c-d}{\sqrt{a^2+b^2}}\frac{c'-d'}{\sqrt{a'^2+b'^2}}\left[1+\left(\frac{aa'+bb'}{ab'-a'b}\right)^2\right]^{1/2}$$

$$= \frac{(c-d)(c'-d')}{ab'-a'b}.$$

EXERCISE

1. Prove that the equation of the bisectors of the angle between $ax^2+2hxy+by^2=0$ is
$$(a-b)(x^2-y^2)+4hxy=0.$$
2. Prove that two pairs of straight lines $ax^2+2hxy+by^2=0$ and $a^2x^2+2h(a+b)xy+b^2y^2=0$ have the same bisectors.
3. Prove that two pairs of straight lines $ax^2+2hxy+by^2=0$ and $a'x^2+2h'xy+b'y^2=0$ have the same bisectors if $h(a'-b')=h'(a-b)$.
4. If two pairs of straight lines $x^2-2pxy-y^2=0$ and $x^2-2qxy-y^2=0$ be such that each bisects the angle between the other pairs then prove that $pq=-1$.
5. Show that lines joining origin to the point of intersection of $ax^2+2hxy+by^2+2gx=0$ and $a'x^2+2h'xy+b'y^2+2g'x=0$ will be at right angle if $g'(a+b)=g(a'+b')$.
6. Show that lines joining origin to the point of intersection of the parabola $y^2=4ax$ and the straight line $y=mx+c$ are (a) at right angles if $c+4am=0$, (b) coincident if $c=\frac{a}{m}$.
7. Find the angle between the lines joining origin to the point of intersection of $\frac{x}{a}+\frac{y}{b}=1$ and the circle $x^2+y^2=ax+by$.
8. Show that the two straight lines $x^2(\tan^2\theta+\cos^2\theta)-2xy\tan\theta+y^2\sin^2\theta=0$ make with the x-axis angles such that the difference of their tangents is 2.
9. Show that the equation of the lines through the origin, each of which makes an angle α with the line $y=x$ is $x^2-2xy\sec 2\alpha+y^2=0$.
10. Show that four lines given by $(y-mx)^2=c^2(1+m^2)$ and $(y-nx)^2=c^2(1+n^2)$ form a rhombus.
11. Show that the pair of lines $(a-b)(x^2-y^2)+4hxy=0$ and the pair of lines $h(x^2-y^2)-(a-b)xy=0$ are such that each pair bisects the angle between the other pair.

12. Show that the area of the triangle formed by the straight lines $ax^2 + 2hxy + by^2 = 0$ and $lx + my = 1$ is $\dfrac{\sqrt{h^2 - ab}}{am^2 - 2hlm + bl^2}$.
13. Show that the triangle formed by the straight lines $ax^2+2hxy+by^2 = 0$ and $lx+my = 1$ is (a) right-angled if $(a+b)(al^2+2hlm+bm^2)$, (b) isosceles if $h(l^2-m^2) = (a-b)lm$.
14. If the equation $ax^2 + 2hxy + by^2 + 2gx + 2fy + c = 0$ represents two intersecting straight lines, then show that the square of the distance of their point of intersection from origin is

$$\frac{c(a+b) - f^2 - g^2}{ab - h^2}.$$

15. Prove that the straight lines $ax^2 + 2hxy + by^2 + \lambda(x^2 + y^2) = 0$ have the same pair of bisector. Interpret the case when $\lambda = -(a+b)$.

[*Hints*: Equation of the bisectors of the given lines is given by

$$\frac{x^2 - y^2}{(a+\lambda) - (b+\lambda)} = \frac{xy}{h}$$

$$\text{or,} \quad \frac{x^2 - y^2}{a - b} = \frac{xy}{h} \tag{1}$$

which is independent of λ and hence for all values of λ the given equation has the same pair of bisectors. When $\lambda = -(a+b)$, the given equation becomes $bx^2-2hxy+ay^2 = 0$ which has the same bisector given by Eqn. (1).]

16. A triangle has the lines $ax^2 + 2hxy + by^2 = 0$ for two of its sides and the point (p, q) for its orthocentre. Prove that the equation of the third side is $(a+b)(xp+yq) = bp^2 - 2hpq + aq^2$.

[*Hints*: Taking the equation of third line as $\dfrac{x}{\alpha} + \dfrac{y}{\beta} = 1$ and using the results of Example 8 we get

$$p\alpha = q\beta = \frac{a+b}{\dfrac{b}{\alpha^2} - \dfrac{2h}{\alpha\beta} + \dfrac{a}{\beta^2}}$$

$$\Rightarrow \quad p = \frac{(a+b)\alpha\beta^2}{a\alpha^2 - 2h\alpha\beta + b\beta^2} \quad \text{and} \quad q = \frac{(a+b)\alpha^2\beta}{a\alpha^2 - 2h\alpha\beta + b\beta^2}$$

$$\Rightarrow \quad \frac{p}{\beta} = \frac{q}{\alpha} = \lambda \ \text{(say)} \Rightarrow p\lambda = \frac{(a+b)qp^2}{aq^2 - 2hqp + bp^2}$$

$$\Rightarrow \quad \alpha = \frac{aq^2 - 2hqp + bp^2}{(a+b)p} \quad \text{and} \quad \beta = \frac{aq^2 - 2hqp + bp^2}{(a+b)q}$$

Putting these values in $\dfrac{x}{\alpha} + \dfrac{y}{\beta} = 1$ we get the required result.]

17. If the lines $ax^2 + 2hxy + by^2 = 0$ form an equilateral triangle with the line $x\cos\alpha + y\sin\alpha = p$, show that

$$\frac{a}{1 - 2\cos 2\alpha} = \frac{h}{2\sin 2\alpha} = \frac{b}{1 + 2\cos 2\alpha}.$$

[*Hints*: Putting $A = \cos\alpha$; $B = \sin\alpha$; $C = -p$ in the Eqn. (1) of Example 10 we get

$$(\cos^2\alpha - 3\sin^2\alpha)x^2 + 8\cos\alpha\sin\alpha xy + (\sin^2\alpha - 3\cos^2\alpha)y^2 = 0$$
$$\Rightarrow \quad (1 - 2\cos 2\alpha)x^2 - 4xy\sin 2\alpha + (1 + 2\cos 2\alpha)y^2 = 0.$$

Now comparing this result with $ax^2 + 2hxy + by^2 = 0$ we obtain given result which was to prove.]

18. If the equation $ax^2 + 2hxy + by^2 + 2gx + 2fy + c = 0$ represents a pair of straight lines, then show that the equation to the third pair of straight lines passing through the points where these meet the axes is $ax^2 - 2hxy + by^2 + 2gx + 2fy + c + \dfrac{4fgxy}{c} = 0$.

[*Hints*: Given that the equation

$$ax^2 + 2hxy + by^2 + 2gx + 2fy + c = 0 \tag{1}$$

represents a pair of straight lines. Equation of coordinate axes is

$$xy = 0. \tag{2}$$

Now equation of the third pair of straight lines passing through the point of intersection of Eqns. (1) and (2) can be written as

$$ax^2 + 2hxy + by^2 + 2gx + 2fy + c + 2\lambda xy = 0$$
$$\text{or,} \quad ax^2 + 2(h + \lambda)xy + by^2 + 2gx + 2fy + c = 0. \tag{3}$$

Since Eqn. (3) represents a pair of straight lines, we have

$$abc + 2fg(h + \lambda) - af^2 - bg^2 - c(h + \lambda)^2 = 0$$

which on simplification gives

$$\lambda = \frac{2(fg - ch)}{c}. \tag{4}$$

Using Eqn. (4) in Eqn. (3) we get the result.]

19. If the equation $ax^2 + 2hxy + by^2 + 2gx + 2fy + c = 0$ represents two straight lines equidistant from origin then show that $f^4 - g^4 = c(bf^2 - ag^2)$.

[*Hints*: If $l_1x + m_1y + c_1 = 0$ and $l_2x + m_2y + c_2 = 0$ be the individual equations of the lines given by

$$ax^2 + 2hxy + by^2 + 2gx + 2fy + c = 0$$

then we have

$$l_1l_2 = a; \quad m_1m_2 = b; \quad c_1c_2 = c;$$
$$l_2m_1 + l_1m_2 = 2h; \quad l_2c_1 + l_1c_2 = 2g; \quad c_2m_1 + c_1m_2 = 2f. \tag{1}$$

Since the lines are equidistant from origin,

$$\frac{c_1}{\sqrt{l_1^2 + m_1^2}} = \frac{c_2}{\sqrt{l_2^2 + m_2^2}}$$
$$\Rightarrow \quad c_1^2(l_2^2 + m_2^2) = c_2^2(l_1^2 + m_1^2)$$
$$\Rightarrow \quad c_1^2l_2^2 - c_2^2l_1^2 = c_2^2m_1^2 - c_1^2m_2^2 \tag{2}$$

Using Eqn. (1) in Eqn. (2) we get the required result.]

20. Show that the pair of straight lines $ax^2 + 2hxy + by^2 + 2gx + 2fy + c = 0$ form a rhombus with the straight lines $ax^2 + 2hxy + by^2 = 0$ if $h(g^2 - f^2) = gf(a - b)$.
[*Hints*: Equation of the diagonal AC can be written as

$$ax^2 + 2hxy + by^2 + 2gx + 2fy + c + \lambda(ax^2 + 2hxy + by^2) = 0.$$

Since equation of AC is a linear equation in x, y, there would be no second degree term in the above equation which leads that $\lambda = -1$.
Hence equation of AC is

$$2gx + 2fy + c = 0. \tag{1}$$

The point of intersection of the lines given by the equation $ax^2 + 2hxy + by^2 + 2gx + 2fy + c = 0$ is $\left(\frac{hf - bg}{ab - h^2}, \frac{hg - af}{ab - h^2}\right)$ and the point of intersection of the lines given by the equation $ax^2 + 2hxy + by^2 = 0$ is $(0, 0)$. So the slope of the diagonal OB is $\left(\frac{hg - af}{hf - bg}\right)$ and that of AC is $-\frac{g}{f}$. As the diagonals of a rhombus are perpendicular to each other, we have

$$-\frac{g}{f}\left(\frac{hg - af}{hf - bg}\right) = -1$$

$$\Rightarrow \quad h(g^2 - f^2) = gf(a - b).]$$

21. Show that the pair of straight lines $(ab - h^2)(ax^2 + 2hxy + by^2 + 2gx + 2fy) + af^2 + bg^2 - 2fgh = 0$ represents a pair of straight lines and form a rhombus with the straight lines $ax^2 + 2hxy + by^2 = 0$, if $h(g^2 - f^2) = gf(a - b)$.
[*Hints*: We know that

$$ax^2 + 2hxy + by^2 + 2gx + 2fy + c = 0 \tag{1}$$

represents a pair of straight lines if

$$af^2 + bg^2 + ch^2 - 2fgh - abc = 0$$

$$\text{i.e., if} \quad af^2 + bg^2 - 2fgh - c(ab - h^2) = 0$$

$$\text{i.e., if} \quad af^2 + bg^2 - 2fgh = c(ab - h^2)$$

$$\text{i.e., if} \quad c = \frac{af^2 + bg^2 - 2fgh}{ab - h^2}. \tag{2}$$

Using Eqn. (2) in Eqn. (1) we get the first result and for the second part see problem no. Ex. 20.]

22. Prove that if one of the lines $ax^2 + 2hxy + by^2 = 0$ be perpendicular to one of the lines of $a'x^2 + 2h'xy + b'y^2 = 0$, then

$$(aa' - bb')^2 + 4(ah' + hb')(a'h + h'b) = 0.$$

23. Prove that if one of the bisectors of the angle between $ax^2 + 2hxy + by^2 = 0$ coincides with one of the lines of $Ax^2 + 2Hxy + By^2 = 0$, then

$$h^2(A + B)^2 = \{2hH - B(a - b)\}\{2hH + A(a - b)\}.$$

24. Prove that if one of the lines $ax^2 + 2hxy + by^2 = 0$ coincides with one of the lines of $a'x^2 + 2h'xy + b'y^2 = 0$, then

$$(ab' - ba')^2 = 4(bh' - hb')(a'h - h'a).$$

25. A parallelogram is formed by the pair of straight lines $ax^2 + 2hxy + by^2 = 0$ and the straight lines through (p, q) parallel to those. Show that the equation of the diagonal, which does not pass through the origin, is

$$(2x - p)(ap + hq) + (2y - q)(bq + hp) = 0.$$

[*Hints*: The equation of the pair of straight lines through the point (p, q) and parallel to the lines

$$ax^2 + 2hxy + by^2 = 0 \tag{1}$$

is

$$a(x - p)^2 + 2h(x - p)(y - q) + b(y - q)^2 = 0. \tag{2}$$

As the diagonal, which does not pass through the origin, passes through the point of intersections of Eqns. (1) and (2), we can write its equation as

$$a(x - p)^2 + 2h(x - p)(y - q) + b(y - q)^2 + \lambda(ax^2 + 2hxy + by^2) = 0. \tag{3}$$

Since equation of a straight line is a linear equation in x, y, value of λ should be such that there would be no second degree term in x, y in Eqn. (3) which leads that $\lambda = -1$. So from Eqn. (3) we get

$$\begin{aligned} & -2apx + ap^2 - 2hyp - 2hxq + 2hpq + bq^2 - 2bqy = 0 \\ \Rightarrow \quad & 2apx + 2hxq - ap^2 - hpq + 2hyp + 2bqy - bq^2 - hpq = 0 \\ \Rightarrow \quad & (2x - p)(ap + hq) + (2y - q)(bq + hp) = 0.] \end{aligned}$$

26. Show that the equation of the line joining feet of the perpendiculars from the point $(d, 0)$ on the straight lines $ax^2 + 2hxy + by^2 = 0$ is $(a - b)x + 2hy + bd = 0$.

[*Hints*: Let $y = m_1x$ and $y = m_2x$ be the individual equations of the lines given by

$$ax^2 + 2hxy + by^2 = 0.$$

Then we have

$$m_1 + m_2 = -\frac{2h}{b} \quad \text{and} \quad m_1m_2 = \frac{a}{b}. \tag{1}$$

Equation of the perpendicular drawn from $(d, 0)$ to the line $y = m_1x$ is

$$x + m_1y = d.$$

Solving the equations $y = m_1x$ and $x + m_1y = d$ we get

$$A: \left(\frac{d}{1 + m_1^2}, \frac{dm_1}{1 + m_1^2}\right). \tag{2}$$

Similarly, the coordinate of B is : $\left(\dfrac{d}{1 + m_2^2}, \dfrac{dm_2}{1 + m_2^2}\right)$. (3)

Therefore, equation of AB is

$$\frac{x-\dfrac{d}{1+m_1^2}}{\dfrac{dm_1}{y-\dfrac{1+m_1^2}{}}}=\frac{\dfrac{d}{1+m_1^2}-\dfrac{d}{1+m_2^2}}{\dfrac{dm_1}{1+m_1^2}-\dfrac{dm_2}{1+m_2^2}}$$

$$\Rightarrow \quad \frac{(1+m_1^2)x-d}{(1+m_1^2)y-dm_1}=\frac{m_1+m_2}{m_1m_2-1}=\frac{2h}{b-a}$$

$$\Rightarrow \quad (a-b)(1+m_1^2)x-d(a-b) + 2h(1+m_1^2)y-2dhm_1=0. \tag{4}$$

Since $y=m_1x$ is one of the lines given by $ax^2+2hxy+by^2=0$, we have

$$a+2hm_1+bm_1^2=0$$

$$b(1+m_1^2)=b-a-2hm_1. \tag{5}$$

Rewriting Eqn. (4) as $(1+m_1^2)[(a-b)x+2hy]=d[a-b+2hm_1]$ and using Eqn. (5) we get

$$(1+m_1^2)[(a-b)x+2hy]=-db(1+m_1^2)$$

$$\Rightarrow \quad (a-b)x+2hy+bd=0.]$$

27. Find the area of the triangle formed by the x-axis and the bisectors of the lines $ax^2+2hxy+by^2+2gx+2fy+c=0$.

[*Hints*: The equation of the bisectors of the pair of straight lines $ax^2+2hxy+by^2+2gx+2fy+c=0$ is given by

$$\frac{(x-\alpha)^2-(y-\beta)^2}{a-b}=\frac{(x-\alpha)(y-\beta)}{h}$$

where the point of intersection (α,β) of the lines $ax^2+2hxy+by^2+2gx+2fy+c=0$ is given by

$$\alpha=\sqrt{\frac{f^2-bc}{h^2-ab}};\quad \beta=\sqrt{\frac{g^2-ac}{h^2-ab}}. \tag{1}$$

If the bisectors meet x-axis at $(x_1,0)$ and $(x_2,0)$, then $x_1-\alpha$ and $x_2-\alpha$ are roots of

$$\frac{(x-\alpha)^2-\beta^2}{a-b}=\frac{(x-\alpha)(-\beta)}{h}$$

i.e. of the equation

$$h(x-\alpha)^2+(a-b)(x-\alpha)\beta-h\beta^2=0.$$

So, we have

$$(x_1-\alpha)+(x_2-\alpha)=\frac{-\beta(a-b)}{h} \quad \text{and} \quad (x_1-\alpha)(x_2-\alpha)=-\beta^2. \tag{2}$$

Now,

$$(x_1 - x_2) = (x_1 - \alpha) - (x_2 - \alpha) = \beta\sqrt{\left(\frac{a-b}{h}\right)^2 + 4}.$$

So, area of the triangle

$$= (x_1 - x_2)\beta/2 = \beta^2 \frac{\sqrt{[(a-b)^2 + 4h^2]}}{2h}$$

$$= \frac{\sqrt{[(a-b)^2 + 4h^2]}}{2h} \frac{g^2 - ac}{h^2 - ab}.$$

28. Prove that the equation $ax^2 + 2hxy + by^2 + 2gx + 2fy + c = 0$ represents a pair of parallel straight lines if $\frac{a}{h} = \frac{h}{b} = \frac{g}{f}$. Find the distance between them.
[*Hints*: Let

$$ax^2 + 2hxy + by^2 + 2gx + 2fy + c = (\sqrt{a}x + \sqrt{b}y + c_1)(\sqrt{a}x + \sqrt{b}y + c_2).$$

Comparing both sides we get

$$2\sqrt{(ab)} = 2h \tag{1}$$

$$\sqrt{a}(c_1 + c_2) = 2g \tag{2}$$

$$\sqrt{b}(c_1 + c_2) = 2f \tag{3}$$

$$c_1 c_2 = c \tag{4}$$

From Eqn. (1) we get

$$ab = h^2 \quad \Rightarrow \quad \frac{h}{b} = \frac{a}{h}. \tag{5}$$

Dividing Eqn. (2) by Eqn. (3) we get

$$\sqrt{\frac{a}{b}} = \frac{g}{f} \quad \Rightarrow af^2 = bg^2 \Rightarrow abf^2 = b^2g^2$$

$$\Rightarrow \quad h^2 f^2 = b^2 g^2 \qquad \text{[by (5)]}$$

$$\Rightarrow \quad hf = bg$$

$$\Rightarrow \quad \frac{h}{b} = \frac{g}{f}. \tag{6}$$

By Eqns. (5) and (6) we get

$$\frac{a}{h} = \frac{h}{b} = \frac{g}{f}.$$

Now, difference between two parallel straight lines = difference of their distances measured from origin $= \frac{c_1}{\sqrt{a+b}} \sim \frac{c_2}{\sqrt{a+b}}$

$$= \sqrt{\frac{(c_1 + c_2)^2 - 4c_1c_2}{a+b}}$$

$$= 2\sqrt{\frac{g^2 - ac}{a(a+b)}} \qquad \text{\{by Eqns. (2) and (4) \}.]}$$

29. A point moves so that the distance between the feet of perpendiculars from it on the lines $ax^2 + 2hxy + by^2 = 0$ is a constant distance $2d$. Show that its locus is

$$(x^2 + y^2)(h^2 - ab) = d^2[(a - b)^2 + 4h^2].$$

[*Hints*: Let OA and OB be the lines represented by $ax^2 + 2hxy + by^2 = 0$ and $P(\alpha, \beta)$ be the moving point. PA, PB are the perpendiculars draw from P on OA and OB; then O, A, P, B lie on a circle having OP as diameter. Let $\angle AOB = \theta$ and $\angle OBA = \phi$. Then we have

$$\tan\theta = 2\frac{\sqrt{h^2 - ab}}{a + b}. \tag{1}$$

From the triangle ΔAOB we get

$$\frac{AB}{\sin\theta} = \frac{OA}{\sin\phi}. \tag{2}$$

Again, from triangle ΔAOP we get

$$\frac{OP}{\sin 90°} = \frac{OA}{\sin\phi}. \tag{3}$$

From Eqns. (2) and (3)

$$OP^2 = AB^2(1 + \cot^2\theta)$$

$$\Rightarrow \quad \alpha^2 + \beta^2 = 4d^2\left[1 + \frac{(a + b)^2}{4(h^2 - ab)}\right] \qquad \text{[using Eqn. (1)]}$$

$$\Rightarrow \quad (h^2 - ab)(\alpha^2 + \beta^2) = d^2[4(h^2 - ab) + (a + b)^2]$$

$$\Rightarrow \quad (h^2 - ab)(\alpha^2 + \beta^2) = d^2[4h^2 + (a - b)^2]$$

Generalising we get the result.]

30. A parallelogram is formed by the pair of straight lines $ax^2 + 2hxy + by^2 = 0$ and the straight lines through (p, q) parallel to those. Show that the area of the parallelogram is

$$\frac{ap^2 + 2hpq + bq^2}{2\sqrt{h^2 - ab}}.$$

[*Hints*: If $y = m_1x$ and $y = m_2x$ be the individual equations of the lines given by

$$ax^2 + 2hxy + by^2 = 0$$

then

$$m_1 + m_2 = -\frac{2h}{b} \quad \text{and} \quad m_1m_2 = \frac{a}{b}. \tag{1}$$

Again, if θ be the angle between the given pair of straight lines, then we have

$$\tan\theta = 2\frac{\sqrt{h^2 - ab}}{a + b}. \tag{2}$$

Now, area of the parallelogram

$$= 2\Delta ABC = CD \cdot p_1 = \frac{p_1 p_2}{\sin\theta} \tag{3}$$

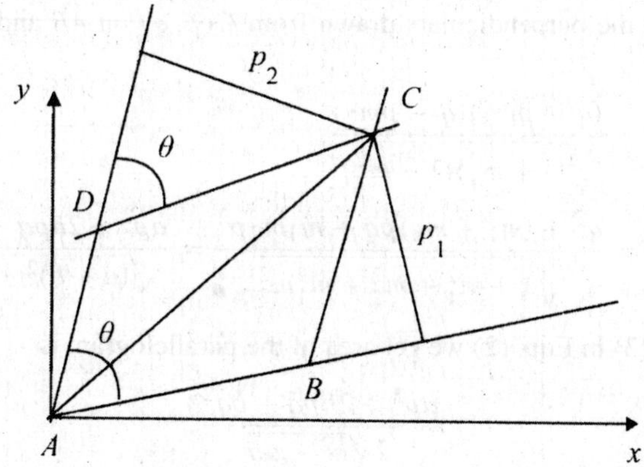

where p_1, p_2 are the perpendiculars drawn from C on AB and on AD produced respectively. Now

$$p_1 p_2 = \frac{(q - pm_1)(q - pm_2)}{\sqrt{(1+m_1^2)(1+m_2^2)}}$$

$$= \frac{q^2 - (m_1 + m_2)pq + m_1 m_2 p^2}{\sqrt{1 + m_1^2 + m_2^2 + m_1^2 m_2^2}} = \frac{ap^2 + 2hpq + bq^2}{\sqrt{(a-b)^2 + 4h^2}}. \quad (4)$$

Using Eqns. (4), (3) in Eqn. (2) we get area of the parallelogram is

$$\frac{ap^2 + 2hpq + bq^2}{2\sqrt{h^2 - ab}}.]$$

31. If $ax^2 + 2hxy + by^2 + 2gx + 2fy + c = 0$ represents a pair of straight lines, show that the area of the parallelogram formed by them and the pair of parallel straight lines through the origin is $\dfrac{c}{2\sqrt{h^2 - ab}}$.

[*Hints*: If $y = m_1 x$ and $y = m_2 x$ be the individual equations of the lines given by

$$ax^2 + 2hxy + by^2 = 0$$

then

$$m_1 + m_2 = -\frac{2h}{b} \quad \text{and} \quad m_1 m_2 = \frac{a}{b}. \quad (1)$$

Again, if θ be the angle between the given pair of straight lines, then we have

$$\tan\theta = 2\frac{\sqrt{h^2 - ab}}{a + b}. \quad (2)$$

Now, area of the parallelogram

$$= 2\Delta ABC = CD \cdot p_1 = \frac{p_1 p_2}{\sin\theta} \quad (3)$$

where p_1, p_2 are the perpendiculars drawn from $C(p, q)$ on AB and on AD produced respectively. Now

$$p_1 p_2 = \frac{(q - pm_1)(q - pm_2)}{\sqrt{(1 + m_1^2)(1 + m_2^2)}}$$

$$= \frac{q^2 - (m_1 + m_2)pq + m_1 m_2 p^2}{\sqrt{1 + m_1^2 + m_2^2 + m_1^2 m_2^2}} = \frac{ap^2 + 2hpq + bq^2}{\sqrt{(a - b)^2 + 4h^2}}. \tag{4}$$

Using Eqns. (4), (3) in Eqn. (2) we get area of the parallelogram is

$$\frac{ap^2 + 2hpq + bq^2}{2\sqrt{h^2 - ab}}. \tag{5}$$

Now as (p, q) satisfy the equation $ax^2 + 2hxy + by^2 + 2gx + 2fy + c = 0$, we have

$$ap^2 + 2hpq + bq^2 + 2gp + 2fq + c = 0 \quad \text{and} \quad gp + fq + c = 0. \tag{6}$$

Using Eqn. (6) in (5) we get area of the parallelogram $= \dfrac{c}{2\sqrt{h^2 - ab}}$.]

32. If the equation $ax^2 + 2hxy + by^2 + 2gx + 2fy + c = 0$ represents two perpendicular straight lines, show that the distance of their point of intersecion from origin is $\dfrac{f^2 + g^2}{h^2 - ab}$.

33. If each of the equations $ax^2 + 2hxy + by^2 + 2gx + 2fy + c = 0$ and $ax^2 + 2hxy + by^2 - 2gx - 2fy - c = 0$ represents a pair of straight lines, show that the area of the parallelogram formed by them is $\dfrac{2c}{\sqrt{h^2 - ab}}$.

[*Hints*: If $l_1 x + m_1 y + c_1 = 0$ and $l_2 x + m_2 y + c_2 = 0$ be the individual equations of the lines given by

$$ax^2 + 2hxy + by^2 + 2gx + 2fy + c = 0$$

then we have

$$l_1 l_2 = a; \quad m_1 m_2 = b; \quad c_1 c_2 = c;$$

$$l_2 m_1 + l_1 m_2 = 2h; \quad l_2 c_1 + l_1 c_2 = 2g; \quad c_2 m_1 + c_1 m_2 = 2f. \tag{1}$$

Again, if θ be the angle between the given pair of straight lines, then we have

$$\tan\theta = 2\frac{\sqrt{h^2 - ab}}{a + b}. \tag{2}$$

Now area of the parallelogram

$$= 2\Delta ABC = CD \cdot p_1 = \frac{p_1 p_2}{\sin\theta} \tag{3}$$

where p_1, p_2 are the perpendiculars drawn from $C(p, q)$ on AB and on AD produced respectively. Now

$$p_1 p_2 = \frac{(pl_1 + qm_1 - c_1)(pl_2 + qm_2 - c_2)}{\sqrt{(l_1^2 + m_1^2)(l_2^2 + m_2^2)}}$$

$$= \frac{ap^2 + 2hpq + bq^2 - 2gp - 2fq + c}{\sqrt{(a - b)^2 + 4h^2}}. \tag{4}$$

Using Eqns. (4), (3) in Eqn. (2) we get area of the parallelogram is

$$\frac{ap^2 + 2hpq + bq^2 - 2gp - 2fq + c}{2\sqrt{h^2 - ab}}. \tag{5}$$

Now as (p, q) satisfy the equation $ax^2 + 2hxy + by^2 + 2gx + 2fy + c = 0$, we have

$$ap^2 + 2hpq + bq^2 + 2gp + 2fq + c = 0 \text{ and } gp + fq + c = 0. \tag{6}$$

Using Eqn. (6) in Eqn. (5) we get area of the parallelogram $= \dfrac{2c}{\sqrt{h^2 - ab}}$.]

34. The straight lines joining the origin to the common points of the curve $ax^2 + 2hxy + by^2 = c$ and the straight line $lx + my = 1$ are at right angles. Show that the locus of the foot of the perpendicular from origin on the straight line is $(a + b)(x^2 + y^2) = c$.

[*Hints*: Making homogeneous $ax^2 + 2hxy + by^2 = c$ with the aid of $lx + my = 1$ we get

$$ax^2 + 2hxy + by^2 = c(lx + my)^2$$
$$\text{or,} \quad (a - cl^2)x^2 + 2(h - clm)xy + (b - cm^2)y^2 = 0.$$

If this represents a pair of perpendicular straight lines, then we have

$$(a - cl^2) + (b - cm^2) = 0$$
$$\text{or,} \quad a + b = c(l^2 + m^2). \tag{1}$$

Equation of the straight line passing through the origin and perpendicular to $lx + my = 1$ is

$$mx = ly.$$

Solving $lx + my = 1$ and $mx = ly$ we get $x = \frac{l}{l^2+m^2}$; $y = \frac{m}{l^2+m^2}$.

Hence locus of the foot of the perpendicular is

$$x^2 + y^2 = \frac{1}{l^2 + m^2} = \frac{c}{a + b}$$
$$\text{or,} \quad (a + b)(x^2 + y^2) = c.]$$

35. Show that the distance between the point of intersection of the straight line $x \cos\alpha + y \sin\alpha = p$ with the lines $ax^2 - 2hxy + by^2 = 0$ is

$$\frac{2p\sqrt{h^2 - ab}}{b\cos^2\alpha - 2h\sin 2\alpha + a\sin^2\alpha}.$$

36. Prove that two of the lines represented by the equation $ax^4 + bx^3y + cx^2y^2 + dxy^3 + ay^4 = 0$ will bisect the angle between other two, if $c + 6a = 0$ and $b + d = 0$.

[*Hints*: We know that the bisectors are perpendicular to each other and hence the given equation should represent two pairs of mutually perpendicular lines. Hence we can assume

$$ax^4 + bx^3y + cx^2y^2 + dxy^3 + ay^4 = (x^2 + pxy - y^2)(ax^2 + qxy - ay^2).$$

Comparing both sides we get

$$ap + q = b \quad \text{and} \quad -aq - p = d, \quad \text{i.e. } b + d = 0 \tag{1}$$

$$\text{and} \quad pq = c + 2a. \tag{2}$$

As bisectors of $x^2 + pxy - y^2 = 0$ are given by $ax^2 + qxy - ay^2 = 0$, so comparing the equation $\dfrac{x^2 - y^2}{1+1} = \dfrac{xy}{p/2}$, i.e. $p(x^2 - y^2) = 4xy$ with $ax^2 + qxy - ay^2 = 0$ we get $pq = -4a$, i.e. $c + 2a = -4a$ or, $c + 6a = 0$.]

37. Show that one of the bisectors of the angle between the lines $ax^2 + 2hxy + by^2 = 0$ will pass through the point of intersection of the straight lines $ax^2 + 2hxy + by^2 + 2gx + 2fy + c = 0$ if $h(g^2 - f^2) = fg(a - b)$.

Chapter 3

Circle

3.1 Definition

A circle is a locus of a moving point, which is always at a fixed distance from a fixed point. The fixed point is called its *centre* and the fixed distance is called its *radius*.

3.2 Equation of a Circle

If we take the point (h, k) as the centre and r as the radius of the circle, then equation of the circle is given by

$$(x - h)^2 + (y - k)^2 = r^2 \tag{1}$$

which can be simplified to

$$x^2 + y^2 + 2gx + 2fy + c = 0$$

where $g = -h$, $f = -k$ and $c = h^2 + k^2 - r^2$ and this is taken as the general equation of the circle. So comparing the general second degree equation in x, y with it we see that the general equation $ax^2 + 2hxy + by^2 + 2gx + 2fy + c = 0$ will represent a circle if

$$a = b \quad \text{and} \quad h = 0.$$

If the centre of the circle be origin, then the Eqn. (1) reduces to

$$x^2 + y^2 = r^2.$$

If the circle (1) passes through origin, then its equation becomes

$$x^2 + y^2 - 2hx - 2ky = 0.$$

Equation of a circle whose centre is at the point (h, k) and which touches the x-axis is

$$(x - h)^2 + (y - k)^2 = k^2$$

$$\text{i.e.,} \quad x^2 + y^2 - 2hx - 2ky = h^2.$$

Similarly, equation of a circle whose centre is at the point (h, k) and which touches the y-axis is

$$(x-h)^2 + (y-k)^2 = h^2$$
$$\text{i.e.,} \quad x^2 + y^2 - 2hx - 2ky = k^2$$

and equation of a circle, which touches both the axes, is

$$(x-h)^2 + (y-h)^2 = h^2$$
$$\text{i.e.,} \quad x^2 + y^2 - 2hx - 2hy + h^2 = 0.$$

If the points (x_1, y_1) and (x_2, y_2) be taken as the end points of a diameter of a circle, then for any point (x, y) on the perimeter of the circle we have

$$\frac{y-y_1}{x-x_1} \times \frac{y-y_2}{x-x_2} = -1$$
$$\text{or,} \quad (x-x_1)(x-x_2) + (y-y_1)(y-y_2) = 0$$

which is the equation of a circle whose end points are (x_1, y_1) and (x_2, y_2).

3.3 Position of a Point with Respect to a Circle

Any point $P(h, k)$ will be outside, on or inside the circle

$$x^2 + y^2 + 2gx + 2fy + c = 0$$

according as the distance of it from the centre $(-g, -f)$ is greater, equal to or less than the radius $\sqrt{(g^2 + f^2 - c)}$.

$$\text{i.e., according as} \quad (h+g)^2 + (k+f)^2 >, \; = \; \text{or} \; < g^2 + f^2 - c$$
$$\text{i.e., according as} \quad h^2 + k^2 + 2gh + 2fk + c >, \; = \; \text{or} \; < 0.$$

3.4 Equation of Tangent and Normal at any Point $P(h, k)$ on a Circle

Consider the circle

$$x^2 + y^2 + 2gx + 2fy + c = 0.$$

As the normal to a circle at any point passes through its centre, the tangent at that point is perpendicular to the radius. Now slope of the radius CP, C being the centre of the circle, is $(k+f)/(h+g)$ and hence the slope of the tangent at that point is $-(h+g)/(k+f)$. So the equation of the tangent at $P(h,k)$ is

$$y - k = -\frac{h+g}{k+f}(x-h)$$
$$\text{or,} \quad (y-k)(k+f) + (x-h)(h+g) = 0$$
$$\text{or,} \quad xh + yk + gx + fy = h^2 + k^2 + hg + kf$$
$$\text{or,} \quad xh + yk + gx + fy = h^2 + k^2 + 2hg + 2kf + c - hg - kf - c$$
$$\text{or,} \quad xh + yk + g(x+h) + f(y+k) + c = 0,$$

[as (h, k) lies on the circle]

and the equation of normal is

$$y - k = \frac{k+f}{h+g}(x-h)$$

or, $$(y-k)(h+g) - (x-h)(k+f) = 0$$

or, $$y(h+g) - kh - kg - x(k+f) + hk + hf = 0$$

or, $$y(h+g) - x(k+f) + hf - kg = 0.$$

Cor. 1. *If we take the equation of the circle as $x^2 + y^2 = a^2$, then the line $y = mx + c$ will touch it if the perpendicular distance from the centre of the circle to the line $y = mx + c$ is equal to the radius of the circle, which gives*

$$\frac{c}{\pm\sqrt{1+m^2}} = a$$

i.e., $y = mx \pm a\sqrt{1+m^2}$ touches the circle $x^2 + y^2 = a^2$ for all values of m.

Again, if the line $y = mx + c$ touches the circle $x^2 + y^2 = a^2$ at (h, k), then the equations $xh + yk = a^2$ and $y = mx + c$ must be identical, which gives

$$\frac{-m}{h} = \frac{1}{k} = \frac{c}{a^2}, \quad \text{i.e. } h = \frac{-ma^2}{c} \quad \text{and } k = \frac{a^2}{c}.$$

So, the line $y = mx + c$ touches the circle $x^2 + y^2 = a^2$ at $(-ma^2/c, a^2/c)$.

3.5 Length of the Tangent Drawn from (h, k) to the Circle $x^2 + y^2 + 2gx + 2fy + c = 0$

The centre of the circle is the point $C : (-g, -f)$ and radius of the circle is $\sqrt{(g^2 + f^2 - c)}$. If the tangent PT to the circle is drawn from the point $P(h, k)$, then from the triangle CPT, we have

$$CP^2 = PT^2 + CT^2$$

or, $$(h+g)^2 + (k+f)^2 = PT^2 + g^2 + f^2 - c$$

or, $$PT^2 = h^2 + k^2 + 2gh + 2fk + c$$

or, $$PT = \sqrt{h^2 + k^2 + 2gh + 2fk + c}.$$

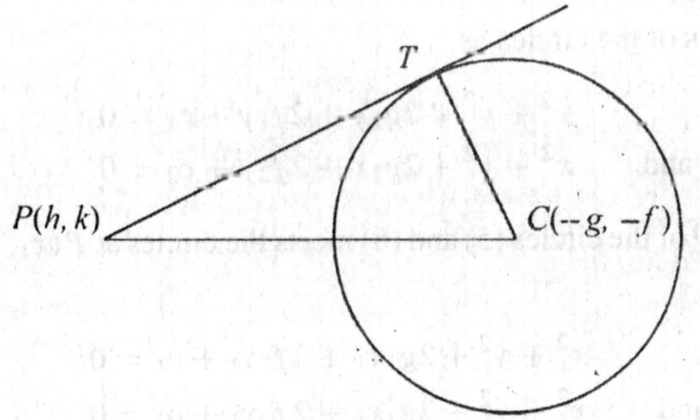

3.6 Equation of Chord of Contact of the Point $P(h, k)$ with Respect to the Circle $x^2 + y^2 = a^2$

Let the tangents drawn from the point (h, k) to the circle $x^2 + y^2 = a^2$ touch the circle at (x_1, y_1) and (x_2, y_2). Equation of tangent at the point (x_1, y_1) to the circle $x^2 + y^2 = a^2$ is

$$xx_1 + yy_1 = a^2. \tag{2}$$

If it passes through the point (h, k), then

$$hx_1 + ky_1 = a^2. \tag{3}$$

Again, equation of tangent at the point (x_2, y_2) to the circle $x^2 + y^2 = a^2$ is

$$xx_2 + yy_2 = a^2.$$

As it passes through the point (h, k), then

$$hx_2 + ky_2 = a^2. \tag{4}$$

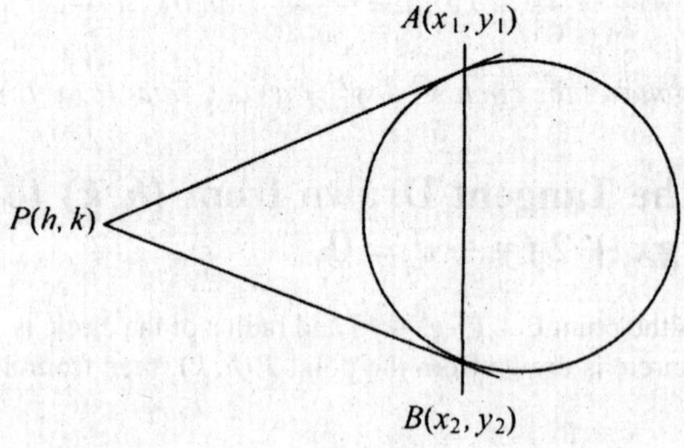

From Eqns. (3) and (4) we observe that the line $hx + ky = a^2$ passes through the points (x_1, y_1) and (x_2, y_2), and this is the equation of the chord of contact of the point $P(h, k)$ with respect to the circle $x^2 + y^2 = a^2$.

3.7 Equation of the Common Chord of Two Circles

Let us take the equations of the circles as

$$x^2 + y^2 + 2g_1x + 2f_1y + c_1 = 0 \tag{5}$$

$$\text{and} \quad x^2 + y^2 + 2g_2x + 2f_2y + c_2 = 0. \tag{6}$$

If the common chord PQ of the circles (5) and (6) meets the circles at $P(x_1, y_1)$ and $Q(x_2, y_2)$, then we have

$$x_1^2 + y_1^2 + 2g_1x_1 + 2f_1y_1 + c_1 = 0$$

$$\text{and} \quad x_1^2 + y_1^2 + 2g_2x_1 + 2f_2y_1 + c_2 = 0.$$

From these equations we get

$$2x_1(g_1 - g_2) + 2y_1(f_1 - f_2) + c_1 - c_2 = 0. \tag{7}$$

Similarly, we also get

$$2x_2(g_1 - g_2) + 2y_2(f_1 - f_2) + c_1 - c_2 = 0. \tag{8}$$

From Eqns. (7) and (8) we observe that the line

$$2x(g_1 - g_2) + 2y(f_1 - f_2) + c_1 - c_2 = 0 \tag{9}$$

passes through the points P: (x_1, y_1) and Q: (x_2, y_2). As the circles are not concentric and also different, we cannot have

$$g_1 = g_2 \quad \text{and} \quad f_1 = f_2 \quad \text{simultaneously}$$

which shows that the line (9) is the equation of the common chord of the circles.

3.8 Equation of Pair of Tangents from a Given Point $P(h, k)$

Let the pair of tangents from the point $P(h, k)$ be drawn to the circle $x^2 + y^2 = a^2$. Equation of any line through the point (h, k) is of the form

$$y - k = m(x - h). \tag{10}$$

If this line touches the circle, then we have

$$\frac{mh - k}{\sqrt{1 + m^2}} = a$$

$$\text{or,} \qquad (mh - k)^2 = a^2(1 + m^2). \tag{11}$$

Eliminating m from Eqn. (11) with the help of Eqn. (10), we get

$$\left(h\frac{y-k}{x-h} - k\right)^2 = a^2\left[1 + \left(\frac{y-k}{x-h}\right)^2\right]$$

$$\text{or,} \quad (hy - xk)^2 = a^2[(x - h)^2 + (y - k)^2]$$

$$\text{or,} \quad x^2(k^2 - a^2) + y^2(h^2 - a^2) - a^2(h^2 + k^2) = 2hkxy - 2a^2hx - 2a^2ky.$$

Adding $x^2h^2 + y^2k^2 + a^4$ to both sides we get

$$x^2(h^2 + k^2 - a^2) + y^2(h^2 + k^2 - a^2) - a^2(h^2 + k^2 - a^2)$$

$$= x^2h^2 + y^2k^2 + a^4 + 2hkxy - 2a^2hx - 2a^2ky$$

$$\text{or,} \quad (x^2 + y^2 - a^2)(h^2 + k^2 - a^2) = (xh + ky - a^2)^2.$$

This is the equation of pair of tangents from a given point $P(h, k)$ to the circle $x^2 + y^2 = a^2$ and of the form

$$SS_1 = T^2$$

where $S = x^2 + y^2 - a^2$, $S_1 = S$ at (h, k) and $T = xh + ky - a^2$.

Similarly, equation of the pair of tangents to the circle at $x^2 + y^2 + 2gx + 2fy + c = 0$, drawn from the point (x_1, y_1) is

$$(x^2 + y^2 + 2gx + 2fy + c)(x_1^2 + y_1^2 + 2gx_1 + 2fy_1 + c)$$

$$= [xx_1 + yy_1 + g(x + x_1) + f(y + y_1) + c]^2.$$

3.9 Parametric Equation of a Circle

If any point (x, y) on a curve can be expressed as $x = \psi(t)$ and $y = \phi(t)$, where t is an independent variable. Then $x = \psi(t)$, $y = \phi(t)$ are called *parametric equation* of the curve and t is called *parameter*, from which t can be eliminated to get $f(x, y) = 0$. If a radius of the circle $x^2 + y^2 = a^2$ makes an angle θ with x-axis, then we can write

$$x = a\cos\theta \quad \text{and} \quad y = a\sin\theta$$

and as θ varies from 0 to 2π, the point (x, y) describes the circle. So the above relations are called *parametric equation* of the circle $x^2 + y^2 = a^2$.

3.10 Orthogonal Circles

Angle of intersection of two circles is measured as the angle between the tangents to the circles at the point of intersection. If the angle between two circles be a right angle, then we say that the circles intersect orthogonally.

Consider two circles:

$$S = x^2 + y^2 + 2gx + 2fy + c = 0$$
$$\text{and} \quad S_1 = x^2 + y^2 + 2g_1x + 2f_1y + c_1 = 0.$$

Their centres are given by $C: (-g, -f)$ and $C_1: (-g_1, -f_1)$. If P be the point of intersection of the circles, then the tangent T_1PC_1 to the circle $S = 0$ will pass through the centre C_1 of the circle $S_1 = 0$ and similarly, the tangent TPC to the circle $S_1 = 0$ will pass through the centre C of the circle $S = 0$, which imply that the triangle ΔCPC_1 is a right-angled triangle and for this

$$CP^2 + C_1P^2 = CC_1^2$$
$$\text{or,} \quad (g - g_1)^2 + (f - f_1)^2 = g^2 + f^2 - c + g_1^2 + f_1^2 - c_1$$
$$\text{or,} \quad 2gg_1 + 2ff_1 = c + c_1.$$

This is the required condition of orthogonality of the circles $S = 0$ and $S_1 = 0$.

3.11 Radical Axis of Two Circles

The radical axis of two circles is defined to be the locus of a point which moves in such a way that the lengths of the tangents drawn from it to the circles are of equal length.

Consider two circles:

$$S = x^2 + y^2 + 2gx + 2fy + c = 0$$
$$\text{and} \quad S_1 = x^2 + y^2 + 2g_1x + 2f_1y + c_1 = 0.$$

Let (h, k) be a moving point such that the tangents drawn from it to the circles $S = 0$ and $S_1 = 0$ are equal. Then

$$h^2 + k^2 + 2gh + 2fk + c = h^2 + k^2 + 2g_1h + 2f_1k + c_1$$
$$\text{or,} \quad 2(g - g_1)h + 2(f - f_1)k + c - c_1 = 0.$$

Hence, generalizing, we get the locus of the point (h, k), i.e. the equation of the radical axis is

$$2(g - g_1)x + 2(f - f_1)y + c - c_1 = 0$$

which is a straight line and of the form $S - S_1 = 0$, where $S = 0$ and $S_1 = 0$ represent two circles with coefficients of x^2 and y^2 unity.

3.12 Properties of Radical Axis

P1 : *The radical axis of two circles is perpendicular to the line joining their centres.*

If we take the equation of circles as

$$x^2 + y^2 + 2g_1x + 2f_1y + c_1 = 0$$
$$\text{and} \quad x^2 + y^2 + 2g_2x + 2f_2y + c_2 = 0$$

then their radical axis is

$$2(g_2 - g_1)x + 2(f_2 - f_1)y + c_2 - c_1 = 0$$

whose slope is $-(g_2 - g_1)/(f_2 - f_1)$ and slope of the line joining centres of the circles, i.e. of the line joining $(-g_1, -f_1)$ and $(-g_2, -f_2)$ is $(f_2 - f_1)/(g_2 - g_1)$. Now we see that product of the slopes is -1. So, the radical axis of two circles is perpendicular to the line joining their centres.

P2 : *The radical axes of three circles taken in pairs are concurrent. The common point of intersection is called the* **radical centre**.

If we take the equation of circles as

$$x^2 + y^2 + 2g_1x + 2f_1y + c_1 = 0$$
$$x^2 + y^2 + 2g_2x + 2f_2y + c_2 = 0$$
$$x^2 + y^2 + 2g_3x + 2f_3y + c_3 = 0$$

then equations of the radical axes of the circles taken in pairs are

$$2(g_2 - g_1)x + 2(f_2 - f_1)y + c_2 - c_1 = 0$$
$$2(g_3 - g_2)x + 2(f_3 - f_2)y + c_3 - c_2 = 0$$
$$2(g_1 - g_3)x + 2(f_1 - f_3)y + c_1 - c_3 = 0.$$

The coefficient determinant is

$$\begin{vmatrix} 2(g_2 - g_1) & 2(f_2 - f_1) & c_2 - c_1 \\ 2(g_3 - g_2) & 2(f_3 - f_2) & c_3 - c_2 \\ 2(g_1 - g_3) & 2(f_1 - f_3) & c_1 - c_3 \end{vmatrix}.$$

Performing $R_1 + R_2 + R_3 \to R_1$, we see that the determinant vanishes, which shows that the radical axes are concurrent.

P3 : *The radical axis of two circles bisect their common tangent.*

P4 : *The radical centre of three circles described on sides of a triangle as diameters is the orthocentre of the triangle.*

P5 : *If two circles cut a third circle orthogonally, the radical axis of the two circles will pass through the centre of the third circle.*

3.13 Coaxial Circles and Its Various Forms

A system of circles, every pair of which has the same radical axis is called *coaxial circles.*

A system of circles can be coaxial with two given circles $S_2 = 0$ and $S_1 = 0$, and the form of the coaxial circles is $S_1 - kS_2 = 0$, for $k \neq 1$.

A system of circles can be coaxial with a circle $S = 0$ and a straight line $P = 0$, and the form of the coaxial circle is $S - kP = 0$.

Examples

▶▷▶ 1. *Prove that the locus of the centre of the circles which intercept a chord of length 2a on x-axis and passes through point* $(0, b)$ *is the curve*

$$x^2 - 2yb + b^2 = a^2.$$

Solution: Consider the circle

$$(x - \alpha)^2 + (y - \beta)^2 = r^2.$$

Putting $y = 0$ we get

$$(x - \alpha)^2 + (0 - \beta)^2 = r^2$$

or, $$(x - \alpha)^2 = r^2 - \beta^2$$

or, $$x = \alpha \pm \sqrt{r^2 - \beta^2}.$$

By the given condition, we have

$$2\sqrt{r^2 - \beta^2} = 2a$$

or, $$r^2 = \beta^2 + a^2. \qquad (1)$$

Again it is given that the circle passes through $(0, b)$. So, we get

$$(0 - \alpha)^2 + (b - \beta)^2 = r^2 = \beta^2 + a^2 \quad \text{[by Eqn. (1)]}$$

or, $$\alpha^2 + b^2 - 2b\beta + \beta^2 = \beta^2 + a^2$$

or, $$\alpha^2 + b^2 - 2b\beta = a^2.$$

Generalizing we get the locus of the centre of the circle is

$$x^2 - 2yb + b^2 = a^2. \quad \textbf{[Proved]}$$

2. *Prove that the equation of the circle on the line joining the intersection of $ax^2 + 2hxy + by^2 = 0$ and $lx + my = 1$ as diameter is*

$$(x^2 + y^2)(am^2 - 2hlm + bl^2) + 2x(hm - bl) + 2y(hl - am) + a + b = 0.$$

Solution: If $y = m_1x$ and $y = m_2x$ be the individual equations of the line given by $ax^2 + 2hxy + by^2 = 0$, then we have

$$m_1 + m_2 = -\frac{2h}{b} \quad \text{and} \quad m_1m_2 = \frac{a}{b}. \tag{1}$$

Solving $y = m_1x$ with $lx + my = 1$ we get $x = \dfrac{1}{l + mm_1}$; $y = \dfrac{m_1}{l + mm_1}$; and solving $y = m_2x$ with $lx + my = 1$ we get $x = \dfrac{1}{l + mm_2}$; $y = \dfrac{m_2}{l + mm_2}$.

Hence, the equation of the circle describing the line segment joining $\left(\dfrac{1}{l + mm_1}, \dfrac{m_1}{l + mm_1}\right)$ and $\left(\dfrac{1}{l + mm_2}, \dfrac{m_2}{l + mm_2}\right)$ as diameter is

$$\left(x - \frac{1}{l + mm_1}\right)\left(x - \frac{1}{l + mm_2}\right) + \left(y - \frac{m_1}{l + mm_1}\right)\left(y - \frac{m_2}{l + mm_2}\right) = 0$$

or, $$x^2 + y^2 - x\left(\frac{1}{l + mm_1} + \frac{1}{l + mm_2}\right) - y\left(\frac{m_1}{l + mm_1} + \frac{m_2}{l + mm_2}\right) + \frac{1 + m_1m_2}{(l + mm_1)(l + mm_2)} = 0$$

or, $$x^2 + y^2 - \frac{2l + m(m_1 + m_2)}{l^2 + lm(m_1 + m_2) + m^2m_1m_2}x - \frac{2mm_1m_2 + l(m_1 + m_2)}{l^2 + lm(m_1 + m_2) + m^2m_1m_2}y + \frac{m_1m_2 + 1}{l^2 + lm(m_1 + m_2) + m^2m_1m_2} = 0.$$

Now, using Eqn. (1) and $l^2 + lm(m_1 + m_2) + m^2m_1m_2 = \dfrac{am^2 - 2hlm + bl^2}{b}$ we get the equation of the circle as

$$(x^2 + y^2)(am^2 - 2hlm + bl^2) + 2x(hm - bl) + 2y(hl - am) + a + b = 0. \qquad \textbf{[Proved]}$$

3. *Show that the equation of the straight line meeting the circle $x^2 + y^2 = a^2$ in two points at equal distance d from (x_1, y_1) on the circle is $xx_1 + yy_1 - a^2 + d^2/2 = 0$.*

Solution: Let the straight line meets the circle at $B: (x_2, y_2)$ and $C: (x_3, y_3)$. Then

$$AB^2 = (x_1 - x_2)^2 + (y_1 - y_2)^2 = d^2$$

or, $$(x_1^2 + y_1^2) + (x_2^2 + y_2^2) - 2(x_1x_2 + y_1y_2) = d^2$$

or, $$2(x_1x_2 + y_1y_2) = 2a^2 - d^2. \tag{1}$$

Similarly, $$2(x_1x_3 + y_1y_3) = 2a^2 - d^2. \tag{2}$$

Now, the triangle ABC is isosceles triangle, so the perpendicular from $A: (x_1, y_1)$ on BC will bisect the chord and will pass through the centre $(0, 0)$.

Slope of the bisector is y_1/x_1. So slope of BC is $-x_1/y_1$. Hence equation of BC is

$$y - y_2 = -\frac{x_1}{y_1}(x - x_2)$$

$$\text{or,}\quad yy_1 + xx_1 = y_2y_1 + x_2x_1 = a^2 - d^2/2. \qquad \textbf{[Proved]}$$

▶▶ **4**. *Find the equation of the circle circumscribing the triangle formed by the lines* $a_1x + b_1y + c_1 = 0$*;* $a_2x + b_2y + c_2 = 0$*;* $a_3x + b_3y + c_3 = 0$.

Solution: Let us consider the equation

$$(a_1x + b_1y + c_1)(a_2x + b_2y + c_2) + \lambda(a_3x + b_3y + c_3)(a_2x + b_2y + c_2) + \mu(a_1x + b_1y + c_1)(a_3x + b_3y + c_3) = 0. \tag{1}$$

Above equation is satisfied by the points obtained by taking any two of three given equations, which means that the curve given by Eqn. (1) passes through the vertices of the triangle formed by the given lines.

Now the Eqn. (1) will represent a circle if

$$\text{coefficient of } x^2 = \text{coefficient of } y^2 \quad \text{and} \quad \text{coefficient of } xy = 0,$$

which give

$$a_1a_2 - b_1b_2 + \lambda(a_3a_2 - b_3b_2) + \mu(a_1a_3 - b_1b_3) = 0 \tag{2}$$

$$\text{and}\quad a_1b_2 + b_1a_2 + \lambda(b_3a_2 + a_3b_2) + \mu(a_1b_3 + b_1a_3) = 0. \tag{3}$$

From Eqn. (1) we get

$$\frac{1}{a_3x + b_3y + c_3} + \frac{\lambda}{a_1x + b_1y + c_1} + \frac{\mu}{a_2x + b_2y + c_2} = 0. \tag{4}$$

Eliminating λ and μ from the Eqns. (4), (3) and (2) we get

$$\begin{vmatrix} \dfrac{1}{a_3x + b_3y + c_3} & \dfrac{\lambda}{a_1x + b_1y + c_1} & \dfrac{\mu}{a_2x + b_2y + c_2} \\ a_1a_2 - b_1b_2 & a_3a_2 - b_3b_2 & a_1a_3 - b_1b_3 \\ a_1b_2 + b_1a_2 & b_3a_2 + a_3b_2 & a_1b_3 + b_1a_3 \end{vmatrix} = 0$$

$$\text{or,}\quad \sum \frac{(a_3a_2 - b_3b_2)(a_1b_3 + b_1a_3) - (a_3a_1 - b_3b_1)(a_2b_3 + b_2a_3)}{a_3x + b_3y + c_3} = 0$$

$$\text{or,}\quad \sum \frac{(a_3^2 + b_3^2)(a_1b_2 - b_1a_2)}{a_3x + b_3y + c_3} = 0$$

$$\text{or,}\quad \begin{vmatrix} \dfrac{a_3^2 + b_3^2}{a_3x + b_3y + c_3} & a_3 & b_3 \\ \dfrac{a_2^2 + b_2^2}{a_2x + b_2y + c_2} & a_2 & b_2 \\ \dfrac{a_1^2 + b_1^2}{a_1x + b_1y + c_1} & a_1 & b_1 \end{vmatrix} = 0$$

which is the equation of the circle circumscribing the triangle formed by the given lines.

▶▷▶ 5. *The straight line $Ax + By = 1$ makes with $ax^2 + 2hxy + by^2 = c$ a chord which subtends a right angle at the origin, show that $c(A^2 + B^2) = a + b$ and that the chord envelopes a circle of radius $\sqrt{\dfrac{c}{a+b}}$.*

Solution: Making homogeneous $ax^2 + 2hxy + by^2 = c$ with the aid of $Ax + By = 1$ we get

$$ax^2 + 2hxy + by^2 = c(Ax + By)^2$$

$$\text{or,} \quad (a - cA^2)x^2 + 2(h - cAB)xy + (b - cB^2)y^2 = 0$$

which is a second degree homogeneous equation in x, y and represents a pair of straight lines passing through the origin.

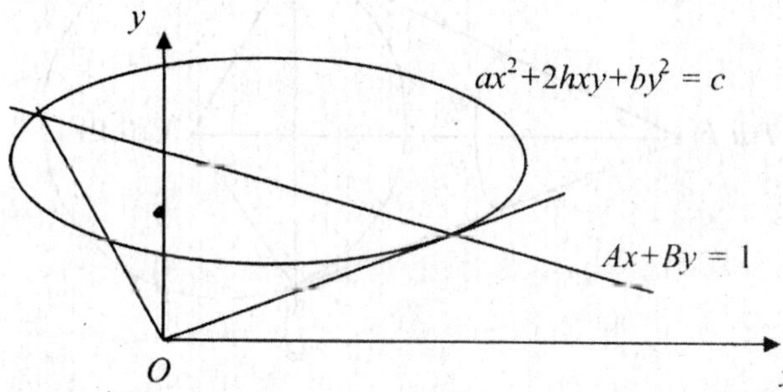

If the angle between them is 90°, then we have

$$(a - cA^2) + (b - cB^2) = 0$$

$$\text{or,} \quad c(A^2 + B^2) = a + b. \qquad (1)$$

Now, if the straight line $Ax + By = 1$ touches the circle $x^2 + y^2 = r^2$, then we have

$$r = \frac{1}{\sqrt{A^2 + B^2}} = \sqrt{\frac{c}{a+b}} \quad \text{[by Eqn. (1)].} \qquad \textbf{[Proved]}$$

▶▷▶ 6. *Show that the locus of the points, chords of contact of tangents from which to the circle $x^2 + y^2 = a^2$ subtends a right angle at the centre is a concentric circle whose radius is $\sqrt{2}$ times the radius of the given circle.*

Solution: Equation of the chord of contact of the point (h, k) with respect to the circle $x^2 + y^2 = a^2$ is

$$xh + yk = a^2 \qquad (1)$$

Now, making homogeneous the equation $x^2 + y^2 = a^2$ with the aid of Eqn. (1) we get

$$x^2 + y^2 = a^2\left(\frac{xh + yk}{a^2}\right)^2$$

$$\text{or,} \quad x^2(a^2 - h^2) + y^2(a^2 - k^2) - 2hkxy = 0$$

which is a second degree homogeneous equation in x, y and represents a pair of straight lines passing through the origin.

If the angle between them is 90°, then we have

$$(a^2 - h^2) + (a^2 - k^2) = 0$$

$$\text{or,} \quad h^2 + k^2 = 2a^2.$$

Generalizing we get the locus of the point (h, k) is

$$x^2 + y^2 = 2a^2. \quad \textbf{[Proved]}$$

▶▷▶ 7. *Find the angle between the pair of tangents drawn from the point $T(h, k)$ to the circle $x^2 + y^2 = a^2$. Find also the area of the triangle formed by the tangents and their chord of contact.*

Solution: Let the tangents drawn from $T(h, k)$ to the circle touch at P and Q and angle $\angle PTQ = 2\theta$. Then due to symmetry, $\angle PTC = \theta = \angle QTC$.

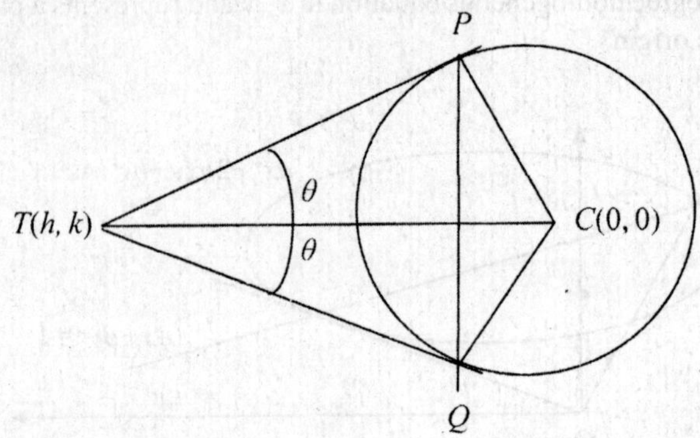

The length of the tangents $= \sqrt{h^2 + k^2 - a^2}$.
From the triangle ΔCPT, we get

$$\sin\theta = \frac{CP}{CT} = \frac{a}{\sqrt{h^2 + k^2}}, \quad \cos\theta = \frac{PT}{CT} = \frac{\sqrt{h^2 + k^2 - a^2}}{\sqrt{h^2 + k^2}}$$

i.e., $$\tan\theta = \frac{a}{\sqrt{h^2 + k^2 - a^2}}.$$

So, the angle between the tangents $= 2\theta = 2\tan^{-1}\left(\frac{a}{\sqrt{h^2 + k^2 - a^2}}\right)$.

Now that area of the triangle

$$\begin{aligned} \Delta PTQ &= \frac{1}{2} TP \times TQ \times \sin 2\theta \\ &= TP \times TQ \times \sin\theta \times \cos\theta \\ &= (h^2 + k^2 - a^2) \times \frac{a}{\sqrt{h^2 + k^2}} \times \frac{\sqrt{h^2 + k^2 - a^2}}{\sqrt{h^2 + k^2}} \\ &= \frac{a(h^2 + k^2 - a^2)^{3/2}}{h^2 + k^2}. \end{aligned}$$

▶▷▶ 8. *If $S_1 = 0$ and $S_2 = 0$ represent two circles, then find the equation of the curve which passes through the points of intersection of $S_1 = 0$ and $S_2 = 0$.*

Solution: Let $S_1 = x^2 + y^2 + 2gx + 2fy + c$ and $S_2 = x^2 + y^2 + 2g'x + 2f'y + c'$. Then

$$\begin{aligned} S_1 + kS_2 &= x^2 + y^2 + 2gx + 2fy + c + k(x^2 + y^2 + 2g'x + 2f'y + c') \\ &= (1 + k)x^2 + (1 + k)y^2 + 2(g + kg')x + 2(f + kf')y + (c + kc'). \end{aligned}$$

So, $S_1 + kS_2 = 0$,

$$\text{i.e.,} \quad x^2 + y^2 + 2\left(\frac{g + kg'}{1 + k}\right)x + 2\left(\frac{f + kf'}{1 + k}\right)y + \left(\frac{c + kc'}{1 + k}\right)x = 0.$$

As the equation $S_1 + kS_2 = 0$ is satisfied by those points which satisfy both $S_1 = 0$ and $S_2 = 0$, this is the equation of the curve which passes through the points of intersection of $S_1 = 0$ and $S_2 = 0$, and represents a circle if $k \neq -1$.

When $k = -1$, then $S_1 - S_2 = 0$ gives $2(g - g')x + 2(g - f')y + c - c' = 0$, which represents the common chord of the circles $S_1 = 0$ and $S_2 = 0$.

▶▷▶ **9.** *If $S = 0$ and $P = 0$ represent a circle and a straight line respectively, then find the equation of the circle described the chord $P = 0$ of the circle $S = 0$ as diameter.*

Solution: Let $S = x^2 + y^2 + 2gx + 2fy + c$ and $P = lx + my + n$, then

$$\begin{aligned} S + kP &= x^2 + y^2 + 2gx + 2fy + c + 2k(lx + my + n) \\ &= x^2 + y^2 + 2(g + kl)x + 2(f + km)y + c + 2kn. \end{aligned}$$

As the equation $S + kP = 0$ is satisfied by those points which satisfy both $S = 0$ and $P = 0$, the equation of the circle passes through the points of intersection $S = 0$ and $P = 0$ is

$$x^2 + y^2 + 2(g + kl)x + 2(f + km)y + c + 2kn = 0$$

whose centre is the point $\{-(g + kl), -(f + km)\}$ and this point must lie on the diameter

$$\begin{aligned} & lx + my + n = 0 \\ \text{i.e.,} \quad & l(g + kl) + m(f + km) = n \\ \text{or,} \quad & k(l^2 + m^2) = n - lg - mf \\ \text{or,} \quad & k = \frac{n - lg - mf}{l^2 + m^2}. \end{aligned}$$

So, the equation of the circle is

$$x^2 + y^2 + 2gx + 2fy + c + 2\frac{n - lg - mf}{l^2 + m^2}(lx + my + n) = 0.$$

▶▷▶ **10.** *A chord of circle $x^2 + y^2 - 2ax = 0$ is drawn to pass through the origin. Show that the locus of the centres of the circles described on this chord as diameter is a circle passing through the centre of the given circle.*

Solution: Let us take $y = mx$ as chord of the given circle, where m is a variable. Then by Problem No. 9, the equation of the circle on this chord as diameter is given by

$$\begin{aligned} & x^2 + y^2 - 2ax + k(y - mx) = 0 \\ \text{or,} \quad & x^2 + y^2 - (2a + km)x + ky = 0 \end{aligned}$$

whose centre is $(\alpha, \beta) = (a + km/2, -k/2)$. If this point lies on the line $y = mx$, then

$$-\frac{k}{2} = \frac{m(2a + km)}{2} \Rightarrow k(1 + m^2) = -2am \Rightarrow k = -\frac{2am}{1 + m^2}.$$

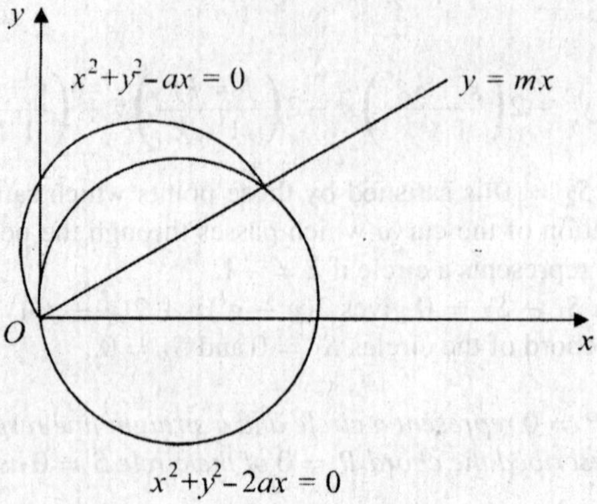

Now, $\alpha = a + km/2 = a - \dfrac{am^2}{1+m^2} = \dfrac{a}{1+m^2}$ and $\beta = -k/2 = \dfrac{am}{1+m^2}$.

Eliminating m we get

$$\alpha^2 + \beta^2 = \left(\frac{a}{1+m^2}\right)^2 + \left(\frac{am}{1+m^2}\right)^2 = \frac{a^2}{1+m^2} = a\alpha$$

$$\text{or,} \quad \alpha^2 + \beta^2 - a\alpha = 0.$$

Hence, locus of the centre, i.e. of the point (α, β) is $x^2 + y^2 - ax = 0$, which is a circle passing through the point $(a, 0)$, i.e. the centre of the given circle.

[Proved]

▶▷ **11**. *Prove that the equation of the circle on the common chord of the circles*

$$(x-h)^2 + y^2 = h^2 \quad \textit{and} \quad x^2 + (y-k)^2 = k^2$$

as diameter is $(x^2+y^2)(h^2+k^2) = 2hk(kx+hy)$.

Solution: Equation of the common chord is

$$(x-h)^2 + y^2 - h^2 - [x^2 + (y-k)^2 - k^2] = 0$$

$$\text{or,} \quad xh - yk = 0.$$

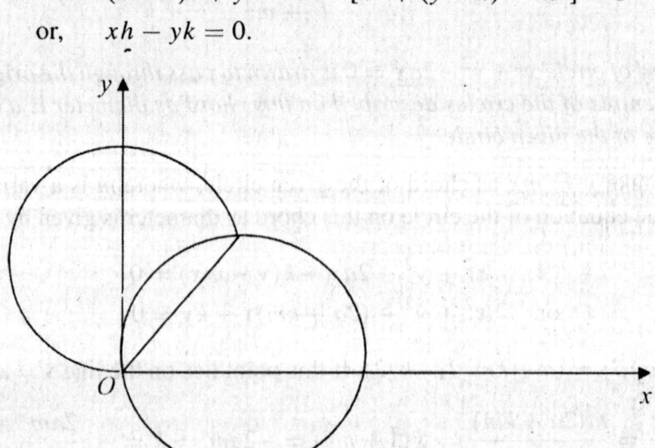

Now, equation of the circle, which passes through the points of intersection of the circle $(x-h)^2+y^2=h^2$ and the line $xh-yk=0$ is

$$(x-h)^2+y^2-h^2+\lambda(xh-yk)=0$$
$$\text{or,}\quad x^2+y^2+(\lambda-2)xh-\lambda yk=0$$

whose centre is $\left(\left(1-\frac{\lambda}{2}\right)h, \frac{\lambda k}{2}\right)$. If this point lies on the line $xh-yk=0$, then we have

$$h^2\left(1-\frac{\lambda}{2}\right)-\frac{\lambda k^2}{2}=0$$
$$\text{or,}\quad \lambda(h^2+k^2)=2h^2$$
$$\text{or,}\quad \lambda=\frac{2h^2}{h^2+k^2}.$$

Hence, the equation of the circle is

$$x^2+y^2+\left(\frac{2h^2}{h^2+k^2}-2\right)xh-\frac{2h^2}{h^2+k^2}yk=0$$
$$\text{or,}\quad x^2+y^2-\frac{2k^2}{h^2+k^2}xh-\frac{2h^2}{h^2+k^2}yk=0$$
$$\text{or,}\quad (x^2+y^2)(h^2+k^2)=2hk(kx+hy).\qquad \textbf{[Proved]}$$

▶▷▶ **12.** *If two plane curves*

$$ax^2+2hxy+by^2+2gx+2fy+c=0$$
$$\textit{and}\quad Ax^2+2Hxy+By^2+2Gx+2Fy+C=0$$

intersect in four concyclic points, prove that

$$H(a-b)=h(A-B).$$

Solution: Equation of the curve which passes through the point of intersection of the given curves can be written as

$$ax^2+2hxy+by^2+2gx+2fy+c$$
$$+k(Ax^2+2Hxy+By^2+2Gx+2Fy+C)=0$$
$$\text{or,}\quad (a+kA)x^2+2(h+kH)xy+(b+kB)y^2$$
$$+2(g+kG)x+2(f+kF)y+c+kC=0.$$

If the above equation represents a circle, then we have

$$\text{Coeff. of } x^2=\text{Coeff. of } y^2 \quad\text{and}\quad \text{Coeff. of } xy=0$$
$$\text{i.e.,}\quad a+kA=b+kB \quad\text{and}\quad h+kH=0.$$

Eliminating k we get

$$H(a-b)=h(A-B).\qquad \textbf{[Proved]}$$

▶▷▶ **13.** *Prove that two circles each of which passes through two fixed points* $(0,k)$ *and* $(0,-k)$ *and touch the line* $y=mx+b$ *will cut orthogonally, if* $b^2=k^2(2+m^2)$.

Solution: If one of the circles be

$$x^2 + y^2 + 2gx + 2fy + c = 0$$

then we have

$$k^2 + 2fk + c = 0 \quad \text{and} \quad k^2 - 2fk + c = 0.$$

From these equations, we get $f = 0$ and $c = -k^2$.

So, the equation of the above circle becomes

$$x^2 + y^2 + 2gx - k^2 = 0 \tag{1}$$

whose centre is $(-g, 0)$, radius $= \sqrt{g^2 + k^2}$.

Now, if the above circle touches the line $y = mx + b$, then we have

$$b - mg = \sqrt{(g^2 + k^2)(1 + m^2)}$$

or, $\quad b^2 - 2mg + m^2g^2 = g^2(1 + m^2) + k^2(1 + m^2)$

or, $\quad g^2 + 2mg + k^2(1 + m^2) - b^2 = 0.$

If g_1, g_2 be its two roots, then

$$g_1g_2 = k^2(1 + m^2) - b^2.$$

Now, two circles of the form (1) will cut orthogonally if

$$2g_1g_2 + 0 = -k^2 - k^2$$

or, $\quad k^2(1 + m^2) - b^2 = -k^2$

or, $\quad b^2 = k^2(2 + m^2).$ **[Proved]**

▶▷▶ **14**. *If $S = 0$, $S' = 0$ be two circles of radii r and R, then prove that the circles $\dfrac{S}{r} \pm \dfrac{S'}{R} = 0$ will intersect orthogonally.*

Solution: Without loss of generality, we assume the equation of the circles as

$$S = (x - a)^2 + y^2 - r^2 = 0$$

and $\quad S' = (x + a)^2 + y^2 - R^2 = 0.$

Now, the individual equations of the circles $\dfrac{S}{r} \pm \dfrac{S'}{R} = 0$ are

(i) $RS + rS' = 0$,

i.e., $\quad R[(x - a)^2 + y^2 - r^2] + r[(x + a)^2 + y^2 - R^2] = 0$

or, $\quad (R + r)x^2 + (R + r)y^2 - 2a(R - r)x + (R + r)a^2 - rR(R + r) = 0$

or, $\quad x^2 + y^2 - 2a[(R - r)/(R + r)]x + a^2 - rR = 0$

and

(ii) $RS - rS' = 0$, i.e.

$$R[(x-a)^2+y^2-r^2]-r[(x+a)^2+y^2-R^2]=0$$
or, $$(R-r)x^2+(R-r)y^2-2a(R+r)x+(R-r)a^2+rR(R-r)=0$$
or, $$x^2+y^2-2a[(R+r)/(R-r)]x+a^2+rR=0.$$

Now, we see that the condition of orthogonality, i.e.

$$2g_1g_2+2f_1f_2=c_1+c_2$$

is satisfied for the circles (i) and (ii). Hence the circles $\frac{S}{r}\pm\frac{S'}{R}=0$ will intersect orthogonally.

[Proved]

▶▷▶ 15. *Find the circumcentre of the triangle formed by the lines $x+y=0$, $x-y=0$ and $lx+my=1$. If l and m vary such that $l^2+m^2=1$, show that the locus of the circumcentre of the triangle is $(x^2-y^2)^2=x^2+y^2$.*

Solution: Solving the equations $x+y=0$, $x-y=0$ and $lx+my=1$, taking two at a time, we get the vertices of the triangle are

$$A(0,0);\quad B\left(\frac{1}{l+m},\frac{1}{l+m}\right)\quad\text{and}\quad C\left(\frac{1}{l-m},\frac{-1}{l-m}\right).$$

If S: (h,k) be the circumcentre then we have $SA^2=SB^2=SC^2$ and hence

$$h^2+k^2=\left(h-\frac{1}{l+m}\right)^2+\left(k-\frac{1}{l+m}\right)^2=\left(h-\frac{1}{l-m}\right)^2+\left(k+\frac{1}{l-m}\right)^2.$$

From $h^2+k^2=\left(h-\frac{1}{l+m}\right)^2+\left(k-\frac{1}{l+m}\right)^2$, we get $h+k=\frac{1}{l+m}$ and from $h^2+k^2=\left(h-\frac{1}{l-m}\right)^2+\left(k+\frac{1}{l-m}\right)^2$, we get $h-k=\frac{1}{l-m}$. Solving these we get the circumcentre is $\left(\frac{l}{l^2-m^2},\frac{-m}{l^2-m^2}\right)$.

Again, given that $l^2+m^2=1$, using this, we get

$$h^2+k^2=\frac{l^2+m^2}{(l^2-m^2)^2}=\frac{1}{(l^2-m^2)^2}=(h^2-k^2)^2.$$

Generalizing, we get the required locus is $(x^2-y^2)^2=x^2+y^2$.

EXERCISE

1. Show that the lengths of the intercepts on the axes by the circle $x^2+y^2+2gx+2fy+c=0$, are $2\sqrt{g^2-c}$ and $2\sqrt{f^2-c}$.
2. Show that if a circle $x^2+y^2=a^2$ cuts off from the straight line $y=mx+c$, a chord of length $2b$, then $c^2=(1+m^2)(a^2-b^2)$.

   [*Hints*: Putting $y=mx+c$ in $x^2+y^2=a^2$, we get

   $$(1+m^2)x^2+2mxc+c^2-a^2=0.$$

If the line $y = mx + c$ cuts the circle $x^2 + y^2 = a^2$ in (x_1, y_1) and (x_2, y_2), then

$$x_2 + x_1 = -\frac{2mc}{1+m^2} \quad \text{and} \quad x_2 x_1 = \frac{c^2 - a^2}{1+m^2}.$$

Now, given that

$$(x_2 - x_1)^2 + (y_2 - y_1)^2 = 4b^2$$

or, $$(x_2 - x_1)^2[1 + m^2] = 4b^2$$

or, $$[(x_2 + x_1)^2 - 4x_1x_2][1 + m^2] = 4b^2$$

or, $$4[m^2c^2 - (c^2 - a^2)(1 + m^2)] = 4b^2(1 + m^2)$$

or, $$m^2c^2 - c^2 - m^2c^2 = -(1 + m^2)(a^2 - b^2)$$

or, $$c^2 = (1 + m^2)(a^2 - b^2).]$$

3. Prove that the equation of the circle described on the chord $x \cos\alpha + y \sin\alpha = p$ of the circle $x^2 + y^2 = a^2$ as diameter is $x^2 + y^2 - a^2 - 2p(x\cos\alpha + y\sin\alpha - p) = 0$.

[*Hints*: Equation of the circle described on the chord $x\cos\alpha + y\sin\alpha = p$ of the circle $x^2 + y^2 = a^2$ as diameter can be written as

$$x^2 + y^2 - a^2 + 2k(x\cos\alpha + y\sin\alpha - p) = 0$$

whose centre $(-k\cos\alpha, -k\sin\alpha)$ lies on the line $x\cos\alpha + y\sin\alpha = p$, etc.]

4. Show that two circles $x^2 + y^2 + 2gx + 2fy = 0$ and $x^2 + y^2 + 2g'x + 2f'y = 0$ will touch each other, if $fg' = f'g$.

[*Hints*: If the circles $x^2 + y^2 + 2gx + 2fy = 0$ and $x^2 + y^2 + 2g'x + 2f'y = 0$ touch each other, then at the point of contact tangents to the circles will be identical. Further, we note that the circles pass through (0, 0), etc.]

5. Find the value of λ so that the equation $\dfrac{x^2}{25} + \dfrac{y^2}{16} - 1 - \lambda(x-1)^2 = 0$ represents a circle.

[*Hints*: Express the equation as $ax^2 + by^2 + 2gx + 2fy + c = 0$ and put $a = b$.]

6. A circle touches the x-axis and cuts off an intercept of length $2a$ from y-axis. Prove that the locus of the centre of the circle is $x^2 - y^2 + a^2 = 0$.

[*Hints*: If the circle touches the x-axis at $(c, 0)$, then its equation will be

$$(x - c)^2 + (y - b)^2 = b^2.$$

Now, putting $x = 0$ in the equation of circle we get

$$(y - b)^2 = b^2 - c^2$$

or, $$y = b \pm \sqrt{b^2 - c^2}.$$

By the given condition

$$2\sqrt{b^2 - c^2} = 2a$$

or, $$b^2 - c^2 = a^2.$$

Generalizing, we get the locus of the centre, i.e. of (c, b) is

$$x^2 - y^2 + a^2 = 0.$$

7. Show that the circumcircle of the triangle formed by the straight lines $ax+by+c=0$, $bx+cy+a=0$ and $cx+ay+b=0$ passes through the origin, if

$$(b^2+c^2)(a^2+c^2)(b^2+a^2)=abc(a+b)(b+c)(c+a).$$

[*Hints*: Proceeding as Example 4, as the circle passes through origin, we get

$$\begin{vmatrix} \dfrac{a^2+b^2}{c} & a & b \\ \dfrac{c^2+b^2}{a} & b & c \\ \dfrac{a^2+c^2}{b} & c & a \end{vmatrix} = 0.$$

Now expand it.]

8. Show that the circumcircle of the triangle formed by the straight lines $x\cos\alpha+y\sin\alpha=p$, $x\cos\beta+y\sin\beta=q$ and $x\cos\gamma+y\sin\gamma=r$ passes through the origin, if

$$qr\sin(\beta-\gamma)+rp\sin(\gamma-\alpha)+pq\sin(\alpha-\beta)=0.$$

[*Hints*: Put $a_1=\cos\alpha$, $a_2=\cos\beta$, $a_3=\cos\gamma$, $b_1=\sin\alpha$, $b_2=\sin\beta$, $b_3=\sin\gamma$ and $c_1=-p$, $c_2=-q$, $c_3=-r$ in Example 4 and put $x=0$, $y=0$.]

9. Prove that the equation of the circle of which the points (x_1,y_1) and (x_2,y_2) are the end points of a chord of a segment containing the angle θ is

$$(x-x_1)(x-x_2)+(y-y_1)(y-y_2)$$
$$\pm\cot\theta[(x-x_1)(y-y_2)-(y-y_1)(x-x_2)]=0.$$

[*Hints*: Take any point $A(x,y)$ in the perimeter of the circle. If points $B(x_1,y_1)$ and $C(x_2,y_2)$ are such that the chord BC subtends an angle θ at the point A, then

$$\tan(\pm\theta)=\frac{(y-y_1)/(x-x_1)-(y-y_2)/(x-x_2)}{1+\dfrac{y-y_2}{x-x_2}\dfrac{y-y_1}{x-x_1}}.$$

Simplifying we get

$$(x-x_1)(x-x_2)+(y-y_1)(y-y_2)$$
$$\pm\cot\theta[(x-x_1)(y-y_2)-(y-y_1)(x-x_2)]=0.]$$

10. If from any point on the circle $x^2+y^2+2gx+2fy+c=0$ tangents be drawn to the circle $x^2+y^2+2gx+2fy+c\sin^2\alpha+(g^2+f^2)\cos^2\alpha=0$, show that the angle between them is 2α.

[*Hints*: Given circles are concentric and common centre is $(-g,-f)$. Now, if the radius of the first circle be r, then the radius of the second circle is $R=r\sin\alpha$. If 2θ be the angle between the tangents, then $\sin\theta=R/r=\sin\alpha$, i.e. $\theta=\alpha$.]

11. The distances from origin of the centre of three circles of the form $x^2+y^2-2\lambda x=c^2$, where c is a constant and λ is a variable, are in G.P. Prove that the lengths of the tangents drawn to them from any point on the circle $x^2+y^2=c^2$ are also in G.P.

12. Find the locus of the centre of the circles $x^2 + y^2 + 2gx + 2fy + c = 0$, if the tangents drawn from origin to those are perpendicular.

[*Hints*: Equation of pair of tangents drawn from (0, 0) to the circle $x^2 + y^2 + 2gx + 2fy + c = 0$ is

$$(x^2 + y^2 + 2gx + 2fy + c)c = (gx + fy + c)^2$$

or, $$(x^2 + y^2 + 2gx + 2fy + c)c = g^2x^2 + f^2y^2 + c^2 + 2gcx + 2fcy + 2fgxy$$

or, $$(c - g^2)x^2 - 2ghxy + (c - f^2)y^2 = 0.$$

Now apply Coeff. of x^2 + Coeff. of $y^2 = 0$.]

13. Tangents PQ and PR are drawn to the circle $x^2 + y^2 = c^2$ from the point $P(h, k)$. Show that equation of the circumcircle of the triangle PQR is

$$x^2 + y^2 - xh - yk = 0.$$

[*Hints*: Equation of QR is $xh + yk = c^2$. Equation of the circle describing QR as a chord is

$$x^2 + y^2 - c^2 + \lambda(xh + yk - c^2) = 0.$$

As it passes through $P(h, k)$, we get

$$h^2 + k^2 - c^2 + \lambda(h^2 + k^2 - c^2) = 0 \Rightarrow \lambda = -1.]$$

14. Find the general equation of the circle which cuts the circle $x^2 + y^2 = c^2$ orthogonally and show that if it passes through the point (a, b), then it will also passes through $\left(\dfrac{ac^2}{a^2 + b^2}, \dfrac{bc^2}{a^2 + b^2}\right)$.

15. Prove that if the four points of intersection of the circles $x^2 + y^2 + ax + by + c = 0$ and $x^2+y^2+a'x+b'y+c' = 0$ by the straight lines $Ax+By+C = 0$ and $A'x+B'y+C' = 0$ respectively are concyclic, then

$$\begin{vmatrix} a - a' & b - b' & c - c' \\ A & B & C \\ A' & B' & C' \end{vmatrix} = 0.$$

[*Hints*: Let the four points lie on the circle $S \equiv x^2 + y^2 + 2gx + 2fy + d = 0$. Denoting the given equations as $S_1 = 0$, $S_2 = 0$, $L_1 = 0$ and $L_2 = 0$ respectively we see that equation of radical axes of $S_1 = 0$, $S = 0$ is $L_1 = 0$, of $S_2 = 0$, $S = 0$ is $L_2 = 0$ and of $S_1 = 0$, $S_2 = 0$ is $S_1 - S_2 = 0$, i.e. $Ax + By + C = 0$; $A'x + B'y + C' = 0$ and $(a - a')x + (b - b')y + (c - c') = 0$. But these lines are concurrent, so eliminating x, y we get the required condition.]

16. Show that the equation of the circle which has for its diameter the chord cut off on the straight line $ax + by + c = 0$ by the circle $(a^2 + b^2)(x^2 + y^2) = 2c^2$ is

$$(a^2 + b^2)(x^2 + y^2) + 2c(ax + by) = 0.$$

17. If for the conic $ax^2 + 2hxy + by^2 + 2gx + 2fy + c = 0$, the characteristic roots of the matrix $\begin{pmatrix} a & h \\ h & b \end{pmatrix}$ are equal, show that the conic is either a circle or a point circle or an imaginary circle.

[*Hints*: Characteristic roots of the given matrix are the roots of the equation $(a-\lambda)(b-\lambda)-h^2=0$ or, $\lambda^2-(a+b)\lambda+ab-h^2=0$. If the roots of this equation are equal, then we have

$$\begin{aligned} & (a+b)^2 = 4(ab-h^2) \\ \Rightarrow\quad & (a-b)^2+4h^2=0 \\ \Rightarrow\quad & a=b \quad\text{and}\quad h=0. \end{aligned}$$

So, the given equation becomes

$$x^2+y^2+2\frac{g}{a}x+2\frac{f}{a}y+\frac{c}{a}=0$$

which is a circle or a point circle or an imaginary circle according as

$$g^2+f^2 >=< ac.]$$

18. Show that the locus of the feet of the perpendiculars drawn from the point $(a, 0)$ on the tangents to the circle $x^2+y^2=a^2$ is $(x^2+y^2-ax)^2=(x-a)^2+y^2$.

Chapter 4

Parabola

4.1 Definition of Conic

The locus of a point which moves in such a way that its distance from a fixed point bears a constant ratio to its distance from a fixed straight line is called *conic*. The fixed point is called its *focus*, the fixed straight line is called its *directrix* and the constant is denoted by e and is called its *eccentricity*. The nature of the conics depend on e. The conic is a (i) circle if $e = 0$, (ii) ellipse if $e < 1$, (iii) parabola if $e = 1$ and (iv) hyperbola if $e > 1$.

4.2 Parabola

The locus of a point which moves in such a way that its distance from a fixed point is equal to its distance from a fixed straight line is called *parabola*. If we choose the point $(a, 0)$ as focus and the line $x + a = 0$ as directrix, then the equation of parabola becomes

$$y^2 = 4ax$$

which is the standard/basic/normal/canonical form of parabola.

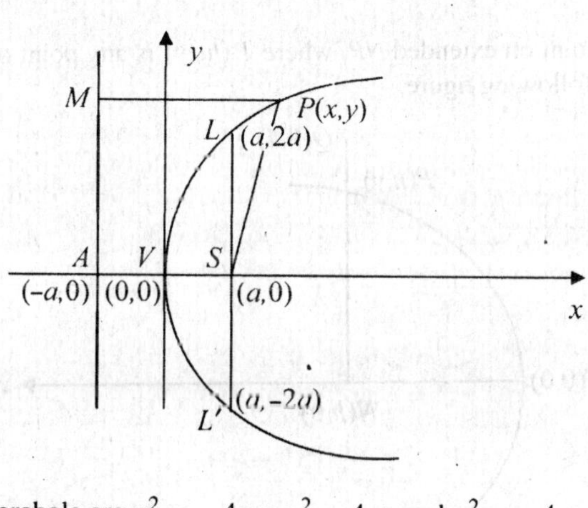

Other forms of parabola are $y^2 = -4ax$; $x^2 = 4ay$ and $x^2 = -4ay$.

4.3 Some Important Results Associated with the Standard Form of Parabola

(i) Vertex: $V(0,0)$; (ii) Focus: $S(a,0)$; (iii) Axis of the parabola: x-axis; (iv) Tangent at the vertex: y-axis; (v) Equation of the directrix: $x+a=0$; (vi) Latus rectum $=4a$; (vii) End points of latus rectum: $L(a,2a)$ and $L'(a,-2a)$; (vii) Focal distance of any point $P(x,y)$: $a+x$.

4.4 General Equation of Parabola

If we take the line $ax+by+c=0$ as directrix and the point (h,k) as focus, then for a point $P(x,y)$ in the parabola

$$SP = \sqrt{(x-h)^2+(y-k)^2}$$

$$\text{and} \quad PM = \frac{ax+by+c}{\sqrt{a^2+b^2}}.$$

Using $SP = PM$, we get

$$[(x-h)^2+(y-k)^2][a^2+b^2] = (ax+by+c)^2$$

$$\text{or,} \quad [a^2+b^2][x^2+y^2-2xh-2yk+h^2+k^2]$$

$$= [a^2x^2+b^2y^2+2abxy+2axc+2bcy+c^2]$$

$$\text{or,} \quad (bx-ay)^2+2Gx+2Fy+C=0$$

where $G=-(a^2+b^2)h-ac$; $F=-(a^2+b^2)k-bc$ and $C=(a^2+b^2)(h^2+k^2)-c^2$.

4.5 Position of a Point $P(h;k)$ with Respect to the Parabola $y^2=4ax$

Let $Q(h,k)$ be any point on extended NP, where $P(h,y)$ is any point on the parabola $y^2 = 4ax$ as shown in the following figure.

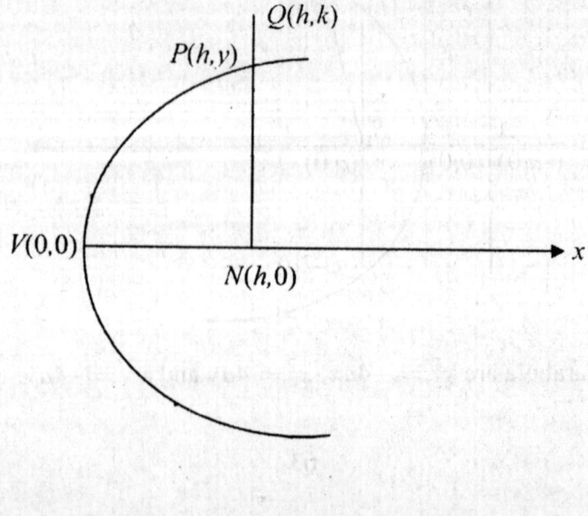

Now, the point Q is outside, on or inside the parabola according as

$$\begin{aligned} & NQ >=< NP \\ \text{or,}\quad & NQ^2 >=< NP^2 \\ \text{or,}\quad & k^2 >=< y^2 = 4ah \\ \text{or,}\quad & k^2 - 4ah >=< 0. \end{aligned}$$

4.6 Equation of Tangent and Normal at any Point $P(h, k)$ on a Parabola

Let us take the parabola $y^2 = 4ax$. Any line through the point (h, k) can be taken as

$$y - k = m(x - h) \tag{1}$$

m being the slope of the line.

Now, eliminating y from the above equations we get

$$\begin{aligned} & [k + m(x - h)]^2 = 4ax \\ \text{or,}\quad & m^2(x - h)^2 + 2km(x - h) + k^2 = 4ax \\ \text{or,}\quad & m^2x^2 + 2x[km - hm^2 - 2a] + k^2 - 2kmh + m^2h^2 = 0. \end{aligned}$$

The line given by Eqn. (1) will touch the parabola if the above equation has two equal roots, i.e. if

$$\begin{aligned} & [km - hm^2 - 2a]^2 = m^2[k^2 - 2kmh + m^2h^2] \\ \text{or,}\quad & k^2m^2 + h^2m^4 + 4a^2 + 4ahm^2 - 4akm - 2khm^3 \\ & = m^2k^2 - 2km^3h + m^4h^2 \\ \text{or,}\quad & a + hm^2 - km = 0 \\ \text{or,}\quad & a + hm^2 - 2m\surd(ah) = 0 \qquad [k^2 = 4ah] \\ \text{or,}\quad & (m\surd h - \surd a)^2 = 0 \\ \text{or,}\quad & m = \sqrt{\frac{a}{h}}. \end{aligned}$$

Putting the value of m in Eqn. (1) we obtain

$$\begin{aligned} & y - k = \sqrt{\frac{a}{h}}(x - h) \\ \text{or,}\quad & yk - k^2 = k\sqrt{\frac{a}{h}}x - k\sqrt{ah} \\ \text{or,}\quad & yk - 4ah = 2\sqrt{ah}\sqrt{\frac{a}{h}}x - 2ah \\ \text{or,}\quad & yk = 2a(x + h). \end{aligned}$$

This is the equation of tangent at (h, k) to the parabola $y^2 = 4ax$.

Equation of normal at (h, k) to the parabola $y^2 = 4ax$ is

$$2ay + kx = 2ak + hk.$$

Cor. 1. *If the line $y = mx + c$ touches the parabola $y^2 = 4ax$ at (h, k), then tangent to the parabola at (h, k), i.e. $yk = 2a(x+h)$ and $y = mx+c$ should be identical. Comparing we get*

$$m = \frac{2a}{k} \quad \text{and} \quad c = \frac{2ah}{k}$$

$$\text{or} \quad k = \frac{2a}{m} \quad \text{and} \quad h = \frac{ck}{2a} = \frac{c}{m}.$$

Since (h, k) lies on the parabola, we have

$$\left(\frac{2a}{m}\right)^2 = \frac{4ac}{m}$$

$$\text{or,} \quad c = \frac{a}{m}.$$

Hence, the line $y = mx + \frac{a}{m}$ touches the parabola $y^2 = 4ax$ for all values of m and the point of contact is $\left(\frac{a}{m^2}, \frac{2a}{m}\right)$. Changing m by $-\frac{1}{m}$, we get $y = -\frac{1}{m}x - am$ is the tangent to the parabola at $(am^2, -2am)$ and at that point the equation of the normal is

$$y = mx - am^3 - 2am.$$

4.7 Equation of Chord of Contact of the Point $P(h, k)$ with Respect to the Parabola $y^2 = 4ax$

Let the tangents drawn from the point (h, k) to the parabola $y^2 = 4ax$ touch the parabola at (x_1, y_1) and (x_2, y_2). Equation of tangent at the point (x_1, y_1) to the parabola $y^2 = 4ax$ is

$$yy_1 = 2a(x + x_1).$$

If it passes through the point (h, k), then

$$ky_1 = 2a(h + x_1). \tag{2}$$

Again, equation of tangent at the point (x_2, y_2) to the parabola $y^2 = 4ax$ is

$$yy_2 = 2a(x + x_2).$$

As it passes through the point (h, k), then

$$ky_2 = 2a(h + x_2). \tag{3}$$

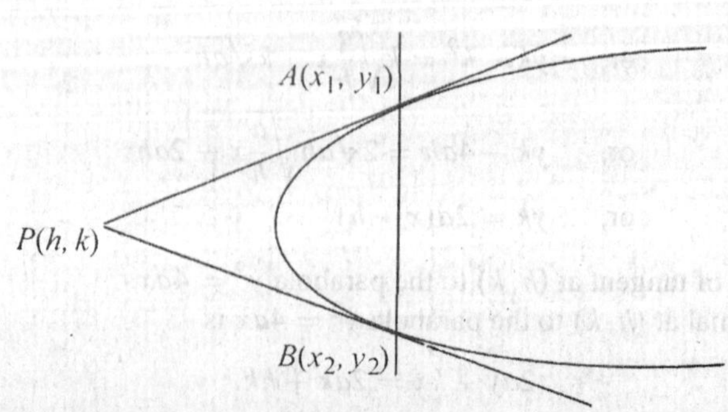

From Eqns. (2) and (3) we observe that the line $ky = 2a(x+h)$ passes through the points (x_1, y_1) and (x_2, y_2), and this is the equation of the chord of contact of the point $P(h, k)$ with respect to the parabola $y^2 = 4ax$.

4.8 Equation of Pair of Tangents from a Given Point $P(h, k)$

Let the pair of tangents from the point $P(h, k)$ be drawn to the parabola $y^2 = 4ax$. Equation of any line through the point (h, k) is of the form

$$\frac{x-h}{l} = \frac{y-k}{m} = r. \tag{4}$$

Any point on this line can be taken as $(h + lr, k + mr)$. If this point lies on the parabola $y^2 = 4ax$, then we have

$$(k+mr)^2 = 4a(h+lr)$$
$$\text{or,} \quad m^2r^2 + 2r(km - 2al) + k^2 - 4ah = 0.$$

Now, if the line given by Eqn. (4) touches the parabola, the above equation in r should have only one root, which gives that point of contact and this gives

$$(km - 2al)^2 = m^2(k^2 - 4ah)$$
$$\text{or,} \quad \left[k - 2a\frac{x-h}{y-k}\right]^2 = k^2 - 4ah$$
$$\text{or,} \quad [ky - k^2 - 2ax + 2ah]^2 = (k^2 - 4ah)(y-k)^2$$
$$\text{or,} \quad [ky - 2a(x+h) - (k^2 - 4ah)]^2 = (k^2 - 4ah)(y-k)^2$$
$$\text{or,} \quad [ky - 2a(x+h)]^2 + (k^2 - 4ah)^2 - 2[ky - 2a(x+h)](k^2 - 4ah) = (k^2 - 4ah)(y-k)^2$$
$$\text{or,} \quad [ky - 2a(x+h)]^2 + (k^2 - 4ah)^2 = (k^2 - 4ah)[(y-k)^2 + 2ky - 4a(x+h)]$$
$$\text{or,} \quad [ky - 2a(x+h)]^2 + (k^2 - 4ah)^2 = (k^2 - 4ah)[y^2 + k^2 - 4a(x+h)]$$
$$\text{or,} \quad T^2 + S_1^2 = S_1[S + S_1]$$
$$\text{or,} \quad SS_1 = T^2$$

where $S = y^2 - 4ax$, $S_1 = k^2 - 4ah$ and $T = ky - 2a(x+h)$.

4.9 Parametric Equation of a Parabola

For any value of t, the point $(at^2, 2at)$ lies on the parabola $y^2 = 4ax$. So, $x = at^2$, $y = 2at$, where t is a variable, is the parametric equation of the parabola $y^2 = 4ax$. Hence, any point on the parabola $y^2 = 4ax$ can be taken as $(at^2, 2at)$. Similarly, the parametric equation of the parabola $x^2 = 4ay$ is $x = 2at$ and $y = at^2$ and any point on the parabola $x^2 = 4ay$ can be taken as $(2at, at^2)$.

4.10 Conormal Points

Points through which the normals drawn pass through a common point are called *conormal points*.

We know that the equation of the normal to the parabola $y^2 = 4ax$ at $(am^2, -2am)$ is

$$y = mx - am^3 - 2am.$$

If it passes through (h, k), then

$$k = mh - am^3 - 2am$$

$$\text{or,} \quad am^3 + m(2a - h) + k = 0.$$

The above cubic equation in m implies that from a point (h, k) three normals can be drawn to the parabola $y^2 = 4ax$. Now, if the normals cut the parabola at $(am_1^2, -2am_1)$, $(am_2^2, -2am_2)$ and $(am_3^2, -2am_3)$, then we have

$$m_1 + m_2 + m_3 = 0 \tag{5}$$

$$m_1m_2 + m_2m_3 + m_1m_3 = \frac{2a - h}{a} \tag{6}$$

$$m_1m_2m_3 = -\frac{k}{a}. \tag{7}$$

Examples

▶▷ 1. *Prove that the tangents at the end point of a focal chord meet at right angles on the directrix.*

Solution: Equation of the tangent at $P(at^2, 2at)$ to the parabola $y^2 = 4ax$ is

$$yt = x + at^2.$$

If PQ be a focal chord, then the equation of tangent at Q is

$$-\frac{y}{t} = x + \frac{a}{t^2}$$

which is perpendicular to the tangent at P. Solving the above equation we get

$$y = a\left(t^2 - \frac{1}{t^2}\right) \Big/ \left(t + \frac{1}{t}\right) = a\left(t - \frac{1}{t}\right)$$

$$\text{and} \quad x = a\left(t - \frac{1}{t}\right)t - at^2 = -a$$

which is the equation of the directrix, this shows that the tangents at the end point of a focal chord intersect at right angle on the directrix.

▶▷ 2. *Prove that the length of the focal chord of a parabola varies inversely as the square of the distance from the vertex.*

Solution: Equation of the focal chord joining $(at^2, 2at)$ and $\left(\frac{a}{t^2}, -\frac{2a}{t}\right)$ is

$$\frac{y + 2a/t}{2at + 2a/t} = \frac{x - a/t^2}{at^2 - a/t^2}$$

$$\text{or,} \quad (y + 2a/t)(t - 1/t) = 2(x - a/t^2)$$

$$\text{or,} \quad y(t - 1/t) - 2x + 2a = 0$$

and the length of the focal chord

$$L = a\sqrt{\left(t^2 - \frac{1}{t^2}\right)^2 + 4\left(t + \frac{1}{t}\right)^2}$$

$$= a\left(t + \frac{1}{t}\right)\sqrt{\left(t - \frac{1}{t}\right)^2 + 4} = a\left(t + \frac{1}{t}\right)^2.$$

If p be the length of the perpendicular drawn from vertex (0, 0) to the focal chord, then

$$p = \frac{2a}{\sqrt{\left(t - \frac{1}{t}\right)^2 + 4}} = \frac{2a}{t + \frac{1}{t}} = 2a\sqrt{\frac{a}{L}}$$

or, $\quad L = \dfrac{4a^3}{p^2}.\quad$ **[Proved]**

▶▷▶ 3. *Prove that the portion of a tangent to a parabola cut off between the directrix and the curve subtends a right angle at the focus.*

Solution: Equation of the tangent at $P(at^2, 2at)$ to the parabola $y^2 = 4ax$ is

$$yt = x + at^2.$$

Putting $x = -a$ in the above equation we get

$$y = a\left(t - \frac{1}{t}\right).$$

So, the tangent at $P(at^2, 2at)$ cuts the directrix $x = -a$ at $Q\{-a, a(t - 1/t)\}$. Now, slope of SQ, $S(a, 0)$ being the focus, is $m_1 = \dfrac{a\left(t - \frac{1}{t}\right)}{-2a} = \dfrac{1 - t^2}{2t}$
and slope of SP is

$$m_2 = \frac{2at}{a(t^2 - 1)} = \frac{2t}{t^2 - 1}.$$

From these results we get $m_1 m_2 = -1$, i.e. $\angle PSQ = 90°$. **[Proved]**

▶▷▶ 4. *Prove that the area of the triangle inscribed in a parabola is twice the area of the triangle formed by the tangents at the vertices.*

Solution: If $P(at_1^2, 2at_1)$, $Q(at_2^2, 2at_2)$ and $R(at_3^2, 2at_3)$ be any three points on the parabola $y^2 = 4ax$, the tangents at P, Q, R are

$$yt_1 = x + at_1^2, \tag{1}$$

$$yt_2 = x + at_2^2, \tag{2}$$

$$yt_3 = x + at_3^2 \tag{3}$$

respectively. Now, solving Eqns. (1) and (2), we get

$$y(t_1 - t_2) = a\left(t_1^2 - t_2^2\right)$$

or, $\quad y = a(t_1 + t_2) \quad$ and $\quad x = at_1t_2.$

So, the coordinates of the points of intersection of the tangents at P and Q; Q and R; P and R are $A:\ \{at_1t_2, a(t_1+t_2)\}$, $B:\ \{at_3t_2, a(t_3+t_2)\}$ and $C:\ \{at_1t_3, a(t_1+t_3)\}$ respectively.

Now, the area of the triangle

$$\begin{aligned}\Delta PQR &= \frac{1}{2}[at_1^2(2at_2 - 2at_3) + at_2^2(2at_3 - 2at_1) + at_3^2(2at_1 - 2at_2)] \\ &= a^2[t_1^2(t_2 - t_3) + t_2^2(t_3 - t_1) + t_3^2(t_1 - t_2)] \\ &= a^2(t_2 - t_3)(t_3 - t_1)(t_1 - t_2)\end{aligned}$$

and the area of the triangle

$$\begin{aligned}\Delta ABC &= \frac{1}{2}[at_1t_2(at_2 - at_1) + at_3t_2(at_3 - at_2) + at_1t_3(at_1 - at_3)] \\ &= \frac{a^2}{2}[t_1t_2(t_2 - t_1) + t_3t_2(t_3 - t_2) + t_1t_3(t_1 - t_3)] \\ &= \frac{a^2}{2}(t_2 - t_3)(t_3 - t_1)(t_1 - t_2).\end{aligned}$$

So, $\Delta PQR = 2\Delta ABC$. **[Proved]**

▶▷ 5. *Prove that the orthocentre of a triangle formed by the tangents to a parabola lies on its directrix.*

Solution: If $P(at_1^2, 2at_1)$, $Q(at_2^2, 2at_2)$ and $R(at_3^2, 2at_3)$ be any three points on the parabola $y^2 = 4ax$, the tangents at P, Q, R are

$$BC:\quad yt_1 = x + at_1^2, \tag{1}$$

$$CA:\quad yt_2 = x + at_2^2, \tag{2}$$

$$AB:\quad yt_3 = x + at_3^2 \tag{3}$$

respectively. Now solving Eqns. (1) and (2), we get

$$y(t_1 - t_2) = a\left(t_1^2 - t_2^2\right)$$

$$\text{or,}\quad y = a(t_1 + t_2) \quad \text{and} \quad x = at_1t_2.$$

So, the coordinates of the points of intersection of the tangents at P and Q; Q and R; P and R are $C:\ \{at_1t_2, a(t_1+t_2)\}$, $A:\ \{at_3t_2, a(t_3+t_2)\}$ and $B:\ \{at_1t_3, a(t_1+t_3)\}$ respectively.

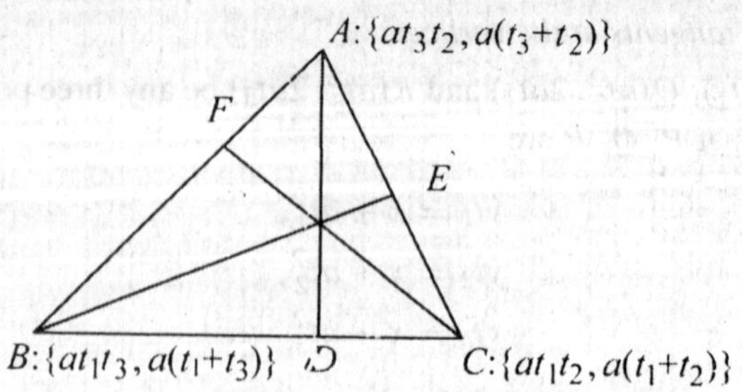

Equation of AD, i.e. of perpendicular from the vertex A to opposite side BC is

$$y - a(t_3 + t_2) = -t_1(x - at_3t_2). \tag{4}$$

Similarly, equation of BE is

$$y - a(t_3 + t_1) = -t_2(x - at_3t_1). \tag{5}$$

Eliminating y from Eqns. (4) and (5) we get $x = -a$ and $y = a(t_1 + t_2 + t_3 + t_1t_2t_3)$ and due to symmetry , this point also lies on the line CF. Which shows that the orthocentre of a triangle formed by the tangents to a parabola $y^2 = 4ax$ lies on its directrix.

▶▷▶ 6. *If the tangents drawn from any point on the line $x + 4a = 0$ to the parabola $y^2 = 4ax$, prove the chord of contact will subtend a right angle at the vertex.*

Solution: Let the tangents are drawn from the point (h, k) to the parabola $y^2 = 4ax$. Then equation of chord of contact is

$$yk = 2a(x + h). \tag{1}$$

Now, making homogeneous the equation $y^2 = 4ax$ with the aid of Eqn. (1) we get

$$y^2 = 4ax\frac{yk - 2ax}{2ah}$$

$$\text{or,} \quad hy^2 - 2kxy + 4ax^2 = 0$$

which is a second degree homogeneous equation in x, y and represents a pair of straight lines passing through the origin. If the lines are perpendicular, then we have

$$\text{Coeff. of } x^2 + \text{Coeff. of } y^2 = 0$$

$$\text{or,} \quad h + 4a = 0.$$

Generalizing we get $x + 4a = 0$, which shows that if tangents drawn from any point on the line $x + 4a = 0$ to the parabola $y^2 = 4ax$, then the chord of contact will subtend a right angle at the vertex.

▶▷▶ 7. *If a normal chord to the parabola subtends a right angle at the vertex, prove that it is inclined to the axis of the parabola at an angle* $\tan^{-1}\sqrt{2}$.

Solution: If we take the parabola as $y^2 = 4ax$, then equation of the normal at $(am^2, -2am)$ is

$$y = mx - am^3 - 2am. \tag{1}$$

Now, making homogeneous $y^2 = 4ax$ with the aid of Eqn. (1) we get

$$y^2 = \frac{4ax(y - mx)}{am^3 + 2am}$$

$$\text{or,} \quad (am^3 + 2am)y^2 + 4axy - 4amx^2 = 0$$

which is a second degree homogeneous equation in x, y and represents a pair of straight lines passing through the origin. If the lines are perpendicular, then we have

$$\text{Coeff. of } x^2 + \text{Coeff. of } y^2 = 0$$

$$\text{or,} \quad am^3 + 2am - 4am = 0$$

$$\text{or,} \quad m^2 = 2 \quad [\text{as } m \neq 0]$$

$$\text{or,} \quad m = \tan\theta = \sqrt{2}$$

$$\text{or,} \quad \theta = \tan^{-1}\sqrt{2}. \qquad \textbf{[Proved]}$$

▶▷▶ **8**. *If tangents are drawn from a point to the parabola $y^2 = 4ax$ are such that the chord of contact is a normal chord, find the locus of the point.*

Solution: Let the moving point be (h, k); then chord of contact of this point with respect to the parabola is

$$yk = 2a(x+h). \tag{1}$$

Again, we know that, for all values of m,

$$y = mx - am^3 - 2am \tag{2}$$

is a normal to the parabola.

Comparing Eqns. (1) and (2), we get

$$\frac{k}{1} = \frac{2a}{m} = \frac{-2ah}{am^3 + 2am}$$

from which we get

$$m = \frac{2a}{k} \quad \text{and} \quad am^2 + 2a = -h.$$

Eliminating m we get

$$\frac{4a^3}{k^2} + 2a + h = 0.$$

Generalizing we get

$$4a^3 + y^2(2a + x) = 0.$$

▶▷▶ **9**. *Find the locus of the point of intersection of two tangents to the parabola $y^2 = 4ax$ such that they intercept a constant length L from the directrix.*

Solution: Let the tangents are drawn from the point (h, k) to the parabola $y^2 = 4ax$ and touch it at $P(at_1^2, 2at_1)$ and $Q(at_2^2, 2at_2)$, then the tangents at P, Q are

$$yt_1 = x + at_1^2 \tag{1}$$

$$\text{and} \quad yt_2 = x + at_2^2. \tag{2}$$

Solving Eqns. (1) and (2) we get

$$h = at_1t_2 \quad \text{and} \quad k = a(t_1 + t_2). \tag{3}$$

Putting $x = -a$ in the above equation we see that the tangents meet the directrix at

$$M\colon \{-a, a(t_1 - 1/t_1)\} \quad \text{and} \quad N\colon \{-a, a(t_2 - 1/t_2)\}.$$

Given that $MN = L$, i.e.

$$a\left[\left(t_1 - \frac{1}{t_1}\right) - \left(t_2 - \frac{1}{t_2}\right)\right] = L$$

$$\text{or,} \quad a(t_1 - t_2)(1 + t_1t_2) = Lt_1t_2$$

$$\text{or,} \quad a\left(1 + \frac{h}{a}\right)\sqrt{\frac{k^2}{a^2} - \frac{4h}{a}} = L\frac{h}{a}$$

$$\text{or,} \quad (k^2 - 4ah)(h+a)^2 = h^2L^2.$$

Generalizing, we get

$$(y^2 - 4ax)(x + a)^2 = x^2L^2.$$

▶▷▶ **10**. *Find the locus of the point of intersection of two tangents to the parabola $y^2 = 4ax$ so that they form a triangle of constant area A with the tangent at the vertex.*

Solution: Let the tangents are drawn from the point $R(h, k)$ to the parabola $y^2 = 4ax$ and touch it at $P(at_1^2, 2at_1)$ and $Q(at_2^2, 2at_2)$; then the tangents at P, Q are

$$yt_1 = x + at_1^2 \qquad (1)$$

$$\text{and} \quad yt_2 = x + at_2^2. \qquad (2)$$

Solving Eqns. (1) and (2) we get

$$h = at_1t_2 \quad \text{and} \quad k = a(t_1 + t_2). \qquad (3)$$

Putting $x = 0$ in the above equation we see that the tangents meet the tangent at the vertex at $M : (0, at_1)$ and $N : (0, at_2)$.

Now, area of the triangle ΔMNR is

$$\frac{1}{2}h(at_1 - at_2) = A \quad \text{(given)}$$

$$\Rightarrow \quad a^2h^2[(t_1 + t_2)^2 - 4t_1t_2] = 4A^2$$

$$\Rightarrow \quad a^2h^2\left(\frac{k^2}{a^2} - \frac{4h}{a}\right) = 4A^2$$

$$\Rightarrow \quad h^2[k^2 - 4ha] = 4A^2.$$

Generalizing, we get

$$x^2[y^2 - 4xa] = 4A^2.$$

▶▷▶ **11**. *Find the area of the triangle formed by the tangents drawn from $R(h, k)$ to the parabola $y^2 = 4ax$ and their chord of contact.*

Solution: Let the tangents are drawn from the point $R(h, k)$ to the parabola $y^2 = 4ax$ and touch it at $P(at_1^2, 2at_1)$ and $Q(at_2^2, 2at_2)$, then the tangents at P, Q are

$$yt_1 = x + at_1^2 \qquad (1)$$

$$\text{and} \quad yt_2 = x + at_2^2. \qquad (2)$$

Solving Eqns. (1) and (2) we get

$$h = at_1t_2 \quad \text{and} \quad k = a(t_1 + t_2). \qquad (3)$$

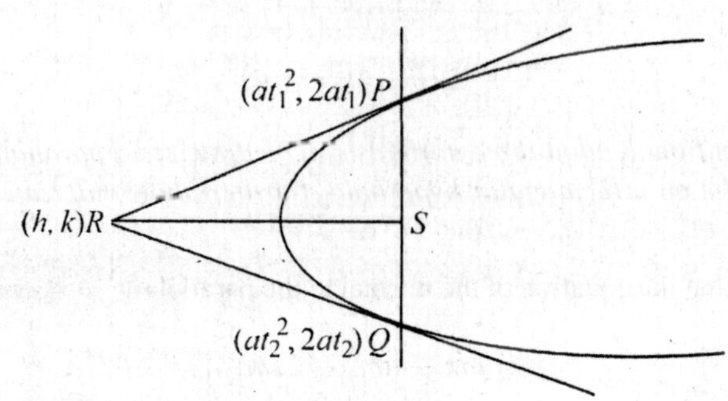

Now,

$$\begin{aligned}\text{length of } PQ &= \sqrt{\left(at_1^2 - at_2^2\right)^2 + (2at_1 - 2at_2)^2}\\ &= a(t_1 - t_2)\sqrt{(t_1+t_2)^2+4}\\ &= a\sqrt{(t_1+t_2)^2 - 4t_1t_2}\sqrt{(t_1+t_2)^2+4}\\ &= a\sqrt{\frac{k^2}{a^2} - \frac{4h}{a}}\sqrt{\frac{k^2}{a^2}+4}\\ &= \frac{\sqrt{(k^2-4ah)(k^2+4a^2)}}{a}.\end{aligned}$$

Now, equation of PQ (being chord of contact of the point (h, k) w.r.t. the parabola) is

$$yk = 2a(x+h).$$

So,

$$RS = \frac{k^2 - 4ah}{\sqrt{k^2+4a^2}}$$

and the area of the triangle

$$\Delta PQR = \frac{1}{2}PQ \times RS = \frac{(k^2-4ah)^{3/2}}{2a}.$$

▶▷▶ **12.** *Find the locus of the middle points of the portion of the tangents intercepted between the curve and directrix.*

Solution: Equation of the tangent at $P(at^2, 2at)$ to the parabola $y^2 = 4ax$ is

$$yt = x + at^2.$$

It meets the directrix $x = -a$ at Q: $\left(-a, a\left(t - \frac{1}{t}\right)\right)$. If (h, k) be the middle point of PQ, then we have

$$h = \frac{a}{2}(t^2-1) \quad \text{and} \quad k = \frac{a}{2}\left(3t - \frac{1}{t}\right).$$

Eliminating t from the above results we get

$$k^2(2h+a) = a(3h+a)^2.$$

Generalizing, we get

$$y^2(2x+a) = a(3x+a)^2.$$

▶▷▶ **13.** *Show that from a point three normals can be drawn to a parabola $y^2 = 4ax$ and feet of the normals lie on a rectangular hyperbola. Further, show that sum of the ordinates of the feet is zero.*

Solution: We know that the equation of the normal to the parabola $y^2 = 4ax$ at $(am^2, -2am)$ is

$$y = mx - am^3 - 2am.$$

If it passes through (h, k), then

$$k = mh - am^3 - 2am$$

$$\text{or,} \quad am^3 + m(2a - h) + k = 0.$$

The above cubic equation in m implies that from a point (h, k) three normals can be drawn to the parabola $y^2 = 4ax$. Now, if the normals cut the parabola at $(am_1^2, -2am_1)$, $(am_2^2, -2am_2)$ and $(am_3^2, -2am_3)$, then we have

$$\text{Sum of the roots} = m_1 + m_2 + m_3 = \frac{0}{a} = 0.$$

Now, sum of the ordinates of the feet $= -2a(m_1 + m_2 + m_3) = 0$.

Again, consider the equation of normal at (α, β), which is

$$y - \beta = -\frac{\beta}{2a}(x - \alpha).$$

If it passes through (h, k), then

$$k - \beta = -\frac{\beta}{2a}(h - \alpha)$$

$$\text{or,} \quad 2ak - 2a\beta = -\beta h + \beta\alpha$$

$$\text{or,} \quad \beta\alpha + (2a - h)\beta - 2ak = 0.$$

Hence, locus of the foot of the normal is

$$xy + (2a - h)y + 2ak = 0$$

which represents a rectangular hyperbola.

▶▷▶ **14**. *Find the locus of the point through which three normals can be drawn to the parabola $y^2 = 4ax$ such that two of the normals coincide.*

Solution: Let (h, k) be the point through which three normals are drawn to the parabola $y^2 = 4ax$ and if the normals cut the parabola at $(am_1^2, -2am_1)$, $(am_2^2, -2am_2)$ and $(am_3^2, -2am_3)$, then we have

$$m_1 + m_2 + m_3 = 0, \tag{1}$$

$$m_1m_2 + m_2m_3 + m_1m_3 = \frac{2a - h}{a} \tag{2}$$

$$\text{and} \quad m_1m_2m_3 = -\frac{k}{a}. \tag{3}$$

Now, putting $m_1 = m_2$ in Eqns. (1)–(3) we get

$$m_3 = -2m_1,$$

$$-2m_1^3 = -\frac{k}{a}$$

$$\text{and} \quad m_1^2 - m_3^2 = \frac{2a - h}{a}$$

$$\text{or,} \quad -3m_1^2 = \frac{2a - h}{a}.$$

Now, eliminating m_1 we get

$$\frac{k^2}{4a^2} = -\frac{(2a-h)^3}{27a^3}$$

$$\text{or,} \quad 27ak^2 + 4(2a-h)^3 = 0.$$

Hence, the locus of the point (h, k) is

$$27ay^2 + 4(2a-x)^3 = 0.$$

▶▷▶ **15.** *Three normals are drawn from a point to the parabola $y^2 = 4ax$ meet the axis in points whose distances from the vertex are in A.P. Show that the locus of the point is*

$$27ay^2 + 2(2a-x)^3 = 0.$$

Solution: Equation of the normal to the parabola $y^2 = 4ax$ at $(am^2, -2am)$ is

$$y = mx - am^3 - 2am.$$

It meets the axis at $(am^2 + 2a, 0)$.

Let (h, k) be the point through which three normals are drawn to the parabola $y^2 = 4ax$ and if the normals cut the parabola at $(am_1^2, -2am_1)$, $(am_2^2, -2am_2)$ and $(am_3^2, -2am_3)$, then we have

$$m_1 + m_2 + m_3 = 0, \tag{1}$$

$$m_1m_2 + m_2m_3 + m_1m_3 = \frac{2a-h}{a} \tag{2}$$

$$\text{and} \quad m_1m_2m_3 = -\frac{k}{a}. \tag{3}$$

Given that

$$2(am_2^2 + 2a) = am_1^2 + 2a + am_3^2 + 2a$$

$$\text{or,} \quad 2m_2^2 = m_1^2 + m_3^2$$

$$\text{or,} \quad 3m_2^2 = m_1^2 + m_2^2 + m_3^2 = 0 - 2\frac{2a-h}{a}$$

$$\text{or,} \quad m_2^2 = -2\frac{2a-h}{3a}.$$

Again

$$2m_2^2 = m_1^2 + m_3^2 = m_2^2 - 2m_1m_3$$

$$\text{or,} \quad m_2^2 = \frac{2k}{am_2}$$

$$\text{or,} \quad m_2^3 = \frac{2k}{a}.$$

Eliminating m_2, we get

$$\frac{4k^2}{a^2} = -\frac{8(2a-h)^3}{27a^3}$$

$$\text{or,} \quad 27ak^2 + 2(2a-h)^3 = 0.$$

Generalizing, we get

$$27ay^2 + 2(2a - x)^3 = 0. \qquad \textbf{[Proved]}$$

▶▷▶ **16**. *Find the locus of the point through which three normals can be drawn to the parabola $y^2 = 4ax$ such that two of the normals make complementary angles with the axis of the parabola.*

Solution: Let one of the normals makes an angle θ, then another will make an angle $90° - \theta$ with the axis. So, $m_1 m_2 = \tan\theta \times \cot\theta = 1$. Now using the following results

$$m_1 + m_2 + m_3 = 0 \qquad (1)$$

$$m_1 m_2 + m_2 m_3 + m_1 m_3 = \frac{2a - h}{a} \qquad (2)$$

$$m_1 m_2 m_3 = -\frac{k}{a} \qquad (3)$$

we obtain

$$m_3 = -\frac{k}{a}; \quad 1 - m_3^2 = \frac{2a - h}{a}.$$

Eliminating m_3 we get

$$1 - \frac{k^2}{a^2} = \frac{2a - h}{a}$$

$$\text{or,} \quad \frac{k^2}{a^2} = \frac{h - a}{a}.$$

Generalizing, we get

$$y^2 = a(x - a).$$

▶▷▶ **17**. *If the normals at P, Q, R on the parabola $y^2 = 4ax$ meet in the point (h, k), then show that the centroid of the triangle ΔPQR is $\left(\frac{2(h - 2a)}{3}, 0\right)$ and orthocentre of the triangle ΔPQR is $\left(h - 6a, -\frac{k}{2}\right)$.*

Solution: If (h, k) be the point where the three normals to the parabola $y^2 = 4ax$ meet and if the feet of the normals are $P : (am_1^2, -2am_1)$, $Q : (am_2^2, -2am_2)$ and $R : (am_3^2, -2am_3)$, then we have

$$m_1 + m_2 + m_3 = 0 \qquad (1)$$

$$m_1 m_2 + m_2 m_3 + m_1 m_3 = \frac{2a - h}{a} \qquad (2)$$

$$m_1 m_2 m_3 = -\frac{k}{a}. \qquad (3)$$

Now, centroid of the triangle (α, β) is given by

$$\alpha = \frac{a}{3}\sum m_1^2 = \frac{a}{3}\left[\left(\sum m_1\right)^2 - 2\sum m_1 m_2\right]$$

$$= \frac{a}{3}\left(0 - 2\frac{2a - h}{a}\right)$$

$$= -2\frac{2a - h}{3}$$

and $\beta = -\frac{2a}{3}\sum m_1 = 0$. So, the centroid of the triangle ΔPQR is $\left(\frac{2(h-2a)}{3}, 0\right)$.

Now, equation of the line through P and perpendicular to QR is

$$y + 2am_1 = \frac{1}{2}(m_2 + m_3)(x - am_1^2)$$

and equation of the line through Q and perpendicular to PR is

$$y + 2am_2 = \frac{1}{2}(m_1 + m_3)(x - am_2^2).$$

Solving the above equations we get

$$4a(m_1 - m_2) = x(m_2 - m_1) - a(m_1 - m_2)(m_1m_2 + m_2m_3 + m_1m_3)$$

$$\text{or,} \quad x = -4a - a\frac{2a-h}{a} = h - 6a$$

$$\text{and} \quad (y + 2am_1)(m_1 + m_3) - (y + 2am_2)(m_2 + m_3)$$
$$= \frac{a}{2}(m_2^2 - m_1^2)(m_2 + m_3)(m_1 + m_3)$$

$$\text{or,} \quad (m_1 - m_2)y = 2a[m_2(m_2 + m_3) - m_1(m_1 + m_3)]$$
$$+ \frac{a}{2}(m_2^2 - m_1^2)(m_2 + m_3)(m_1 + m_3)$$

$$\text{or,} \quad (m_1 - m_2)y = 2a[m_2^2 - m_1^2 + m_3(m_2 - m_1)]$$
$$+ \frac{a}{2}(m_2^2 - m_1^2)(m_2 + m_3)(m_1 + m_3)$$

$$\text{or,} \quad y = -2a(m_1 + m_2 + m_3) - \frac{a}{2}(m_2 + m_1)(m_2 + m_3)(m_1 + m_3)$$

$$\text{or,} \quad y = 0 + \frac{a}{2}m_1m_2m_3 = -\frac{k}{2}.$$

Hence, the orthocentre of the triangle ΔPQR is $(h - 6a, -k/2)$.

▶▷ **18.** *Find the locus of the foot of the perpendicular to tangent to the parabola $y^2 = 4ax$ from its vertex.*

Solution: We know that for all values of m the line $y = mx + \frac{a}{m}$ touches the parabola $y^2 = 4ax$. Equation of the perpendicular to the tangent from the vertex, i.e. (0, 0) is $y = -\frac{1}{m}x$. To get the locus of the foot of the perpendicular, i.e. of the point of intersection of

$$y = mx + \frac{a}{m} \quad \text{and} \quad y = -\frac{1}{m}x$$

we eliminate the parameter m from the above equations and get

$$y = -\frac{x}{y}x - \frac{ay}{x}$$

$$\text{or,} \quad (x + a)y^2 + x^3 = 0.$$

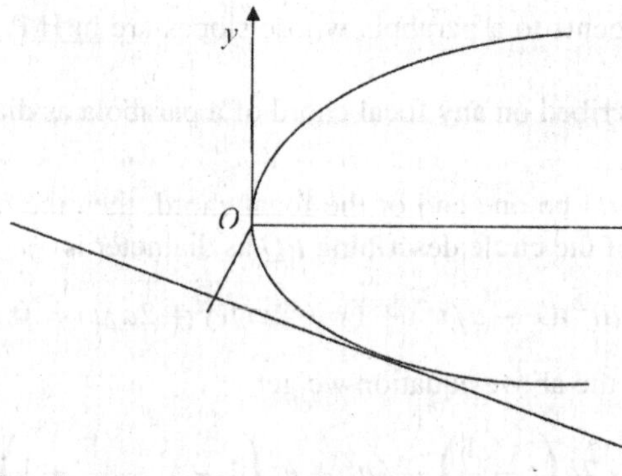

19. *Show that the length of a side of an equilateral triangle, two vertices of which lie on a parabola $y^2 = 4ax$ and the remaining vertex is at the focus, is either $4a(2+\sqrt{3})$ or $4a(2-\sqrt{3})$.*

Solution: Let the vertices of the equilateral triangle be $S(a, 0)$, $P(at_1^2, 2at_1)$, $Q(at_2^2, 2at_2)$. Then using $SP = SQ$ we get

$$\begin{aligned} & (t_1^2-1)^2+4t_1^2=(t_2^2-1)^2+4t_2^2 \\ \Rightarrow\ & (t_1^2+1)^2=(t_2^2+1)^2 \\ \Rightarrow\ & t_1^2=t_2^2 \\ \Rightarrow\ & t_1=-t_2, \quad \text{since } t_1 \neq t_2. \end{aligned}$$

Again, using $SP = PQ$ we obtain

$$\begin{aligned} & (t_1^2-1)^2+4t_1^2=(t_1^2-t_2^2)^2+4(t_1-t_2)^2=16t_1^2 \\ \Rightarrow\ & t_1^2 \pm 2\sqrt{3}t_1-1=0 \\ \Rightarrow\ & t_1=\frac{\pm 2\sqrt{3}+\sqrt{12+4}}{2}=2\pm\sqrt{3}. \end{aligned}$$

Hence, length of the sides of the equilateral triangle is

$$a(t_1^2+1)=a(4+3\pm 4\sqrt{3}+1)=4a(2\pm\sqrt{3}).$$

EXERCISE

1. If $P : (at^2, 2at)$ be the coordinates of one end of a focal chord of the parabola $y^2 = 4ax$, find the coordinates of the other end.

   [*Hints*: If $Q : (at_1^2, 2at_1)$ be the other end of the focal chord, then use the result that area of the triangle $\Delta PSQ = 0$, $S(a, 0)$ being the focus of the parabola.]
2. PN is any ordinate of the parabola $y^2 = 4ax$; M is a point on PN which divides it in the ratio $m : n$. Show that the locus of M is a parabola.
3. A circle and a parabola intersect in four points. Prove that the algebraic sum of the ordinates of the four points is zero.

   [*Hints*: Equation of the circle is $x^2 + y^2 + 2gx + 2fy + c = 0$. Equation of the parabola is $y^2 = 4ax$. Put $x = \frac{y^2}{4a}$ in the equation of circle. Simplifying it we get $\sum y_1 = 0$.]
4. Show that the tangents at the end points of a focal chord intersect at right angle on the directrix.

5. Prove that any three tangents to a parabola whose slopes are in H.P. form a triangle of constant area.
6. Prove that the circle described on any focal chord of a parabola as diameter touches the directrix.

   [*Hints*: Let $P: \ (at^2, 2at)$ be one end of the focal chord, then the other end is $(a/t^2, -2a/t)$. Now equation of the circle describing PQ as diameter is

$$(x - at^2)(x - a/t^2) + (y - 2at)(y + 2a/t) = 0.$$

   Now, putting $x = -a$ in the above equation we get

$$y^2 - 2a\left(t - \frac{1}{t}\right) - 4a^2 + a^2\left(1 + t^2 + \frac{1}{t^2} + 1\right) = 0$$

$$\text{or,} \quad y^2 - 2a\left(t - \frac{1}{t}\right) + \left[a\left(t - \frac{1}{t}\right)\right]^2 = 0$$

$$\text{or,} \quad \left[y - a\left(t - \frac{1}{t}\right)\right]^2 = 0.]$$

7. Prove that the locus of the point of intersection of two perpendicular normals is

$$y^2 = a(x - 3a).$$

8. Tangents are drawn to the parabola $y^2 = 4ax$ from the point (h, k); find the point of intersection of corresponding normals.
9. Prove that the locus of the point of intersections of the normal at the end points of a focal chord of the parabola is another parabola.
10. Prove that the length of the intercept of the normal at $P(at^2, 2at)$ of a parabola $y^2 = 4ax$ made by the circle on PS as diameter, S being the focus, is $a\sqrt{1 + t^2}$.
11. The two parabolas $y^2 = 4a(x - k)$ and $x^2 = 4a(y - k')$ always touch one another, k, k' being parameters. Prove that the point of contact lies on the curve $xy = 4a^2$.

    [*Hints*: Write the equations of tangents at (α, β) to both the parabola and compare the equations.]
12. Show that the locus of the point through which three normals can be drawn to the parabola $y^2 = 4ax$ such that one bisects the angle between other two is

$$27ay^2 = (x - 5a)^2(2x - a).$$

13. Tangents are drawn to the parabola $y^2 = 4ax$ at the points whose abscissae are in the ratio $p : 1$. Show that the locus of their point of intersection is a parabola.

    [*Hints*: Point of intersection of the tangents at $P(at^2, 2at)$ and $Q(at_1^2, 2at_1)$ is given by

$$h = att_1 \quad \text{and} \quad k = a(t + t_1).$$

    Given that $t = \sqrt{p}t_1$. Then $h = a\sqrt{p}t_1^2$ and $k = at_1(1 + \sqrt{p})$. Eliminating t_1, we obtain

$$h = a\sqrt{p}\left(\frac{k}{a(1 + \sqrt{p})}\right)^2$$

$$\text{or,} \quad \sqrt{p}k^2 = a(1 + \sqrt{p})^2 h.$$

    Generalizing, we get

$$\sqrt{p}y^2 = a(1 + \sqrt{p})^2 x.]$$

14. Tangents are drawn to the parabola $y^2 = 4ax$ at the points whose ordinates are in the ratio $p^2 : q^2$. Show that the locus of their point of intersection is a parabola.

[*Hints*: Point of intersection of the tangents at $P(at^2, 2at)$ and $Q(at_1^2, 2at_1)$ is given by

$$h = att_1 \quad \text{and} \quad k = a(t + t_1).$$

Given that $t = \frac{p^2 t_1}{q^2}$. Then $h = \frac{ap^2 t_1^2}{q^2}$ and $k = at_1\left(1 + \frac{p^2}{q^2}\right)$. Eliminating t_1, we obtain

$$h = \frac{ap^2}{q^2}\left(\frac{k}{a(1 + p^2/q^2)}\right)^2$$

$$\text{or,} \quad k^2 = a\left(1 + \frac{p^2}{q^2}\right)^2 \frac{q^2 h}{p^2}.$$

Generalizing, we get

$$y^2 = a\left(\frac{p^2}{q^2} + \frac{q^2}{p^2} + 2\right)x.]$$

15. Prove that if the line $lx + my + n = 0$ touches the parabola $y^2 = 4px - 4pq$, then

$$l^2 q + nl = pm^2.$$

[*Hints*: If the $lx + my + n = 0$ touches the parabola $y^2 = 4px - 4pq$ at (h, k), then it would be identical with $yk = 2p(h+x) - 4pq$. Compare this equation with $lx+my+n = 0$ to get the values of h and k. Now put these values in $k^2 = 4ph - 4pq$.]

16. A circle cuts the parabola $y^2 = 4ax$ at right angles and passes through the focus. Show that the centre lies on the curve $y^2(2x + a) = a(3x + a)^2$.

[*Hints*: Equation of the circle having centre at the point (h, k) and which passes through the focus $(a, 0)$ of the parabola $y^2 = 4ax$ is

$$(x - h)^2 + (y - k)^2 = (a - h)^2 + k^2. \tag{1}$$

Since, the circle cuts the parabola $y^2 = 4ax$ at right angles, tangent at any point $(at^2, 2at)$ to the parabola will pass through (h, k), i.e.

$$at^2 - kt + h = 0. \tag{2}$$

Again, the circle passes through $(at^2, 2at)$, so we have

$$(at^2 - h)^2 + (2at - k)^2 = (a - h)^2 + k^2.$$

Using Eqn. (2) and simplifying we get

$$(kt + a - h)(kt - a - 3h) = 0$$

which gives $t = \frac{h-a}{k}$ or $\frac{3h+a}{k}$ and using these results in Eqn. (2) we get

$$k^2(2h + a) = a(3h + a)^2 \quad \text{or} \quad k = 0, \text{ which is not true in general.}$$

So, the locus of the centre (h, k) is $y^2(2x + a) = a(3x + a)^2$.]

17. Find the locus of the point through which three normals can be drawn to the parabola $y^2 = 4ax$ such that two of the normals are at right angles.

[*Hints*: Proceed as Example 14 and use $m_1m_2 = -1$.]

18. Prove that the locus of the middle point of the portion of a normal to the parabola $y^2 = 4ax$ intercepted between the curve and axis is another parabola.

[*Hints*: Equation of the normal to the parabola $y^2 = 4ax$ at $(am^2, -2am)$ is

$$y = mx - am^3 - 2am.$$

It meets the axis at $(am^2 + 2a, 0)$. Let (h, k) be the middle point; then $h = a + am^2$ and $k = -am$, eliminating m we get $k^2 = a(h - a)$. Hence, locus of the middle point is $y^2 = a(x - a)$.]

19. The tangents at the extremities of a normal chord of the parabola meet in a point T. Find the locus of the point T.
20. Find the locus of a point which is such that two of the normals drawn from it to the parabola $y^2 = 4ax$ are at right angles.
21. The two tangents from a point P to the parabola $y^2 = 4ax$ make angles α, β with x-axis. Find the locus of the point when
    (i) $\tan\alpha + \tan\beta = k$ and
    (ii) $\tan^2\alpha + \tan^2\beta = k$.
22. Tangents are drawn at the ends of a normal chord of the parabola $y^2 = 4ax$. Find the locus of their point of intersection.
23. The normals at the extremities of a chord of the parabola $y^2 = 4ax$ meet on the parabola. Prove that the middle point of the chord lies on the parabola

$$y^2 = 2a(x + 2a).$$

24. Prove that the area of the triangle inscribed in the parabola $y^2 = 4ax$ is

$$\frac{1}{8a}(y_1 - y_2)(y_2 - y_3)(y_3 - y_1)$$

where y_i, $i = 1, 2, 3$ are ordinates of the vertices.

[*Hints*: From Example 4, we see that the area of the triangle inscribed in the parabola $y^2 = 4ax$ is equal to

$$\begin{aligned} & a^2(t_2 - t_3)(t_3 - t_1)(t_1 - t_2) \\ &= \frac{a^2}{2a \times 2a \times 2a}(2at_2 - 2at_3)(2at_3 - 2at_1)(2at_1 - 2at_2) \\ &= \frac{1}{8a}(y_1 - y_2)(y_2 - y_3)(y_3 - y_1).] \end{aligned}$$

Chapter 5

Ellipse

5.1 Definition

The locus of a point which moves in such a way that its distance from a fixed point, called focus, is less than to its distance from a fixed straight line, called directrix, is called *ellipse*. If we choose the points $(\pm ae, 0)$ as foci and the lines $x \pm \frac{a}{e} = 0$ as directrices, $e(< 1)$ being the eccentricity, then the equation of ellipse becomes

$$\frac{x^2}{a^2} + \frac{y^2}{b^2} = 1$$

where $b^2 = a^2(1 - e^2)$, which is the standard/basic/normal/canonical form of ellipse.

Another Definition

Ellipse is the locus of a point which moves in such a way that sum of its distances from two fixed points, called foci, is constant $2a$, length of the major axis.

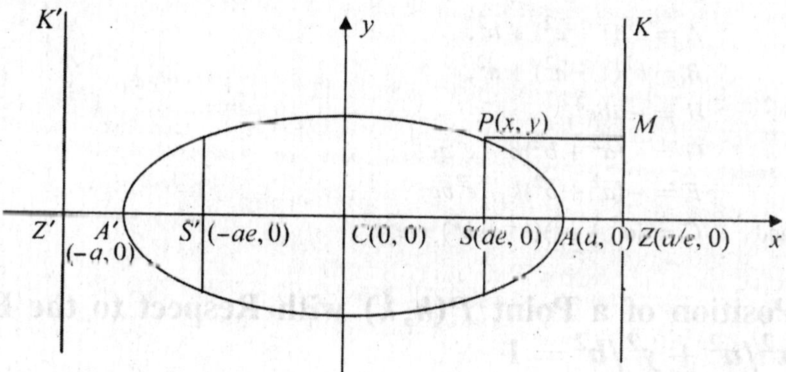

5.2 Some Important Results Associated with the Standard Form of Ellipse

(i) Centre: $C(0,0)$,
(ii) Foci: $S(ae,0)$, $S'(-ae,0)$,
(iii) Major axis of the ellipse: x-axis,
(iv) Minor axis of the ellipse: y-axis,
(v) End points of major axis $(\pm a, 0)$,
(vi) End points of minor axis $(0, \pm b)$,
(vii) Eccentricity: $e = \sqrt{1 - b^2/a^2}$,
(viii) Equation of the directrices: $x = \pm a/e$,
(ix) Latus rectum $= \dfrac{2b^2}{a}$,
(x) End points of latus rectums: $L\left(ae, \pm\dfrac{b^2}{a}\right)$ and $L'\left(-ae, \pm\dfrac{b^2}{a}\right)$,
(xi) Focal distance of any point (x, y): $a \pm ex$.

5.3 General Equation of Ellipse

If we take the line $ax + by + c = 0$ as directrix and the point (h, k) as focus, then for a point $P(x, y)$ in the ellipse

$$SP = \sqrt{(x-h)^2 + (y-k)^2}$$

and $$PM = \frac{ax + by + c}{\sqrt{a^2 + b^2}}.$$

Using $SP = ePM$, we get

$$[(x-h)^2 + (y-k)^2][a^2 + b^2] = e^2(ax + by + c)^2$$

or, $$[a^2 + b^2][x^2 + y^2 - 2xh - 2yk + h^2 + k^2]$$
$$= e^2[a^2x^2 + b^2y^2 + 2abxy + 2axc + 2bcy + c^2]$$

or, $$Ax^2 + 2Hxy + By^2 + 2Gx + 2Fy + C = 0$$

where

$$A = a^2(1 - e^2) + b^2,$$
$$B = b^2(1 - e^2) + a^2,$$
$$H = -abe^2,$$
$$G = -(a^2 + b^2)h - e^2ac,$$
$$F = -(a^2 + b^2)k - e^2bc,$$

and $$C = (a^2 + b^2)(h^2 + k^2) - e^2c^2.$$

5.4 Position of a Point $P(h, k)$ with Respect to the Ellipse $x^2/a^2 + y^2/b^2 = 1$

Let $Q(h, k)$ be any point on extended NP, where $P(h, y)$ is any point on the ellipse $x^2/a^2 + y^2/b^2 = 1$ as shown in the following figure.

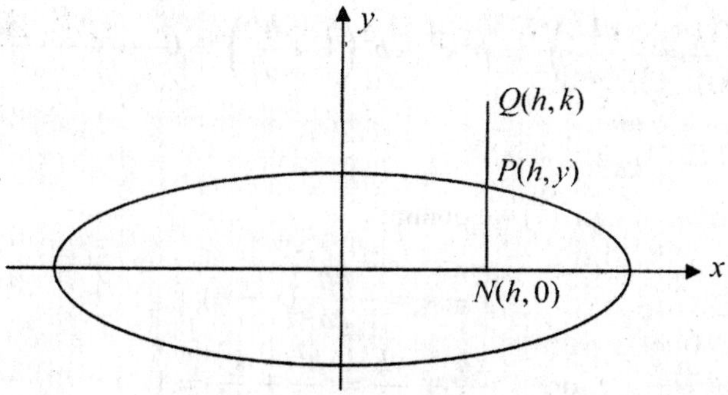

Now the point Q is outside, on or inside the ellipse according as

$$
\begin{aligned}
& NQ >=< NP \\
\text{or,} \quad & NQ^2 >=< NP^2 \\
\text{or,} \quad & k^2 >=< y^2 = b^2(1 - h^2/a^2) \\
\text{or,} \quad & h^2/a^2 + k^2/b^2 - 1 >=< 0.
\end{aligned}
$$

5.5 Equation of Tangent and Normal at any Point $P(h, k)$ on an Ellipse

Any line through the point (h, k) can be taken as

$$y - k = m(x - h) \tag{1}$$

m being the slope of the line. Equation of the ellipse is

$$\frac{x^2}{a^2} + \frac{y^2}{b^2} = 1.$$

Now, eliminating y from the above equations we get

$$
\begin{aligned}
& \frac{x^2}{a^2} + \frac{[k + m(x - h)]^2}{b^2} = 1 \\
\text{or,} \quad & a^2[m^2(x - h)^2 + 2km(x - h) + k^2] + b^2x^2 - a^2b^2 = 0 \\
\text{or,} \quad & (a^2m^2 + b^2)x^2 + 2x[km - hm^2]a^2 \\
& \quad + a^2[m^2h^2 + k^2 - b^2 - 2kmh] = 0.
\end{aligned}
$$

The line given by Eqn. (1) will touch the ellipse if the above equation has two equal roots, i.e. if

$$
\begin{aligned}
& a^4[km - hm^2]^2 = a^2[k^2 - b^2 + m^2h^2 - 2kmh](a^2m^2 + b^2) \\
\text{or,} \quad & a^2[k^2m^2 + h^2m^4 - 2khm^3] = a^2m^2k^2 - a^2m^2b^2 + a^2m^4h^2 \\
& \quad + b^2k^2 - b^4 + b^2m^2h^2 - 2kmh(a^2m^2 + b^2) \\
\text{or,} \quad & a^2k^2m^2 = b^2k^2 - b^4 - 2kmhb^2 \qquad \text{[as } (h, k) \text{ lies on the ellipse]} \\
\text{or,} \quad & \left(akm + \frac{hb^2}{a}\right)^2 = \frac{h^2b^4}{a^2} + b^2k^2 - b^4
\end{aligned}
$$

or, $\left(akm + \frac{hb^2}{a}\right)^2 = b^2k^2 - b^4\left(1 - \frac{h^2}{a^2}\right) = 0$

or, $m = -\frac{hb^2}{ka^2}$.

Putting the value of m in Eqn. (1) we obtain

$$y - k = -\frac{hb^2}{ka^2}(x - h)$$

$$\text{or,} \quad \frac{xh}{a^2} + \frac{yk}{b^2} = \frac{h^2}{a^2} + \frac{k^2}{b^2} = 1.$$

This is the equation of tangent at (h, k) to the ellipse $\frac{x^2}{a^2} + \frac{y^2}{b^2} = 1$.

Equation of normal at (h, k) to the ellipse $\frac{x^2}{a^2} + \frac{y^2}{b^2} = 1$ is

$$\frac{xk}{b^2} - \frac{yh}{a^2} = hk\left(\frac{1}{b^2} - \frac{1}{a^2}\right)$$

$$\text{or,} \quad \frac{x - h}{h/a^2} = \frac{y - k}{k/b^2}.$$

Cor. 3. *If the line $y = mx + c$ touches the ellipse $\frac{x^2}{a^2} + \frac{y^2}{b^2} = 1$ at (h, k), then tangent to the ellipse at (h, k), i.e. $\frac{xh}{a^2} + \frac{yk}{b^2} = 1$ and $y = mx + c$ should be identical. Comparing we get*

$$m = -hb^2; \quad ka^2 = 1 \quad \text{and} \quad c = a^2b^2$$

$$\text{or,} \quad k = 1/a^2 \quad \text{and} \quad h = -m/b^2.$$

Since (h, k) lies on the ellipse we have

$$\frac{m^2}{a^2b^4} + \frac{1}{a^4b^2} = 1$$

$$\text{or,} \quad a^2m^2 + b^2 = a^4b^4 = c^2, \quad \text{i.e.} \quad c = \sqrt{a^2m^2 + b^2}.$$

Hence, the line $y = mx + \sqrt{a^2m^2 + b^2}$ touches the ellipse $\frac{x^2}{a^2} + \frac{y^2}{b^2} = 1$ for all values of m.

5.6 Equation of Chord of Contact of the Point $P(h, k)$ with Respect to the Ellipse $\frac{x^2}{a^2} + \frac{y^2}{b^2} = 1$

Let the tangents drawn from the point (h, k) to the ellipse $\frac{x^2}{a^2} + \frac{y^2}{b^2} = 1$ touch the ellipse at (x_1, y_1) and (x_2, y_2). Equation of tangent at the point (x_1, y_1) to the ellipse $\frac{x^2}{a^2} + \frac{y^2}{b^2} = 1$ is

$$\frac{xx_1}{a^2} + \frac{yy_1}{b^2} = 1.$$

If it passes through the point (h, k), then

$$\frac{hx_1}{a^2} + \frac{ky_1}{b^2} = 1. \tag{2}$$

Again, equation of tangent at the point (x_2, y_2) to the ellipse $\frac{x^2}{a^2} + \frac{y^2}{b^2} = 1$ is

$$\frac{xx_2}{a^2} + \frac{yy_2}{b^2} = 1$$

As it passes through the point (h, k), then

$$\frac{hx_2}{a^2} + \frac{ky_2}{b^2} = 1. \tag{3}$$

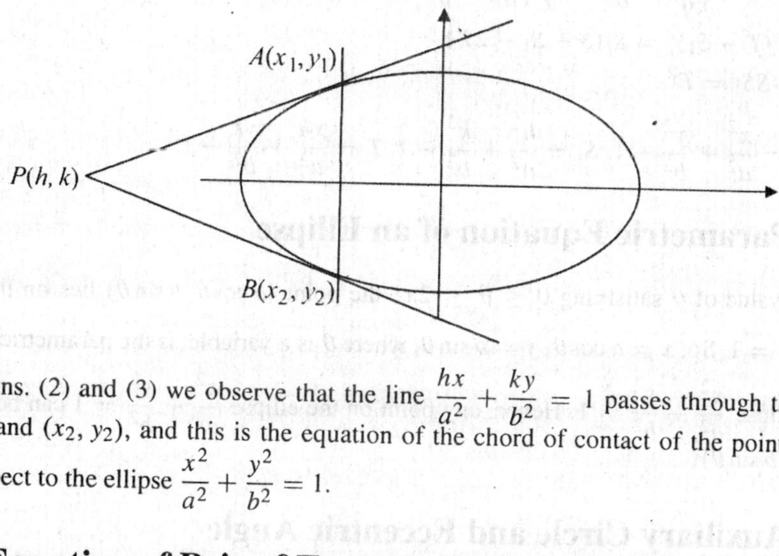

From Eqns. (2) and (3) we observe that the line $\frac{hx}{a^2} + \frac{ky}{b^2} = 1$ passes through the points (x_1, y_1) and (x_2, y_2), and this is the equation of the chord of contact of the point $P(h, k)$ with respect to the ellipse $\frac{x^2}{a^2} + \frac{y^2}{b^2} = 1$.

5.7 Equation of Pair of Tangents from a Given Point $P(h, k)$

Let the pair of tangents from the point $P(h, k)$ be drawn to the ellipse $\frac{x^2}{a^2} + \frac{y^2}{b^2} = 1$. Equation of any line through the point (h, k) is of the form

$$\frac{x - h}{l} = \frac{y - k}{m} = r. \tag{4}$$

Any point on this line can be taken as $(h + lr, k + mr)$. If this point lies on the ellipse $\frac{x^2}{a^2} + \frac{y^2}{b^2} = 1$, then we have

$$\frac{(h + lr)^2}{a^2} + \frac{(k + lm)^2}{b^2} = 1$$

$$\text{or,} \quad \left(\frac{l^2}{a^2} + \frac{m^2}{b^2}\right)r^2 + 2r\left(\frac{lh}{a^2} + \frac{mk}{b^2}\right) + \left(\frac{h^2}{a^2} + \frac{k^2}{b^2} - 1\right) = 0.$$

Now, if the line given by (4) touches the ellipse, the above equation in r should have only one root, which gives the point of contact and this gives

$$\left(\frac{lh}{a^2}+\frac{mk}{b^2}\right)^2=\left(\frac{h^2}{a^2}+\frac{k^2}{b^2}-1\right)\left(\frac{l^2}{a^2}+\frac{m^2}{b^2}\right)$$

or, $$\left(\frac{(x-h)h}{(y-k)a^2}+\frac{k}{b^2}\right)^2=\left(\frac{h^2}{a^2}+\frac{k^2}{b^2}-1\right)\left(\frac{(x-h)^2}{(y-k)^2a^2}+\frac{1}{b^2}\right)$$

or, $$\left(\frac{xh}{a^2}+\frac{yk}{b^2}-\frac{h^2}{a^2}-\frac{k^2}{b^2}\right)^2=\left(\frac{h^2}{a^2}+\frac{k^2}{b^2}-1\right)\left(\frac{(x-h)^2}{a^2}+\frac{(y-k)^2}{b^2}\right)$$

or, $$\left(\frac{xh}{a^2}+\frac{yk}{b^2}-1-\frac{h^2}{a^2}-\frac{k^2}{b^2}+1\right)^2$$
$$=\left(\frac{h^2}{a^2}+\frac{k^2}{b^2}-1\right)\left(\frac{x^2}{a^2}+\frac{y^2}{b^2}-1+\frac{h^2}{a^2}+\frac{k^2}{b^2}-1-2\left(\frac{xh}{a^2}+\frac{yk}{b^2}-1\right)\right)$$

or, $(T-S_1)^2=S_1(S+S_1-2T)$

or, $SS_1=T^2$

where $S=\frac{x^2}{a^2}+\frac{y^2}{b^2}-1$, $S_1=\frac{h^2}{a^2}+\frac{k^2}{b^2}-1$, $T=\frac{xh}{a^2}+\frac{yk}{b^2}-1$.

5.8 Parametric Equation of an Ellipse

For any value of θ satisfying $0\le\theta\le 2\pi$, the point $(a\cos\theta, b\sin\theta)$ lies on the ellipse $\frac{x^2}{a^2}+\frac{y^2}{b^2}=1$. So, $x=a\cos\theta$, $y=b\sin\theta$, where θ is a variable, is the parametric equation of the ellipse $\frac{x^2}{a^2}+\frac{y^2}{b^2}=1$. Hence, any point on the ellipse $\frac{x^2}{a^2}+\frac{y^2}{b^2}=1$ can be taken as $(a\cos\theta, b\sin\theta)$.

5.9 Auxiliary Circle and Eccentric Angle

The circle describing the major axis of the ellipse as diameter is called *auxiliary circle* of the ellipse. The auxiliary circle of the ellipse $\frac{x^2}{a^2}+\frac{y^2}{b^2}=1$ $(a>b)$ is $x^2+y^2=a^2$.

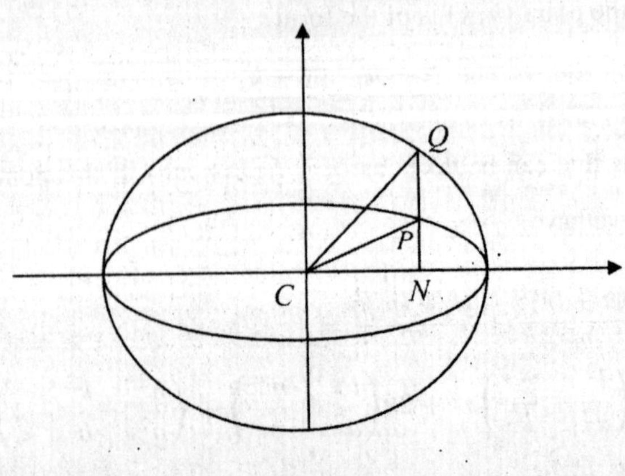

Let P be any point on the ellipse and PN be a perpendicular to the major axis. If NP is extended to meet the auxiliary circle at Q, then clearly abscissae of both P and Q are same. As the points (x, PN) and (x, QN) lies on the ellipse and the auxiliary circle respectively, we have

$$\frac{x^2}{a^2} + \frac{PN^2}{b^2} = 1$$

$$\text{and} \quad x^2 + QN^2 = a^2$$

which give

$$PN = \frac{b}{a}QN$$

this shows that any ordinate on the ellipse is $\frac{b}{a}$ times the ordinate of the corresponding point on the auxiliary circle.

If the line joining center of the ellipse and the point Q on the auxiliary circle makes an angle ϕ with the major axis of the ellipse, the angle ϕ is called the *eccentric angle* of the point P on the ellipse. Obviously, the coordinates of the point Q are $(a\cos\phi, a\sin\phi)$ and hence the coordinates of the point P are $(a\cos\phi, b\sin\phi)$.

5.10 Conormal Points

Equation of normal to the ellipse $\frac{x^2}{a^2} + \frac{y^2}{b^2} = 1$ at the point (x_1, y_1) is

$$\frac{x - x_1}{x_1/a^2} = \frac{y - y_1}{y_1/b^2}.$$

If it passes through (h, k), then

$$\frac{h - x_1}{x_1/a^2} = \frac{k - y_1}{y_1/b^2} = \lambda \quad \text{(say)}.$$

Simplifying, we get

$$x_1 = \frac{a^2h}{a^2 + \lambda} \quad \text{and} \quad y_1 = \frac{b^2k}{b^2 + \lambda}.$$

As the point (x_1, y_1) lies on the ellipse, we have

$$\frac{a^2h^2}{(a^2 + \lambda)^2} + \frac{b^2k^2}{(b^2 + \lambda)^2} = 1$$

from which we get four values of λ and this indicates that four normals can be drawn to the ellipse from the point (h, k) and feet of the normals are given by

$$\left(\frac{ha^2}{a^2 + \lambda_i}, \frac{kb^2}{b^2 + \lambda_i}\right), \quad i = 1, 2, 3, 4.$$

Now, the results $a^2 + \lambda = \frac{a^2h}{x_1}$ and $b^2 + \lambda = \frac{b^2k}{y_1}$ give

$$a^2 - b^2 = \frac{a^2h}{x_1} - \frac{b^2k}{y_1}$$

which shows that the feet of the normals lie on the curve

$$a^2 - b^2 = \frac{a^2h}{x} - \frac{b^2k}{y}$$

$$\text{or,} \quad xy(a^2 - b^2) = a^2hy - b^2kx$$

which is a rectangular hyperbola.

Another Method

Equation of normal to the ellipse $\frac{x^2}{a^2} + \frac{y^2}{b^2} = 1$ at the point $(a\cos\theta, b\sin\theta)$ is given by

$$xa\sec\theta - yb\operatorname{cosec}\theta = a^2 - b^2.$$

If it passes through (h, k), then we have

$$ha\sec\theta - kb\operatorname{cosec}\theta = a^2 - b^2.$$

Introducing the variable $t = \tan(\theta/2)$ in the above equation we get

$$ah\frac{1+t^2}{1-t^2} - bk\frac{1+t^2}{2t} = a^2 - b^2$$

$$\text{or,} \quad 2aht(1+t^2) - bk(1-t^4) = 2(a^2 - b^2)t(1-t^2)$$

from which we get four values of t, i.e. of θ and this indicates that four normals can be drawn to the ellipse from the point (h, k).

Simplifying, the above equation in t, we get

$$bkt^4 + 2(ah + a^2 - b^2)t^3 + 2(ah - a^2 + b^2)t - bk = 0.$$

If the feet of the normals are given by $(a\cos\theta_i, b\sin\theta_i)$, where $t_i = \tan\left(\frac{\theta_i}{2}\right)$, then we have by theory of equation

$$\sum t_1 = \frac{-2(ah + a^2 - b^2)}{bk}$$

$$\sum t_1t_2 = 0$$

$$\sum t_1t_2t_3 = \frac{-2(ah - a^2 + b^2)}{bk}$$

$$t_1t_2t_3t_4 = -1.$$

Now, by basic trigonometry, we have

$$\tan\left(\frac{\theta_1}{2} + \frac{\theta_2}{2} + \frac{\theta_3}{2} + \frac{\theta_4}{2}\right) = \frac{\sum t_1 - \sum t_1t_2t_3}{1 - \sum t_1t_2 + t_1t_2t_3t_4} = \infty$$

i.e. $(\theta_1 + \theta_2 + \theta_3 + \theta_4)$ is an odd multiple of π.

So, four normals can be drawn from a point to the ellipse and the sum of the eccentric angles of their feet is an odd multiple of π.

Examples

▶▶ 1. *Prove that the length of the focal chord of the ellipse which is inclined to the major axis at an angle θ is* $\dfrac{2ab^2}{a^2\sin^2\theta + b^2\cos^2\theta}$.

Solution: Let PQ be the focal chord which is inclined to the major axis of the ellipse at an angle θ. Then equation of PQ is

$$y = m(x - ae), \quad \text{where } m = \tan\theta.$$

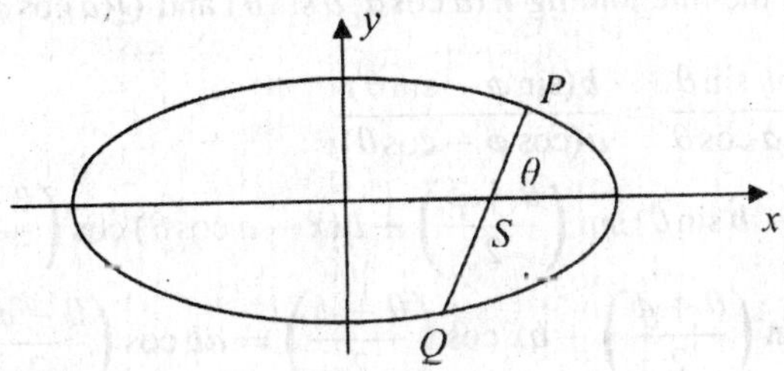

Now putting this result in equation of ellipse we get

$$\frac{x^2}{a^2} + \frac{m^2(x - ae)^2}{b^2} = 1$$

$$\text{or,} \quad (a^2m^2 + b^2)x^2 - 2a^3m^2ex + a^2(a^2m^2e^2 - b^2) = 0.$$

Now, if the chord PQ cuts the ellipse at (x_1, y_1) and (x_2, y_2), then

$$x_1 + x_2 = \frac{2a^3m^2e}{a^2m^2 + b^2}$$

$$\text{and} \quad x_1x_2 = \frac{a^2(a^2m^2e^2 - b^2)}{a^2m^2 + b^2}.$$

So,

$$\begin{aligned}(x_1 - x_2)^2 &= (x_1 + x_2)^2 - 4x_1x_2 \\ &= \frac{4a^2[a^4m^4e^2 - (a^2m^2e^2 - b^2)(a^2m^2 + b^2)]}{(a^2m^2 + b^2)^2} \\ &= \frac{4a^2b^2(b^2 + a^2m^2 - a^2m^2e^2)}{(a^2m^2 + b^2)^2} \\ &= \frac{4a^2b^4(1 + m^2)}{(a^2m^2 + b^2)^2}.\end{aligned}$$

Hence, the length of the chord

$$\begin{aligned}&= \sqrt{(x_1 - x_2)^2 + (y_1 - y_2)^2} \\ &= \sqrt{(x_1 - x_2)^2 + m^2(x_1 - x_2)^2}\end{aligned}$$

$$= \frac{2ab^2(1+m^2)}{a^2m^2+b^2}$$
$$= \frac{2ab^2(1+\tan^2\theta)}{a^2\tan^2\theta+b^2}$$
$$= \frac{2ab^2}{a^2\sin^2\theta+b^2\cos^2\theta}. \quad \textbf{[Proved]}$$

▶▷▶ **2.** *Find the equation of the chord of an ellipse joining two points whose eccentric angles are θ and ϕ.*

Solution: Equation of the line joining $P(a\cos\theta, b\sin\theta)$ and $Q(a\cos\phi, b\sin\phi)$ is

$$\frac{y-b\sin\theta}{x-a\cos\theta} = \frac{b(\sin\phi-\sin\theta)}{a(\cos\phi-\cos\theta)}$$

or, $$a(y-b\sin\theta)\sin\left(\frac{\theta+\phi}{2}\right)+b(x-a\cos\theta)\cos\left(\frac{\theta+\phi}{2}\right)=0$$

or, $$ay\sin\left(\frac{\theta+\phi}{2}\right)+bx\cos\left(\frac{\theta+\phi}{2}\right)=ab\cos\left(\frac{\theta-\phi}{2}\right).$$

▶▷▶ **3.** *Prove that the line joining two points on an ellipse such that the difference of their eccentric angles is constant, touches another ellipse.*

Solution: Equation of the line joining $P(a\cos\theta, b\sin\theta)$ and $Q(a\cos\phi, b\sin\phi)$ is

$$ay\sin\left(\frac{\theta+\phi}{2}\right)+bx\cos\left(\frac{\theta+\phi}{2}\right)=ab\cos\left(\frac{\theta-\phi}{2}\right).$$

Given that $\phi-\theta=2\alpha$ (constant). So the above equation becomes

$$ay\sin\lambda+bx\cos\lambda=ab\cos\alpha, \quad \text{where it is assumed that } \phi+\theta=2\lambda$$

or, $$\frac{x\cos\lambda}{a\cos\alpha}+\frac{y\sin\lambda}{b\cos\alpha}=1 \tag{1}$$

which is of the form of a tangent to an ellipse

$$\frac{x^2}{a^2\cos^2\alpha}+\frac{y^2}{b^2\cos^2\alpha}=1.$$

So, the chord touches the ellipse

$$\frac{x^2}{a^2}+\frac{y^2}{b^2}=\cos^2\alpha.$$

▶▷▶ **4.** *Find the locus of the middle point of the line joining two points on an ellipse such that the difference of their eccentric angles is constant.*

Solution: Let (h, k) be the middle point of the line joining $P(a\cos\theta, b\sin\theta)$ and $Q(a\cos\phi, b\sin\phi)$. Then

$$2h=a(\cos\theta+\cos\phi) \quad \text{and} \quad 2k=b(\sin\theta+\sin\phi)$$

which gives

$$\frac{4h^2}{a^2} + \frac{4k^2}{b^2} = 2 + 2(\cos\theta\cos\phi + \sin\theta\sin\phi)$$
$$= 2[1 + \cos 2\lambda], \quad \text{[given that } \phi - \theta = 2\lambda \text{ (constant)]}$$
$$= 4\cos^2\lambda.$$

Hence, locus of the middle point of the chord is

$$\frac{x^2}{a^2} + \frac{y^2}{b^2} = \cos^2\lambda.$$

▶▷▶ 5. *If the tangent and normal to the ellipse at a point meet the minor axis at Q and R respectively, show that QR subtends a right angle at the foci of the ellipse.*

Solution: Equations of tangent and normal at the point $P(a\cos\theta, b\sin\theta)$ to the ellipse $\frac{x^2}{a^2} + \frac{y^2}{b^2} = 1$ are

$$\frac{x}{a}\cos\theta + \frac{y}{b}\sin\theta = 1 \tag{1}$$

and

$$\frac{x}{b}\sin\theta - \frac{y}{a}\cos\theta = \left(\frac{a}{b} - \frac{b}{a}\right)\sin\theta\cos\theta. \tag{2}$$

They meet the minor axis at Q: $(0, b\,\mathrm{cosec}\,\theta)$ and R: $\{0, (b^2 - a^2)\sin\theta/b\}$. If S: $(ae, 0)$ be one of the foci, then slope of $SQ = -\frac{b\,\mathrm{cosec}\,\theta}{ae}$ and slope of $SR = -\frac{b^2 - a^2}{bae}\sin\theta$.

Now, product of the slopes

$$= \frac{b\,\mathrm{cosec}\,\theta}{ae} \times \frac{b^2 - a^2}{bae}\sin\theta$$
$$= \frac{b^2 - a^2}{a^2e^2} = -1$$

which is also true if we change e by $-e$. **[Proved]**

▶▷▶ 6. *Any ordinate NP of the ellipse $\frac{x^2}{a^2} + \frac{y^2}{b^2} = 1$ cuts the auxiliary circle at Q. Find the locus of the point of intersection of the normals at P and Q.*

Solution: Equation of the normal at $P(a\cos\theta, b\sin\theta)$ to the ellipse is

$$\frac{x}{b}\sin\theta - \frac{y}{a}\cos\theta = \left(\frac{a}{b} - \frac{b}{a}\right)\sin\theta\cos\theta.$$

Putting $a = b$, we get the equation of the normal to auxiliary circle at Q as

$$x\sin\theta - y\cos\theta = 0.$$

To get the locus of the point of intersection of the above normals, we eliminate θ as follows:

$$x\sin\theta - y\cos\theta = 0$$
$$\Rightarrow \quad \frac{x}{\cos\theta} = \frac{y}{\sin\theta} = \frac{\sqrt{x^2 + y^2}}{1}$$
$$\Rightarrow \quad \cos\theta = \frac{x}{\sqrt{x^2 + y^2}} \quad \text{and} \quad \sin\theta = \frac{y}{\sqrt{x^2 + y^2}}.$$

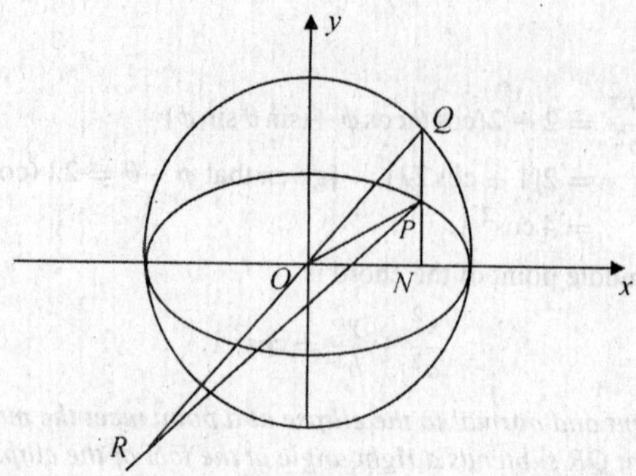

Putting these results in

$$\frac{x}{b}\sin\theta - \frac{y}{a}\cos\theta = \left(\frac{a}{b} - \frac{b}{a}\right)\sin\theta\cos\theta$$

we get

$$\frac{xy}{b\sqrt{x^2+y^2}} - \frac{xy}{a\sqrt{x^2+y^2}} = \frac{xy(a^2-b^2)}{ab(x^2+y^2)}$$

$$\Rightarrow \quad (a-b)\sqrt{(x^2+y^2)} = (a^2-b^2)$$

$$\Rightarrow \quad x^2+y^2 = (a+b)^2.$$

▶▶ 7. *If the normal at P to the ellipse* $\frac{x^2}{a^2}+\frac{y^2}{b^2}=1$ *cuts the major axis of the ellipse at Q, find the locus of the middle point of PQ.*

Solution: Equation of the normal at $P(a\cos\theta, b\sin\theta)$ to the ellipse is

$$\frac{x}{b}\sin\theta - \frac{y}{a}\cos\theta = \left(\frac{a}{b} - \frac{b}{a}\right)\sin\theta\cos\theta.$$

Putting $y=0$, we get $x = \frac{a^2-b^2}{a}\cos\theta$.

If (h, k) be the middle point of PQ, then

$$2h = a\cos\theta + \frac{a^2-b^2}{a}\cos\theta \quad \text{and} \quad 2k = b\sin\theta.$$

Eliminating θ, we get

$$\left(\frac{2h}{2a-b^2/a}\right)^2 + \left(\frac{2k}{b}\right)^2 = 1.$$

Hence, the locus of the middle point is

$$\left(\frac{ax}{2a^2-b^2}\right)^2 + \left(\frac{y}{b}\right)^2 = \frac{1}{4}.$$

▶▷▶ **8**. *The tangent at the point θ of the ellipse* $\frac{x^2}{a^2}+\frac{y^2}{b^2}=1$ *cuts the auxiliary circle at P and Q. If PQ subtends a right angle at the centre of the ellipse, then show that eccentricity of the ellipse is* $\frac{1}{\sqrt{1+\sin^2\theta}}$.

Solution: Equation of the tangent at $(a\cos\theta, b\sin\theta)$ is

$$\frac{x}{a}\cos\theta+\frac{y}{b}\sin\theta=1. \tag{1}$$

Now, making homogeneous the equation $x^2+y^2=a^2$ with the aid of Eqn. (1) we get

$$x^2+y^2=a^2\left(\frac{x}{a}\cos\theta+\frac{y}{b}\sin\theta\right)^2$$

$$\text{or,}\quad x^2\sin^2\theta-\frac{2a}{b}xy\sin\theta\cos\theta+y^2\left(1-\frac{a^2}{b^2}\sin^2\theta\right)=0.$$

The above second degree homogeneous equation in x, y represents a pair of straight lines passing through origin, i.e. centre of the ellipse.

If the lines given by above equation are perpendicular, then

$$\sin^2\theta+\left(1-\frac{a^2}{b^2}\sin^2\theta\right)=0$$

$$\text{or,}\quad \left(1-\frac{a^2}{b^2}\right)\sin^2\theta=-1$$

$$\text{or,}\quad \left(1-\frac{1}{1-e^2}\right)\sin^2\theta=-1$$

$$\text{or,}\quad e^2\sin^2\theta=1-e^2$$

$$\text{or,}\quad e^2(1+\sin^2\theta)=1$$

$$\text{or,}\quad e=\frac{1}{\sqrt{1+\sin^2\theta}}.\qquad \textbf{[Proved]}$$

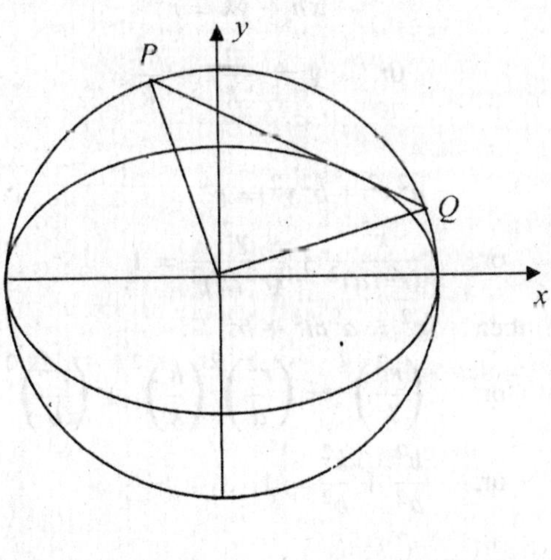

9. *Find the locus of the point of intersection of the tangents if the line joining their points of contact to the centre of the ellipse be at right angle.*

Solution: If the tangents at P and Q on the ellipse $\frac{x^2}{a^2}+\frac{y^2}{b^2}=1$ intersect at (h, k), the equation of PQ is

$$\frac{xh}{a^2}+\frac{yk}{b^2}=1. \tag{1}$$

Now, making homogeneous the equation of ellipse with the aid of Eqn. (1) we get

$$\frac{x^2}{a^2}+\frac{y^2}{b^2}=\left(\frac{xh}{a^2}+\frac{yk}{b^2}\right)^2$$

$$\text{or,}\quad \left(\frac{1}{a^2}-\frac{h^2}{a^4}\right)x^2-\frac{2hkxy}{a^2b^2}+\left(\frac{1}{b^2}-\frac{k^2}{b^4}\right)y^2=0.$$

The above second degree homogeneous equation in x, y represents a pair of straight lines passing through origin i.e., centre of the the ellipse.

If the straight lines given by above equation are perpendicular, then we have

$$\left(\frac{1}{a^2}-\frac{h^2}{a^4}\right)+\left(\frac{1}{b^2}-\frac{k^2}{b^4}\right)=0$$

$$\text{or,}\quad \frac{h^2}{a^4}+\frac{k^2}{b^4}=\frac{1}{a^2}+\frac{1}{b^2}.$$

Hence, locus of the point (h, k) is

$$\frac{x^2}{a^4}+\frac{y^2}{b^4}=\frac{1}{a^2}+\frac{1}{b^2}.$$

10. *Tangents are drawn from any point on the ellipse* $\frac{x^2}{a^2}+\frac{y^2}{b^2}=1$ *to the circle* $x^2+y^2=r^2$. *Prove that the chords of contact are tangents to the ellipse* $a^2x^2+b^2y^2=r^4$.

Solution: Equation of the chord of contact of the point (h, k) with respect to the circle $x^2+y^2=r^2$ is

$$xh+yk=r^2$$

$$\text{or,}\quad y=-\frac{h}{k}x+\frac{r^2}{k}.$$

If it touches the ellipse

$$a^2x^2+b^2y^2=r^4$$

$$\text{or,}\quad \frac{x^2}{(r^2/a)^2}+\frac{y^2}{(r^2/b)^2}=1$$

$$\text{then}\quad c^2=a^2m^2+b^2$$

$$\text{or,}\quad \left(\frac{r^2}{k}\right)^2=\left(\frac{r^2}{a}\right)^2\left(\frac{h}{k}\right)^2+\left(\frac{r^2}{b}\right)^2$$

$$\text{or,}\quad \frac{h^2}{a^2}+\frac{k^2}{b^2}=1.$$

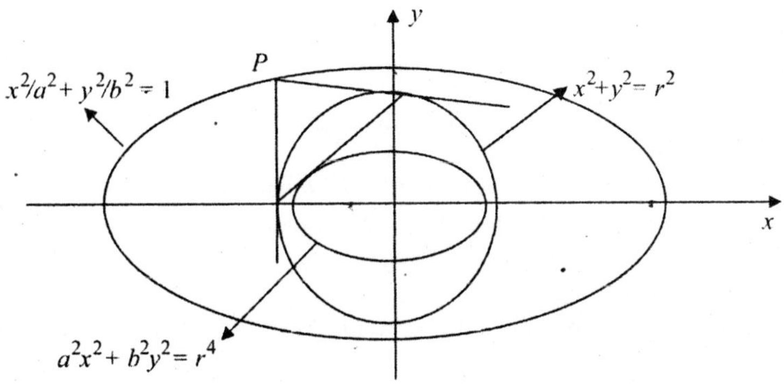

Hence, locus of the point (h, k) is

$$\frac{x^2}{a^2} + \frac{y^2}{b^2} = 1. \qquad \textbf{[Proved]}$$

▶▷▶ **11**. *Find the locus of the point, the chord of contact of the tangents from which to the ellipse* $\frac{x^2}{a^2} + \frac{y^2}{b^2} = 1$ *always touches the circle* $x^2 + y^2 = r^2$.

Solution: Equation of the chord of contact of the point (h, k) with respect to the ellipse $\frac{x^2}{a^2} + \frac{y^2}{b^2} = 1$ is

$$\frac{xh}{a^2} + \frac{yk}{b^2} = 1$$

$$\text{or,} \quad y = -\frac{hb^2}{ka^2}x + \frac{b^2}{k}.$$

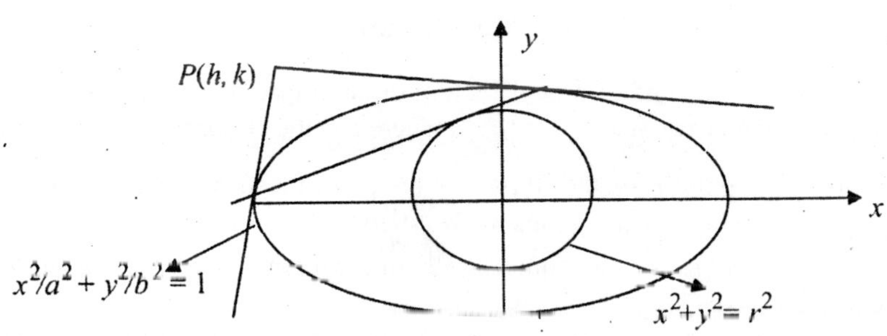

If it touches the circle

$$x^2 + y^2 = r^2$$

then we have

$$c^2 = r^2m^2 + r^2$$

$$\text{or,} \quad \left(\frac{b^2}{k}\right)^2 = \left[\left(\frac{hb^2}{ka^2}\right)^2 + 1\right]r^2$$

$$\text{or,} \quad a^4b^4 = (h^2b^4 + k^2a^4)r^2.$$

F-7

Generalizing, we get the locus of the point (h, k) is

$$\frac{x^2}{a^4} + \frac{y^2}{b^4} = \frac{1}{r^2}.$$

▶▷▶ **12.** *Any tangent to an ellipse is cut off by the tangents at the extremities of major axis at T and T'. Prove that the circle describing TT' as diameter passes through the foci of the ellipse.*

Solution: Equations of the tangents at the extremities of the major axis of the ellipse $\frac{x^2}{a^2} + \frac{y^2}{b^2} = 1$ are $x = \pm a$ and equation of the tangent to the ellipse at the point $(a\cos\theta, b\sin\theta)$ is

$$\frac{x}{a}\cos\theta + \frac{y}{b}\sin\theta = 1.$$

It meets the lines $x = \pm a$ at $T\left(a, \frac{1-\cos\theta}{\sin\theta}b\right)$ and $T'\left(-a, \frac{1+\cos\theta}{\sin\theta}b\right)$. So, equation of the circle describing TT' as diameter is

$$(x-a)(x+a) + \left(y - \frac{1-\cos\theta}{\sin\theta}b\right)\left(y - \frac{1+\cos\theta}{\sin\theta}b\right) = 0$$

$$\Rightarrow \quad x^2 + y^2 - \frac{2b}{\sin\theta}y - a^2 + b^2 = 0$$

$$\Rightarrow \quad x^2 + y^2 - \frac{2b}{\sin\theta}y - a^2e^2 = 0.$$

Clearly, the above circle passes through the foci of the ellipse, i.e. $(\pm ae, 0)$.

EXERCISE

1. If the normal at one end of a latus rectum of an ellipse passes through one end of the minor axis, then prove that $e^4 + e^2 = 1$, where e is the eccentricity.
2. The normal at a point P on the ellipse $\frac{x^2}{a^2} + \frac{y^2}{b^2} = 1$ meets the axes at Q and R. Find the equation of the locus of the middle point of QR.
3. Show that product of the perpendiculars from foci on any tangent is equal to the square of semi-minor axis.
4. The tangent and normal at a point P on the ellipse meet its major axis at T and T'; show that $TT' = a$.
5. Find the locus of a point, the chord of contact of the tangents from which with respect to the ellipse $\frac{x^2}{a^2} + \frac{y^2}{b^2} = 1$ touches the circle $x^2 + y^2 = c^2$.
6. A tangent to the ellipse $\frac{x^2}{a^2} + \frac{y^2}{b^2} = 1$ meets the ellipse $\frac{x^2}{a} + \frac{y^2}{b} = a + b$ at the points P and Q. Prove that the tangents at P and Q are perpendicular.
7. Show that the locus of the foot of the perpendicular drawn from centre of the ellipse $\frac{x^2}{a^2} + \frac{y^2}{b^2} = 1$ on any tangent to it is $(x^2 + y^2)^2 = a^2x^2 + b^2y^2$.

8. Tangents are drawn from the point $T(h, k)$ which meet the ellipse $\frac{x^2}{a^2} + \frac{y^2}{b^2} = 1$ at P and Q. Show that the area of the triangle ΔCPQ, C being the centre of the ellipse, is

$$\frac{a^2b^2\sqrt{b^2h^2 + a^2k^2 - a^2b^2}}{b^2h^2 + a^2k^2}.$$

9. Tangents are drawn from the point $T(h, k)$ which meet the ellipse $\frac{x^2}{a^2} + \frac{y^2}{b^2} = 1$ at P and Q. Show that the area of the triangle ΔTPQ is

$$\frac{\sqrt{b^2h^2 + a^2k^2 - a^2b^2}}{b^2h^2 + a^2k^2}.$$

10. An ellipse is rotated through a right angle in its own plane about the centre, which is fixed. Prove that the locus of the point of intersection of a tangent to the ellipse in its original position with the tangent at the same point of the curve in new position is

$$(x^2 + y^2)(x^2 + y^2 - a^2 - b^2) = 2(b^2 - a^2)xy.$$

[*Hints*: Without loss of generality, we assume the equation of the ellipse as

$$\frac{x^2}{a^2} + \frac{y^2}{b^2} = 1.$$

Equation of tangent to the ellipse at $(a\cos\theta, b\sin\theta)$ is

$$xb\cos\theta + ya\sin\theta - ab = 0.$$

Using the equations of rotation we get

$$x' = x\cos 90° + y\sin 90° = y$$

and $$y' = -x\sin 90° + y\cos 90° = -x.$$

Thus transformed equation of the ellipse is $\frac{x^2}{b^2} + \frac{y^2}{a^2} = 1$ and point of contact of the tangent becomes $(b\sin\theta, -a\cos\theta)$. Hence, equation of the tangent in new position is

$$xa\sin\theta - yb\cos\theta - ab = 0.$$

Solving the equation of tangents for $\cos\theta$ and $\sin\theta$, and eliminating θ, we get the result.]

11. Show that the normals at the four points where the lines $\frac{lx}{a} + \frac{my}{b} = 1$ and $\frac{x}{la} + \frac{y}{mb} + 1 = 0$ cut the ellipse $\frac{x^2}{a^2} + \frac{y^2}{b^2} = 1$ are concurrent.

[*Hints*: Let two points A, B lie on the line $\frac{lx}{a} + \frac{my}{b} = 1$. Equation of the curve through the feet of the four normals of the ellipse $\frac{x^2}{a^2} + \frac{y^2}{b^2} = 1$ drawn from the point (h, k) is

$$(a^2 - b^2)xy = a^2hy - b^2ky. \quad (1)$$

Now, if equation of the line through other two points A', B' be $l'x + m'y = p'$, then

$$\frac{x^2}{a^2} + \frac{y^2}{b^2} - 1 + \lambda\left(\frac{lx}{a} + \frac{my}{b} - 1\right)(l'x + m'y - p') = 0 \quad (2)$$

represents the curve containing the four normals for some values of λ. Comparing Eqns. (1) and (2) we get

$$\lambda ll' + \frac{1}{a} = 0; \quad \lambda mm' + \frac{1}{b} = 0; \quad \text{and} \quad \lambda p' = 1$$

$$\text{or,} \quad l' = -\frac{1}{\lambda al}; \quad m' = -\frac{1}{\lambda bm}; \quad \text{and} \quad p' = \frac{1}{\lambda}.$$

So, equation of the line $A'B'$ is

$$\frac{x}{la} + \frac{y}{mb} + 1 = 0.$$

Hence the result.]

Chapter 6

Hyperbola

6.1 Definition

The locus of a point which moves in such a way that its distance from a fixed point, called focus, is greater than to its distance from a fixed straight line, called directrix, is called *hyperbola*. If we choose the points $(\pm ae, 0)$ as foci and the lines $x \pm \frac{a}{e} = 0$ as directrices, $e(> 1)$ being the eccentricity, then the equation of hyperbola becomes

$$\frac{x^2}{a^2} - \frac{y^2}{b^2} = 1$$

where $b^2 = a^2(e^2 - 1)$, which is the standard/basic/normal/canonical form of hyperbola.

Another Definition

Hyperbola is the locus of a point which moves in such a way that difference of its distances from two fixed points, called foci, is constant $2a$, length of the transverse axis.

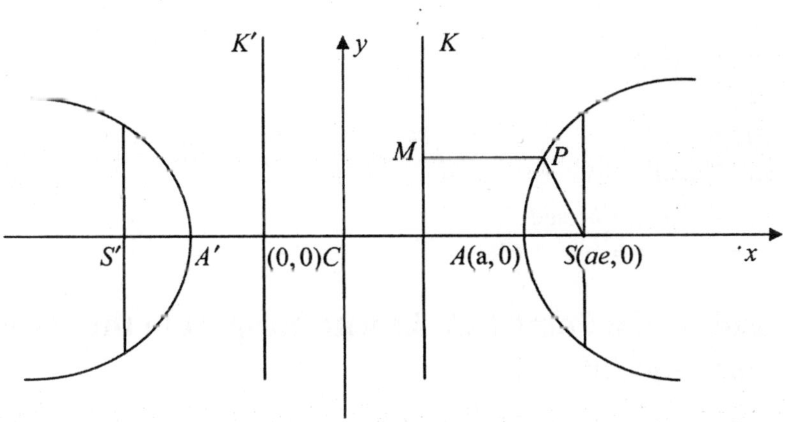

6.2 Some Important Results Associated with the Standard Form of Hyperbola

(i) Centre: $C(0, 0)$,
(ii) Foci: $S(ae, 0)$, $S'(-ae, 0)$
(iii) Transverse axis of the hyperbola: x-axis,
(iv) Conjugate axis of the hyperbola: y-axis,
(v) End points of transverse axis $(\pm a, 0)$,
(vi) End points of conjugate axis $(0, \pm b)$,
(vii) Eccentricity: $e = \sqrt{1 + b^2/a^2}$,
(viii) Equation of the directrices: $x = \pm a/e$,
(ix) Latus rectum $= \dfrac{2b^2}{a}$,
(x) End points of latus rectums: $L\left(ae, \pm\dfrac{b^2}{a}\right)$ and $L'\left(-ae, \pm\dfrac{b^2}{a}\right)$,
(xi) Focal distance of any point (x, y): $a \pm ex$.

6.3 General Equation of Hyperbola

If we take the line $ax + by + c = 0$ as directrix and the point (h, k) as focus, then for a point $P(x, y)$ in the hyperbola

$$SP = \sqrt{(x-h)^2 + (y-k)^2}$$

$$\text{and} \quad PM = \frac{ax + by + c}{\sqrt{a^2 + b^2}}.$$

Using $SP = ePM$, we get

$$[(x-h)^2 + (y-k)^2][a^2 + b^2] = e^2(ax + by + c)^2$$

$$\text{or,} \quad [a^2 + b^2][x^2 + y^2 - 2xh - 2yk + h^2 + k^2]$$
$$= e^2[a^2x^2 + b^2y^2 + 2abxy + 2axc + 2bcy + c^2]$$

$$\text{or,} \quad Ax^2 + 2Hxy + By^2 + 2Gx + 2Fy + C = 0$$

where

$A = a^2(e^2 - 1) - b^2$,
$B = b^2(e^2 - 1) - a^2$,
$H = abe^2$,
$G = (a^2 + b^2)h + e^2ac$,
$F = (a^2 + b^2)k + e^2bc$ and
$C = e^2c^2 - (a^2 + b^2)(h^2 + k^2)$.

6.4 Position of a Point $P(h, k)$ with Respect to the Hyperbola $x^2/a^2 - y^2/b^2 = 1$

Let $Q(h, k)$ be any point on extended NP, where $P(h, y)$ is any point on the hyperbola $\dfrac{x^2}{a^2} - \dfrac{y^2}{b^2} = 1$ as shown in the following figure.

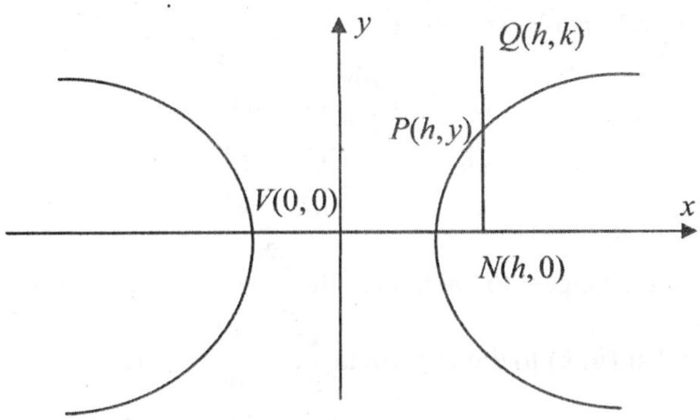

Now the point Q is outside, on or inside the hyperbola according as

$$NQ >=< NP$$

$$\text{or,} \quad NQ^2 >=< NP^2$$

$$\text{or,} \quad k^2 >=< y^2 = -b^2(1 - h^2/a^2)$$

$$\text{or,} \quad h^2/a^2 - k^2/b^2 - 1 >=< 0.$$

6.5 Equation of Tangent and Normal at Any Point $P(h, k)$ on a Hyperbola

Any line through the point (h, k) can be taken as

$$y - k = m(x - h) \tag{1}$$

m being the slope of the line. Equation of the hyperbola is $\dfrac{x^2}{a^2} - \dfrac{y^2}{b^2} = 1$.

Now, eliminating y from the above equations we get

$$\frac{x^2}{a^2} - \frac{[k + m(x - h)]^2}{b^2} = 1$$

$$\text{or,} \quad a^2[m^2(x - h)^2 + 2km(x - h) + k^2] - b^2x^2 + a^2b^2 = 0$$

$$\text{or,} \quad (a^2m^2 - b^2)x^2 + 2x[km - hm^2]a^2 + a^2[m^2h^2 + k^2 + b^2 - 2kmh] = 0.$$

The line given by Eqn. (1) will touch the hyperbola if the above equation has two equal roots

$$\text{i.e., if} \quad a^4[km - hm^2]^2 = a^2[k^2 + b^2 + m^2h^2 - 2kmh](a^2m^2 - b^2)$$

$$\text{or,} \quad a^2[k^2m^2 + h^2m^4 - 2khm^3] = a^2m^2k^2 + a^2m^2b^2 + a^2m^4h^2 - b^2k^2 - b^4 - b^2m^2h^2 - 2kmh(a^2m^2 - b^2)$$

$$\text{or,} \quad a^2k^2m^2 = -b^2k^2 - b^4 + 2kmhb^2 \quad [\text{as } (h, k) \text{ lies on the hyperbola}]$$

$$\text{or,} \quad \left(akm - \frac{hb^2}{a}\right)^2 = \frac{h^2b^4}{a^2} - b^2k^2 - b^4$$

$$\text{or,} \quad \left(akm - \frac{hb^2}{a}\right)^2 = -b^2k^2 - b^4\left(1 - \frac{h^2}{a^2}\right) = 0$$

$$\text{or,} \quad m = \frac{hb^2}{ka^2}.$$

Putting the value of m in Eqn. (1) we obtain

$$y - k = \frac{hb^2}{ka^2}(x - h)$$

$$\text{or,} \qquad \frac{xh}{a^2} - \frac{yk}{b^2} = \frac{h^2}{a^2} - \frac{k^2}{b^2} = 1.$$

This is the equation of the tangent to the hyperbola $\frac{x^2}{a^2} - \frac{y^2}{b^2} = 1$ at (h, k).

Equation of normal at (h, k) to the hyperbola $\frac{x^2}{a^2} - \frac{y^2}{b^2} = 1$ is

$$\frac{xk}{b^2} + \frac{yh}{a^2} = hk\left(\frac{1}{b^2} + \frac{1}{a^2}\right).$$

Cor. 4. *If the line $y = mx + c$ touches the hyperbola $\frac{x^2}{a^2} - \frac{y^2}{b^2} = 1$ at (h, k), then tangent to the hyperbola at (h, k), i.e. $\frac{xh}{a^2} - \frac{yk}{b^2} = 1$ or $xhb^2 - yka^2 - a^2b^2 = 0$ and $y = mx + c$ should be identical. Comparing we get*

$$m = hb^2; \quad ka^2 = 1 \quad \text{and} \quad c = -a^2b^2$$

$$\text{or,} \quad k = 1/a^2 \quad \text{and} \quad h = m/b^2.$$

Since (h, k) lies on the hyperbola, we have

$$\frac{m^2}{a^2b^4} - \frac{1}{a^4b^2} = 1$$

$$\text{or,} \quad a^2m^2 - b^2 = a^4b^4 = c^2, \quad \text{i.e. } c = \sqrt{a^2m^2 - b^2}.$$

Hence, the line $y = mx + \sqrt{a^2m^2 - b^2}$ touches the hyperbola $\frac{x^2}{a^2} - \frac{y^2}{b^2} = 1$ for all values of m.

6.6 Equation of Chord of Contact of the Point $P(h, k)$ with Respect to the Hyperbola $x^2/a^2 - y^2/b^2 = 1$

Let the tangents drawn from the point (h, k) to the hyperbola $\frac{x^2}{a^2} - \frac{y^2}{b^2} = 1$ touch the hyperbola at (x_1, y_1) and (x_2, y_2). Equation of tangent at the point (x_1, y_1) to the hyperbola $\frac{x^2}{a^2} - \frac{y^2}{b^2} = 1$ is

$$\frac{xx_1}{a^2} - \frac{yy_1}{b^2} = 1. \tag{2}$$

If it passes through the point (h, k), then

$$\frac{hx_1}{a^2} - \frac{ky_1}{b^2} = 1. \tag{3}$$

Again, equation of tangent at the point (x_2, y_2) to the hyperbola $\frac{x^2}{a^2} - \frac{y^2}{b^2} = 1$ is

$$\frac{xx_2}{a^2} - \frac{yy_2}{b^2} = 1.$$

As it passes through the point (h, k), then

$$\frac{hx_2}{a^2} - \frac{ky_2}{b^2} = 1. \tag{4}$$

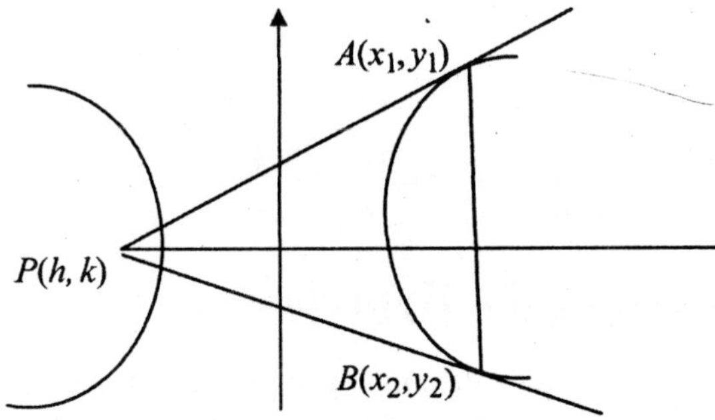

From Eqns. (3) and (4) we observe that the line $\frac{hx}{a^2} - \frac{ky}{b^2} = 1$ passes through the points (x_1, y_1) and (x_2, y_2), and this is the equation of the chord of contact of the point $P(h, k)$ with respect to the hyperbola $\frac{x^2}{a^2} - \frac{y^2}{b^2} = 1$.

6.7 Equation of Pair of Tangents from a Given Point $P(h, k)$

Let the pair of tangents from the point $P(h, k)$ be drawn to the hyperbola $\frac{x^2}{a^2} - \frac{y^2}{b^2} = 1$. Equation of any line through the point (h, k) is of the form

$$\frac{x-h}{l} = \frac{y-k}{m} = r. \tag{5}$$

Any point on this line can be taken as $(h + lr, k + mr)$. If this point lies on the hyperbola $\frac{x^2}{a^2} - \frac{y^2}{b^2} = 1$, then we have

$$\frac{(h+lr)^2}{a^2} - \frac{(k+lm)^2}{b^2} = 1$$

$$\text{or,} \quad \left(\frac{l^2}{a^2} - \frac{m^2}{b^2}\right)r^2 + 2r\left(\frac{lh}{a^2} - \frac{mk}{b^2}\right) + \left(\frac{h^2}{a^2} - \frac{k^2}{b^2} - 1\right) = 0.$$

Now, if the line given by Eqn. (5) touches the hyperbola, the above equation in r should have only one root, which gives the point of contact and this gives

$$\left(\frac{lh}{a^2} - \frac{mk}{b^2}\right)^2 = \left(\frac{h^2}{a^2} - \frac{k^2}{b^2} - 1\right)\left(\frac{l^2}{a^2} - \frac{m^2}{b^2}\right)$$

$$\text{or,} \quad \left(\frac{(x-h)h}{(y-k)a^2} - \frac{k}{b^2}\right)^2 = \left(\frac{h^2}{a^2} - \frac{k^2}{b^2} - 1\right)\left(\frac{(x-h)^2}{(y-k)^2a^2} - \frac{1}{b^2}\right)$$

$$\text{or,} \quad \left(\frac{xh}{a^2} - \frac{yk}{b^2} - \frac{h^2}{a^2} + \frac{k^2}{b^2}\right)^2 = \left(\frac{h^2}{a^2} - \frac{k^2}{b^2} - 1\right)\left(\frac{(x-h)^2}{a^2} - \frac{(y-k)^2}{b^2}\right)$$

$$\text{or,}\quad \left(\frac{xh}{a^2}-\frac{yk}{b^2}-1-\frac{h^2}{a^2}+\frac{k^2}{b^2}+1\right)^2$$

$$=\left(\frac{h^2}{a^2}-\frac{k^2}{b^2}-1\right)\left(\frac{x^2}{a^2}-\frac{y^2}{b^2}-1+\frac{h^2}{a^2}-\frac{k^2}{b^2}-1-2\left(\frac{xh}{a^2}-\frac{yk}{b^2}-1\right)\right)$$

$$\text{or,}\quad (T-S_1)^2 = S_1(S+S_1-2T)$$

$$\text{or,}\quad SS_1 = T^2$$

where $S=\frac{x^2}{a^2}-\frac{y^2}{b^2}-1$, $S_1=\frac{h^2}{a^2}-\frac{k^2}{b^2}-1$, $T=\frac{xh}{a^2}-\frac{yk}{b^2}-1$.

6.8 Parametric Equation of a Hyperbola

For any value of θ satisfying $0 \le \theta \le 2\pi$, the point $(a\sec\theta, b\tan\theta)$ lies on the hyperbola $\frac{x^2}{a^2}-\frac{y^2}{b^2}=1$. So, $x=a\sec\theta$, $y=b\tan\theta$, where θ is a variable, is the parametric equation of the hyperbola $\frac{x^2}{a^2}-\frac{y^2}{b^2}=1$. Hence, any point on the hyperbola $\frac{x^2}{a^2}-\frac{y^2}{b^2}=1$ can be taken as $(a\sec\theta, b\tan\theta)$.

6.9 Rectangular Hyperbola

A hyperbola, whose transverse and conjugate axes are of equal length, is called *rectangular hyperbola.* So the basic form of a rectangular hyperbola is $x^2-y^2=a^2$ whose eccentricity is $\sqrt{2}$. The parametric equation of rectangular hyperbola is $x=a\sec\theta$, $y=a\tan\theta$, i.e. any point on the rectangular hyperbola $x^2-y^2=a^2$ can be taken as $(a\sec\theta, a\tan\theta)$. Rotating coordinate axes, i.e. its axes suitably this equation of rectangular hyperbola can be transformed to $xy=c^2$, whose parametric equation is $x=ct$, $y=c/t$.

6.10 Chord, Tangent, Normal of the Rectangular Hyperbola $xy=c^2$

If $P(ct_1, c/t_1)$ and $Q(ct_2, c/t_2)$ be any two points on the hyperbola $xy=c^2$, the equation of PQ is

$$y-\frac{c}{t_2}=\frac{c/t_2-c/t_1}{ct_2-ct_1}(x-ct_2)$$

$$\text{or,}\quad y-\frac{c}{t_2}=-\frac{x-ct_2}{t_2t_1}$$

$$\text{or,}\quad yt_2t_1+x-c(t_1+t_2)=0.$$

Now, taking limit $Q\to P$, i.e. $t_2\to t_1$ in the above equation we get, the equation of tangent at $(ct_1, c/t_1)$ is

$$yt_1^2+x-2ct_1=0.$$

From it we also get, the equation of normal is

$$xt_1^2-y=ct_1^3-\frac{c}{t_1}.$$

6.11 Conjugate Hyperbola

The hyperbola whose transverse and conjugate axes are the conjugate and transverse axes of the hyperbola $\frac{x^2}{a^2}-\frac{y^2}{b^2}=1$ is called its *conjugate*. So the equation of the conjugate hyperbola of the hyperbola $\frac{x^2}{a^2}-\frac{y^2}{b^2}=1$ is $\frac{y^2}{b^2}-\frac{x^2}{a^2}=1$, i.e. $\frac{x^2}{a^2}-\frac{y^2}{b^2}=-1$ and the equation of the conjugate hyperbola of the hyperbola $x^2-y^2=a^2$ is $x^2-y^2=-a^2$.

6.12 Auxiliary Circle

The circle describing the transverse axis of the hyperbola as diameter is called *auxiliary circle* of the hyperbola. The auxiliary circle of the hyperbola $\frac{x^2}{a^2}-\frac{y^2}{b^2}=1$ $(a>b)$ is $x^2+y^2=a^2$.

6.13 Conormal Points

Equation of normal to the hyperbola $\frac{x^2}{a^2}-\frac{y^2}{b^2}=1$ at the point (x_1, y_1) is

$$\frac{x-x_1}{x_1/a^2}=-\frac{y-y_1}{y_1/b^2}.$$

If it passes through (h, k), then

$$\frac{h-x_1}{x_1/a^2}=-\frac{k-y_1}{y_1/b^2}=\lambda \text{ (say)}.$$

Simplifying, we get

$$x_1=\frac{a^2h}{a^2+\lambda} \quad \text{and} \quad y_1=\frac{-b^2k}{-b^2+\lambda}.$$

As the point (x_1, y_1) lies on the hyperbola, we have

$$\frac{a^2h^2}{(a^2+\lambda)^2}-\frac{b^2k^2}{(-b^2+\lambda)^2}=1$$

from which we get four values of λ and which indicates that four normals can be drawn to the ellipse from the point (h, k) and feet of the normals are given by

$$\left(\frac{ha^2}{a^2+\lambda_i}, \frac{kb^2}{-b^2+\lambda_i}\right), \qquad i=1,2,3,4$$

Now, the results $a^2+\lambda=\frac{a^2h}{x_1}$ and $b^2-\lambda=\frac{b^2k}{y_1}$ give

$$a^2+b^2=\frac{a^2h}{x_1}+\frac{b^2k}{y_1}$$

which shows that the feet of the normals lie on the curve

$$a^2+b^2=\frac{a^2h}{x}+\frac{b^2k}{y}$$

$$\text{or,} \quad xy(a^2+b^2)=a^2hy+b^2kx$$

which is a rectangular hyperbola.

Another Method

Equation of normal to the hyperbola $\frac{x^2}{a^2} - \frac{y^2}{b^2} = 1$ at the point $(a\sec\theta, b\tan\theta)$ is given by

$$xa\cos\theta + yb\cot\theta = a^2 + b^2.$$

If it passes through (h, k), then we have

$$ha\cos\theta + kb\cot\theta = a^2 + b^2.$$

Introducing the variable $t = \tan(\theta/2)$ in the above equation we get

$$ah\frac{1-t^2}{1+t^2} + bk\frac{1-t^2}{2t} = a^2 + b^2$$

$$\text{or,} \quad 2aht(1-t^2) + bk(1-t^4) = 2(a^2+b^2)t(1+t^2)$$

from which we get four values of t, i.e. of θ and which indicates that four normals can be drawn to the hyperbola from the point (h, k).

Simplifying the above equation in t, we get

$$bkt^4 + 2(ah + a^2 + b^2)t^3 + 2(-ah + a^2 + b^2)t - bk = 0.$$

If the feet of the normals are given by $(a\cos\theta_i, b\sin\theta_i)$, where $t_i = \tan\left(\frac{\theta_i}{2}\right)$, then we have by theory of equation

$$\sum t_1 = \frac{-2(ah + a^2 + b^2)}{bk}$$

$$\sum t_1t_2 = 0$$

$$\sum t_1t_2t_3 = \frac{2(ah - a^2 - b^2)}{bk}$$

$$t_1t_2t_3t_4 = -1.$$

Now, by basic trigonometry, we have

$$\tan\left(\frac{\theta_1}{2} + \frac{\theta_2}{2} + \frac{\theta_3}{2} + \frac{\theta_4}{2}\right) = \frac{\sum t_1 - \sum t_1t_2t_3}{1 - \sum t_1t_2 + t_1t_2t_3t_4} = \infty$$

i.e. $(\theta_1 + \theta_2 + \theta_3 + \theta_4)$ is an odd multiple of π.

So, four normals can be drawn from a point to the hyperbola and the sum of the eccentric angles of their fect is an odd multiple of π.

Examples

►► 1. *In a rectangular hyperbola, prove that $SA \times SA' = a^2$, where S is one of the foci and A, A' are the vertices of the hyperbola.*

Solution: Let the equation of the rectangular hyperbola be $x^2 - y^2 = a^2$.

Now, $\quad SA = ea - a = \sqrt{2}a - a \quad$ and $\quad SA' = ea + a = \sqrt{2}a + a$.

So, $\quad SA \times SA' = (\sqrt{2}a - a)(\sqrt{2}a + a) = 2a^2 - a^2 = a^2$. **[Proved]**

▶▷▶ 2. *In a rectangular hyperbola, prove that $SP \times S'P = CP^2$, where P is any point on the hyperbola, S, S' are the foci and C is the centre of the hyperbola.*

Solution: Let the equation of the rectangular hyperbola be $x^2 - y^2 = a^2$. If $P(h, k)$ be any point of the hyperbola, then

$$SP = eh - a = \sqrt{2}h - a \quad \text{and} \quad S'P = eh + a = \sqrt{2}h + a.$$

So,
$$SP \times S'P = (\sqrt{2}h - a)(\sqrt{2}h + a) = 2h^2 - a^2$$
$$= 2h^2 - (h^2 - k^2) = h^2 + k^2 = CP^2. \qquad \textbf{[Proved]}$$

▶▷▶ 3. *Find the equation of the chord of a hyperbola joining two points whose eccentric angles are θ and ϕ.*

Solution: Equation of the line joining $P(a\sec\theta, b\tan\theta)$ and $Q(a\sec\phi, b\tan\phi)$ is

$$\frac{y - b\tan\theta}{x - a\sec\theta} = \frac{b(\tan\phi - \tan\theta)}{a(\sec\phi - \sec\theta)}$$

or,
$$\frac{y - b\tan\theta}{x - a\sec\theta} = \frac{b\cos\left(\frac{\phi-\theta}{2}\right)}{a\sin\left(\frac{\phi+\theta}{2}\right)}$$

or,
$$a(y - b\sin\theta)\sin\left(\frac{\phi+\theta}{2}\right) = b(x - a\sec\theta)\cos\left(\frac{\phi-\theta}{2}\right)$$

or,
$$ay\sin\left(\frac{\phi+\theta}{2}\right) - bx\cos\left(\frac{\phi-\theta}{2}\right) + ab\cos\left(\frac{\phi+\theta}{2}\right) = 0.$$

▶▷▶ 4. *Show that the foot of the perpendicular drawn from the focus on any tangent to the hyperbola $\frac{x^2}{a^2} - \frac{y^2}{b^2} = 1$ lies on a circle.*

Solution: Equation of the tangent at $(a\sec\theta, b\tan\theta)$ is

$$\frac{x\sec\theta}{a} - \frac{y\tan\theta}{b} = 1. \qquad (1)$$

Then equation of the perpendicular from $(ae, 0)$ to the tangent is

$$\frac{y\sec\theta}{a} + \frac{x\tan\theta}{b} = \frac{ae\tan\theta}{b}. \qquad (2)$$

To get the locus of the point of intersection of Eqns. (1) and (2) we eliminate θ from Eqns. (1) and (2). Squaring and adding, we get

$$x^2\left(\frac{\sec^2\theta}{a^2} + \frac{\tan^2\theta}{b^2}\right) + y^2\left(\frac{\sec^2\theta}{a^2} + \frac{\tan^2\theta}{b^2}\right) = 1 + \frac{a^2e^2\tan^2\theta}{b^2}$$

or,
$$(x^2 + y^2)\left(\frac{\sec^2\theta}{a^2} + \frac{\tan^2\theta}{b^2}\right) = 1 + \frac{(a^2 + b^2)\tan^2\theta}{b^2}$$

or,
$$x^2 + y^2 = a^2.$$

▶▷▶ 5. *Show that the locus of the middle points of the chords of constant length 2d of the rectangular hyperbola $xy = c^2$ is*

$$(x^2 + y^2)(xy - c^2) = d^2xy.$$

Solution: Let (h, k) be the middle point of the chord of the rectangular hyperbola $xy = c^2$ which is inclined to the x-axis at an angle θ, then coordinates of the extremities are

$$(h + d\cos\theta, k + d\sin\theta) \quad \text{and} \quad (h - d\cos\theta, k - d\sin\theta)$$

which satisfy the equation $xy = c^2$.

So,

$$(h + d\cos\theta)(k + d\sin\theta) = c^2$$

and $$(h - d\cos\theta)(k - d\sin\theta) = c^2.$$

Adding and subtracting we get

$$hk + d^2\cos\theta\sin\theta = c^2 \quad \text{and} \quad h\sin\theta + k\cos\theta = 0.$$

From the second equation

$$\frac{\sin\theta}{k} = \frac{\cos\theta}{-h} = \frac{1}{\sqrt{h^2 + k^2}}.$$

Eliminating θ, we get

$$hk + d^2\left(-\frac{hk}{h^2 + k^2}\right) = c^2$$

or, $$(hk - c^2)(h^2 + k^2) = d^2hk.$$

Hence, locus of the point (h, k) is

$$(x^2 + y^2)(xy - c^2) = d^2xy.$$

EXERCISE

1. If the straight line $y = mx + \sqrt{a^2m^2 - b^2}$ touches the hyperbola $\frac{x^2}{a^2} - \frac{y^2}{b^2} = 1$ at $(a\sec\theta, b\tan\theta)$, then show that $\theta = \sin^{-1}\left(\frac{b}{am}\right)$.
2. Find the condition that the straight line $x\cos\alpha + y\sin\alpha = p$ may touch the hyperbola $\frac{x^2}{a^2} - \frac{y^2}{b^2} = 1$.
3. If the tangents at (x_1, y_1) and (x_2, y_2) on the hyperbola $\frac{x^2}{a^2} - \frac{y^2}{b^2} = 1$ are perpendicular, then show that $\frac{x_1x_2}{y_1y_2} + \frac{a^4}{b^4} = 0$.
4. If the normal to the hyperbola $\frac{x^2}{a^2} - \frac{y^2}{b^2} = 1$ at P meets the transverse axis at G and conjugate axis at g and if CF is the perpendicular from the centre C on the normal, show that (i) $PF \times PG = b^2$ and (ii) $PF \times Pg = a^2$.

5. Find the locus of the point from which two perpendicular tangents can be drawn to the hyperbola $\frac{x^2}{a^2} - \frac{y^2}{b^2} = 1$.
6. A normal to the hyperbola $\frac{x^2}{a^2} - \frac{y^2}{b^2} = 1$ meets the axes at Q and R. Show that the locus of the point of intersection of the perpendiculars to the axes through Q and R is

$$a^2x^2 - b^2y^2 = (a^2 + b^2)^2.$$

7. If the tangent and normal to the rectangular hyperbola $xy = c^2$ cut off intercepts a_1 and a_2 from one axis and b_1 and b_2 from other axis, then show that $a_1a_2 + b_1b_2 = 0$.
8. If P, Q be two variable points on the hyperbola $xy = c^2$ such that the tangent at Q passes through the foot of the ordinate of P, show that the locus of the point of intersection of the tangents at P and Q is a hyperbola.
9. Show that if the normal to the rectangular hyperbola $xy = c^2$ at the point t_1 meets the curve again at t_2, then $t_1^3t_2 = -1$.
10. If the normal at P to the rectangular hyperbola $xy = c^2$ meets the curve again in Q, show that circle on PQ as diameter cuts the curve at the other end of the diameter through P.
11. A line through the origin meets the circle $x^2+y^2 = a^2$ at P and the hyperbola $x^2-y^2 = a^2$ at Q. Prove that the locus of the point of intersection of the tangent at P to the circle with the tangent at Q to the hyperbola is the curve $(a^4 + 4y^4)x^2 = a^6$.
12. The tangents at two points P, Q to the rectangular hyperbola $xy = c^2$ intersect at T. Prove that the line joining T and origin bisects the chord PQ.

Chapter 7

Pole and Polars

7.1 Definition

If the chords be drawn from a point to a curve, then the locus of the point of intersection of the tangents to the curve drawn at the extremities of the chords is a straight line. This straight line is called *polar* of the point with respect to the curve and the point is called *pole* of the *polar*. So, the polar of a point with respect to a curve is the locus of the point of intersection of tangents at the extremities of the chords through that point.

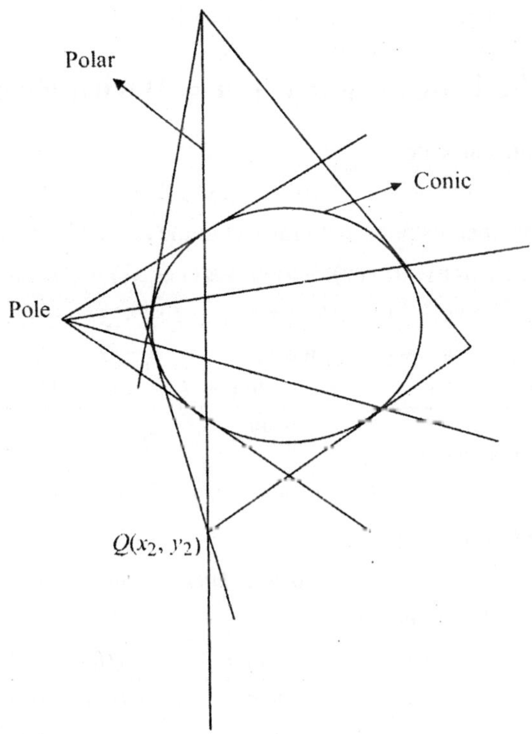

F-8

Let $P(x_1, y_1)$ be the pole and the tangents at the extremities of a chord drawn from $P(x_1, y_1)$ meet at $Q(x_2, y_2)$; then equation of the chord of contact of $Q(x_2, y_2)$ with respect to the curve

$$ax^2 + 2hxy + by^2 + 2gx + 2fy + c = 0 \tag{1}$$

is $$axx_2 + h(xy_2 + x_2y) + byy_2 + g(x + x_2) + f(y + y_2) + c = 0.$$

According to definition of polar it will pass through $P(x_1, y_1)$, i.e.

$$ax_1x_2 + h(x_1y_2 + x_2y_1) + by_1y_2 + g(x_1 + x_2) + f(y_1 + y_2) + c = 0$$

which shows that $Q(x_2, y_2)$ lies on the line

$$axx_1 + h(xy_1 + x_1y) + byy_1 + g(x + x_1) + f(y + y_1) + c = 0.$$

This is the equation of polar of the point (x_1, y_1) with respect to the curve given by Eqn. (1).

Particular Cases

If the pole be (h, k), then its polar with respect to

(1) Circle: $x^2 + y^2 = a^2$ is $xh + yk = a^2$,
(2) Circle: $x^2 + y^2 + 2gx + 2fy + c = 0$ is $xh + yk + g(x + h) + f(y + k) + c = 0$,
(3) Parabola: $y^2 = 4ax$ is $yk = 2a(x + h)$,
(4) Ellipse : $\frac{x^2}{a^2} + \frac{y^2}{b^2} = 1$ is $\frac{xh}{a^2} + \frac{yk}{b^2} = 1$,
(5) Hyperbola: $\frac{x^2}{a^2} - \frac{y^2}{b^2} = 1$ is $\frac{xh}{a^2} - \frac{yk}{b^2} = 1$,
(6) Rectangular hyperbola: $xy = c^2$ is $xk + yh = 2c^2$.

7.2 Pole of the Line $lx + my + n = 0$ with Respect to a Curve

Let the equation of the curve be

$$ax^2 + 2hxy + by^2 + 2gx + 2fy + c = 0. \tag{2}$$

If (x_1, y_1) be the pole, then its polar with respect to the curve given by Eqn. (2) is

$$axx_1 + h(xy_1 + x_1y) + byy_1 + g(x + x_1) + f(y + y_1) + c = 0$$

or, $$(ax_1 + hy_1 + g)x + (hx_1 + by_1 + f)y + gx_1 + fy_1 + c = 0.$$

Comparing this with $lx + my + n = 0$, we get

$$\frac{ax_1 + hy_1 + g}{l} = \frac{hx_1 + by_1 + f}{m} = \frac{gx_1 + fy_1 + c}{n}.$$

From 1st and 2nd ratios we get

$$(am - hl)x_1 + (hm - bl)y_1 = fl - mg \tag{3}$$

and from 2nd and 3rd ratios we get

$$(nh - mg)x_1 + (nb - fm)y_1 = mc - nf. \tag{4}$$

Solving Eqns. (3) and (4) we get

$$x_1 = \frac{n(hf - bg) + m(fg - hc) + l(bc - f^2)}{n(ab - h^2) + m(hg - af) + l(hf - bg)}$$

$$y_1 = \frac{n(hg - af) + m(ac - g^2) + l(fg - hc)}{n(ab - h^2) + m(hg - af) + l(hf - bg)}.$$

Particular Cases

The pole of the line $lx + my + n = 0$ with respect to the curve

(1) Circle: $x^2 + y^2 = a^2$ is $\left(-\dfrac{a^2 l}{n}, -\dfrac{a^2 m}{n}\right)$,

(2) Circle: $x^2 + y^2 + 2gx + 2fy + c = 0$ is

$$\left(\frac{l(g^2 + f^2 - c)}{gl + fm - n} - g, \frac{m(g^2 + f^2 - c)}{gl + fm - n} - f\right),$$

(3) Parabola: $y^2 = 4ax$ is $\left(\dfrac{n}{l}, -\dfrac{2am}{l}\right)$,

(4) Ellipse: $\dfrac{x^2}{a^2} + \dfrac{y^2}{b^2} = 1$ is $\left(-\dfrac{a^2 l}{n}, -\dfrac{b^2 m}{n}\right)$,

(5) Hyperbola: $\dfrac{x^2}{a^2} - \dfrac{y^2}{b^2} = 1$ is $\left(-\dfrac{a^2 l}{n}, \dfrac{b^2 m}{n}\right)$,

(6) Rectangular hyperbola: $xy = c^2$ is $\left(-\dfrac{2c^2 m}{n}, -\dfrac{2c^2 l}{n}\right)$.

7.3 Properties of Pole and Polar

(a) If the polar of a point $P(x_1, y_1)$ with respect to a curve passes through $Q(x_2, y_2)$, then the polar of Q will pass through P and such points are called *conjugate points* with respect to the curve.

Proof: Equation of the polar of the point $P(x_1, y_1)$ with respect to the curve

$$ax^2 + 2hxy + by^2 + 2gx + 2fy + c = 0 \tag{5}$$

is $axx_1 + h(xy_1 + x_1 y) + byy_1 + g(x + x_1) + f(y + y_1) + c = 0.$

If it passes through $Q(x_2, y_2)$, then

$$ax_2 x_1 + h(x_2 y_1 + x_1 y_2) + by_2 y_1 + g(x_2 + x_1) + f(y_2 + y_1) + c = 0. \tag{6}$$

Again the equation of the polar of the point $Q(x_2, y_2)$ with respect to the curve is

$$axx_2 + h(xy_2 + x_2 y) + byy_2 + g(x + x_2) + f(y + y_2) + c = 0. \tag{7}$$

Comparing Eqn. (7) with Eqn. (6) we see that the line given by Eqn. (7) passes through the point $P(x_1, y_1)$.

(b) If pole of the line $lx+my+n = 0$ with respect to a curve lies on the line $Lx+My+N = 0$, then pole of the second line will lie on the first line and such lines are called *conjugate lines* with respect to the curve.

Proof: Let the equation of the curve be

$$ax^2 + 2hxy + by^2 + 2gx + 2fy + c = 0. \tag{8}$$

If (x_1, y_1) be the pole of the line $lx + my + n = 0$ with respect to the curve given by Eqn. (8), then we have

$$x_1 = \frac{n(hf - bg) + m(fg - hc) + l(bc - f^2)}{n(ab - h^2) + m(hg - af) + l(hf - bg)}$$

$$y_1 = \frac{n(hg - af) + m(ac - g^2) + l(fg - hc)}{n(ab - h^2) + m(hg - af) + l(hf - bg)}.$$

If it lies on the line $Lx + My + N = 0$, then

$$\begin{aligned} &L[n(hf - bg) + m(fg - hc) + l(bc - f^2)] \\ &+ M[n(hg - af) + m(ac - g^2) + l(fg - hc)] \\ &+ N[n(ab - h^2) + m(hg - af) + l(hf - bg)] = 0. \end{aligned} \tag{9}$$

Again, if (x_2, y_2) be the pole of the line $Lx + My + N = 0$ with respect to the curve given by Eqn. (8), then we have

$$x_2 = \frac{N(hf - bg) + M(fg - hc) + L(bc - f^2)}{N(ab - h^2) + M(hg - af) + L(hf - bg)}$$

$$y_2 = \frac{N(hg - af) + M(ac - g^2) + L(fg - hc)}{N(ab - h^2) + M(hg - af) + L(hf - bg)}.$$

If it lies on the line $lx + my + n = 0$, then

$$\begin{aligned} &l[N(hf - bg) + M(fg - hc) + L(bc - f^2)] \\ &+ m[N(hg - af) + M(ac - g^2) + L(fg - hc)] \\ &+ n[N(ab - h^2) + M(hg - af) + L(hf - bg)] = 0 \\ \text{or,}\quad &L[l(bc - f^2) + m(fg - hc) + n(hf - bg)] \\ &+ M[n(hg - af) + m(ac - g^2) + l(fg - hc)] \\ &+ N[n(ab - h^2) + m(hg - af) + l(hf - bg)] = 0 \end{aligned} \tag{10}$$

which is same as equation (9). Hence the proposition is proved.

(c) The polar of the point of intersection of any two straight lines with respect to a curve is the straight line joining the poles of the straight lines.

Proof: Let P, Q be the poles of the lines $L = 0$ and $L' = 0$ with respect to a curve $S = 0$ and the lines intersect at C. Then polar of C will pass through P and also through Q, i.e. polar of C is the line joining P and Q. Hence the proposition is proved.

7.4 Special Properties of Pole and Polar for Circle

(1) The distances of two points from the centre of a circle are proportional to distance of each from the polar of other.

Proof: If C be the centre of the circle $x^2 + y^2 = a^2$ and $P(x_1, y_1)$, $Q(x_2, y_2)$ be two points, then $CP = \sqrt{x_1^2 + y_1^2}$ and $CQ = \sqrt{x_2^2 + y_2^2}$. Now equations of polars with respect to the circles are

$$xx_1 + yy_1 = a^2 \quad \text{and} \quad xx_2 + yy_2 = a^2 \text{ respectively.}$$

Now,

$$d_1 = \text{distance of } P \text{ from the polar of } Q = \frac{x_1x_2 + y_1y_2 - a^2}{\sqrt{x_2^2 + y_2^2}}$$

$$\text{and}\quad d_2 = \text{distance of } Q \text{ from the polar of } P = \frac{x_1x_2 + y_1y_2 - a^2}{\sqrt{x_1^2 + y_1^2}}.$$

Hence,

$$\frac{d_1}{d_2} = \frac{\sqrt{x_1^2 + y_1^2}}{\sqrt{x_2^2 + y_2^2}} = \frac{CP}{CQ}. \qquad \textbf{[Proved]}$$

(2) The polar of a point with respect to a circle is perpendicular to the line joining the centre of the circle and the point.

Proof: Equation of the polar of the point $P(x_1, y_1)$ with respect to the circle $x^2 + y^2 = a^2$ is $xx_1 + yy_1 = a^2$, whose slope is $-x_1/y_1$. Slope of the line joining the centre $(0, 0)$ and $P(x_1, y_1)$ is y_1/x_1, from which we see that product of the slopes $= -1$.

(3) If O be the centre of a circle, P be any point and OP meets the polar at Q, then $OP \times OQ = (\text{radius})^2$.

Proof: Equation of the polar of the point $P(x_1, y_1)$ with respect to the circle $x^2 + y^2 = a^2$ is $xx_1 + yy_1 = a^2$. Since OP is perpendicular to polar of P,

$$OQ = \text{length of the perpendicular from } O \text{ to } xx_1 + yy_1 = a^2$$

$$= \frac{a^2}{\sqrt{x_1^2 + y_1^2}} = \frac{a^2}{OP}$$

i.e., $$OP \times OQ = a^2.$$

7.5 Condition for Conjugate Lines

We have already seen that the lines $lx + my + n = 0$ and $Lx + My + N = 0$ will be conjugate to each other with respect to the curve

$$ax^2 + 2hxy + by^2 + 2gx + 2fy + c = 0$$

if

$$L[n(hf - bg) + m(fg - hc) + l(bc - f^2)]$$
$$+ M[n(hg - af) + m(ac - g^2) + l(fg - hc)]$$
$$+ N[n(ab - h^2) + m(hg - af) + l(hf - bg)] = 0.$$

Particular Cases

The lines $lx + my + n = 0$ and $Lx + My + N = 0$ will be conjugate to each other with respect to

(1) Circle: $x^2 + y^2 = a^2$ if $a^2(lL + mM) = nN$,

(2) Circle: $x^2 + y^2 + 2gx + 2fy + c = 0$ if

$$L\left(\frac{l(g^2 + f^2 - c)}{gl + fm - n} - g\right) + M\left(\frac{m(g^2 + f^2 - c)}{gl + fm - n} - f\right) + N = 0,$$

(3) Parabola: $y^2 = 4ax$ if $lN + nL = 2amM$,

(4) Ellipse: $\dfrac{x^2}{a^2} + \dfrac{y^2}{b^2} = 1$ if $a^2lL + b^2mM = nN$,

(5) Hyperbola: $\dfrac{x^2}{a^2} - \dfrac{y^2}{b^2} = 1$ if $a^2lL - b^2mM = nN$.

7.6 Self-conjugate or Self-polar Triangle

If two of the three points A, B, C are such that polar of a point with respect to a curve, passes through the other two points, i.e. each side of the triangle ABC is polar of opposite vertex, then the triangle ABC is called a self-conjugate or self-polar triangle.

Examples

▶▷▶ 1. *If the polar of a point with respect to the circle $x^2 + y^2 = a^2$ touches the circle $4x^2 + 4y^2 = a^2$, show that the point lies on the circle $x^2 + y^2 = 4a^2$.*

Solution: Equation of the polar of the point (h, k) with respect to the circle $x^2 + y^2 = a^2$ is

$$xh + yk = a^2$$

$$\text{or,} \quad y = -\frac{h}{k}x + \frac{a^2}{k}.$$

If it touches the circle $4x^2 + 4y^2 = a^2$, i.e. $x^2 + y^2 = (a/2)^2$, then we have

$$c = \pm r\sqrt{1 + m^2}$$

$$\text{or,} \quad \frac{a^2}{k} = \pm\frac{a}{2}\sqrt{1 + \frac{h^2}{k^2}}$$

$$\text{or,} \quad h^2 + k^2 = 4a^2.$$

Generalizing, we get

$$x^2 + y^2 = 4a^2. \qquad \textbf{[Proved]}$$

▶▷▶ 2. *If the pole of the straight line with respect to the circle $x^2 + y^2 = a^2$ lies on the circle $x^2 + y^2 = f^2a^2$, then prove that the straight line touches the circle $x^2 + y^2 = \dfrac{a^2}{f^2}$.*

Solution: Equation of the polar of the point (h, k) with respect to the circle $x^2 + y^2 = a^2$ is

$$xh + yk = a^2$$

$$\text{or,} \quad y = -\frac{h}{k}x + \frac{a^2}{k}.$$

If it touches the circle $x^2 + y^2 = (a/f)^2$, then we have

$$c = \pm r\sqrt{1 + m^2}$$

$$\text{or,} \quad \frac{a^2}{k} = \pm\frac{a}{f}\sqrt{1 + \frac{h^2}{k^2}}$$

$$\text{or,} \quad h^2 + k^2 = f^2a^2.$$

Generalizing, we get,

$$x^2 + y^2 = f^2a^2. \qquad \textbf{[Proved]}$$

▶▷▶ 3. *Show that the locus of the poles of the tangents to the parabola $y^2 = 4bx$ with respect to the parabola $y^2 = 4ax$ is another parabola.*

Solution: Equation of the polar of the point (h, k) with respect to the $y^2 = 4ax$ is

$$yk = 2a(x + h)$$

$$\text{or,} \quad y = \frac{2a}{k}x + \frac{2ah}{k}.$$

If it touches the parabola $y^2 = 4bx$, then we have

$$c = \frac{a}{m}$$

$$\text{or,} \quad \frac{2ah}{k} = \frac{bk}{2a}$$

$$\text{or,} \quad k^2 = \frac{4a^2}{b}h.$$

Generalizing, we get $y^2 = \dfrac{4a^2}{b}x$, which is a parabola.

▶▷▶ 4. *Show that the pole of the tangents to the rectangular hyperbola $xy = c^2$ with respect to the circle $x^2 + y^2 = a^2$ lies on $xy = \dfrac{a^4}{4c^2}$.*

Solution. Equation of the polar of the point (h, k) with respect to the circle $x^2 + y^2 = a^2$ is

$$xh + yk = a^2$$

$$\text{or,} \quad y = -\frac{h}{k}x + \frac{a^2}{k}.$$

If it touches the hyperbola $xy = c^2$, then the equation

$$\left(-\frac{h}{k}x + \frac{a^2}{k}\right)x = c^2$$

$$\text{i.e.,} \quad hx^2 - a^2x + kc^2 = 0$$

will have two equal roots, which gives

$$(-a^2)^2 = 4hkc^2$$

$$\text{or,} \quad hk = \frac{a^4}{4c^2}.$$

Generalizing, we get

$$xy = \frac{a^4}{4c^2}. \qquad \textbf{[Proved]}$$

▶▷▶ 5. *Find the locus of the poles of the chords of the ellipse $\dfrac{x^2}{a^2} + \dfrac{y^2}{b^2} = 1$ which subtend a right angle at the centre of the ellipse.*

Solution: Let (h, k) be the pole; then equation of the polar PQ, with respect to the ellipse

$$\frac{x^2}{a^2} + \frac{y^2}{b^2} = 1 \qquad (1)$$

$$\text{is} \quad \frac{xh}{a^2} + \frac{yk}{b^2} = 1. \qquad (2)$$

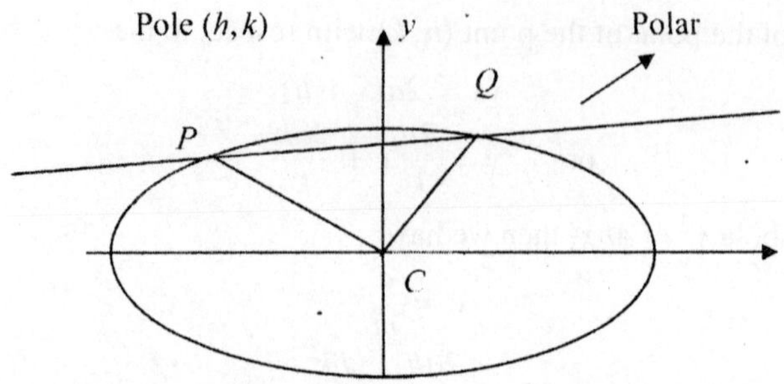

Making homogeneous Eqn. (1) with the aid of Eqn. (2), we get

$$\frac{x^2}{a^2}+\frac{y^2}{b^2}=\left(\frac{xh}{a^2}+\frac{yk}{b^2}\right)^2$$

which is a second degree homogeneous equation in x, y, and represents a pair of straight lines passing through the origin, i.e. through C. Now, if PQ subtends a right angle at the point C, i.e. if the lines CP and CQ are perpendicular, then we have

$$\text{Coeff. of } x^2 + \text{Coeff. of } y^2 = 0$$

or, $$\left(\frac{1}{a^2}-\frac{h^2}{a^4}\right)+\left(\frac{1}{b^2}-\frac{k^2}{b^4}\right)=0$$

or, $$\frac{h^2}{a^4}+\frac{k^2}{b^4}=\frac{1}{a^2}+\frac{1}{b^2}.$$

Generalizing, we get the locus of the point (h, k) is

$$\frac{x^2}{a^4}+\frac{y^2}{b^4}=\frac{1}{a^2}+\frac{1}{b^2}.$$

▶▶ **6.** *Show that the locus of the pole of the normal chords of the parabola* $y^2 = 4ax$ *is* $y^2(x+2a)+4a^3=0$.

Solution: Equation of the polar of the point (h, k) with respect to the parabola $y^2 = 4ax$ is

$$yk = 2a(x+h)$$

or, $$y=\frac{2a}{k}x+\frac{2ah}{k}. \tag{1}$$

Again, equation of the normal at the point $(at^2, -2at)$ to the parabola $y^2 = 4ax$ is

$$y = tx - at^3 - 2at. \tag{2}$$

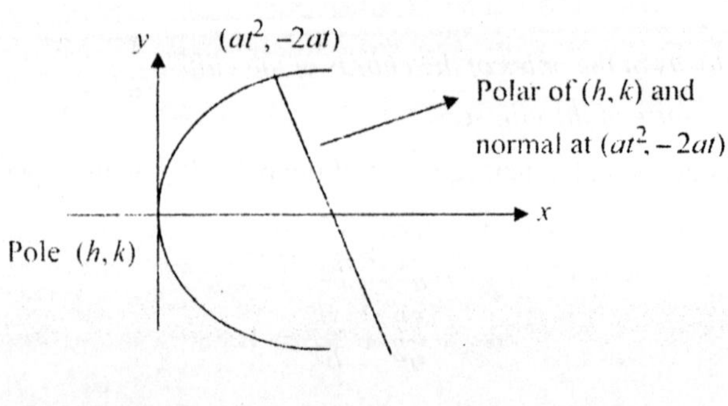

Comparing Eqns. (1) and (2), we get

$$1 = \frac{2a}{kt} = -\frac{2ah}{k(at^3 + 2at)}$$

from which we get

$$t = \frac{2a}{k} \quad \text{and} \quad at^2 + 2a = -h.$$

Eliminating t, we get

$$\frac{4a^3}{k^2} + 2a + h = 0.$$

Generalizing, we get

$$4a^3 + y^2(2a + x) = 0.$$

▶▷▶ 7. *Find the locus of the pole of the normal chords of the ellipse* $\frac{x^2}{a^2} + \frac{y^2}{b^2} = 1$.

Solution: Equation of the polar of the point (h, k) with respect to the ellipse $\frac{x^2}{a^2} + \frac{y^2}{b^2} = 1$ is $\frac{xh}{a^2} + \frac{yk}{b^2} = 1$.

Again, equation of the normal at $(a\cos\theta, b\sin\theta)$ to the ellipse $\frac{x^2}{a^2} + \frac{y^2}{b^2} = 1$ is

$$xa\sec\theta - yb\operatorname{cosec}\theta = a^2 - b^2.$$

Comparing these two equations, we get

$$\frac{h\cos\theta}{a^3} = -\frac{k\sin\theta}{b^3} = \frac{1}{a^2 - b^2}.$$

Eliminating θ, we get

$$\left(\frac{a^3}{h(a^2 - b^2)}\right)^2 + \left(\frac{-b^3}{k(a^2 - b^2)}\right)^2 = 1$$

or, $$\frac{a^6}{h^2} + \frac{b^6}{k^2} = (a^2 - b^2)^2.$$

Generalizing, we get the locus of the point (h, k) is

$$\frac{a^6}{x^2} + \frac{b^6}{y^2} = (a^2 - b^2)^2.$$

▶▷▶ 8. *Find the locus of the pole of the normal chord of the rectangular hyperbola* $xy = c^2$.

Solution: Equation of the polar of the point (h, k) with respect to the $xy = c^2$ is

$$xk + yh = 2c^2. \tag{1}$$

Now equation of the normal at the point $(ct, c/t)$ is

$$xt^2 - y = ct^3 - c/t. \tag{2}$$

As Eqns. (1) and (2) represent the same line, we get

$$\frac{k}{t^2} = -h = \frac{2ct}{t^4 - 1}.$$

Eliminating the parameter t, we get

$$h^2 = \left(\frac{2ct}{t^4 - 1}\right)^2.$$

$$\text{or,} \quad h^2\left(\frac{k^2}{h^2} - 1\right)^2 = -\frac{4c^2k}{h}$$

$$\text{or,} \quad (k^2 - h^2)^2 + 4c^2kh = 0.$$

Generalizing, we get the locus of the point (h, k) is

$$(y^2 - x^2)^2 + 4c^2xy = 0.$$

▶▷▶ **9.** *If the sum of the ordinate of two points on the ellipse* $\frac{x^2}{a^2} + \frac{y^2}{b^2} = 1$ *be b, show that the locus of the pole of the chord which joins them is*

$$\frac{x^2}{a^2} + \frac{y^2}{b^2} = \frac{2y}{b}.$$

Solution: Equation of the polar of the point (h, k) with respect to the ellipse $\frac{x^2}{a^2} + \frac{y^2}{b^2} = 1$ is $\frac{xh}{a^2} + \frac{yk}{b^2} = 1$.

For the point of intersection of the chord and ellipse we solve those equations as follows:

$$\frac{y^2}{b^2} + \frac{a^2}{h^2}\left(\frac{yk}{b^2} - 1\right)^2 = 1$$

$$\text{or,} \quad \left(h^2 + \frac{a^2k^2}{b^2}\right)y^2 - 2yka^2 + b^2(a^2 - h^2) = 0.$$

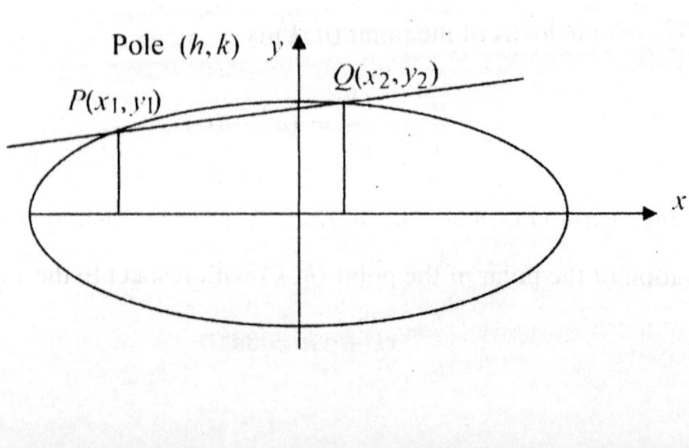

If the chord intersects the ellipse in $P(x_1, y_1)$ and $Q(x_2, y_2)$, then

$$y_1 + y_2 = \frac{2ka^2}{h^2 + \dfrac{a^2k^2}{b^2}} = b \text{ (given)}$$

$$\text{or,} \quad \frac{2ka^2b^2}{b^2h^2 + a^2k^2} = b$$

$$\text{or,} \quad h^2b^2 + a^2k^2 = 2kba^2$$

$$\text{or,} \quad \frac{h^2}{a^2} + \frac{k^2}{b^2} = \frac{2k}{b}.$$

Generalizing, we get the locus of the point (h, k) is

$$\frac{x^2}{a^2} + \frac{y^2}{b^2} = \frac{2y}{b}.$$

▶▷▶ **10**. *Find the locus of the poles of the tangents to the director circle of the ellipse with respect to the ellipse.*

Solution: Equation of the polar of the point (h, k) with respect to the ellipse $\dfrac{x^2}{a^2} + \dfrac{y^2}{b^2} = 1$ is

$$\frac{xh}{a^2} + \frac{yk}{b^2} = 1$$

$$\text{or,} \quad y = -\frac{hb^2}{ka^2}x + \frac{b^2}{k}.$$

If it touches the director circle of the ellipse, i.e. $x^2 + y^2 = a^2 + b^2$, then we have

$$\left(\frac{b^2}{k}\right)^2 = (a^2 + b^2)\left[1 + \left(-\frac{hb^2}{ka^2}\right)^2\right]$$

$$\text{or,} \quad (ab)^4 = (a^2 + b^2)[k^2a^4 + h^2b^4]$$

$$\text{or,} \quad \frac{h^2}{a^4} + \frac{k^2}{b^4} = \frac{1}{a^2 + b^2}.$$

Generalizing, we get the locus of (h, k) is

$$\frac{x^2}{a^4} + \frac{y^2}{b^4} = \frac{1}{a^2 + b^2}.$$

▶▷▶ **11**. *If a point lies on the ellipse* $\dfrac{x^2}{A^2} + \dfrac{y^2}{B^2} = 1$, *then prove that its polar with respect to the ellipse* $\dfrac{x^2}{a^2} + \dfrac{y^2}{b^2} = 1$ *touches the ellipse* $\dfrac{A^2x^2}{a^4} + \dfrac{B^2y^2}{b^4} = 1$.

Solution: Equation of the polar of the point (h, k) with respect to the ellipse $\dfrac{x^2}{a^2} + \dfrac{y^2}{b^2} = 1$ is

$$\frac{xh}{a^2} + \frac{yk}{b^2} = 1$$

$$\text{or,} \quad y = -\frac{hb^2}{ka^2}x + \frac{b^2}{k}.$$

If it touches the ellipse $\dfrac{A^2x^2}{a^4}+\dfrac{B^2y^2}{b^4}=1$, i.e. $\dfrac{x^2}{(a^2/A)^2}+\dfrac{y^2}{(b^2/B)^2}=1$, then we have

$$\left(\frac{b^2}{k}\right)^2=\left(-\frac{hb^2}{ka^2}\right)^2\left(\frac{a^2}{A}\right)^2+\left(\frac{b^2}{B}\right)^2$$

or, $$\frac{b^4}{k^2}=\frac{h^2b^4}{A^2k^2}+\frac{b^4}{B^2}$$

or, $$\frac{h^2}{A^2}+\frac{k^2}{B^2}=1.$$

Generalizing, we see that the point (h, k) lies on the ellipse

$$\frac{x^2}{A^2}+\frac{y^2}{B^2}=1. \qquad \textbf{[Proved]}$$

▶▷▶ **12.** *Find the locus of the poles of the chords of the hyperbola which subtends a right angle at the centre of the hyperbola.*

Solution: Equation of the polar of the point (h, k) with respect to the hyperbola $\dfrac{x^2}{a^2}-\dfrac{y^2}{b^2}=1$ is $\dfrac{xh}{a^2}-\dfrac{yk}{b^2}=1$.

Now, making homogeneous the equation of the hyperbola with the aid of the equation of the chord, we get

$$\frac{x^2}{a^2}-\frac{y^2}{b^2}=\left(\frac{xh}{a^2}-\frac{yk}{b^2}\right)^2$$

or, $$\left(\frac{1}{a^2}-\frac{h^2}{a^4}\right)x^2+\frac{2hkxy}{a^2b^2}-\left(\frac{1}{b^2}+\frac{k^2}{b^4}\right)y^2=0.$$

The above second degree homogeneous equation in x, y represents a pair of straight lines passing through the origin, i.e. centre of the hyperbola: If these lines are perpendicular, then we have

$$\text{Coeff. of } x^2+\text{Coeff. of } y^2=0$$

or, $$\left(\frac{1}{a^2}-\frac{h^2}{a^4}\right)-\left(\frac{1}{b^2}+\frac{k^2}{b^4}\right)=0$$

or, $$\frac{h^2}{a^4}+\frac{k^2}{b^4}=\frac{1}{a^2}-\frac{1}{b^2}.$$

Generalizing, we get the locus of the pole (h, k) is

$$\frac{x^2}{a^4}+\frac{y^2}{b^4}=\frac{1}{a^2}-\frac{1}{b^2}.$$

▶▷▶ **13.** *If A and B be two points on the ellipse* $\dfrac{x^2}{a^2}+\dfrac{y^2}{b^2}=1$ *such that three times the eccentric angle of the one is equal to the supplement of that of the other, then find the locus of the pole of AB with respect to the ellipse.*

Solution: Equation of the line joining A: $(a\cos\theta, b\sin\theta)$ and B: $(a\cos\phi, b\sin\phi)$ is

$$ay\sin\left(\frac{\theta+\phi}{2}\right)+bx\cos\left(\frac{\theta+\phi}{2}\right)=ab\cos\left(\frac{\theta-\phi}{2}\right).$$

Now, given that $\phi+3\theta=\pi$, so the above equation becomes

$$ay\sin\left(\frac{\pi-2\theta}{2}\right)+bx\cos\left(\frac{\pi-2\theta}{2}\right)=ab\cos\left(\frac{\pi-4\theta}{2}\right)$$

or, $$ay\cos\theta+bx\sin\theta=ab\sin 2\theta$$

or, $$\frac{x}{2a\cos\theta}+\frac{y}{2b\sin\theta}=1. \qquad (1)$$

Equation of the polar of the point (h, k) with respect to the ellipse $\frac{x^2}{a^2}+\frac{y^2}{b^2}=1$ is

$$\frac{xh}{a^2}+\frac{yk}{b^2}=1. \qquad (2)$$

Comparing Eqns. (1) and (2), we get

$$\frac{2h\cos\theta}{a}=\frac{2k\sin\theta}{b}=1.$$

Eliminating θ, we get

$$\frac{a^2}{h^2}+\frac{b^2}{k^2}=4.$$

Hence, locus of the point (h, k) is

$$\frac{a^2}{x^2}+\frac{b^2}{y^2}=4.$$

▶▶ **14.** *An ellipse and a hyperbola have the same principal axes. Show that the polar of any point on either curve with respect to the other touches the first curve.*

Solution: Let the equations of ellipse and hyperbola be

$$\frac{x^2}{a^2}+\frac{y^2}{b^2}=1$$

and $$\frac{x^2}{a^2}-\frac{y^2}{b^2}=1.$$

Let (h, k) be a point on the ellipse. Then we have

$$\frac{h^2}{a^2}+\frac{k^2}{b^2}=1. \qquad (1)$$

Equation of the polar of the point (h, k) with respect to the hyperbola

$$\frac{x^2}{a^2}-\frac{y^2}{b^2}=1 \quad \text{is} \quad \frac{xh}{a^2}-\frac{yk}{b^2}=1$$

or, $$y=\frac{hb^2}{ka^2}x-\frac{b^2}{k}. \qquad (2)$$

It touches the ellipse if

$$c^2 = a^2m^2 + b^2$$

or, $$\left(\frac{b^2}{k}\right)^2 = a^2\left(\frac{hb^2}{ka^2}\right)^2 + b^2$$

or, $$a^4b^4 = a^2b^4h^2 + b^2a^4k^2$$

or, $$\frac{h^2}{a^2} + \frac{k^2}{b^2} = 1, \text{ which is true by Eqn. (1).}$$

Hence, the polar of any point on ellipse with respect to the hyperbola touches the ellipse. Similarly, changing b^2 by $-b^2$, it can be shown that the polar of any point on hyperbola with respect to the ellipse touches the hyperbola.

▶▷▶ **15**. *Chords of a rectangular hyperbola are at right angles and they subtend a right angle at the fixed point O; prove that they intersect on the polar of O.*

Solution: Equation of pair of perpendicular straight lines through the point $O(h, k)$ is

$$[y - k - m(x - h)]\left[y - k + \frac{1}{m}(x - h)\right] = 0$$

or, $$(y - k)^2 - (x - h)^2 - \left(m - \frac{1}{m}\right)(x - h)(y - k) = 0. \tag{1}$$

If we take the equation of rectangular hyperbola as $xy = a^2$, then equation of the polar of the point O with respect to it is

$$xk + yh = 2a^2; \tag{2}$$

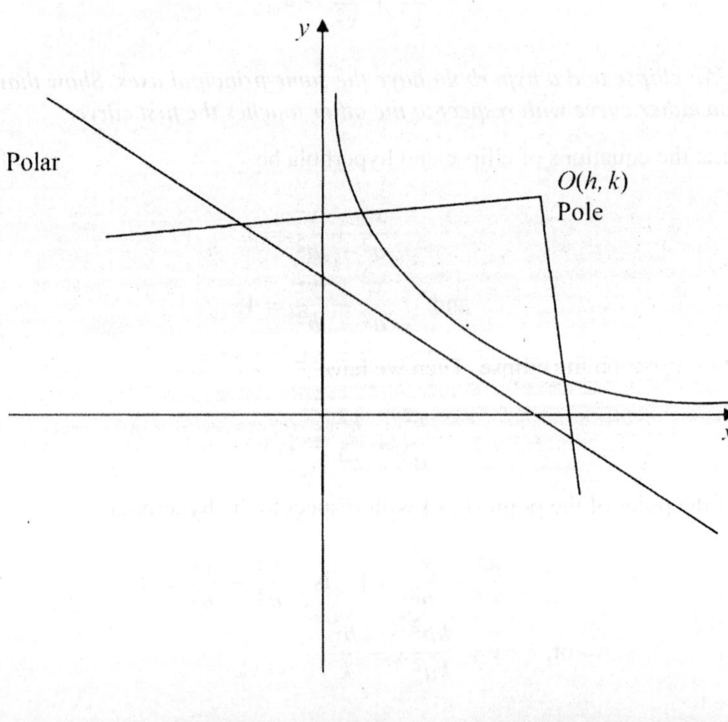

For the points of intersection of the straight lines (1) with the polar (2) we solve (1) and (2). Putting

$$x = \frac{2a^2 - yh}{k}, \quad \text{i.e. } x - h = \frac{2a^2 - 2kh - (y-k)h}{k}$$

in the Eqn. (1) we get

$$(y-k)^2 - \left(\frac{2a^2 - 2kh - (y-k)h}{k}\right)^2 - \left(m - \frac{1}{m}\right)\left(\frac{2a^2 - 2kh - (y-k)h}{k}\right)(y-k) = 0$$

or, $$k^2(y-k)^2 - (2a^2 - 2kh - (y-k)h)^2 - k\left(m - \frac{1}{m}\right)(2a^2 - 2kh - (y-k)h)(y-k) = 0$$

or, $$(y-k)^2\left(k^2 - h^2 + hk\left(m - \frac{1}{m}\right)\right) + 2(a^2 - kh)\left[2h - k\left(m - \frac{1}{m}\right)\right](y-k) - 4(a^2 - hk)^2 = 0. \quad (3)$$

Discriminant of the above quadratic equation is

$$4(a^2 - kh)^2\left[2h - k\left(m - \frac{1}{m}\right)\right]^2 + 16(a^2 - kh)^2\left[k^2 - h^2 + hk\left(m - \frac{1}{m}\right)\right]$$

$$= 4(a^2 - kh)^2\left[4h^2 + k^2\left(m - \frac{1}{m}\right)^2 - 4hk\left(m - \frac{1}{m}\right) + 4k^2 - 4h^2 + 4hk\left(m - \frac{1}{m}\right)\right]$$

$$= 4(a^2 - kh)^2k^2\left(m + \frac{1}{m}\right)^2 > 0.$$

So, the roots of the Eqn. (3) are real and unequal showing that the lines given by Eqn. (1) intersect the polar (2) of the point O in two distinct points.

EXERCISE

1. Prove that the locus of the pole with respect to the ellipse $\frac{x^2}{a^2} + \frac{y^2}{b^2} = 1$ of any tangent to the auxiliary circle is the curve $\frac{x^2}{a^4} + \frac{y^2}{b^4} = \frac{1}{a^2}$.

[*Hints*: Equation of the polar of the point (h, k) with respect to the ellipse

$$\frac{x^2}{a^2} + \frac{y^2}{b^2} = 1.$$

is

$$\frac{xh}{a^2} + \frac{yk}{b^2} = 1$$

or, $$y = -\frac{hb^2}{ka^2}x + \frac{b^2}{k}.$$

It touches the auxiliary circle of the ellipse, i.e. $x^2+y^2=a^2$ if

$$c^2=a^2(m^2+1)$$

$$\text{or,}\quad \left(\frac{b^2}{k}\right)^2=a^2\left[\left(\frac{hb^2}{ka^2}\right)^2+1\right]$$

$$\text{or,}\quad a^2b^4=b^4h^2+a^4k^2$$

$$\text{or,}\quad \frac{h^2}{a^4}+\frac{k^2}{b^4}=\frac{1}{a^2}.$$

Generalizing, we get the locus of the point (h,k) is

$$\frac{x^2}{a^4}+\frac{y^2}{b^4}=\frac{1}{a^2}.]$$

2. Find the locus of the pole of the normal chords of the hyperbola

$$\frac{x^2}{a^2}-\frac{y^2}{b^2}=1.$$

3. Find the locus of the poles of the chords of the ellipse which subtends a right angle at the centre of the ellipse.
4. Show that polar of a point on $x^2+4ay=0$ with respect to $xy=2a^2$ will touch the parabola $y^2=4ax$.
5. Show that the polar of any point on the circle $x^2+y^2-2ax-3a^2=0$ with respect to the circle $x^2+y^2+2ax-3a^2=0$ will touch the parabola $y^2=-4ax$.
6. Show that the locus of the poles of the tangents to the parabola $y^2=4ax$ with respect to the circle $x^2+y^2=2ax$ is the circle $x^2+y^2=ax$.
7. Find the locus of the poles of the tangents to the auxiliary circle of the hyperbola $\frac{x^2}{a^2}-\frac{y^2}{b^2}=1$ with respect to the hyperbola itself.
8. Find the locus of the poles if the polar with respect to the hyperbola $\frac{x^2}{a^2}-\frac{y^2}{b^2}=1$ be equally inclined to the coordinate axes.

[*Hints*: Equation of the polar of the point (h,k) with respect to the hyperbola

$$\frac{x^2}{a^2}-\frac{y^2}{b^2}=1 \quad\text{is}\quad \frac{xh}{a^2}-\frac{yk}{b^2}=1.$$

If this line is equally inclined to the coordinate axes, then

$$\frac{h}{a^2}=\frac{k}{b^2}.$$

Hence, locus of the point (h,k) is $ya^2=xb^2$.]

9. The polars of $P(x_1,y_1)$ and $Q(x_2,y_2)$ with respect to the ellipse $\frac{x^2}{a^2}+\frac{y^2}{b^2}=1$ meet the ellipse in four points. Show that the equation of the conic which passes through the six points is

$$\left(\frac{x^2}{a^2}+\frac{y^2}{b^2}-1\right)\left(\frac{x_1x_2}{a^2}+\frac{y_1y_2}{b^2}-1\right)=\left(\frac{xx_1}{a^2}+\frac{yy_1}{b^2}-1\right)\left(\frac{xx_2}{a^2}+\frac{yy_2}{b^2}=1\right).$$

[*Hints*: Equations of the polar of $P(x_1, y_1)$ and $Q(x_2, y_2)$ with respect to the ellipse $\frac{x^2}{a^2}+\frac{y^2}{b^2}=1$ are $\frac{xx_1}{a^2}+\frac{yy_1}{b^2}=1$ and $\frac{xx_2}{a^2}+\frac{yy_2}{b^2}=1$. These lines meet the ellipse in four points. Equation of the curve which passes through these four points is

$$\frac{x^2}{a^2}+\frac{y^2}{b^2}-1+k\left(\frac{xx_1}{a^2}+\frac{yy_1}{b^2}-1\right)\left(\frac{xx_2}{a^2}+\frac{yy_2}{b^2}-1\right)=0.$$

If this curve passes through P and Q, then

$$\frac{x_1^2}{a^2}+\frac{y_1^2}{b^2}-1+k\left(\frac{x_1^2}{a^2}+\frac{y_1^2}{b^2}-1\right)\left(\frac{x_1x_2}{a^2}+\frac{y_1y_2}{b^2}-1\right)=0$$

and $$\frac{x_2^2}{a^2}+\frac{y_2^2}{b^2}-1+k\left(\frac{x_2^2}{a^2}+\frac{y_2^2}{b^2}-1\right)\left(\frac{x_1x_2}{a^2}+\frac{y_1y_2}{b^2}-1\right)=0$$

which gives

$$k=-\frac{1}{\frac{x_1x_2}{a^2}+\frac{y_1y_2}{b^2}-1}.$$

Hence, equation of the curve is

$$\left(\frac{x^2}{a^2}+\frac{y^2}{b^2}-1\right)\left(\frac{x_1x_2}{a^2}+\frac{y_1y_2}{b^2}-1\right)=\left(\frac{xx_1}{a^2}+\frac{yy_1}{b^2}-1\right)\left(\frac{xx_2}{a^2}+\frac{yy_2}{b^2}-1\right).]$$

10. Show that the locus of the poles of the chords of the hyperbola $\frac{x^2}{a^2}-\frac{y^2}{b^2}=1$ which touch the circle on the line joining the foci as diameter, is

$$\frac{x^2}{a^4}+\frac{y^2}{b^4}=\frac{1}{a^2+b^2}.$$

[*Hints*: Equation of the polar of the point (h, k) with respect to the hyperbola

$$\frac{x^2}{a^2}-\frac{y^2}{b^2}=1 \quad \text{is} \quad \frac{xh}{a^2}-\frac{yk}{b^2}=1$$

or, $$y=\frac{hb^2}{ka^2}x-\frac{b^2}{k}. \tag{1}$$

Equation of the circle on the line joining the foci $(\pm ae, 0)$, as diameter, is

$$(x-ae)(x+ae)+y^2=0$$

or, $$x^2+y^2=a^2e^2=a^2+b^2. \tag{2}$$

If the straight line (1) touches the circle (2), then we have

$$\left(\frac{b^2}{k}\right)^2=(a^2+b^2)\left[1+\left(\frac{hb^2}{ka^2}\right)^2\right]$$

or, $$(ab)^4=(a^2+b^2)[k^2a^4+h^2b^4]$$

or, $$\frac{h^2}{a^4}+\frac{k^2}{b^4}=\frac{1}{a^2+b^2}.$$

Generalizing, we get the locus of (h, k) is

$$\frac{x^2}{a^4}+\frac{y^2}{b^4}=\frac{1}{a^2+b^2}.]$$

11. Find the locus of the pole of the chords of the ellipse $\dfrac{x^2}{a^2}+\dfrac{y^2}{b^2}=1$ whose middle point lies on the line $x+y=k$.

12. If the product of the perpendiculars from the foci on a polar of a point with respect to the ellipse $\dfrac{x^2}{a^2}+\dfrac{y^2}{b^2}=1$ be c^2, then prove that the locus of the point is $b^4x^2(c^2+a^2e^2)+c^2a^4y^2=a^4b^4$.

[*Hints*: Equation of the polar of the point (h, k) with respect to the ellipse

$$\frac{x^2}{a^2}+\frac{y^2}{b^2}=1 \quad \text{is} \quad \frac{xh}{a^2}+\frac{yk}{b^2}=1.$$

Foci of the ellipse are given by $(\pm ae, 0)$. Given that

$$\frac{\dfrac{eh}{a}-1}{\sqrt{\dfrac{h^2}{a^4}+\dfrac{k^2}{b^4}}}\times\frac{-\dfrac{eh}{a}-1}{\sqrt{\dfrac{h^2}{a^4}+\dfrac{k^2}{b^4}}}=c^2$$

$$\text{or,}\quad 1-\frac{e^2h^2}{a^2}=c^2\left(\frac{h^2}{a^4}+\frac{k^2}{b^4}\right)$$

$$\text{or,}\quad a^4b^4-e^2h^2a^2b^4=c^2(h^2b^4+k^2a^4)$$

$$\text{or,}\quad b^4h^2(c^2+a^2e^2)+c^2a^4k^2=a^4b^4.$$

Generalizing, we get

$$b^4x^2(c^2+a^2e^2)+c^2a^4y^2=a^4b^4.]$$

13. Show that the locus of the poles of the chords of the parabola $y^2 = 4ax$ which are bisected by a fixed line $lx+my+n=0$ is

$$l(y^2-2ax)+2a(my+n)=0.$$

[*Hints*: Equation of the polar of the point with respect to the parabola $y^2=4ax$ is

$$yk=2a(x+h).$$

If it intersects the parabola at A: (x_1, y_1) and B: (x_2, y_2), then x_1, x_2 are the roots of the equation

$$\left(\frac{2a(x+h)}{k}\right)^2=4ax$$

$$\text{or,}\quad a(x^2+2xh+h^2)=xk^2$$

$$\text{or,}\quad ax^2+x(2ah-k^2)+ah^2=0.$$

Then

$$x_1 + x_2 = \frac{k^2}{a} - 2h$$

and $$y_1 + y_2 = \frac{2a(x_1 + x_2)}{k} + \frac{4ah}{k} = 2a\left(\frac{k}{a} - \frac{2h}{k}\right) + \frac{4ah}{k}.$$

If the chord is bisected by the fixed line $lx + my + n = 0$, then we have

$$l(x_1 + x_2) + m(y_1 + y_2) + 2n = 0$$

or, $$l\left(\frac{k^2}{a} - 2h\right) + m\left[2a\left(\frac{k}{a} - \frac{2h}{k}\right) + \frac{4ah}{k}\right] + 2n = 0$$

or, $$l\left(\frac{k^2 - 2ah}{a}\right) + m\left[2\left(\frac{k^2 - 2ah}{k}\right) + \frac{4ah}{k}\right] + 2n = 0$$

or, $$l(k^2 - 2ah) + 2amk + 2na = 0$$

or, $$l(k^2 - 2ah) + 2a(mk + n) = 0.$$

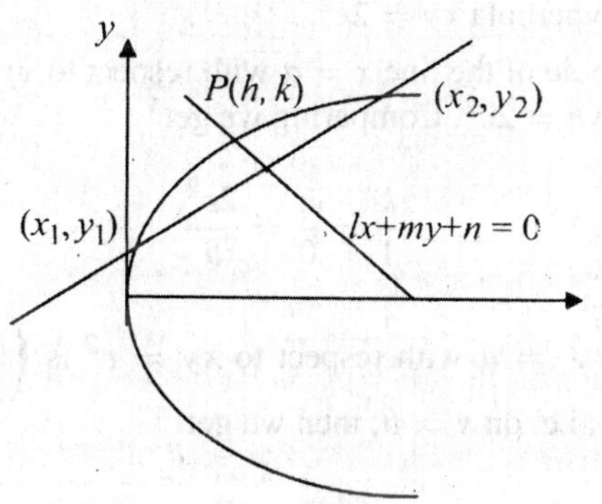

Generalizing, we get

$$l(y^2 - 2ax) + 2a(my + n) = 0.]$$

14. If three points be collinear, then prove that their polars with respect to an ellipse are concurrent.
15. Find the condition that the two straight lines $lx + my = 1$ and $Lx + My = 1$ will conjugate with respect to the conic $ax^2 + 2hxy + by^2 = 1$.
16. Find the condition that the lines $lx + my + n = 0$ and $Lx + My + N = 0$ will be conjugate to each other with respect to

    (a) $x^2 + y^2 = a^2$;
    (b) $x^2 + y^2 + 2gx + 2fy + c = 0$;
    (c) $y^2 = 4ax$;
    (d) $\dfrac{x^2}{a^2} + \dfrac{y^2}{b^2} = 1$;
    (e) $\dfrac{x^2}{a^2} - \dfrac{y^2}{b^2} = 1$.

17. Find the locus of the points which are such that the chords of contact of tangents drawn from them to the conic $ax^2 + by^2 = 1$ form a triangle of constant area with coordinate axes.

[*Hints*: Equation of chord of contact of the point (h, k) w.r.t. the conic $ax^2 + by^2 = 1$ is $axh + byk = 1$ or $\frac{x}{1/ah} + \frac{y}{1/bk} = 1$. Now area of the triangle $= \frac{1}{2}\frac{1}{ahbk} = \frac{Q^2}{2} =$ constant. Hence, locus of (h, k) is $xy = \frac{1}{abQ^2} = c^2$.]

18. Chords of the ellipse $\frac{x^2}{a^2} + \frac{y^2}{b^2} = 1$ touch the parabola $ay^2 = -2b^2x$; prove that the locus of their poles is the parabola $ay^2 = 2b^2x$.

19. Find the locus of the pole of the line $\frac{x}{a} + \frac{y}{b} = 1$ with respect to the circle which touches the coordinate axes.

20. Prove that the locus of the poles of the chords of the parabola $y^2 = 4ax$ which subtend a constant angle α at the vertex is the curve

$$(x + 4a)^2 = 4(y^2 - 4ax)\cot^2 \alpha.$$

21. Two lines $x = \alpha$ and $y = \beta$ are conjugate with respect to the hyperbola $xy = c^2$. Prove that (α, β) lies on the hyperbola $xy = 2c^2$.

[*Hints*: If (h, k) be the pole of the line $x = \alpha$ with respect to $xy = c^2$, then the line must be identical with $xk + yh = 2c^2$. Comparing we get

$$\frac{k}{1} = \frac{h}{0} = \frac{2c^2}{\alpha}.$$

So, the pole of the line $x = \alpha$ with respect to $xy = c^2$ is $\left(0, \frac{2c^2}{\alpha}\right)$. If it lies on the conjugate line of $x = \alpha$, i.e. on $y = \beta$, then we get

$$\frac{2c^2}{\alpha} = \beta.$$

Hence, locus of the point (α, β) is $xy = 2c^2$.]

Chapter 8

Diameters

The locus of the middle points of a series of parallel chords of a conic is called the *diameter* of the conic. So it is necessary to know the equation of the chord of a conic whose middle point (x_1, y_1) is given.

8.1 Equation of a Chord of a Conic in Terms of Its Middle Point

Let the equation of the conic be

$$ax^2 + 2hxy + by^2 + 2gx + 2fy + c = 0.$$

Equation of a straight line through the point (x_1, y_1) is

$$\frac{x - x_1}{l} = \frac{y - y_1}{m} = r.$$

Any point on the line can be taken as $(x_1 + lr, y_1 + mr)$. If the straight line intersect the conic, for some value of r, the above point will lie on the conic. Putting $x = x_1 + lr$ and $y = y_1 + mr$ in the equation of conic, we get

$$\begin{aligned} & a(x_1 + lr)^2 + 2h(x_1 + lr)(y_1 + mr) + b(y_1 + mr)^2 \\ & \quad + 2g(x_1 + lr) + 2f(y_1 + mr) + c = 0 \\ \text{or,} \quad & (al^2 + 2hlm + bm^2)r^2 + 2[(ax_1 + hy_1 + g)l + (hx_1 + by_1 + f)m]r \\ & \quad + ax_1^2 + 2hx_1y_1 + by_1^2 + 2gx_1 + 2fy_1 + c = 0. \end{aligned}$$

The above second degree equation in r gives two values of r, this indicates that the straight line intersects the conic in two points, which are equidistant from the middle point of the chord, i.e. from (x_1, y_1). In this case, the above equation in r must have two roots which are equal in magnitude but opposite in sign, which gives

$$\begin{aligned} & (ax_1 + hy_1 + g)l + (hx_1 + by_1 + f)m = 0 \\ \text{or,} \quad & (ax_1 + hy_1 + g)(x - x_1) + (hx_1 + by_1 + f)(y - y_1) = 0 \\ \text{or,} \quad & axx_1 + h(xy_1 + x_1y) + byy_1 + gx + fy = ax_1^2 + 2hx_1y_1 + by_1^2 + gx_1 + fy_1. \end{aligned}$$

Adding $gx_1 + fy_1 + c$ to both sides, we get

$$\begin{aligned} axx_1 + h(xy_1 + x_1y) + byy_1 + g(x + x_1) + f(y + y_1) + c \\ = ax_1^2 + 2hx_1y_1 + by_1^2 + 2gx_1 + 2fy_1 + c. \end{aligned}$$

This is the equation of the chord of the conic whose middle point is (x_1, y_1) and of the form $T = S_1$, where

$$T = axx_1 + h(xy_1 + x_1y) + byy_1 + g(x + x_1) + f(y + y_1) + c$$

and $$S_1 = ax_1^2 + 2hx_1y_1 + by_1^2 + 2gx_1 + 2fy_1 + c.$$

Particular Cases

Equation of the chord, whose middle point is (x_1, y_1), of

(a) $x^2 + y^2 = a^2$ is $xx_1 + yy_1 - a^2 = x_1^2 + y_1^2 - a^2$, i.e. $xx_1 + yy_1 = x_1^2 + y_1^2$;
(b) $x^2 + y^2 + 2gx + 2fy + c = 0$ is $xx_1 + yy_1 + g(x + x_1) + f(y + y_1) + c = x_1^2 + y_1^2 + 2gx_1 + 2fy_1 + c$, i.e. $xx_1 + yy_1 + gx + fy = x_1^2 + y_1^2 + gx_1 + fy_1$;
(c) $y^2 = 4ax$ is $yy_1 - 2a(x + x_1) = y_1^2 - 4ax_1$, i.e. $yy_1 - 2ax = y_1^2 - 2ax_1$;
(d) $\dfrac{x^2}{a^2} + \dfrac{y^2}{b^2} = 1$ is $\dfrac{xx_1}{a^2} + \dfrac{yy_1}{b^2} - 1 = \dfrac{x_1^2}{a^2} + \dfrac{y_1^2}{b^2} - 1$, i.e. $\dfrac{xx_1}{a^2} + \dfrac{yy_1}{b^2} = \dfrac{x_1^2}{a^2} + \dfrac{y_1^2}{b^2}$;
(e) $\dfrac{x^2}{a^2} - \dfrac{y^2}{b^2} = 1$ is $\dfrac{xx_1}{a^2} - \dfrac{yy_1}{b^2} - 1 = \dfrac{x_1^2}{a^2} - \dfrac{y_1^2}{b^2} - 1$, i.e. $\dfrac{xx_1}{a^2} - \dfrac{yy_1}{b^2} = \dfrac{x_1^2}{a^2} - \dfrac{y_1^2}{b^2}$;
(f) $xy = c^2$ is $xy_1 + yx_1 - 2c^2 = 2x_1y_1 - 2c^2$, i.e. $xy_1 + yx_1 = 2x_1y_1$.

8.2 Diameters, Conjugate Diameters and Principal Diameters

We have already defined that the locus of the middle points of a series of parallel chords of a conic is called the *diameter* of the conic.

If (x_1, y_1) be the middle point of one of these chords of the conic

$$ax^2 + 2hxy + by^2 + 2gx + 2fy + c = 0 \tag{1}$$

then its equation is

$$\begin{aligned} axx_1 + h(xy_1 + x_1y) + byy_1 + g(x + x_1) + f(y + y_1) + c \\ = ax_1^2 + 2hx_1y_1 + by_1^2 + 2gx_1 + 2fy_1 + c \\ \text{or,} \quad (ax_1 + hy_1 + g)x + (hx_1 + by_1 + f)y \\ = ax_1^2 + 2hx_1y_1 + by_1^2 + gx_1 + fy_1 \end{aligned}$$

Now, if the chords be parallel to the line

$$\frac{x}{l} = \frac{y}{m} \tag{2}$$

then we have

$$\frac{ax_1 + hy_1 + g}{m} = -\frac{hx_1 + by_1 + f}{l}$$

or, $$(al + mh)x_1 + (hl + mb)y_1 + gl + mf = 0.$$

Hence, locus of the point (x_1, y_1) is

$$(al + mh)x + (hl + mb)y + gl + mf = 0 \quad (3)$$

which is the equation of the diameter of the conic given by Eqn. (1).

Diameters such that one bisects the chords parallel to other are called *conjugate diameters*. So, if the diameter given by Eqn. (3) is parallel to the line

$$\frac{x}{L} = \frac{y}{M} \quad (4)$$

then we have

$$\frac{al + mh}{M} = -\frac{hl + mb}{L}$$

$$\text{or,} \quad alL + h(mL + lM) + bmM = 0.$$

Hence, the diameters parallel to the chords (2) and (4) will be conjugate to each other if the above condition is satisfied.

If the angle between two conjugate diameters is 90°, i.e. if the two conjugate diameters are perpendicular to each other, then these are called *principal diameters*. For example, any diameter of a circle is principal diameter, but in case of ellipse there is only one pair of principal diameters, major axis and minor axis.

8.3 Diameters of Conic Sections

(a) *Parabola*: Equation of the chord of the parabola $y^2 = 4ax$, whose middle point is (h, k), is

$$yk - 2a(x + h) = k^2 - 4ah$$

$$\text{or,} \quad y = \frac{2a}{k}x + \frac{k^2 - 2ah}{k}.$$

If it is parallel to the line $y = mx + c$, then

$$m = \frac{2a}{k}, \quad \text{i.e. } k = \frac{2a}{m}.$$

Hence, the locus of (h, k), i.e. equation of the diameter is $y = \frac{2a}{m}$, which is a straight line parallel to $y = 0$, i.e. axis of the parabola.

(b) *Ellipse*: Equation of the chord of the ellipse $\frac{x^2}{a^2} + \frac{y^2}{b^2} = 1$, whose middle point is (h, k), is

$$\frac{xh}{a^2} + \frac{yk}{b^2} = \frac{h^2}{a^2} + \frac{k^2}{b^2}$$

$$\text{or,} \quad y = -\frac{hb^2}{ka^2}x + \frac{b^2h^2}{ka^2} + k.$$

If it is parallel to the line $y = mx + c$, then

$$m = -\frac{hb^2}{ka^2}, \quad \text{i.e. } k = -\frac{b^2}{ma^2}h.$$

Hence, the locus of (h, k), i.e. equation of the diameter is $y = -\frac{b^2}{ma^2}x$, which is a straight line passing through the origin, i.e. centre of the ellipse and of the form $y = Mx$, where

$$M = -\frac{b^2}{ma^2}, \quad \text{i.e. } Mm = -\frac{b^2}{a^2}$$

which does not change if m and M are interchanged. Hence, two diameters $y = mx$ and $y = Mx$ are conjugate to each other if

$$Mm = -\frac{b^2}{a^2}.$$

(c) *Hyperbola*: Equation of the chord of the hyperbola $\frac{x^2}{a^2} - \frac{y^2}{b^2} = 1$, whose middle point is (h, k), is

$$\frac{xh}{a^2} - \frac{yk}{b^2} = \frac{h^2}{a^2} - \frac{k^2}{b^2}$$

$$\text{or,} \quad y = \frac{hb^2}{ka^2}x - \frac{b^2h^2}{ka^2} + k.$$

If it is parallel to the line $y = mx + c$, then

$$m = \frac{hb^2}{ka^2}, \quad \text{i.e. } k = \frac{b^2}{ma^2}h.$$

Hence, the locus of (h, k), i.e. equation of the diameter is $y = \frac{b^2}{ma^2}x$, which is a straight line passing through the origin, i.e. centre of the hyperbola and of the form $y = Mx$, where

$$M = \frac{b^2}{ma^2}, \quad \text{i.e. } Mm = \frac{b^2}{a^2}$$

which does not change if m and M are interchanged. Hence, two diameters $y = mx$ and $y = Mx$ are conjugate to each other if

$$Mm = \frac{b^2}{a^2}.$$

(d) *Rectangular hyperbola*: Equation of the chord of the rectangular hyperbola $xy = c^2$, whose middle point is (h, k), is

$$xk + yh = 2hk$$

$$\text{or,} \quad y = -\frac{k}{h}x + 2k.$$

If it is parallel to the line $y = mx + c$, then

$$m = -\frac{k}{h}, \quad \text{i.e. } k = -mh.$$

Hence, the locus of (h, k), i.e. equation of the diameter is $y = -mx$, which is a straight line passing through the origin, i.e. centre of the rectangular hyperbola and of the form $y = Mx$, where $M = -m$, i.e. $M + m = 0$, which does not change if m and M are interchanged. Hence, two diameters $y = mx$ and $y = Mx$ are conjugate to each other if $M + m = 0$.

8.4 Properties of Diameters of a Conic

P1 : *The tangents at an extremity of any diameter of a conic is parallel to the system of parallel chords bisected by the diameter.*

Proof: Equation of the diameter bisecting all chords parallel to the straight line $y = mx + c$ of the conic

$$ax^2 + 2hxy + by^2 + 2gx + 2fy + c = 0$$

is

$$(a + mh)x + (h + mb)y + g + mf = 0.$$

If the diameter meets the conic at (x_1, y_1), then

$$m = -\frac{ax_1 + hy_1 + g}{hx_1 + by_1 + f}.$$

Now, equation of the tangent to the conic at (x_1, y_1) is

$$axx_1 + h(xy_1 + x_1y) + byy_1 + g(x + x_1) + f(y + y_1) + c = 0$$

or, $$(ax_1 + hy_1 + g)x + (hx_1 + by_1 + f)y + gx_1 + fy_1 + c = 0$$

whose slope is $-\dfrac{ax_1 + hy_1 + g}{hx_1 + by_1 + f}$, which shows that the tangent is parallel to the straight line $y = mx + c$.

P2 : *The tangents at the ends of any chord of a conic meet on the diameter which bisects the chord.*

Proof: Let $y = mx + c$ be a chord of the conic

$$ax^2 + 2hxy + by^2 + 2gx + 2fy + c = 0.$$

If (x_1, y_1) be the point of intersection of the tangents drawn at the extremities of the chord, then the chord will be the chord of contact of the point (x_1, y_1) with respect to the conic whose equation is

$$axx_1 + h(xy_1 + x_1y) + byy_1 + g(x + x_1) + f(y + y_1) + c = 0$$

or, $$(ax_1 + hy_1 + g)x + (hx_1 + by_1 + f)y + gx_1 + fy_1 + c = 0.$$

Comparing this with the equation $y = mx + c$, we get

$$m = -\frac{ax_1 + hy_1 + g}{hx_1 + by_1 + f}$$

or, $$(a + mh)x_1 + (h + mb)y_1 + g + mf = 0$$

which shows that the point (x_1, y_1) lies on

$$(a + mh)x + (h + mb)y + g + mf = 0$$

which is a diameter of the conic bisecting the system of chords parallel to $y = mx + c$.

8.5 Properties of Conjugate Diameters of Ellipse

P1 : *The eccentric angles of the extremities of a pair of semi-conjugate diameters of an ellipse differ by an odd multiple of $\pi/2$.*

Proof: If CP and CD be two semi-conjugate diameters of the ellipse $\frac{x^2}{a^2}+\frac{y^2}{b^2}=1$, whose centre is C and P, D are the extremities of the diameters. If the eccentric angles of P, D are θ and ϕ, then coordinates of P and D are respectively $(a\cos\theta, b\sin\theta)$ and $(a\cos\phi, b\sin\phi)$ respectively.

Now, slope of $CP = \frac{b}{a}\tan\theta$ and slope of $CD = \frac{b}{a}\tan\phi$. Since CP, CD are semi-conjugate diameters, product of their slopes $= -\frac{b^2}{a^2}$, which gives

$$\tan\theta\tan\phi = -1$$

$$\text{or,}\quad \cos(\theta-\phi)=0$$

$$\text{or,}\quad \theta-\phi = \text{odd multiple of } \pi/2.$$

P2 : *The sum of squares of two semi-conjugate diameters of an ellipse is constant.*

Proof: Let CP and CD be two semi-conjugate diameters of the ellipse $\frac{x^2}{a^2}+\frac{y^2}{b^2}=1$ and coordinates of P be $(a\cos\theta, b\sin\theta)$; then coordinates of D will be $(-a\sin\theta, b\cos\theta)$.

So, $CP^2 = a^2\cos^2\theta + b^2\sin^2\theta$ and $CD^2 = a^2\sin^2\theta + b^2\cos^2\theta$. Adding, we get

$$\begin{aligned} CP^2 + CD^2 &= a^2\cos^2\theta + b^2\sin^2\theta + a^2\sin^2\theta + b^2\cos^2\theta \\ &= a^2 + b^2 = \text{constant.} \end{aligned}$$

P3 : *The tangents at the extremities of a pair of conjugate diameters of an ellipse form a parallelogram of constant area.*

Proof: Let PCP' and DCD' be two conjugate diameters of the ellipse $\frac{x^2}{a^2}+\frac{y^2}{b^2}=1$ and coordinates of P be $(a\cos\theta, b\sin\theta)$; then coordinates of P', D and D' will be $(-a\cos\theta, -b\sin\theta)$, $(-a\sin\theta, b\cos\theta)$ and $(a\sin\theta, -b\cos\theta)$ respectively.

Equations of tangents at P, P', D and D' are

$$\frac{x}{a}\cos\theta + \frac{y}{b}\sin\theta = 1,$$

$$-\frac{x}{a}\cos\theta + \frac{y}{b}\sin\theta = -1,$$

$$-\frac{x}{a}\sin\theta + \frac{y}{b}\cos\theta = 1,$$

$$\text{and}\quad -\frac{x}{a}\sin\theta + \frac{y}{b}\cos\theta = -1 \text{ respectively.}$$

From above we see that the tangents at P, P' are parallel and also the tangents at D and D' are parallel, so they form a parallelogram.

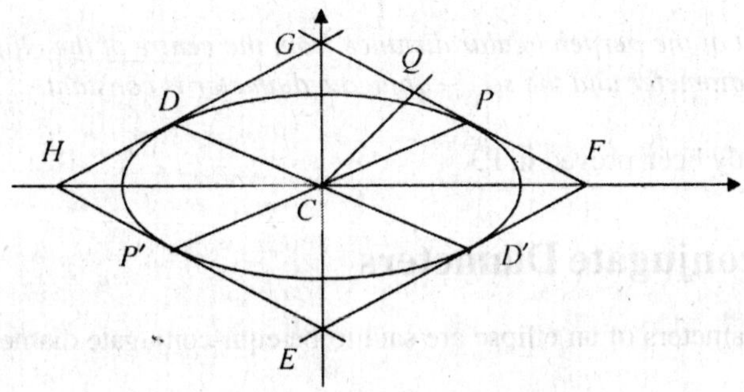

$$\text{Area of the parallelogram } EFGH = 4 \times \text{area of } CPGD$$
$$= 4 \times 2(\text{area of } \Delta CPG)$$
$$= 4CQ \times PG$$
$$= 4CQ \times CD.$$

Now,

$$CQ = \text{length of the perpendicula. from origin to the tangent } \frac{x}{a}\cos\theta + \frac{y}{b}\sin\theta = 1$$

$$= \frac{1}{\sqrt{\dfrac{\cos^2\theta}{a^2} + \dfrac{\sin^2\theta}{b^2}}} = \frac{ab}{\sqrt{a^2\sin^2\theta + b^2\cos^2\theta}} = \frac{ab}{CD}.$$

So, area of the parallelogram $EFGH = 4ab =$ constant.

P4 : *If CP and CD are two semi-conjugate diameters of an ellipse whose foci are S and S′, then* $PS \times PS' = CD^2$.

Proof: Let CP and CD be two semi-conjugate diameters of the ellipse $\dfrac{x^2}{a^2} + \dfrac{y^2}{b^2} = 1$ and coordinates of P be $(a\cos\theta, b\sin\theta)$; then coordinates of D will be $(-a\sin\theta, b\cos\theta)$. Foci of the ellipse are given by $(\pm ae, 0)$. So

$$PS = \sqrt{(a\cos\theta - ae)^2 + b^2\sin^2\theta}$$
$$= \sqrt{a^2\cos^2\theta + a^2e^2 - 2a^2e\cos\theta + b^2\sin^2\theta}$$
$$= \sqrt{a^2\cos^2\theta + a^2 - b^2 - 2a^2e\cos\theta + b^2\sin^2\theta}$$
$$= \sqrt{a^2\cos^2\theta + a^2 - 2a^2e\cos\theta - b^2\cos^2\theta}$$
$$= \sqrt{a^2e^2\cos^2\theta + a^2 - 2a^2e\cos\theta}$$
$$= a - ae\cos\theta$$

Changing e by $-e$ we get, $PS' = a + ae\cos\theta$.

Now,

$$PS \times PS' = (a - ae\cos\theta)(a + ae\cos\theta)$$
$$= a^2 - a^2e^2\cos^2\theta$$
$$= a^2 - (a^2 - b^2)\cos^2\theta$$
$$= b^2\cos^2\theta + a^2\sin^2\theta$$
$$= CD^2.$$

P5 : *The product of the perpendicular distance from the centre of the ellipse to the tangent at one end of the diameter and the semi-conjugate diameter is constant.*

This has already been proved in P3.

8.6 Equi-conjugate Diameters

Two conjugate diameters of an ellipse are said to be equi-conjugate diameters if they are of equal length.

Let CP and CD be two semi-conjugate diameters of the ellipse $\frac{x^2}{a^2}+\frac{y^2}{b^2}=1$ and coordinates of P be $(a\cos\theta, b\sin\theta)$ then coordinates of D will be $(-a\sin\theta, b\cos\theta)$. If P and D are the end points of two equi-conjugate diameters, then

$$\begin{aligned} & CP = CD \\ \text{or,}\quad & b^2\cos^2\theta + a^2\sin^2\theta = a^2\cos^2\theta + b^2\sin^2\theta \\ \text{or,}\quad & (a^2-b^2)\sin^2\theta = (a^2-b^2)\cos^2\theta \\ \text{or,}\quad & \tan^2\theta = 1 \\ \text{or,}\quad & \theta = \pi/4 \text{ or } 3\pi/4. \end{aligned}$$

i.e., two equi-conjugate diameters are equally inclined to the axes in opposite sense and their equations are $y = \pm\frac{b}{a}x$. In this case, the end points are P: $\left(\frac{a}{\sqrt{2}}, \frac{b}{\sqrt{2}}\right)$ and D: $\left(-\frac{a}{\sqrt{2}}, \frac{b}{\sqrt{2}}\right)$, and the length of equi-conjugate diameter $= 2\sqrt{\frac{a^2}{2}+\frac{b^2}{2}} = \sqrt{2(a^2+b^2)}$.

8.7 Properties of Conjugate Diameters of Hyperbola

P1 : *If a pair of diameters be conjugate with respect to a hyperbola, then it is also conjugate with respect to the conjugate hyperbola.*

Proof: We have that if $y = mx$ and $y = Mx$ be the pair of conjugate diameters of the hyperbola $\frac{x^2}{a^2}-\frac{y^2}{b^2}=1$, then

$$Mm = \frac{b^2}{a^2}$$

which remains unchanged if b^2 and a^2 are changed by $-b^2$ and $-a^2$. Hence, a pair of conjugate diameters of a hyperbola is also conjugate diameters of conjugate hyperbola.

P2 : *If a diameter meets a hyperbola in real points, then it does not meet the conjugate hyperbola in real points.*

Proof: Let $y = mx$ be a diameter of the hyperbola $\frac{x^2}{a^2}-\frac{y^2}{b^2}=1$ whose conjugate hyperbola is $\frac{x^2}{a^2}-\frac{y^2}{b^2}=-1$. The x-coordinates of the point of intersections of the line with the hyperbolas are

$$x = \pm \frac{ab}{\sqrt{b^2 - a^2m^2}}$$

$$\text{and} \quad x = \pm \sqrt{\frac{-a^2b^2}{b^2 - a^2m^2}} \quad \text{respectively.}$$

Obviously, both the values of x cannot be real, i.e. if the first equation gives real value of x, then second equation gives imaginary value of x.

P3 : *One of a pair of conjugate diameters of a hyperbola meet the curve in real points.*

Proof: If $y = mx$ and $y = Mx$ be a pair of conjugate diameters of the hyperbola $\frac{x^2}{a^2} - \frac{y^2}{b^2} = 1$, then we have $mM = \frac{b^2}{a^2}$.

The x-coordinates of the point of intersections of the lines with the hyperbola are

$$x = \pm \frac{ab}{\sqrt{b^2 - a^2m^2}}$$

$$\begin{aligned} \text{and} \quad x &= \pm \frac{ab}{\sqrt{b^2 - a^2M^2}} \\ &= \pm \frac{ab}{\sqrt{b^2 - b^4/a^2m^2}} \\ &= \pm \frac{a^2m}{\sqrt{a^2m^2 - b^2}} \quad \text{respectively.} \end{aligned}$$

Obviously, both the values of x cannot be real, i.e. if the first equation gives real value of x, then second equation gives imaginary value of x.

P4 : *If a pair of conjugate diameters cut the hyperbola and its conjugate at P, P' and D, D' respectively, then $CP^2 - CD^2 = a^2 - b^2$.*

Proof: Let P: $(a \sec\theta, b \tan\theta)$ be a point on the hyperbola $\frac{x^2}{a^2} - \frac{y^2}{b^2} = 1$; then coordinates of D will be $(a \tan\theta, b \sec\theta)$.

So,

$$\begin{aligned} CP^2 &= a^2 \sec^2\theta + b^2 \tan^2\theta \\ \text{and} \quad CD^2 &= a^2 \tan^2\theta + b^2 \sec^2\theta. \end{aligned}$$

Subtracting, we get

$$\begin{aligned} CP^2 - CD^2 &= a^2 \sec^2\theta + b^2 \tan^2\theta - a^2 \tan^2\theta - b^2 \sec^2\theta \\ &= a^2 - b^2 = \text{constant.} \end{aligned}$$

P5 : *The tangents at the extremities of a pair of conjugate diameters of an hyperbola form a parallelogram of constant area.*

Proof: Let PCP' and DCD' be two conjugate diameters of the hyperbola $\frac{x^2}{a^2} - \frac{y^2}{b^2} = 1$ and coordinates of P be $(a\sec\theta, b\tan\theta)$; then coordinates of P', D and D' will be $(-a\sec\theta, -b\tan\theta)$, $(a\tan\theta, b\sec\theta)$ and $(-a\tan\theta, -b\sec\theta)$ respectively.

Equations of tangents at P, P', D and D' are

$$\frac{x}{a}\sec\theta - \frac{y}{b}\tan\theta = 1,$$
$$\frac{x}{a}\sec\theta - \frac{y}{b}\tan\theta = -1,$$
$$\frac{x}{a}\tan\theta - \frac{y}{b}\sec\theta = 1,$$
$$\text{and} \quad \frac{x}{a}\tan\theta - \frac{y}{b}\sec\theta = -1 \text{ respectively.}$$

From above we see that the tangents at P, P' are parallel and also the tangents at D and D' are parallel, so they form a parallelogram.

$$\begin{aligned}\text{Area of the parallelogram } EFGH &= 4 \times \text{area of } CPGD \\ &= 4 \times 2(\text{area of } \Delta CPG) \\ &= 4CQ \times PG \\ &= 4CQ \times CD.\end{aligned}$$

Now,

$$CQ = \text{length of the perpendicular from origin to the tangent } \frac{x}{a}\sec\theta - \frac{y}{b}\tan\theta = 1$$

$$= \frac{1}{\sqrt{\frac{\sec^2\theta}{a^2} + \frac{\tan^2\theta}{b^2}}} = \frac{ab}{\sqrt{a^2\tan^2\theta + b^2\sec^2\theta}} = \frac{ab}{CD}.$$

So, area of the parallelogram $EFGH = 4ab =$ constant.

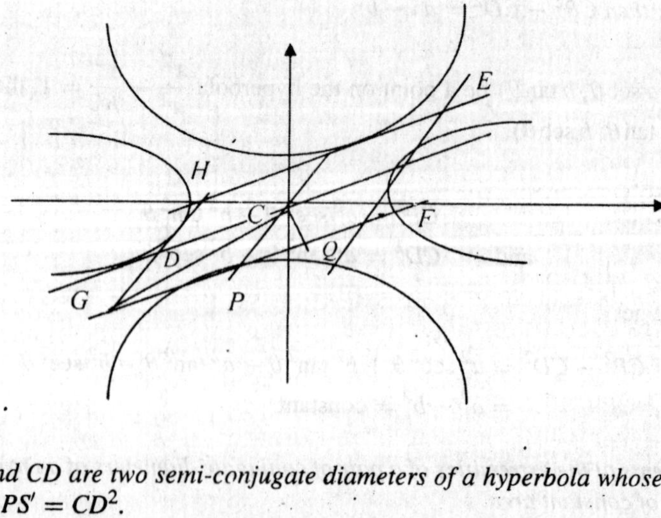

P6 : *If CP and CD are two semi-conjugate diameters of a hyperbola whose foci are S and S', then $PS \times PS' = CD^2$.*

Proof: Let CP and CD be two semi-conjugate diameters of the hyperbola $\frac{x^2}{a^2} - \frac{y^2}{b^2} = 1$ and coordinates of P be $(a\sec\theta, b\tan\theta)$; then coordinates of D will be $(a\tan\theta, b\sec\theta)$. Foci of the hyperbola are given by $(\pm ae, 0)$. So

$$
\begin{aligned}
PS &= \sqrt{(a\sec\theta - ae)^2 + b^2\tan^2\theta} \\
&= \sqrt{a^2\sec^2\theta + a^2e^2 - 2a^2e\sec\theta + b^2\tan^2\theta} \\
&= \sqrt{a^2\sec^2\theta + a^2 + b^2 - 2a^2e\sec\theta + b^2\tan^2\theta} \\
&= \sqrt{a^2\sec^2\theta + a^2 - 2a^2e\sec\theta + b^2\sec^2\theta} \\
&= \sqrt{a^2e^2\sec^2\theta + a^2 - 2a^2e\sec\theta} \\
&= ae\sec\theta - a.
\end{aligned}
$$

Similarly, $PS' = a + ae\sec\theta$. Now

$$
\begin{aligned}
PS \times PS' &= (ae\sec\theta - a)(a + ae\sec\theta) \\
&= a^2e^2\sec^2\theta - a^2 \\
&= (a^2 + b^2)\sec^2\theta - a^2 \\
&= b^2\sec^2\theta + a^2\tan^2\theta \\
&= CD^2.
\end{aligned}
$$

P7 : *The product of the perpendicular distance from the centre of the hyperbola to the tangent at one end of the diameter and the semi-conjugate diameter is constant.*

This has already been proved in P5.

Examples

▶▷▶ **1.** *Prove that the locus of the middle points of focal chords of a parabola is another parabola.*

Solution: Equation of the chord of the parabola $y^2 = 4ax$, whose middle point is (h, k), is

$$yk - 2ax = k^2 - 2ah.$$

If it passes through the focus of the parabola, i.e. through the point $(a, 0)$, then

$$0 - 2a^2 = k^2 - 2ah.$$

Hence, locus of the point (h, k) is

$$y^2 = 2a(x - a), \text{ which is another parabola.}$$

▶▷▶ **2.** *Prove that the locus of the middle points of all chords, which are drawn through the vertex of the parabola is another parabola.*

Solution: Equation of the chord of the parabola $y^2 = 4ax$, whose middle point is (h, k), is

$$yk - 2ax = k^2 - 2ah.$$

If it passes through the vertex of the parabola, i.e. through the point (0, 0), then

$$0 = k^2 - 2ah.$$

Hence, locus of the point (h, k) is

$$\therefore y^2 = 2ax, \text{ which is another parabola.}$$

▶▷▶ **3.** *Find the locus of the middle points of normal chords of the parabola $y^2 = 4ax$.*

Solution: Equation of the chord of the parabola $y^2 = 4ax$, whose middle point is (h, k), is

$$yk - 2ax = k^2 - 2ah.$$

Equation of the normal at $(at^2, 2at)$ to the parabola is

$$tx + y = 2at + at^3.$$

As both the equations represent same line, comparing we get

$$\frac{k}{1} = \frac{2a}{-t} = \frac{k^2 - 2ah}{2at + at^3}$$

which gives

$$t = -\frac{2a}{k} \quad \text{and} \quad k(2at + at^3) = k^2 - 2ah$$

$$\text{or,} \quad k\left[2a\left(-\frac{2a}{k}\right) + a\left(-\frac{2a}{k}\right)^3\right] = k^2 - 2ah$$

$$\text{or,} \quad -4a^2k^2 - 8a^4 = k^2(k^2 - 2ah)$$

$$\text{or,} \quad k^4 - 2ahk^2 + 4a^2k^2 + 8a^4 = 0.$$

Hence, locus of the point (h, k) is

$$y^4 - 2axy^2 + 4a^2y^2 + 8a^4 = 0.$$

▶▷▶ **4.** *Find the locus of the middle points of the chords of the ellipse which subtend a right angle at the centre.*

Solution: Equation of the chord of the ellipse $\frac{x^2}{a^2} + \frac{y^2}{b^2} = 1$, whose middle point is (h, k), is

$$\frac{xh}{a^2} + \frac{yk}{b^2} = \frac{h^2}{a^2} + \frac{k^2}{b^2}. \tag{1}$$

Making homogeneous the equation of the ellipse with aid of Eqn. (1), we get

$$\frac{x^2}{a^2} + \frac{y^2}{b^2} = \left[\frac{\frac{xh}{a^2} + \frac{yk}{b^2}}{\frac{h^2}{a^2} + \frac{k^2}{b^2}}\right]^2$$

$$\text{or,} \quad \left(\frac{x^2}{a^2} + \frac{y^2}{b^2}\right)\left(\frac{h^2}{a^2} + \frac{k^2}{b^2}\right)^2 = \left[\frac{xh}{a^2} + \frac{yk}{b^2}\right]^2$$

$$\text{or,} \quad x^2\left[\frac{1}{a^2}\left(\frac{h^2}{a^2} + \frac{k^2}{b^2}\right)^2 - \frac{h^2}{a^4}\right] - \frac{2xyhk}{a^2b^2} + y^2\left[\frac{1}{b^2}\left(\frac{h^2}{a^2} + \frac{k^2}{b^2}\right)^2 - \frac{k^2}{b^4}\right] = 0.$$

The above second degree homogeneous equation in x, y represents a pair of straight lines which pass through the origin, i.e. centre of the ellipse. If these lines are perpendicular, then we have

$$\text{Coeff. of } x^2 + \text{Coeff. of } y^2 = 0$$

i.e., $$\left[\frac{1}{a^2}\left(\frac{h^2}{a^2}+\frac{k^2}{b^2}\right)^2 - \frac{h^2}{a^4}\right] + \left[\frac{1}{b^2}\left(\frac{h^2}{a^2}+\frac{k^2}{b^2}\right)^2 - \frac{k^2}{b^4}\right] = 0$$

or, $$\left(\frac{1}{a^2}+\frac{1}{b^2}\right)\left(\frac{h^2}{a^2}+\frac{k^2}{b^2}\right)^2 = \frac{h^2}{a^4}+\frac{k^2}{b^4}.$$

Hence, locus of the point (h, k) is

$$\left(\frac{x^2}{a^2}+\frac{y^2}{b^2}\right)^2\left(\frac{1}{a^2}+\frac{1}{b^2}\right) = \frac{x^2}{a^4}+\frac{y^2}{b^4}.$$

►► **5.** *Prove that the locus of the middle points of the chords of constant length c of the parabola $y^2 = 4ax$ is the curve*

$$(y^2 + 4a^2)(y^2 - 4ax) + a^2c^2 = 0.$$

Solution: Let (h, k) be the middle point of the chord which intersects the parabola at P: $(at_1^2, 2at_1)$ and Q: $(at_2^2, 2at_2)$. Then

$$h = \frac{a(t_1^2 + t_2^2)}{2} \quad \text{and} \quad k = a(t_1 + t_2)$$

which gives $2t_1t_2 = (t_1 + t_2)^2 - (t_1^2 + t_2^2) = \left(\frac{k}{a}\right)^2 - \frac{2h}{a}$.

Then,

$$\begin{aligned}
\text{length of } PQ &= \sqrt{(at_1^2 - at_2^2)^2 + (2at_1 - 2at_2)^2} \\
&= a(t_1 - t_2)\sqrt{(t_1 + t_2)^2 + 4} \\
&= a\sqrt{(t_1 + t_2)^2 - 4t_1t_2}\sqrt{(t_1 + t_2)^2 + 4} \\
&= a\sqrt{(k/a)^2 - 2(k^2/a^2 - 2h/a)}\sqrt{(k/a)^2 + 4}
\end{aligned}$$

or, $$ac = \sqrt{(4ah - k^2)(k^2 + 4a^2)}$$

or, $$(k^2 + 4a^2)(k^2 - 4ah) + a^2c^2 = 0.$$

Hence, locus of the point (h, k) is

$$(y^2 + 4a^2)(y^2 - 4ax) + a^2c^2 = 0.$$

►► **6.** *Show that the locus of the middle points of the chords of constant length 2d of the rectangular hyperbola $xy = c^2$ is*

$$(x^2 + y^2)(xy - c^2) = d^2xy.$$

Solution: Let (h, k) be the middle point of the chord which intersects the hyperbola $xy = c^2$ at P: $(ct_1, c/t_1)$ and Q: $(ct_2, c/t_2)$. Then

$$h = \frac{c(t_1 + t_2)}{2} \quad \text{and} \quad k = \frac{c(t_1 + t_2)}{2t_1t_2}$$

which gives $t_1t_2 = h/k$.

Then,

$$\text{length of } PQ = \sqrt{(ct_1 - ct_2)^2 + (c/t_1 - c/t_2)^2}$$

$$= c(t_1 - t_2)\sqrt{1 + \frac{1}{t_1^2 t_2^2}}$$

$$= c\sqrt{(t_1 + t_2)^2 - 4t_1 t_2}\sqrt{1 + \frac{1}{t_1^2 t_2^2}}$$

$$= c\sqrt{\frac{4h^2}{c^2} - \frac{4h}{k}}\sqrt{1 + \frac{k^2}{h^2}}$$

or, $$2dh = 2\sqrt{\frac{(kh^2 - hc^2)(h^2 + k^2)}{k}}$$

or, $$kd^2h^2 = h(hk - c^2)(h^2 + k^2)$$

or, $$(h^2 + k^2)(hk - c^2) = d^2hk.$$

Hence, locus of the point (h, k) is

$$(x^2 + y^2)(xy - c^2) = d^2xy.$$

▶▷ **7.** *Prove that the locus of the middle points of the chords of contact of tangents to the hyperbola $x^2 - y^2 = a^2$ from points on auxiliary circle is*

$$(x^2 - y^2)^2 = a^2(x^2 + y^2).$$

Solution: Equation of the chord of the hyperbola $x^2 - y^2 = a^2$ whose middle point is (h, k), is

$$xh - yk = h^2 - k^2.$$

Equation of the auxiliary circle of the hyperbola $x^2 - y^2 = a^2$ is $x^2 + y^2 = a^2$. Any point on it can be taken as $(a\cos\theta, a\sin\theta)$. Equation of chord of contact of this point with respect to the hyperbola $x^2 - y^2 = a^2$ is

$$x\cos\theta - y\sin\theta = a.$$

Since, the above two equations represent the same line, comparing we get

$$\frac{\cos\theta}{h} = \frac{\sin\theta}{k} = \frac{a}{h^2 - k^2}.$$

Eliminating θ, we get

$$\left(\frac{ah}{h^2 - k^2}\right)^2 + \left(\frac{ak}{h^2 - k^2}\right)^2 = 1$$

or, $$a^2(h^2 + k^2) = (h^2 - k^2)^2.$$

So, locus of the point (h, k) is

$$(x^2 - y^2)^2 = a^2(x^2 + y^2).$$

▶▷▶ **8.** *If (h, k) be the pole of the chord with respect to the ellipse $\dfrac{x^2}{a^2}+\dfrac{y^2}{b^2}=1$ whose middle point is (x_0, y_0), then show that*

$$h=\frac{x_0}{\left(\dfrac{x_0^2}{a^2}+\dfrac{y_0^2}{b^2}\right)} \quad \textit{and} \quad k=\frac{y_0}{\left(\dfrac{x_0^2}{a^2}+\dfrac{y_0^2}{b^2}\right)}.$$

Solution: If (h, k) be the pole, then its polar with respect to the ellipse

$$\frac{x^2}{a^2}+\frac{y^2}{b^2}=1$$

is

$$\frac{xh}{a^2}+\frac{yk}{b^2}=1.$$

Equation of the chord of $\dfrac{x^2}{a^2}+\dfrac{y^2}{b^2}=1$ whose middle point is (x_0, y_0), is

$$\frac{xx_0}{a^2}+\frac{yy_0}{b^2}=\frac{x_0^2}{a^2}+\frac{y_0^2}{b^2}.$$

Since, both the equations represent same line, comparing we get

$$\frac{x_0}{h}=\frac{y_0}{k}=\frac{x_0^2}{a^2}+\frac{y_0^2}{b^2}$$

or, $$h=\frac{x_0}{\left(\dfrac{x_0^2}{a^2}+\dfrac{y_0^2}{b^2}\right)} \quad \text{and} \quad k=\frac{y_0}{\left(\dfrac{x_0^2}{a^2}+\dfrac{y_0^2}{b^2}\right)}.$$

▶▷▶ **9.** *If d be the length of the chord joining two points (h, k) and (h', k') on the ellipse $\dfrac{x^2}{a^2}+\dfrac{y^2}{b^2}=1$ and D be the length of the diameter parallel to the chord, then show that*

$$\frac{4d^2}{D^2}=\left(\frac{h-h'}{a}\right)^2+\left(\frac{k-k'}{b}\right)^2.$$

Solution: Let $P(a\cos\theta, b\sin\theta)$ be one of the extremities of a diameter which is parallel to the chord joining (h, k) and (h', k'), then

$$\frac{k-k'}{h-h'}=\frac{b}{a}\tan\theta. \tag{1}$$

Given that $$d^2=(h-h')^2+(k-k')^2. \tag{2}$$

We also have $$D^2=4\times CP^2=4[a^2\cos^2\theta+b^2\sin^2\theta], \tag{3}$$

C being the centre of the ellipse.

From Eqns. (2) and (3), we get

$$\frac{4d^2}{D^2}=\frac{(h-h')^2+(k-k')^2}{a^2\cos^2\theta+b^2\sin^2\theta}$$

$$
\begin{aligned}
&= (1+\tan^2\theta)\frac{(h-h')^2+(k-k')^2}{a^2+b^2\tan^2\theta} \\
&= \left(1+\frac{a^2(k-k')^2}{b^2(h-h')^2}\right)\frac{(h-h')^2+(k-k')^2}{a^2+b^2\tan^2\theta} \\
&= \left((h-h')^2+\frac{a^2(k-k')^2}{b^2}\right)\frac{1+(k-k')^2/(h-h')^2}{a^2+b^2\tan^2\theta} \\
&= \left(\frac{(h-h')^2}{a^2}+\frac{(k-k')^2}{b^2}\right)\frac{1+(k-k')^2/(h-h')^2}{1+(b^2/a^2)\tan^2\theta} \\
&= \frac{(h-h')^2}{a^2}+\frac{(k-k')^2}{b^2} \quad \text{[by the Eqn. (1)].}
\end{aligned}
$$

▶▶ **10.** *Two conjugate diameters of the ellipse* $\frac{x^2}{a^2}+\frac{y^2}{b^2}=1$ *cut the circle* $x^2+y^2=r^2$ *at P and Q. Prove that the locus of the pole of PQ with respect to the ellipse is* $\frac{x^2}{a^2}+\frac{y^2}{b^2}=\frac{a^2+b^2}{r^2}$.

Solution: Equation of the polar of $P(h,k)$ with respect to the ellipse is

$$\frac{xh}{a^2}+\frac{yk}{b^2}=1. \tag{1}$$

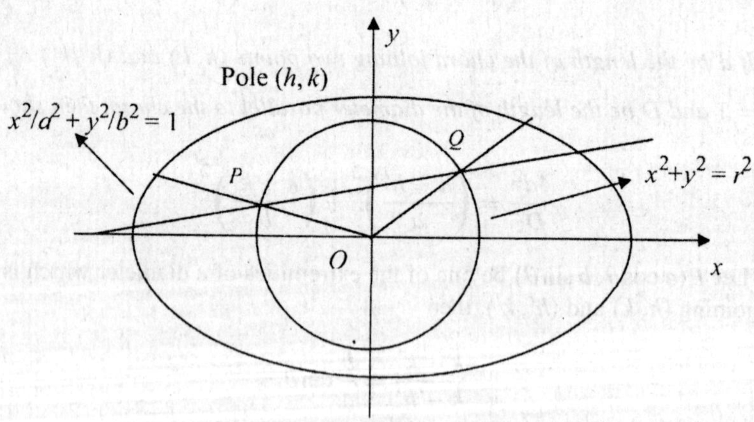

Making homogeneous the equation $x^2+y^2=r^2$ with the aid of Eqn. (1), we get

$$x^2+y^2=r^2\left(\frac{xh}{a^2}+\frac{yk}{b^2}\right)^2$$

or,
$$\left(1-\frac{r^2h^2}{a^4}\right)x^2-\frac{2hkr^2xy}{a^2b^2}+\left(1-\frac{r^2k^2}{b^4}\right)y^2=0. \tag{2}$$

If the individual equations of the above pair of straight lines are $y = mx$ and $y = m'x$, then

$$mm' = \frac{\left(1 - \frac{r^2h^2}{a^4}\right)}{\left(1 - \frac{r^2k^2}{b^4}\right)}. \tag{3}$$

Again, if the pair of straight lines represents the conjugate diameters of the ellipse, then

$$mm' = -\frac{b^2}{a^2}. \tag{4}$$

From Eqns. (3) and (4) we get

$$\frac{\left(1 - \frac{r^2h^2}{a^4}\right)}{\left(1 - \frac{r^2k^2}{b^4}\right)} = -\frac{b^2}{a^2}$$

$$\text{or,} \quad a^2 - \frac{r^2h^2}{a^2} = -b^2 + \frac{r^2k^2}{b^2}$$

$$\text{or,} \quad r^2\left(\frac{h^2}{a^2} + \frac{k^2}{b^2}\right) = a^2 + b^2.$$

Hence, locus of the point (h, k) is

$$\frac{x^2}{a^2} + \frac{y^2}{b^2} = \frac{a^2 + b^2}{r^2}.$$

▶▶ **11.** *Two conjugate diameters of the ellipse* $\frac{x^2}{a^2} + \frac{y^2}{b^2} = 1$ *cut the circle* $x^2 + y^2 = r^2$ *at P and Q. Prove that the locus of the middle point of PQ is*

$$a^2\{(x^2 + y^2)^2 - r^2x^2\} + b^2\{(x^2 + y^2)^2 - r^2y^2\} = 0.$$

Solution: Equation of the chord of the circle $x^2 + y^2 = r^2$, whose middle point is (h, k), is

$$xh + yk = h^2 + k^2. \tag{1}$$

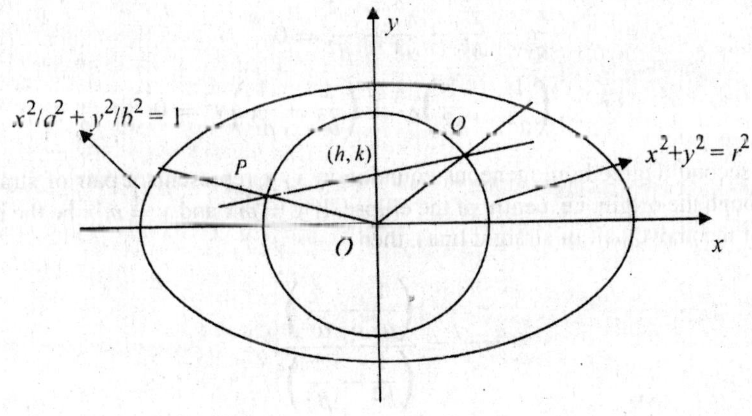

Making homogeneous the equation $x^2 + y^2 = r^2$ with the aid of Eqn. (1), we get

$$x^2 + y^2 = r^2\left(\frac{xh + yk}{h^2 + k^2}\right)^2$$

$$\text{or,}\quad x^2[(h^2 + k^2)^2 - r^2h^2] - 2xyhkr^2 + y^2[(h^2 + k^2)^2 - r^2k^2] = 0. \tag{2}$$

If the individual equations of the above pair of straight lines are $y = mx$ and $y = m'x$, then

$$mm' = \frac{(h^2 + k^2)^2 - r^2h^2}{(h^2 + k^2)^2 - r^2k^2}. \tag{3}$$

Again, if the pair of straight lines represents the conjugate diameters of the ellipse, then

$$mm' = -\frac{b^2}{a^2}. \tag{4}$$

From Eqns. (3) and (4) we get

$$\frac{(h^2 + k^2)^2 - r^2h^2}{(h^2 + k^2)^2 - r^2k^2} = -\frac{b^2}{a^2}$$

$$\text{or,}\quad a^2[(h^2 + k^2)^2 - r^2h^2] = -b^2[(h^2 + k^2)^2 - r^2k^2] = 0$$

$$\text{or,}\quad a^2[(h^2 + k^2)^2 - r^2h^2] + b^2[(h^2 + k^2)^2 - r^2k^2] = 0.$$

Hence, locus of the point (h, k) is

$$a^2\{(x^2 + y^2)^2 - r^2x^2\} + b^2\{(x^2 + y^2)^2 - r^2y^2\} = 0.$$

▶▷▶ **12.** *If the points of intersection of the ellipses* $\frac{x^2}{a^2} + \frac{y^2}{b^2} = 1$ *and* $\frac{x^2}{\alpha^2} + \frac{y^2}{\beta^2} = 1$ *be the end points of the conjugate diameters of the former, then prove that*

$$\frac{a^2}{\alpha^2} + \frac{b^2}{\beta^2} = 2.$$

Solution: Making homogeneous $\frac{x^2}{a^2} + \frac{y^2}{b^2} = 1$ with the aid of $\frac{x^2}{\alpha^2} + \frac{y^2}{\beta^2} = 1$, we get

$$\frac{x^2}{a^2} + \frac{y^2}{b^2} - \frac{x^2}{\alpha^2} - \frac{y^2}{\beta^2} = 0$$

$$\text{or,}\quad \left(\frac{1}{a^2} - \frac{1}{\alpha^2}\right)x^2 + \left(\frac{1}{b^2} - \frac{1}{\beta^2}\right)y^2 = 0.$$

The above second degree homogeneous equation in x, y represents a pair of straight lines passing through the origin, i.e. centre of the ellipse. If $y = mx$ and $y = m'x$ be the individual equations of the above pair of straight lines, then

$$mm' = \frac{\left(\frac{1}{a^2} - \frac{1}{\alpha^2}\right)}{\left(\frac{1}{b^2} - \frac{1}{\beta^2}\right)}. \tag{1}$$

Again, if the above pair of straight lines represents a pair of conjugate diameters of the ellipse $\frac{x^2}{a^2}+\frac{y^2}{b^2}=1$, then

$$mm' = -\frac{b^2}{a^2}. \tag{2}$$

From Eqns. (1) and (2) we get

$$\frac{\left(\frac{1}{a^2}-\frac{1}{\alpha^2}\right)}{\left(\frac{1}{b^2}-\frac{1}{\beta^2}\right)} = -\frac{b^2}{a^2}$$

$$\text{or,}\quad \left(1-\frac{a^2}{\alpha^2}\right) = -\left(1-\frac{b^2}{\beta^2}\right)$$

$$\text{or,}\quad \frac{a^2}{\alpha^2}+\frac{b^2}{\beta^2} = 2.$$

▶▷▶ **13.** *Show that the lines joining origin and points of intersection of the line* $y = mx + \sqrt{\frac{a^2m^2+b^2}{2}}$ *with the ellipse* $\frac{x^2}{a^2}+\frac{y^2}{b^2}=1$ *are the conjugate diameters.*

Solution: Making homogeneous the equation $\frac{x^2}{a^2}+\frac{y^2}{b^2}=1$ with the aid of $y = mx + \sqrt{\frac{a^2m^2+b^2}{2}}$, we get

$$\frac{x^2}{a^2}+\frac{y^2}{b^2} = \frac{2(y-mx)^2}{(a^2m^2+b^2)}$$

$$\text{or,}\quad \left(m^2+\frac{b^2}{a^2}-2m^2\right)x^2 + 4mxy + \left(1+\frac{a^2m^2}{b^2}-2\right)y^2 = 0.$$

If $y = Mx$ and $y = M'x$ be the individual equations of the above pair of straight lines, then

$$MM' = \left(\frac{b^2}{a^2}-m^2\right)\Big/\left(\frac{a^2m^2}{b^2}-1\right) = -\frac{b^2}{a^2}.$$

Thus the lines joining origin and points of intersection of the line $y = mx + \sqrt{\frac{a^2m^2+b^2}{2}}$ with the ellipse $\frac{x^2}{a^2}+\frac{y^2}{b^2}=1$ are the conjugate diameters.

▶▷▶ **14.** *Any tangent to the ellipse* $\frac{x^2}{a^2}+\frac{y^2}{b^2}=1$ *whose centre is C meets its director circle at P, Q. Show that CP and CQ are the conjugate diameters of the ellipse.*

Solution: We know that the line $y = mx + \sqrt{a^2m^2+b^2}$ touches the ellipse $\frac{x^2}{a^2}+\frac{y^2}{b^2}=1$ for all values of m.

Equation of the director circle of the ellipse $\frac{x^2}{a^2} + \frac{y^2}{b^2} = 1$ is

$$x^2 + y^2 = a^2 + b^2.$$

Making homogeneous the equation of the circle with the aid of the equation of the tangent, we get

$$x^2 + y^2 = (a^2 + b^2)\frac{(y - mx)^2}{(a^2m^2 + b^2)}$$

$$\text{or,} \quad (a^2m^2 + b^2)(x^2 + y^2) = (a^2 + b^2)(y^2 - 2mxy + m^2x^2)$$

$$\text{or,} \quad b^2(1 - m^2)x^2 - 2(a^2 + b^2)mxy - a^2(1 - m^2)y^2 = 0.$$

If $y = Mx$ and $y = M'x$ be the individual equations of the above pair of straight lines, then

$$MM' = -\frac{b^2}{a^2}.$$

Hence, the line joining the centre of the ellipse to the points of intersection of the tangent to the ellipse with its director circle are the conjugate diameters of the ellipse.

▶▷▶ **15.** *Show that the tangents at the extremities of conjugate diameters of the ellipse* $\frac{x^2}{a^2} + \frac{y^2}{b^2} = 1$ *intersect on the ellipse* $\frac{x^2}{a^2} + \frac{y^2}{b^2} = 2$.

Solution: If CP, CD be the semi-conjugate diameters of the ellipse $\frac{x^2}{a^2} + \frac{y^2}{b^2} = 1$, then equation of the tangents at P, D are

$$\frac{x\cos\theta}{a} + \frac{y\sin\theta}{b} = 1$$

$$\text{and} \quad \frac{x\cos(\pi/2 + \theta)}{a} + \frac{y\sin(\pi/2 + \theta)}{b} = 1$$

$$\text{or,} \quad -\frac{x\sin\theta}{a} + \frac{y\cos\theta}{b} = 1.$$

To get the locus of the point of intersection of the tangents at P, D, we eliminate θ from the above equations by squaring and adding and get

$$\left(\frac{x\cos\theta}{a} + \frac{y\sin\theta}{b}\right)^2 + \left(-\frac{x\sin\theta}{a} + \frac{y\cos\theta}{b}\right)^2 = 1 + 1 = 2$$

$$\text{or,} \quad \frac{x^2}{a^2} + \frac{y^2}{b^2} = 2.$$

▶▷▶ **16.** *If P, D be the end points of conjugate semi-diameters of ellipse, then show that the locus of the foot of the perpendicular from centre on the line PD is*

$$2(x^2 + y^2)^2 = a^2x^2 + b^2y^2.$$

Solution: Equation of the line PD is

$$\frac{x\cos(\pi/4 + \theta)}{a} + \frac{y\sin(\pi/4 + \theta)}{b} = \frac{1}{\sqrt{2}}.$$

Equation of the line through (0, 0) and perpendicular to PD is

$$\frac{x\sin(\pi/4+\theta)}{b}-\frac{y\cos(\pi/4+\theta)}{a}=0$$

$$\text{or,}\quad \frac{\sin(\pi/4+\theta)}{by}=\frac{\cos(\pi/4+\theta)}{ax}=\frac{1}{\sqrt{a^2x^2+b^2y^2}}.$$

To get the locus of the foot of the perpendicular, we eliminate θ from the above equations and get

$$\frac{x^2}{\sqrt{a^2x^2+b^2y^2}}+\frac{y^2}{\sqrt{a^2x^2+b^2y^2}}=\frac{1}{\sqrt{2}}$$

$$\text{or,}\quad 2(x^2+y^2)^2=a^2x^2+b^2y^2.$$

▶▷▶ **17.** *If P, D be the end points of conjugate semi-diameters of ellipse, then find the locus of the point of intersection of the circles on CP and CD as diameters.*

Solution: Equation of the circle describing CP as diameter is

$$x(x-a\cos\theta)+y(y-b\sin\theta)=0$$

$$\text{or,}\quad x^2+y^2=ax\cos\theta+yb\sin\theta.$$

Similarly, equation of the circle describing CD as diameter is

$$x^2+y^2=-ax\sin\theta+yb\cos\theta.$$

To get the locus of the point of intersection of the above circles, we eliminate θ from the above equations by squaring and adding, and get

$$2(x^2+y^2)^2=a^2x^2+b^2y^2.$$

▶▷▶ **18.** *Find the locus of the middle points of the chords of the conic*

$$\frac{x^2}{a^2+\lambda}+\frac{y^2}{b^2+\lambda}=1$$

which touches the ellipse $\frac{x^2}{a^2}+\frac{y^2}{b^2}=1.$

Solution: Equation of the chord of the ellipse $\frac{x^2}{a^2+\lambda}+\frac{y^2}{b^2+\lambda}=1$ whose middle point is (h,k), is

$$\frac{xh}{a^2+\lambda}+\frac{yk}{b^2+\lambda}=\frac{h^2}{a^2+\lambda}+\frac{k^2}{b^2+\lambda}$$

$$\text{or,}\quad y=-\frac{hx(b^2+\lambda)}{k(a^2+\lambda)}+\frac{h^2(b^2+\lambda)}{k(a^2+\lambda)}+k.$$

If it touches the ellipse $\frac{x^2}{a^2}+\frac{y^2}{b^2}=1$, then we have

$$\frac{h^2(b^2+\lambda)}{k(a^2+\lambda)}+k=\sqrt{\frac{a^2h^2(b^2+\lambda)^2}{k^2(a^2+\lambda)^2}+b^2}$$

$$\text{or,}\quad \frac{a^2h^2(b^2+\lambda)^2}{k^2(a^2+\lambda)^2}+b^2=\left(\frac{h^2(b^2+\lambda)}{k(a^2+\lambda)}+k\right)^2$$

$$\text{or,}\quad \frac{a^2h^2}{(a^2+\lambda)^2}+\frac{b^2k^2}{(b^2+\lambda)^2}=\left(\frac{h^2}{a^2+\lambda}+\frac{k^2}{b^2+\lambda}\right)^2.$$

Hence, locus of the point (h, k) is

$$\frac{a^2x^2}{(a^2+\lambda)^2}+\frac{b^2y^2}{(b^2+\lambda)^2}=\left(\frac{x^2}{a^2+\lambda}+\frac{y^2}{b^2+\lambda}\right)^2.$$

19. ***A tangent to the parabola $y^2 = -4bx$ meets the parabola $y^2 = 4ax$ in Q, R. Find the locus of the middle point of QR.***

***Solution*: If $P(h, k)$ be the middle point of QR, then equation of QR is**

$$yk - 2ax = k^2 - 2ah$$

$$\text{or,}\quad y = \frac{2a}{k}x + \frac{k^2-2ah}{k}.$$

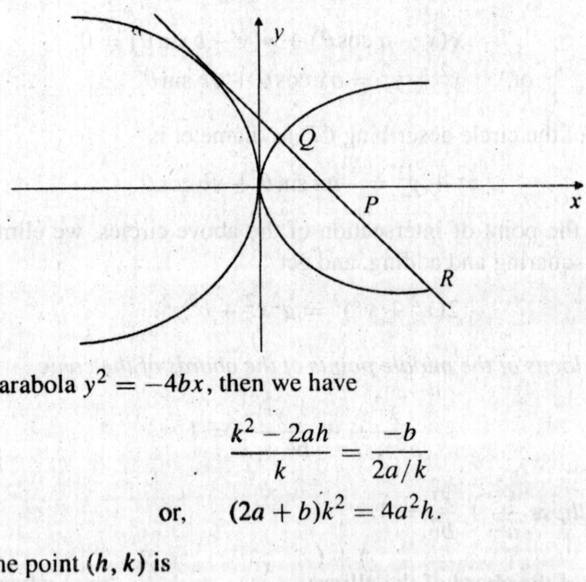

If it touches the parabola $y^2 = -4bx$, then we have

$$\frac{k^2-2ah}{k} = \frac{-b}{2a/k}$$

$$\text{or,}\quad (2a+b)k^2 = 4a^2h.$$

Hence, locus of the point (h, k) is

$$(2a+b)y^2 = 4a^2x.$$

20. *A variable chord of the hyperbola $\frac{x^2}{a^2}-\frac{y^2}{b^2}=1$ is a tangent to the circle $x^2+y^2 = c^2$. Prove that the locus of the mid-point of the chord is the curve*

$$\left(\frac{x^2}{a^2}-\frac{y^2}{b^2}\right)^2 = c^2\left(\frac{x^2}{a^4}+\frac{y^2}{b^4}\right).$$

Solution: Equation of the chord of the hyperbola whose middle point is (h, k) is

$$\frac{xh}{a^2}-\frac{yk}{b^2}=\frac{h^2}{a^2}-\frac{k^2}{b^2}.$$

This will touch the circle if

$$\frac{\dfrac{h^2}{a^2}-\dfrac{k^2}{b^2}}{\sqrt{\dfrac{h^2}{a^4}+\dfrac{k^2}{b^4}}}=c.$$

Generalizing, we get the locus of the point (h, k) is

$$\left(\frac{x^2}{a^2}-\frac{y^2}{b^2}\right)^2=c^2\left(\frac{x^2}{a^4}+\frac{y^2}{b^4}\right).$$

▶▷▶ **21.** *Chords of the rectangular hyperbola $x^2-y^2=a^2$ touch the parabola $y^2=4ax$. Prove that the locus of the middle points is the curve $y^2(x-a)=x^3$.*

Solution: Let (h, k) be the middle point of a chord of the rectangular hyperbola $x^2-y^2=a^2$, then its equation is

$$xh-yk=h^2-k^2$$

$$\text{or,}\quad y=\frac{h}{k}x-\frac{h^2-k^2}{k}.$$

If it touches the parabola $y^2=4ax$, then we have

$$\frac{h^2-k^2}{k}=\frac{ak}{h}$$

$$\text{or,}\quad k^2(h-a)=h^3.$$

Generalizing, we get the locus of the point (h, k) is $y^2(x-a)=x^3$.

▶▷▶ **22.** *Find the locus of the mid points of the chords of $x^2+y^2=a^2$ subtending a right angle at the point (h, k).*

Solution: Let the chord PQ of the circle $x^2+y^2=a^2$ subtending right angle at the point $A(h, k)$. If $R(\alpha, \beta)$ be the middle point of the chord PQ, then its equation is

$$x\alpha+y\beta=\alpha^2+\beta^2=c^2 \quad \text{(say)}.$$

Now, equation of the circle describing the line segment joining $P(x_1, y_1)$ and $Q(x_2, y_2)$ as diameter is

$$(x-x_1)(x-x_2)+(y-y_1)(y-y_2)=0.$$

Since, the chord PQ subtends a right angle at A, the point will lie on the above circle and hence we get

$$(h-x_1)(h-x_2)+(k-y_1)(k-y_2)=0$$

$$\text{or,}\quad h^2+k^2-h(x_1+x_2)-k(y_1+y_2)+x_1x_2+y_1y_2=0. \tag{1}$$

Solving $x^2+y^2=a^2$ and $x\alpha+y\beta=c^2$ for x, we obtain

$$x^2+\frac{(c^2-x\alpha)^2}{\beta^2}=a^2$$

$$\text{or,}\quad (\alpha^2+\beta^2)x^2-2c^2x\alpha+c^4-a^2\beta^2=0$$

$$\text{or,}\quad x^2-2x\alpha+c^2-\left(\frac{a\beta}{c}\right)^2=0.$$

The roots of the above equation are the abscissa of P and Q. Hence, we have

$$x_1 + x_2 = 2\alpha \quad \text{and} \quad x_1x_2 = c^2 - \left(\frac{a\beta}{c}\right)^2. \tag{2}$$

Similarly, solving $x^2 + y^2 = a^2$ and $x\alpha + y\beta = c^2$ for y we get

$$y_1 + y_2 = 2\beta \quad \text{and} \quad y_1y_2 = c^2 - \left(\frac{a\alpha}{c}\right)^2. \tag{3}$$

Using the results (2) and (3) in (1), we get

$$h^2 + k^2 - 2\alpha h - 2\beta k + 2c^2 - a^2 = 0$$

$$\text{or,} \quad 2(\alpha^2 + \beta^2) - 2(\alpha h + \beta k) + h^2 + k^2 - a^2 = 0.$$

Hence, locus of the middle point of PQ is

$$2(x^2 + y^2) - 2(xh + yk) + h^2 + k^2 - a^2 = 0.$$

EXERCISE

1. Tangents are drawn to a circle from points which always lie on a straight line. Prove that the locus of the middle point of the chord of contact is another circle.
2. Show that the locus of the middle point of the chord of contact of tangents to a given circle drawn from points on a given circle is a circle.
3. Find the locus of the middle points of chords passing through a fixed point, the conic being (i) a parabola, (ii) an ellipse and (iii) a hyperbola.
4. A tangent to a parabola $y^2 = 4bx$ meets the parabola $y^2 = 4ax$ in P, Q. Show that the locus of the middle point of PQ is another parabola.

   [*Hints*: If (h, k) be the middle point of PQ, equation of PQ is

   $$yk - 2ax = k^2 - 2ah$$

   $$\text{or,} \quad y = \frac{2a}{k}x + \frac{k^2 - 2ah}{k}.$$

   If it touches the parabola $y^2 = 4bx$, then we have

   $$c = \frac{a}{m}$$

   $$\text{i.e.,} \quad \frac{k^2 - 2ah}{k} = \frac{bk}{2a}$$

   $$\text{or,} \quad 2a(k^2 - 2ah) = bk^2$$

   $$\text{or,} \quad (2a - b)k^2 = 4a^2h.$$

   Hence, locus of the point (h, k) is $(2a - b)y^2 = 4a^2x$, which is another parabola.]
5. A tangent to a parabola $y^2 = 4ax$ meets the circle $x^2 + y^2 = r^2$ in P, Q. Find the locus of the middle point of PQ.
6. Show that the locus of the middle points of the normal chords of the rectangular hyperbola $x^2 - y^2 = a^2$ is $(x^2 - y^2)^3 + 4a^2x^2y^2 = 0$.

7. Find the locus of the middle point of the chords of the parabola $y^2 = 4ax$, which subtend a right angle at the vertex.

[*Hints*: If (h, k) be the middle point of the chord PQ, then equation of PQ is

$$yk - 2ax = k^2 - 2ah.$$

Making homogeneous the equation $y^2 = 4ax$ with the aid of the equation of the chord, we get

$$y^2 = 4ax\left(\frac{yk - 2ax}{k^2 - 2ah}\right)$$

$$\text{or,} \quad (k^2 - 2ah)y^2 - 4akxy + 8a^2x^2 = 0.$$

The above second degree homogeneous equation in x, y represents a pair of straight lines passing through the origin, i.e. vertex of the parabola. If the lines are perpendicular, then we have

$$\text{Coeff. of } x^2 + \text{Coeff. of } y^2 = 0$$

$$\text{or,} \quad k^2 - 2ah + 8a^2 = 0$$

$$\text{or,} \quad k^2 = 2a(h - 4a).$$

Hence, locus of the point (h, k) is

$$y^2 = 2a(x - 4a).]$$

8. Prove that the locus of the middle point of the portion of the tangents to the ellipse $\frac{x^2}{a^2} + \frac{y^2}{b^2} = 1$ included between the axes is $\frac{a^2}{x^2} + \frac{b^2}{y^2} = 4$.

9. Find the locus of the middle point of the chords of ellipse $\frac{x^2}{a^2} + \frac{y^2}{b^2} = 1$, which subtend a right angle at the centre of the ellipse.

10. Find the locus of the middle points of the chords of the ellipse whose poles are on the auxiliary circle of the ellipse.

[*Hints*: If (h, k) be the pole, then its polar with respect to the ellipse

$$\frac{x^2}{a^2} + \frac{y^2}{b^2} = 1$$

is

$$\frac{xh}{a^2} + \frac{yk}{b^2} = 1.$$

Equation of the chord of $\frac{x^2}{a^2} + \frac{y^2}{b^2} = 1$ whose middle point is (x_0, y_0), is

$$\frac{xx_0}{a^2} + \frac{yy_0}{b^2} = \frac{x_0^2}{a^2} + \frac{y_0^2}{b^2}.$$

Since, both the equations represent same line, comparing we get

$$\frac{x_0}{h} = \frac{y_0}{k} = \frac{x_0^2}{a^2} + \frac{y_0^2}{b^2}.$$

$$\text{or,} \quad h = \frac{x_0}{\left(\frac{x_0^2}{a^2} + \frac{y_0^2}{b^2}\right)} \quad \text{and} \quad k = \frac{y_0}{\left(\frac{x_0^2}{a^2} + \frac{y_0^2}{b^2}\right)}.$$

Now since (h, k) lies on the circle

$$x^2 + y^2 = a^2$$

we get

$$\frac{x_0^2}{\left(\frac{x_0^2}{a^2} + \frac{y_0^2}{b^2}\right)^2} + \frac{y_0^2}{\left(\frac{x_0^2}{a^2} + \frac{y_0^2}{b^2}\right)^2} = a^2$$

or, $$x_0^2 + y_0^2 = a^2\left(\frac{x_0^2}{a^2} + \frac{y_0^2}{b^2}\right)^2.$$

Hence, locus of the point (x_0, y_0) is

$$x^2 + y^2 = a^2\left(\frac{x^2}{a^2} + \frac{y^2}{b^2}\right)^2.]$$

11. Perpendicular tangents are drawn to the ellipse $\frac{x^2}{a^2} + \frac{y^2}{b^2} = 1$. Find the locus of the middle point of chord of contact.

[*Hints*: Let the tangents are drawn from the point (h, k), then their joint equation is

$$\left(\frac{x^2}{a^2} + \frac{y^2}{b^2} - 1\right)\left(\frac{h^2}{a^2} + \frac{k^2}{b^2} - 1\right) = \left(\frac{hx}{a^2} + \frac{ky}{b^2} - 1\right)^2.$$

If they are at right angles, then we have

$$\text{Coeff. of } x^2 + \text{Coeff. of } y^2 = 0$$

or, $$h^2 + k^2 = a^2 + b^2. \tag{1}$$

Equation of the chord of contact of (h, k) w.r.t. the ellipse is

$$\frac{hx}{a^2} + \frac{ky}{b^2} = 1 \tag{2}$$

If (α, β) be the middle point of the chord, then its equation is

$$\frac{\alpha x}{a^2} + \frac{\beta y}{b^2} = \frac{\alpha^2}{a^2} + \frac{\beta^2}{b^2}. \tag{3}$$

Comparing Eqns. (2) and (3) we get

$$h = \frac{\alpha}{\frac{\alpha^2}{a^2} + \frac{\beta^2}{b^2}} \quad \text{and} \quad k = \frac{\beta}{\frac{\alpha^2}{a^2} + \frac{\beta^2}{b^2}}.$$

Using these results in Eqn. (1) we get

$$\alpha^2 + \beta^2 = (a^2 + b^2)\left(\frac{\alpha^2}{a^2} + \frac{\beta^2}{b^2}\right)^2.$$

Hence, locus of the point (α, β) is

$$x^2 + y^2 = (a^2 + b^2)\left(\frac{x^2}{a^2} + \frac{y^2}{b^2}\right)^2.]$$

12. Find the locus of the middle points of the chords of constant length $2d$ of the ellipse $\frac{x^2}{a^2} + \frac{y^2}{b^2} = 1$.

[*Hints*: Let (h, k) be the middle point of the chord of the ellipse $\frac{x^2}{a^2}+\frac{y^2}{b^2}=1$ which is inclined to the x-axis at an angle θ, then coordinates of the extremities are $(h+d\cos\theta, k+d\sin\theta)$ and $(h-d\cos\theta, k-d\sin\theta)$ which satisfy the equation

$$\frac{x^2}{a^2}+\frac{y^2}{b^2}=1.$$

So,

$$\frac{(h+d\cos\theta)^2}{a^2}+\frac{(k+d\sin\theta)^2}{b^2}=1$$

and $$\frac{(h-d\cos\theta)^2}{a^2}+\frac{(k-d\sin\theta)^2}{b^2}=1.$$

Adding and subtracting we get two equations in θ. Eliminating θ from these equations we get the required locus.]

13. Prove that the locus of the middle points of all tangents drawn from points on the directrix of the parabola $y^2 = 4ax$ is $y^2(2x+a)=a(3x+a)^2$.

[*Hints*: Equation of the tangent at $(at^2, 2at)$ to the parabola $y^2 = 4ax$ is

$$yt = x + at^2.$$

It intersects the directrix of the parabola, i.e. $x = -a$ at $\left(-a, \frac{a(t^2-1)}{t}\right)$. If (h, k) be the middle point of the line joining $(at^2, 2at)$ and $\left(-a, \frac{a(t^2-1)}{t}\right)$, then

$$2h = a(t^2-1) \quad \text{and} \quad 2k = a\left(3t - \frac{1}{t}\right).$$

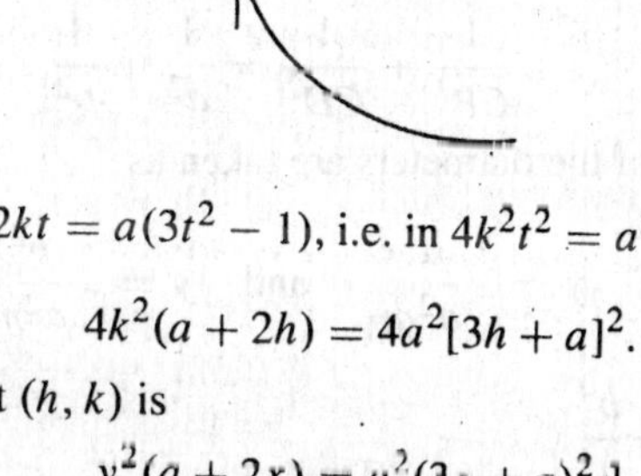

Putting $t^2 = 1 + \frac{2h}{a}$ in $2kt = a(3t^2-1)$, i.e. in $4k^2t^2 = a^2(3t^2-1)^2$, we get

$$4k^2(a+2h) = 4a^2[3h+a]^2.$$

Hence, locus of the point (h, k) is

$$y^2(a+2x) = a^2(3x+a)^2.]$$

14. If P, Q be the extremities of two conjugate diameters of the ellipse $\frac{x^2}{a^2}+\frac{y^2}{b^2}=1$, show that the locus of the middle point of the chord PQ is $\frac{x^2}{a^2}+\frac{y^2}{b^2}=\frac{1}{2}$ and also PQ touches the ellipse.

15. If (x_1, y_1) and (x_2, y_2) be the extremities of a pair of conjugate diameters of an ellipse $\frac{x^2}{a^2}+\frac{y^2}{b^2}=1$, prove that (a) $\frac{x_1x_2}{a^2}+\frac{y_1y_2}{b^2}=0$; (b) $x_1y_2-x_2y_1=\pm ab$.

16. The normal at a variable point P on the ellipse $\frac{x^2}{a^2}+\frac{y^2}{b^2}=1$ meets the diameter CD conjugate to CP at Q. Show that the locus of the point Q is

$$\frac{a^2}{x^2}+\frac{b^2}{y^2}=\left(\frac{a^2-b^2}{x^2+y^2}\right)^2.$$

[*Hints*: Equation of the normal at $P(a\cos\theta, b\sin\theta)$ is

$$ax\sec\theta-by\operatorname{cosec}\theta=a^2-b^2.$$

Equation of CD which is conjugate to diameter CP, is

$$bx\cos\theta+ay\sin\theta=0.$$

To get the point of intersection of the above lines we eliminate θ from the above equations as follows:

From the equation of CD: $\frac{\cos\theta}{ay}=-\frac{\sin\theta}{bx}=\frac{1}{\sqrt{a^2y^2+b^2x^2}}$.

Using these results in the equation of the normal at P, we get

$$ax\sec\theta-by\operatorname{cosec}\theta=a^2-b^2$$

or, $$\frac{x}{y}\sqrt{a^2y^2+b^2x^2}+\frac{y}{x}\sqrt{a^2y^2+b^2x^2}=a^2-b^2$$

or, $$\frac{(x^2+y^2)}{xy}\sqrt{a^2y^2+b^2x^2}=a^2-b^2$$

or, $$\frac{a^2}{x^2}+\frac{b^2}{y^2}=\left(\frac{a^2-b^2}{x^2+y^2}\right)^2.]$$

17. If PCP' and DCD' be two mutually perpendicular diameters of the ellipse $\frac{x^2}{a^2}+\frac{y^2}{b^2}=1$, prove that

$$\frac{1}{CP^2}+\frac{1}{CD^2}=\frac{1}{a^2}+\frac{1}{b^2}.$$

[*Hints*: If the equations of the diameters are taken as

$$y=-\frac{b^2x}{a^2m_1} \quad\text{and}\quad y=-\frac{b^2x}{a^2m_2}$$

then we have $m_1m_2=-\frac{b^4}{a^4}$.

Solving $y=-\frac{b^2x}{a^2m_1}$ and $\frac{x^2}{a^2}+\frac{y^2}{b^2}=1$, we get

$$x=\frac{a^2m_1}{\sqrt{a^2m_1^2+b^2}} \quad\text{and}\quad y=-\frac{b^2}{\sqrt{a^2m_1^2+b^2}}.$$

Similarly, the line $y = -\dfrac{b^2x}{a^2m_2}$ intersects the ellipse $\dfrac{x^2}{a^2} + \dfrac{y^2}{b^2} = 1$ at

$$\left(\frac{a^2m_2}{\sqrt{a^2m_2^2 + b^2}}, -\frac{b^2}{\sqrt{a^2m_2^2 + b^2}}\right).$$

Now,

$$\frac{1}{CP^2} + \frac{1}{CD^2} = \frac{a^2m_1^2 + b^2}{a^4m_1^2 + b^4} + \frac{a^2m_2^2 + b^2}{a^4m_2^2 + b^4}.$$

Now, use $m_2 = -\dfrac{b^4}{m_1a^4}$, etc.]

18. Lines are drawn through the foci of an ellipse perpendicular respectively to a pair of conjugate diameters and intersect at Q. Show that locus of Q is a concentric ellipse.
19. Find the locus of the foot of the perpendicular drawn from the centre of the hyperbola $\dfrac{x^2}{a^2} - \dfrac{y^2}{b^2} = 1$ on any tangent to it.
20. Find the locus of the foot of the perpendicular drawn from the centre of the hyperbola $\dfrac{x^2}{a^2} - \dfrac{y^2}{b^2} = 1$ on any normal to it.
21. CP and CP' are conjugate semi-diameters of the ellipse $\dfrac{x^2}{a^2} + \dfrac{y^2}{b^2} = 1$. If the normals at P and P' of the ellipse intersect the major axis at G and G' respectively, prove that

$$PG^2 + P'G'^2 = a^2(1 - e^2)(2 - e^2).$$

22. The tangents at the extremities of a pair of semi-conjugate diameters of the ellipse $\dfrac{x^2}{a^2} + \dfrac{y^2}{b^2} = 1$ meet at (h, k). Find the locus of the point (h, k).

[*Hints*: Equation of the chord of contact of the point (h, k) w.r.t. the ellipse $\dfrac{x^2}{a^2} + \dfrac{y^2}{b^2} = 1$ is

$$\frac{xh}{a^2} + \frac{yk}{b^2} = 1. \tag{1}$$

If $P(a\cos\theta, b\sin\theta)$ and $D(-a\sin\theta, b\cos\theta)$ be the end points of the semi-conjugate diameters, then equation of PD is

$$bx(\cos\theta - \sin\theta) + ay(\sin\theta + \cos\theta) = ab. \tag{2}$$

Comparing Eqns. (1) and (2) we get

$$\frac{h/a}{\cos\theta - \sin\theta} = \frac{k/b}{\cos\theta + \sin\theta} = 1$$

$$\text{or,} \quad \frac{h^2/a^2 + k^2/b^2}{(\cos\theta - \sin\theta)^2 + (\cos\theta + \sin\theta)^2} = 1$$

$$\text{or,} \quad \frac{h^2}{a^2} + \frac{k^2}{b^2} = 2.$$

Hence, locus of the point (h, k) is $\dfrac{x^2}{a^2} + \dfrac{y^2}{b^2} = 2$.]

23. A normal to the hyperbola $\frac{x^2}{a^2} - \frac{y^2}{b^2} = 1$ meets the axes in Q and R. The lines QL and RL are drawn perpendicular to the axes and meet at L. Show that locus of L is

$$a^2x^2 - b^2y^2 = (a^2 + b^2)^2.$$

[*Hints*: Equation of the normal at $P(a \sec\theta, b\tan\theta)$ is

$$ax\cos\theta + by\cot\theta = a^2 + b^2.$$

It meets the x-axis and y-axis at

$$Q: \left(\frac{a^2+b^2}{a}\sec\theta, 0\right) \quad \text{and} \quad R: \left(0, \frac{a^2+b^2}{b}\tan\theta\right)$$

so the lines QL and RL meet at the point L whose coordinates are

$$\left(\frac{a^2+b^2}{a}\sec\theta, \frac{a^2+b^2}{a}\tan\theta\right) = (h, k) \text{ (say)}.$$

Eliminating θ, we get

$$\frac{a^2h^2 - b^2k^2}{(a^2+b^2)^2} = \sec^2\theta - \tan^2\theta = 1.$$

Hence, locus of the point (h, k) is

$$a^2x^2 - b^2y^2 = (a^2 + b^2)^2.]$$

24. Prove that the locus of the middle points of the chords of contact of the tangents drawn from any point on the circle $x^2 + y^2 = c^2$ to the hyperbola $\frac{x^2}{a^2} - \frac{y^2}{b^2} = 1$ is $\left(\frac{x^2}{a^2} - \frac{y^2}{b^2}\right)^2 = \frac{x^2+y^2}{c^2}$.

25. Pairs of tangents are drawn to the conic $Ax^2 + By^2 = 1$, so as to be always parallel to conjugate diameters of the conic $ax^2 + 2hxy + by^2 = 1$; show that the locus of their point of intersection is

$$ax^2 + 2hxy + by^2 = \frac{a}{A} + \frac{b}{B}.$$

26. Find the locus of the middle points of chords of the (i) ellipse $\frac{x^2}{a^2} + \frac{y^2}{b^2} = 1$, (ii) hyperbola $\frac{x^2}{a^2} - \frac{y^2}{b^2} = 1$ which subtend a right angle at the centre.

27. If the tangents at the extremities of a pair of semi-conjugate diameters of the ellipse $\frac{x^2}{a^2} + \frac{y^2}{b^2} = 1$ meet at (h, k), find the locus of the point (h, k).

[*Hints*: Equations of the tangents at the extremities $(a\cos\theta, b\sin\theta)$ and $(-a\sin\theta, b\cos\theta)$ of the ellipse $\frac{x^2}{a^2} + \frac{y^2}{b^2} = 1$ are

$$\frac{x}{a}\cos\theta + \frac{y}{b}\sin\theta = 1$$

$$\text{and} \quad \frac{x}{a}\sin\theta - \frac{y}{b}\sin\theta = -1.$$

If the tangents meet at (h, k), then we have

$$\frac{h}{a}\cos\theta + \frac{k}{b}\sin\theta = 1$$

$$\text{and} \quad \frac{h}{a}\sin\theta - \frac{k}{b}\sin\theta = -1.$$

To eliminate the parameter θ from the above equations we square and add the results, and get

$$\frac{h^2}{a^2} + \frac{k^2}{b^2} = 2.$$

Generalizing, we get the locus of (h, k) is $\dfrac{x^2}{a^2} + \dfrac{y^2}{b^2} = 2$.]

28. If the chords of an ellipse pass through a fixed point, prove that locus of their middle points is another ellipse.

[*Hints*: Taking (h, k) as the middle point write the equation of the chord. If it passes through the fixed point (α, β), put it into the equation of the chord and then generalize it.]

29. Find the locus of the middle point of the chords of the ellipse $\dfrac{x^2}{a^2} + \dfrac{y^2}{b^2} = 1$, tangents at the extremities of which are at right angles.

[*Hints*: Equation of the director circle, i.e. locus of the perpendicular tangents to the ellipse $\dfrac{x^2}{a^2} + \dfrac{y^2}{b^2} = 1$ is $x^2 + y^2 = a^2 + b^2$. Let (h, k) be a point on it from which the perpendicular tangents are drawn to the ellipse. Then equation of chord of contact of (h, k) with respect to the ellipse is

$$\frac{xh}{a^2} + \frac{yk}{b^2} = 1. \tag{1}$$

Now equation of the chord of the ellipse, whose middle point is (α, β), is

$$\frac{x\alpha}{a^2} + \frac{y\beta}{b^2} = \frac{\alpha^2}{a^2} + \frac{\beta^2}{b^2} \tag{2}$$

which must be identical with Eqn. (1), so comparing we get

$$\frac{h}{\alpha} = \frac{k}{\beta} = \frac{1}{\dfrac{\alpha^2}{a^2} + \dfrac{\beta^2}{b^2}}$$

$$\text{or,} \quad \alpha = h\left(\frac{\alpha^2}{a^2} + \frac{\beta^2}{b^2}\right) \quad \text{and} \quad \beta = k\left(\frac{\alpha^2}{a^2} + \frac{\beta^2}{b^2}\right). \tag{3}$$

Squaring and adding the equations of (3) we get

$$(h^2 + k^2)\left(\frac{\alpha^2}{a^2} + \frac{\beta^2}{b^2}\right)^2 = \alpha^2 + \beta^2.$$

Hence, locus of (α, β) is $(a^2 + b^2)\left(\dfrac{x^2}{a^2} + \dfrac{y^2}{b^2}\right)^2 = x^2 + y^2$, as (h, k) lies on the director circle.]

30. Prove that the tangents to the ellipse $\frac{x^2}{a}+\frac{y^2}{b}=a+b$ at the points where it cut by the tangents of the ellipse $\frac{x^2}{a^2}+\frac{y^2}{b^2}=1$ are at right angle.

[*Hints*: Equation of the chord of contact of the point (h, k) with respect to the ellipse

$$\frac{x^2}{a}+\frac{y^2}{b}=a+b \tag{1}$$

is, $$\frac{xh}{a(a+b)}+\frac{yk}{b(a+b)}=1.$$

This will touch the ellipse $\frac{x^2}{a^2}+\frac{y^2}{b^2}=1$ if

$$a^2\frac{h^2}{a^2(a+b)^2}+b^2\frac{k^2}{b^2(a+b)^2}=1$$

or, $$h^2+k^2=(a+b)^2$$

which shows that (h, k) lies on the director circle of the ellipse (1). Hence the result.]

31. Find the locus of the middle points of the chords of the ellipse $\frac{x^2}{a^2}+\frac{y^2}{b^2}=1$ whose distance from centre is constant.

32. Prove that the locus of the middle points of the chords of contact of the tangents to the ellipse $\frac{x^2}{a^2}+\frac{y^2}{b^2}=1$ at right angles is the curve

$$\left(\frac{x^2}{a^2}+\frac{y^2}{b^2}\right)^2=\frac{x^2+y^2}{a^2+b^2}.$$

33. Prove that the locus of the middle points of the chords of the ellipse $\frac{x^2}{a^2}+\frac{y^2}{b^2}=1$ passing through a fixed point (h, k) is $\frac{x(x-h)}{a^2}+\frac{y(y-k)}{b^2}=0$.

34. Prove that the locus of the middle points of the normal chords of the ellipse $\frac{x^2}{a^2}+\frac{y^2}{b^2}=1$ is the curve $\left(\frac{x^2}{a^2}+\frac{y^2}{b^2}\right)^2\left(\frac{a^6}{x^2}+\frac{b^6}{y^2}\right)=(a^2-b^2)^2$.

35. A tangent to the parabola $x^2=4by$ meets the hyperbola $xy=c^2$ in two points P and Q. Prove that the locus of the middle point of PQ is a parabola.

[*Hints*: Let (h, k) be the middle point of PQ, then its equation is

$$xk+yh=2c^2.$$

If it touches the parabola $x^2=4by$, then the equation

$$x^2=\frac{4b(xk-2c^2)}{h}$$

i.e., $hx^2+4bkx-8bc^2=0$ will have same root

which gives

$$16b^2k^2 = -32hbc^2$$

$$\text{or,} \quad k^2 = -\frac{2hc^2}{b}.$$

Hence, locus of the point (h, k) is $y^2 = -\dfrac{2xc^2}{b}$.]

36. Find the locus of the middle point of the chords of the hyperbola $\dfrac{x^2}{a^2} - \dfrac{y^2}{b^2} = 1$, tangents at the extremities of which are at right angles.

[*Hints*: Equation of the director circle, i.e. locus of the perpendicular tangents to the hyperbola $\dfrac{x^2}{a^2} - \dfrac{y^2}{b^2} = 1$ is $x^2 + y^2 = a^2 - b^2$. Let (h, k) be a point on it from which the perpendicular tangents are drawn to the hyperbola. Then equation of chord of contact of (h, k) with respect to the hyperbola is

$$\frac{xh}{a^2} - \frac{yk}{b^2} = 1. \tag{1}$$

Now equation of the chord of the hyperbola, whose middle point is (α, β), is

$$\frac{x\alpha}{a^2} - \frac{y\beta}{b^2} = \frac{\alpha^2}{a^2} - \frac{\beta^2}{b^2} \tag{2}$$

which must be identical with Eqn. (1); so comparing we get

$$\frac{h}{\alpha} = \frac{k}{\beta} = \frac{1}{\dfrac{\alpha^2}{a^2} - \dfrac{\beta^2}{b^2}}$$

$$\text{or,} \quad \alpha = h\left(\frac{\alpha^2}{a^2} - \frac{\beta^2}{b^2}\right) \quad \text{and} \quad \beta = k\left(\frac{\alpha^2}{a^2} - \frac{\beta^2}{b^2}\right). \tag{3}$$

Squaring and adding the equations of (3) we get

$$(h^2 + k^2)\left(\frac{\alpha^2}{a^2} - \frac{\beta^2}{b^2}\right)^2 = \alpha^2 + \beta^2.$$

Hence, locus of (α, β) is $(a^2 - b^2)\left(\dfrac{x^2}{a^2} - \dfrac{y^2}{b^2}\right)^2 = x^2 + y^2$, as (h, k) lies on the director circle.]

37. Find the locus of the middle point of the chords of the parabola $y^2 = 4ax$, tangents at the extremities of which are at right angles.

[*Hints*: We know that the perpendicular tangents to the parabola $y^2 = 4ax$ intersects on its directrix. So, if the tangents to the parabola be drawn from the point (h, k), then we have

$$h + a = 0. \tag{1}$$

Equation of chord of contact of the point (h, k) with respect to the parabola $y^2 = 4ax$ is

$$yk - 2ax = 2ah. \tag{2}$$

If (α, β) be the middle point of a chord, then its equation is

$$y\beta - 2ax = \beta^2 - 2a\alpha \tag{3}$$

which must be identical with Eqn. (2). So comparing Eqns. (2) and (3) we get

$$\beta^2 - 2a\alpha = 2ah = -2a^2. \qquad \text{[By Eqn. (1)]}$$

Hence, locus of the middle point is $y^2 - 2ax + 2a^2 = 0$.]

38. The polar of a point P, lies on the line $lx + my + n = 0$, with respect to a parabola $y^2 = 4ax$ meets the parabola in two points Q, R. Show that locus of the middle point of QR is

$$l(y^2 - 4ax) + 2a(lx + my + n) = 0.$$

[*Hints*: Polar of the point $P(h, k)$ w.r.t. $y^2 = 4ax$ is

$$yk = 2a(x + h). \tag{1}$$

As P lies on the given line, we have

$$lh + mk + n = 0. \tag{2}$$

If (α, β) be the middle point of QR then equation of QR is

$$y\beta - 2ax = \beta^2 - 2a\alpha. \tag{3}$$

Comparing Eqns. (1) and (3) we get

$$\frac{k}{\beta} = 1 = \frac{2ah}{\beta^2 - 2a\alpha}. \tag{4}$$

Eliminating h, k from Eqn. (2) using Eqn. (4) we get the required result.]

Chapter 9

Polar Coordinate System

9.1 Introduction

Like Cartesian coordinate system, the position of a point in a plane is determined by two quantities; one is distance measured from a fixed point in the plane, called *pole* and other is an angle which is measured in counter-clockwise direction from a fixed line which passes through the pole, called *initial line*. If we take O as pole and the line OX as the initial line then coordinate of a point in the plane is given by (r, θ), where $OP = r$ is called *radius vector* and $\angle POX = \theta$ is called *vectorial angle*. The coordinates (r, θ) are called *polar coordinate* of P. The radius vector is positive, if it be measured from the pole along the line bounding the vectorial angle and it is negative, if measured in the opposite direction. Vectorial angle is assumed positive, if it is measured in the counter-clockwise direction which are described below.

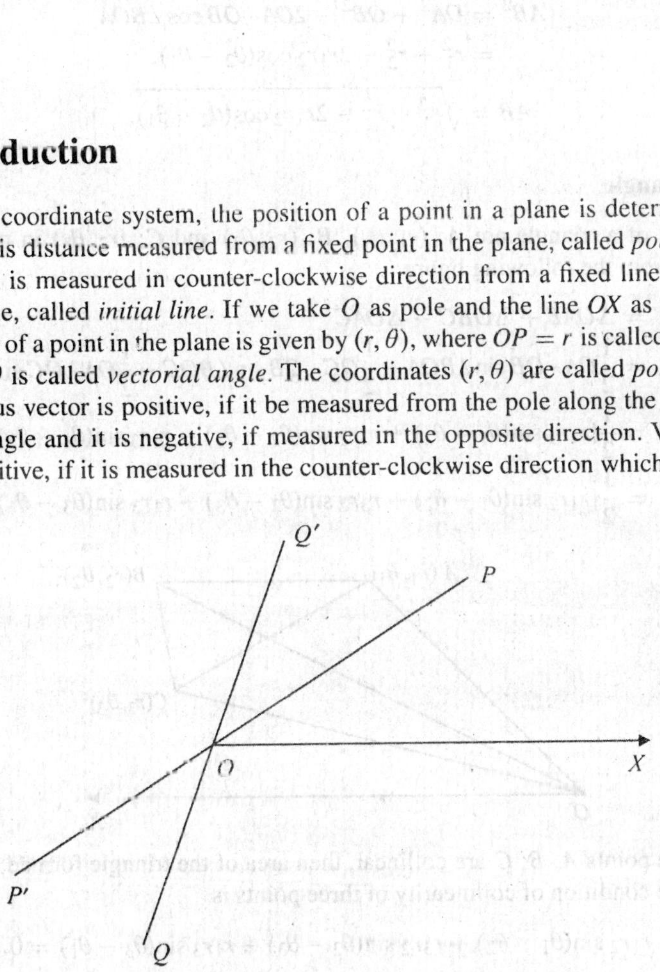

If we take the coordinates of P and Q as (r, θ) and $(R, -\phi)$ respectively, then coordinates of P' and Q' will be $(-r, \theta)$ and $(-R, -\phi)$ respectively.

The polar coordinate of any point (r, θ) may also be expressed as

$$(r, -2\pi + \theta); \quad (-r, \pi + \theta) \quad \text{and} \quad (-r, -\pi + \theta).$$

If we take the origin of Cartesian coordinate system as pole and x-axis of that as the initial line, then we get

$$x = r\cos\theta,\ y = r\sin\theta \text{ and conversely, } r = \sqrt{x^2+y^2},\ \theta = \tan^{-1}\left(\frac{y}{x}\right).$$

9.2 Distance between Two Points (r_1, θ_1) and (r_2, θ_2)

If the coordinates of two points A, B in polar coordinate system are (r_1, θ_1) and (r_2, θ_2), then $OA = r_1$, $OB = r_2$, $\angle AOX = \theta_1$ and $\angle BOX = \theta_2$. So, $\angle BOA = \theta_2 - \theta_1$.

Now

$$\begin{aligned} AB^2 &= OA^2 + OB^2 - 2OA \cdot OB \cos\angle BOA \\ &= r_1^2 + r_2^2 - 2r_1r_2\cos(\theta_2 - \theta_1). \end{aligned}$$

So,

$$AB = \sqrt{r_1^2 + r_2^2 - 2r_1r_2\cos(\theta_2 - \theta_1)}.$$

Area of a Triangle

If the vertices of a triangle are A: (r_1, θ_1), B: (r_2, θ_2) and C: (r_3, θ_3) in polar coordinate system, then from the following figure

$$\begin{aligned} \Delta ABC &= \Delta OAB + \Delta OBC - \Delta OAC \\ &= \frac{1}{2}OA \cdot OB\sin\angle BOA + \frac{1}{2}OC \cdot OB\sin\angle BOC - \frac{1}{2}OA \cdot OC\sin\angle COA \\ &= \frac{1}{2}[r_1r_2\sin(\theta_1 - \theta_2) + r_3r_2\sin(\theta_2 - \theta_3) - r_1r_3\sin(\theta_1 - \theta_3)] \\ &= \frac{1}{2}[r_1r_2\sin(\theta_1 - \theta_2) + r_3r_2\sin(\theta_2 - \theta_3) + r_1r_3\sin(\theta_3 - \theta_1)]. \end{aligned}$$

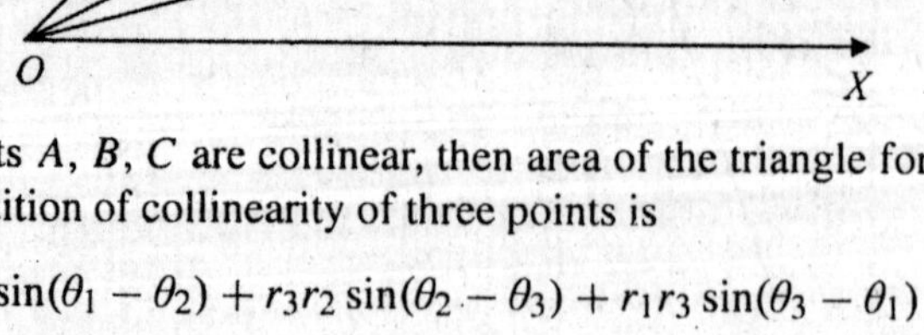

Now, if the points A, B, C are collinear, then area of the triangle formed by these points is zero. So, the condition of collinearity of three points is

$$r_1r_2\sin(\theta_1 - \theta_2) + r_3r_2\sin(\theta_2 - \theta_3) + r_1r_3\sin(\theta_3 - \theta_1) = 0. \tag{1}$$

Equation of a Straight Line

Take O as pole and the line OX as the initial line. If p be the length of the perpendicular from the pole on a straight line and ON, N being the foot of the perpendicular, makes an angle α with the initial line, then coordinate of N is (p, α). Let $Q(r, \theta)$ be any point on the straight line; then $\angle XOQ = \theta$ and $\angle QON = \theta - \alpha$.

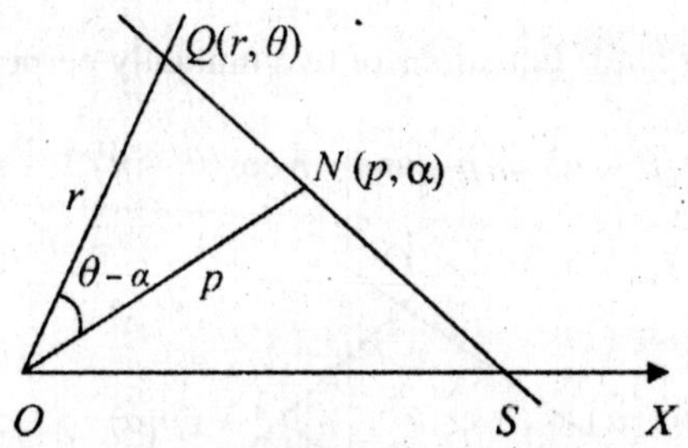

From the above figure

$$\cos(\theta - \alpha) = \frac{p}{r}$$

$$\text{or,} \quad r\cos(\theta - \alpha) = p.$$

This is the polar equation of the straight line QS.

This can also be written as

$$\frac{p}{r} = \cos\alpha\cos\theta + \sin\alpha\sin\theta$$

$$\text{or,} \quad \frac{1}{r} = \frac{\cos\alpha}{p}\cos\theta + \frac{\sin\alpha}{p}\sin\theta$$

$$\text{or,} \quad \frac{1}{r} = A\cos\theta + B\sin\theta.$$

Special Cases

(a) *Straight line parallel to initial line*: Putting $\alpha = \pi/2$ in the above equation, we obtain

$$p = r\sin\theta.$$

(b) *Straight line perpendicular to initial line*: Putting $\alpha = 0$ in the above equation, we obtain

$$p = r\cos\theta.$$

(c) *Straight line passing through the pole*: Putting $p = 0$ in the above equation, we obtain

$$\cos(\theta - \alpha) = 0$$

$$\text{or,} \quad \theta - \alpha = \pi/2$$

$$\text{or,} \quad \theta = \text{constant}.$$

(d) *Parallel straight lines*: As distance between two parallel straight lines is constant and which is the difference of their distances measured from origin, i.e. pole, their equations can be written as

$$r\cos(\theta - \alpha) = p \quad \text{and} \quad r\cos(\theta - \alpha) = p'.$$

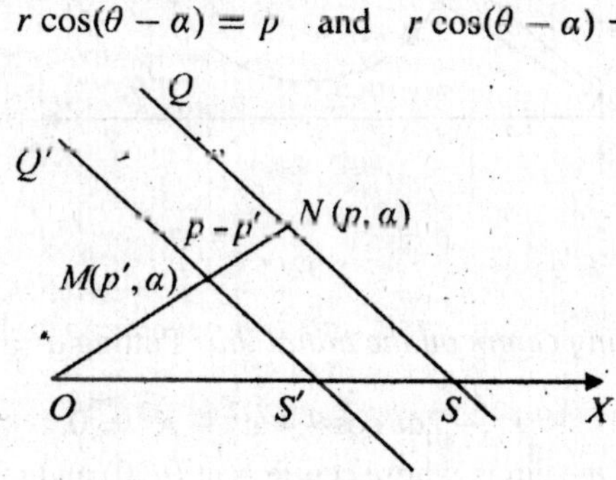

(e) *Perpendicular straight lines*: Equations of two mutually perpendicular straight lines can be written as

$$r\cos(\theta - \alpha) = p \quad \text{and} \quad r\cos(\theta - \pi/2 - \alpha) = p'.$$

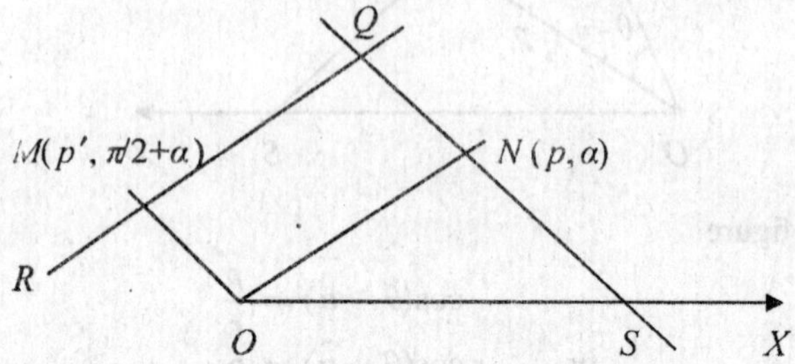

(f) *Equation of a line which passes through* (r_1, θ_1) *and* (r_2, θ_2): If (r, θ) be any point on the line, then the points (r_1, θ_1), (r_2, θ_2) and (r, θ) are collinear. Using the condition of collinearity given by Eqn. (1) we get

$$r_1 r_2 \sin(\theta_1 - \theta_2) + r r_2 \sin(\theta_2 - \theta) + r_1 r \sin(\theta - \theta_1) = 0$$

$$\text{or,} \quad \frac{1}{r}\sin(\theta_1 - \theta_2) + \frac{1}{r_1}\sin(\theta_2 - \theta) + \frac{1}{r_2}\sin(\theta - \theta_1) = 0.$$

This is the equation of the line which passes through (r_1, θ_1) and (r_2, θ_2).

Equation of a Circle

Take O as pole and the line OX as the initial line. If (a, α) be the centre of a circle of radius R, then from the following figure

$$CQ^2 = OC^2 + OQ^2 - 2OC \cdot OQ \cos \angle COQ$$

$$\text{or,} \quad R^2 = a^2 + r^2 - 2ar\cos(\theta - \alpha)$$

$$\text{or,} \quad r^2 - 2ar\cos(\theta - \alpha) + a^2 - R^2 = 0. \tag{2}$$

This is the equation of the circle whose centre is the point (a, α) and radius R.

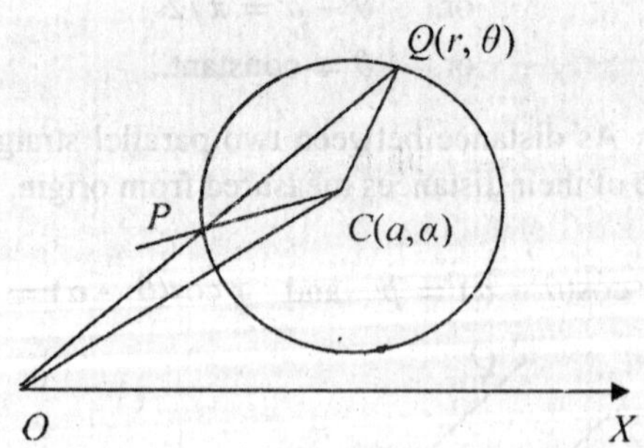

Special Cases

(a) *Equation of circle having centre on the initial line*: Putting $\alpha = 0$ in the above equation we get

$$r^2 - 2ar\cos\theta + a^2 - R^2 = 0.$$

This is the equation of the circle whose centre is at $(a, 0)$ and radius R.

(b) *Equation of circle passing through the pole*: Putting $a = R$ in Eqn. (2) we get

$$r^2 - 2ar\cos(\theta - \alpha) = 0, \quad \text{i.e. } r = 2a\cos(\theta - \alpha)$$

Further, if we put $\alpha = 0$ in the above equation, we get

$$r = 2a\cos\theta$$

which is the equation of the circle which passes through the pole and whose centre lies on initial line.

(c) *Equation of the circle whose centre is the pole*: Putting $a = 0$ in Eqn. (2) we get

$$r = R.$$

(d) *Equation of the circle which touches the initial line at the pole*: Putting $a = R$ and $\alpha = \pi/2$ in the Eqn. (2) we get

$$r^2 - 2ar\cos(\theta - \pi/2) = 0, \quad \text{i.e. } r = 2a\sin\theta.$$

Equation of a Conic

If we take focus, S of the conic as pole and axis of the conic, SX as the initial line, then for any point $P(r, \theta)$ on the conic we have

$$SP = ePM$$

where e is the eccentricity and PM is the length of the perpendicular from P on the directrix ZM.

From the following figure, we see that

$$\begin{aligned} & SP = eRZ \\ \text{or,}\quad & SP = e(SZ - SR) \\ \text{or,}\quad & SP = e(QT - SR) \\ \text{or,}\quad & SP = SQ - eSR \quad [\text{as } Q \text{ is a point on the conic}] \\ \text{or,}\quad & r = l - er\cos\theta \quad [\text{where latus rectum} = 2l] \\ \text{or,}\quad & \frac{l}{r} = 1 + e\cos\theta. \end{aligned}$$

This is the polar equation of the conic with the focus as pole and axis of the conic as initial line.

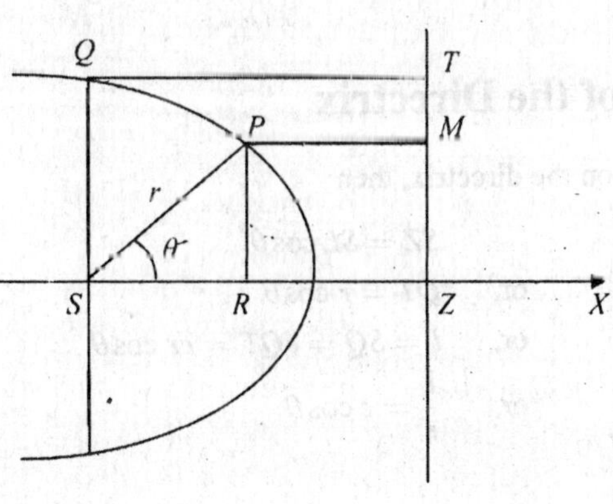

Special Cases

(a) If we take the initial line in the opposite direction, the above equation of conic becomes

$$\frac{l}{r} = 1 + e\cos(\pi - \theta)$$

$$\text{or,} \quad \frac{l}{r} = 1 - e\cos\theta.$$

(b) *When the axis of the conic is inclined to the initial line at an angle α*: From the following figure

$$SP = eRT$$

$$\text{or,} \quad SP = e(ST - SR)$$

$$\text{or,} \quad SP = e(QK - SR)$$

$$\text{or,} \quad SP = SQ - eSR \quad [\text{as } Q \text{ is a point on the conic}]$$

$$\text{or,} \quad r = l - er\cos(\theta - \alpha) \quad [\text{where latus rectum} = 2l]$$

$$\text{or,} \quad \frac{l}{r} = 1 + e\cos(\theta - \alpha).$$

This is the polar equation of the conic when the axis of the conic is inclined to the initial line at an angle α.

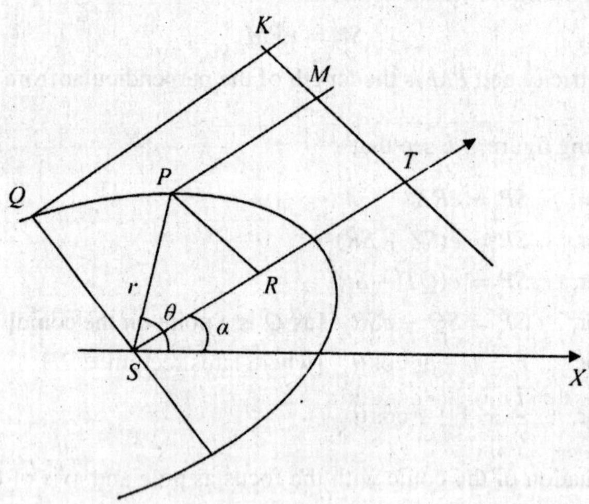

9.3 Equation of the Directrix

If $P(r, \theta)$ be any point on the directrix, then

$$SZ = SL\cos\theta$$

$$\text{or,} \quad QT = r\cos\theta$$

$$\text{or,} \quad l = SQ = eQT = er\cos\theta$$

$$\text{or,} \quad \frac{l}{r} = e\cos\theta.$$

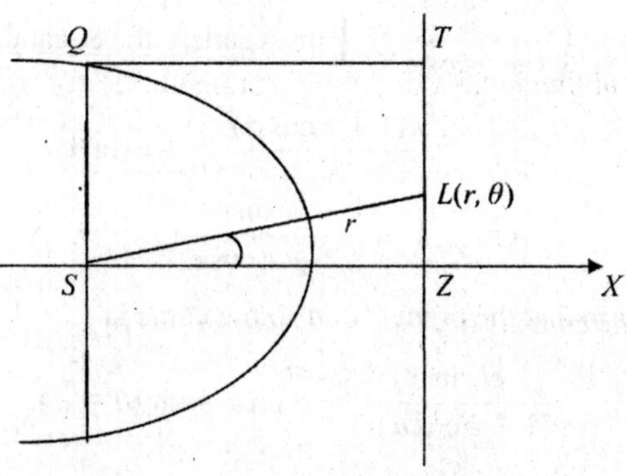

9.4 Equation of Chord, Tangent and Normal

Let the chord $\frac{l}{r} = a\cos\theta + b\sin\theta$ intersects the conic $\frac{l}{r} = 1 + e\cos\theta$ at two points whose vectorial angles are $\alpha + \beta$ and $\alpha - \beta$. For these two points we have

$$1 + e\cos(\alpha+\beta) = a\cos(\alpha+\beta) + b\sin(\alpha+\beta)$$

and $$1 + e\cos(\alpha-\beta) = a\cos(\alpha-\beta) + b\sin(\alpha-\beta)$$

or, $$(a-e)\cos(\alpha+\beta) + b\sin(\alpha+\beta) - 1 = 0$$

and $$(a-e)\cos(\alpha-\beta) + b\sin(\alpha-\beta) - 1 = 0.$$

Solving we get

$$\frac{a-e}{\sin(\alpha-\beta) - \sin(\alpha+\beta)} = \frac{b}{\cos(\alpha+\beta) - \cos(\alpha-\beta)}$$
$$= \frac{1}{\cos(\alpha+\beta)\sin(\alpha-\beta) - \cos(\alpha-\beta)\sin(\alpha+\beta)}$$

or, $$\frac{a-e}{-2\cos\alpha\sin\beta} = \frac{b}{-2\sin\alpha\sin\beta} = \frac{1}{-2\cos\beta\sin\beta}$$

or, $$a - e = \cos\alpha\sec\beta \quad \text{and} \quad b = \sin\alpha\sec\beta.$$

Putting the values of a and b in the equation of chord, we get

$$\frac{l}{r} = (e + \cos\alpha\sec\beta)\cos\theta + \sin\alpha\sec\beta\sin\theta$$

or, $$\frac{l}{r} = e\cos\theta + \sec\beta\cos(\theta-\alpha).$$

If we put $\beta = 0$, which means the chord will touch the conic at the point $\theta = \alpha$, then the above equation becomes

$$\frac{l}{r} = e\cos\theta + \cos(\theta-\alpha)$$

which is the equation of the tangent to the conic at the point $\theta = \alpha$.

As the normal to a conic is perpendicular to the tangent, we assume its equation as

$$\frac{k}{r} = e\cos(\theta + \pi/2) + \cos(\theta + \pi/2 - \alpha)$$

or, $$\frac{k}{r} = -e\sin\theta - \sin(\theta-\alpha).$$

The point of contact, i.e. $\left(\dfrac{l}{1+e\cos\alpha}, \alpha\right)$ must satisfy the equation of normal, which gives

$$\frac{k(1+e\cos\alpha)}{l} = -e\sin\alpha$$

$$\text{or,}\quad k = -\frac{el\sin\alpha}{1+e\cos\alpha}.$$

So, the equation of normal at the point $\theta = \alpha$ to the conic is

$$\frac{el\sin\alpha}{(1+e\cos\alpha)r} = e\sin\theta + \sin(\theta-\alpha).$$

9.5 Chord of Contact

Let the equation of the conic be

$$\frac{l}{r} = 1 + e\cos\theta.$$

If the tangents to the conic at the points $\theta = \alpha+\beta$ and $\theta = \alpha-\beta$ intersect at (r_1, θ_1), then both the tangents

$$\frac{l}{r} = e\cos\theta + \cos(\theta-\alpha-\beta)$$

$$\text{and}\quad \frac{l}{r} = e\cos\theta + \cos(\theta-\alpha+\beta)$$

will pass through (r_1, θ_1), which gives

$$\frac{l}{r_1} = e\cos\theta_1 + \cos(\theta_1-\alpha-\beta) \tag{3}$$

$$\text{and}\quad \frac{l}{r_1} = e\cos\theta_1 + \cos(\theta_1-\alpha+\beta). \tag{4}$$

Subtracting we get

$$\cos(\theta_1-\alpha-\beta) = \cos(\theta_1-\alpha+\beta)$$

$$\text{or,}\quad \theta_1-\alpha-\beta = \pm(\theta_1-\alpha+\beta)$$

$$\text{or,}\quad \theta_1 = \alpha \qquad [\text{as } \beta \neq 0].$$

So, from Eqn. (3) or (4), we get

$$\frac{l}{r_1} - e\cos\theta_1 = \cos\beta.$$

Now, equation of the chord joining two points whose vectorial angles are $\alpha+\beta$ and $\alpha-\beta$ is

$$\frac{l}{r} = e\cos\theta + \sec\beta\cos(\theta-\alpha).$$

Eliminating α, β from the equation of chord we get

$$\left(\frac{l}{r} - e\cos\theta\right)\left(\frac{l}{r_1} - e\cos\theta_1\right) = \cos(\theta-\theta_1).$$

This is the equation of chord of contact of the point (r_1, θ_1) with respect to the conic

$$\frac{l}{r} = 1 + e\cos\theta.$$

Pole and Polar

Let the pole be (r_1, θ_1) and equation of the conic be

$$\frac{l}{r} = 1 + e\cos\theta.$$

If the chord through (r_1, θ_1) cuts the conic at A and B, and the tangents at A and B meet at $Q(\rho, \phi)$, then locus of the point Q is called the polar of P with respect to the conic and AB is the chord of contact of the point Q with respect to the conic. Now equation of AB is

$$\left(\frac{l}{r} - e\cos\theta\right)\left(\frac{l}{\rho} - e\cos\phi\right) = \cos(\theta - \phi).$$

If it passes through the point (r_1, θ_1), then

$$\left(\frac{l}{\rho} - e\cos\phi\right)\left(\frac{l}{r_1} - e\cos\theta_1\right) = \cos(\phi - \theta_1).$$

So, the locus of the point $Q(\rho, \phi)$, i.e. equation of the polar of the point (r_1, θ_1) is

$$\left(\frac{l}{r} - e\cos\theta\right)\left(\frac{l}{r_1} - e\cos\theta_1\right) = \cos(\theta - \theta_1).$$

Examples

▶▶ 1. *Transform the following equations in polar coordinates,*

(a) $x^2 + y^2 = a^2$;
(b) $xy = c^2$;
(c) $(x^2 + y^2)^2 = a^2(x^2 - y^2)$;
(d) $(x^2 + y^2)^2 = ax^2y^2$.

Solution: Putting $x = r\cos\theta$ and $y = r\sin\theta$ in the above equation we get

(a) $r^2 = a^2$, i.e. $r = a$;
(b) $r^2 \sin\theta\cos\theta = c^2$ or, $r^2 \sin 2\theta = 2c^2$;
(c) $r^4 = a^2r^2(\cos^2\theta - \sin^2\theta)$ or, $r^2 = a^2\cos 2\theta$;
(d) $r^4 = ar^4\cos^2\theta\sin^2\theta$ or, $4 = a\sin^2 2\theta$.

▶▶ 2. *Transform the following equations in Cartesian coordinates:*

(a) $r = 2a\cos\theta$;
(b) $\frac{l}{r} = 1 + \cos\theta$;
(c) $\theta = \pi/4$;
(d) $r = a + b\cos\theta$.

Solution: Putting $r^2 = x^2 + y^2$ and $\theta = \tan^{-1}(y/x)$ in the above equations we get

(a) $r^2 = 2ar\cos\theta$ or, $x^2 + y^2 = 2ax$;
(b) $l = r + r\cos\theta$ or, $l = r + x$ or, $r^2 = (l - x)^2$ or, $x^2 + y^2 = l^2 - 2lx + x^2$ or, $y^2 = -2l(x - l/2)$;
(c) $\tan^{-1}(y/x) = \pi/4$ or, $y = x$;
(d) $r^2 = ar + br\cos\theta$ or, $x^2 + y^2 = ar + bx$ or, $(x^2 + y^2 - bx)^2 = a^2(x^2 + y^2)$.

▶▷▶ **3.** *Find the nature of the following conics:*

(a) $\dfrac{2}{r} = 3 - \cos\theta$;
(b) $\dfrac{3}{r} = 2 + 4\cos\theta$;
(c) $\dfrac{2}{r} = 4 + 4\cos\theta$;
(d) $\dfrac{3}{r} = 1 + 2\sin\theta$.

Solution: Comparing the equations with $\dfrac{l}{r} = 1 + e\cos\theta$, we get

(a) $|e| = 1/3 < 1$: ellipse;
(b) $e = 2 > 1$: hyperbola;
(c) $e = 1$: parabola;
(d) $e = 2 > 1$: hyperbola.

▶▷▶ **4.** *Find the points on the conic*

(a) $\dfrac{8}{r} = 3 - \sqrt{2}\cos\theta$, *whose radius vector is 4;*
(b) $\dfrac{l}{r} = 1 - \cos\theta$, *which has smallest radius vector;*
(c) $\dfrac{3}{r} = 1 + 2\cos\theta$, *whose vectorial angle is* $\pi/3$.

Solution:

(a) Putting $r = 4$ in $\dfrac{8}{r} = 3 - \sqrt{2}\cos\theta$, we get $\cos\theta = \dfrac{1}{\sqrt{2}}$, which gives $\theta = \pi/4, \ -\pi/4$.
So, the points are $(4, \pi/4)$ and $(4, -\pi/4)$.

(b) From the equation $\dfrac{l}{r} = 1 - \cos\theta$, we get

$$r = \frac{l}{1 - \cos\theta}.$$

Now r will be minimum when $1 - \cos\theta$ is maximum, i.e. when $\cos\theta$ is minimum. Now minimum value of $\cos\theta = -1$, which gives $\theta = \pi$. So, the minimum value of $r = \dfrac{l}{2}$ and the point is $\left(\dfrac{l}{2}, \pi\right)$.

(c) Putting $\theta = \pi/3$ in $\frac{3}{r} = 1 + 2\cos\theta$, we get

$$\frac{3}{r} = 1 + 2 \times \frac{1}{2} = 2$$

$$\text{or,} \qquad r = \frac{3}{2}.$$

So, the point is $(3/2, \pi/3)$.

▶▷ 5. *Find the point of intersection of the straight lines $r\cos(\theta - \alpha) = p$ and $r\cos(\theta - \beta) = p$.*

Solution: If (ρ, ϕ) be the point of intersection, then at the point of intersection

$$\cos(\phi - \alpha) = \cos(\phi - \beta)$$

$$\text{or,} \qquad \phi - \alpha = \pm(\phi - \beta)$$

$$\text{or,} \qquad \phi - \left(\frac{\alpha + \beta}{2}\right)$$

[taking $(-)$ sign in the r.h.s. as $(+)$ sign does not give any value of ϕ].

Putting the value of ϕ in the following equation, $\rho\cos(\phi - \alpha) = p$ we get

$$\rho\cos\left(\frac{\alpha - \beta}{2}\right) = p$$

$$\text{or,} \qquad \rho = p\sec\left(\frac{\alpha - \beta}{2}\right).$$

So, the point of intersection is $\{p\sec[(\alpha - \beta)/2], (\alpha + \beta)/2\}$.

▶▷ 6. *Find the radius and centre of the circle $r = a\cos\theta + b\sin\theta$.*

Solution:

$$r = a\cos\theta + b\sin\theta = \sqrt{a^2 + b^2}\left(\frac{a}{\sqrt{a^2 + b^2}}\cos\theta + \frac{b}{\sqrt{a^2 + b^2}}\sin\theta\right)$$

$$= \sqrt{a^2 + b^2}[\cos\alpha\cos\theta + \sin\alpha\sin\theta]$$

$$= \sqrt{a^2 + b^2}\cos(\theta - \alpha)$$

where

$$\cos\alpha = \frac{a}{\sqrt{a^2 + b^2}} \quad \text{and} \quad \sin\alpha = \frac{b}{\sqrt{a^2 + b^2}}.$$

Radius of the circle $= \frac{1}{2}\sqrt{a^2 + b^2}$ and centre of the circle is $\left(\frac{1}{2}\sqrt{a^2 + b^2}, \tan^{-1}\left(\frac{b}{a}\right)\right)$.

▶▷ 7. *Show that the perpendicular distance of the point (ρ, ϕ) from the line $r\cos(\theta - \alpha) = p$ is $\rho\cos(\phi - \alpha) - p$.*

Solution: From the following figure, we see that perpendicular distance of the point (ρ, ϕ) from the line $r\cos(\theta - \alpha) = p$ is equal to $PQ = SR = OR - OS = \rho\cos(\phi - \alpha) - p$.

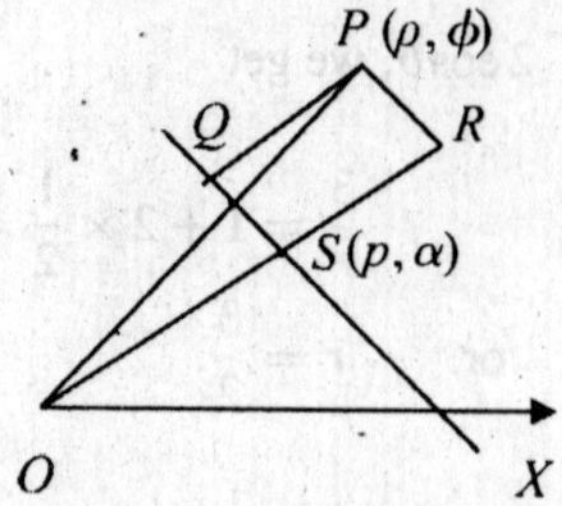

▶▷ **8.** *In any conic prove that sum of the reciprocals of two perpendicular focal chords is constant.*

Solution: Let PSP' and QSQ' be two perpendicular focal chords of the conic

$$\frac{l}{r} = 1 + e\cos\theta.$$

If α be the vectorial angle of the point P, then vectorial angles of Q, P' and Q' will be $\pi/2 + \alpha$, $\pi + \alpha$ and $3\pi/2 + \alpha$ respectively. So, we get

$$\frac{l}{SP} = 1 + e\cos\alpha; \quad \frac{l}{SQ} = 1 - e\sin\alpha;$$

$$\frac{l}{SP'} = 1 - e\cos\alpha; \quad \frac{l}{SQ'} = 1 + e\sin\alpha.$$

From the above results we get

$$PP' = SP + SP' = \frac{l}{1 + e\cos\alpha} + \frac{l}{1 - e\cos\alpha} = \frac{2l}{1 - e^2\cos^2\alpha}.$$

Similarly,

$$QQ' = \frac{2l}{1 - e^2\sin^2\alpha}.$$

Hence,

$$\frac{1}{PP'} + \frac{1}{QQ'} = \frac{2 - e^2}{2l} = \text{constant}.$$

Note: In case of rectangular hyperbola, $e = \sqrt{2}$.

So,

$$PP' = \frac{2l}{1 - 2\cos^2\alpha} = -\frac{2l}{\cos 2\alpha} \quad \text{and} \quad QQ' = \frac{2l}{1 - 2\sin^2\alpha} = \frac{2l}{\cos 2\alpha}$$

thus $PP' = QQ'$ (in length), i.e. of two perpendicular focal chords of a rectangular hyperbola are of equal length.

▶▷ **9.** *If PSQ and $PS'R$ be two focal chords of an ellipse through the foci S and S', then show that $\dfrac{SP}{SQ} + \dfrac{S'P}{S'R}$ is independent of the position of P.*

Solution: If the vectorial angle of the point P be α, then vectorial angle of the point Q will be $\pi + \alpha$ and we get

$$\frac{l}{SP} = 1 + e\cos\alpha; \quad \frac{l}{SQ} = 1 - e\cos\alpha.$$

Adding we get

$$\frac{1}{SP}+\frac{1}{SQ}=\frac{2}{l} \quad \text{or,} \quad \frac{SP}{SQ}=2\frac{SP}{l}-1.$$

Similarly, we get

$$\frac{S'P}{S'R}=2\frac{S'P}{l}-1.$$

Adding these results we get

$$\frac{SP}{SQ}+\frac{S'P}{S'R}=2\frac{SP+S'P}{l}-2=\frac{4a}{l}-2$$

which is a constant, i.e. independent of the position of P.

▶▶ **10.** *If PSP′ and QSQ′ be two perpendicular focal chords of the conic, prove that*

$$\frac{1}{SP\times SP'}+\frac{1}{SQ\times SQ'}=\textit{constant.}$$

Solution: If α be the vectorial angle of the point P, then vectorial angles of Q, P' and Q' will be $\pi/2+\alpha$, $\pi+\alpha$ and $3\pi/2+\alpha$ respectively. So, we get

$$\frac{l}{SP}=1+e\cos\alpha; \quad \frac{l}{SQ}=1-e\sin\alpha;$$
$$\frac{l}{SP'}=1-e\cos\alpha; \quad \frac{l}{SQ'}=1+e\sin\alpha.$$

Now,

$$\frac{1}{SP\times SP'}+\frac{1}{SQ\times SQ'}=\frac{1-e^2\cos^2\alpha}{l^2}+\frac{1-e^2\sin^2\alpha}{l^2}=\frac{2-e^2}{l^2}=\text{constant.}$$

▶▶ **11.** *A focal chord PSP′ of an ellipse makes an angle α with major axis. Show that perpendicular from focus to the tangent at P makes an angle* $\tan^{-1}\left(\frac{\sin\alpha}{e+\cos\alpha}\right)$ *with the axis.*

Solution: Given that the vectorial angle of the point P is α, so the equation of tangent at P is

$$\frac{l}{r}=e\cos\theta+\cos(\theta-\alpha)$$
$$\text{or,} \quad l=r\cos\theta[e+\cos\alpha]+r\sin\theta\sin\alpha.$$

Its slope $=-\left(\frac{e+\cos\alpha}{\sin\alpha}\right)$. So, slope of the line perpendicular to it is $\left(\frac{\sin\alpha}{e+\cos\alpha}\right)$. Hence, the perpendicular from focus to the tangent at P makes an angle $\tan^{-1}\left(\frac{\sin\alpha}{e+\cos\alpha}\right)$ with the axis.

▶▶ **12.** *Prove that portion of the tangent to a conic intercepted between the conic and directrix subtends a right angle at the focus.*

Solution: Let the vectorial angle of the point P on the conic $\frac{l}{r} = 1 + e\cos\theta$, be α and that of the point K, where the tangent meets the directrix $\frac{l}{r} = e\cos\theta$, be θ. Now, equation of tangent at P is

$$\frac{l}{r} = e\cos\theta + \cos(\theta - \alpha).$$

Solving this with $\frac{l}{r} = e\cos\theta$, we get $\cos(\theta - \alpha) = 0$, i.e. $\alpha - \theta = \pi/2$, i.e. $\angle KSP = \pi/2$.

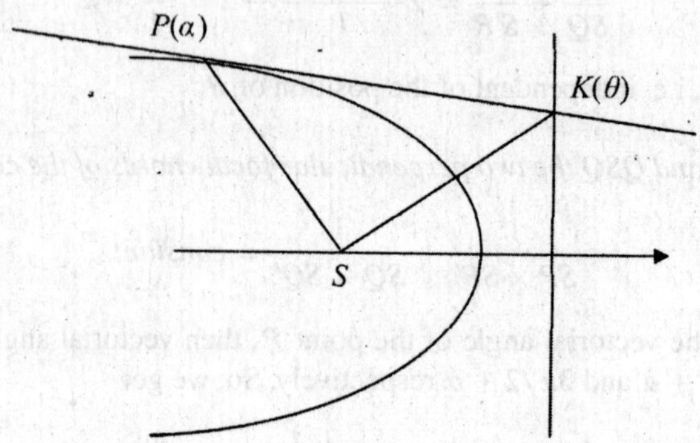

▶▷ **13.** *Show that for the conic $l/r = 1 + e\cos\theta$, equation of the directrix corresponding to the focus other than pole is* $\frac{l}{r} = -\frac{e(1-e^2)}{1+e^2}\cos\theta$.

Solution:

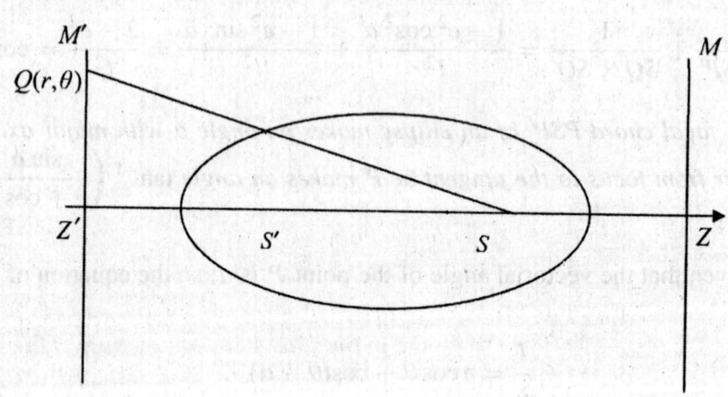

If $Q(r, \theta)$ be a point on the other directrix corresponding to the focus S' and SZ' be the perpendicular from the focus, taken as pole, then

$$SZ' = SQ\cos(\pi - \theta)$$

$$\text{or,} \quad SZ' = -r\cos\theta. \tag{1}$$

Now, $ZZ' = \frac{2a}{e}$ and $SZ = \frac{l}{e}$, so $SZ' = \frac{2a - l}{e}$.

Again from $l = \frac{b^2}{a} = a(1 - e^2)$, we get $a = \frac{l}{1 - e^2}$.

So,

$$SZ' = \left(\frac{2l}{1-e^2} - l\right)\frac{1}{e} = \frac{l(1+e^2)}{e(1-e^2)} = -r\cos\theta \quad \text{[by Eqn. (1)]}$$

$$\text{or,} \quad \frac{l}{r} = -\frac{e(1-e^2)}{1+e^2}\cos\theta$$

which is the equation of the other directrix.

▶▶ **14**. *If the tangents at P and Q of a conic meet at a point T, and S be the focus of the parabola, then prove that $ST^2 = SP \times SQ$.*

Solution: If the vectorial angles of P and Q on the parabola $\frac{l}{r} = 1 + \cos\theta$ be α and β respectively, and S be its focus, then we have

$$SP = \frac{l}{1+\cos\alpha} = \frac{l}{2}\sec^2\left(\frac{\alpha}{2}\right)$$

$$\text{and} \quad SQ = \frac{l}{1+\cos\beta} = \frac{l}{2}\sec^2\left(\frac{\beta}{2}\right)$$

so that

$$SP \times SQ = \frac{l}{2}\sec^2\left(\frac{\alpha}{2}\right) \times \frac{l}{2}\sec^2\left(\frac{\beta}{2}\right).$$

Now, equation of the tangent at $\theta = \alpha$ is

$$\frac{l}{r} = \cos\theta + \cos(\theta - \alpha)$$

and that of the tangent at $\theta = \beta$ is

$$\frac{l}{r} = \cos\theta + \cos(\theta - \beta).$$

If two tangents meet at (ST, γ), then we have

$$\cos(\gamma - \alpha) = \cos(\gamma - \beta)$$

which gives

$$(\gamma - \alpha) = \pm(\gamma - \beta)$$

$$\text{or,} \quad \gamma = \left(\frac{\alpha+\beta}{2}\right). \quad \text{[since + sign in the r.h.s. does not give any value of } \gamma\text{]}$$

Using this in any one of the equations of tangents, we get

$$\frac{l}{ST} = \cos\left(\frac{\alpha+\beta}{2}\right) + \cos\left(\frac{\alpha-\beta}{2}\right)$$

$$\text{or,} \quad \frac{l}{ST} = 2\cos\left(\frac{\alpha}{2}\right)\cos\left(\frac{\beta}{2}\right)$$

$$\text{or,} \quad ST = \frac{l}{2}\sec\left(\frac{\alpha}{2}\right)\sec\left(\frac{\beta}{2}\right).$$

So,

$$ST^2 = SP \times SQ.$$

▶▷▶ **15**. *Find the condition that the line* $\frac{l}{r} = a\cos\theta + b\sin\theta$ *may touch the conic*

$$\frac{l}{r} = 1 + e\cos(\theta - \beta).$$

Solution: Equation of the tangent to the given conic at the point $\theta = \alpha$ is

$$\frac{l}{r} = e\cos(\theta - \beta) + \cos(\theta - \alpha)$$

or, $$\frac{l}{r} = \cos\theta[e\cos\beta + \cos\alpha] + \sin\theta[e\sin\beta + \sin\alpha].$$

Comparing this with the equation

$$\frac{l}{r} = a\cos\theta + b\sin\theta$$

we get

$$\frac{e\cos\beta + \cos\alpha}{a} = \frac{e\sin\beta + \sin\alpha}{b} = 1$$

or, $$\cos\alpha = a - e\cos\beta \quad \text{and} \quad \sin\alpha = b - e\sin\beta.$$

Squaring and adding we get

$$(a - e\cos\beta)^2 + (b - e\sin\beta)^2 = 1.$$

This is the required condition.

▶▷▶ **16**. *Prove that the locus of the middle points of a system of focal chords of a conic is a conic, which is a parabola, ellipse and hyperbola according as the original conic is parabola, ellipse and hyperbola.*

Solution: If the vectorial angle of the end of a focal chord PSP' of the conic $\frac{l}{r} = 1 + e\cos\theta$ be α, then that of other end will be $\pi + \alpha$.

As the points P, P' lie on the conic, we have

$$\frac{l}{SP} = 1 + e\cos\alpha \quad \text{and} \quad \frac{l}{SP'} = 1 - e\cos\alpha.$$

If $Q(\rho, \phi)$ be the middle point of the chord, then from the following figure we see that $\phi = \alpha$, and

$$\begin{aligned} \rho = SQ &= SP - PQ \\ &= SP - \frac{SP + SP'}{2} \\ &= \frac{SP - SP'}{2} \\ &= \frac{1}{2}\left(\frac{l}{1 + e\cos\alpha} - \frac{l}{1 - e\cos\alpha}\right) \\ &= -\frac{el\cos\alpha}{1 - e^2\cos^2\alpha} \end{aligned}$$

or, $$\rho(1 - e^2\cos^2\alpha) + el\cos\alpha = 0.$$

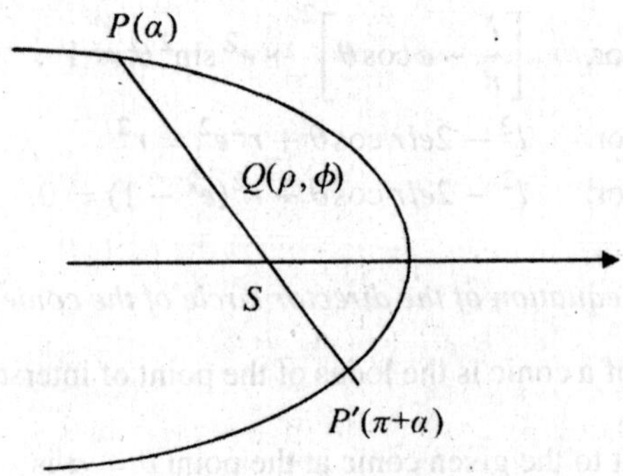

Hence, locus of the point (ρ, ϕ) is

$$r(1 - e^2 \cos^2 \theta) + el \cos \theta = 0.$$

Now, transforming this equation to Cartesian coordinate, we get

$$x^2 + y^2 - e^2x^2 + elx = 0$$

$$\text{or,} \quad x^2(1 - e^2) + y^2 + elx = 0.$$

This will represent ellipse, parabola or hyperbola according as

$$h^2 - ab <, =, > 0$$

$$\text{i.e.,} \quad -(1 - e^2) <,=, > 0$$

$$\text{i.e.,} \quad e^2 <, =, > 1$$

$$\text{i.e.,} \quad e <, =, > 1$$

which is also the condition under which the original conic is ellipse, parabola or hyperbola.

▶▷▶ 17. *Find the polar equation of the auxiliary circle of the conic* $\frac{l}{r} = 1 + e \cos \theta$.

Solution: Auxiliary circle of a conic is the locus of the foot of the perpendicular from focus on any tangent to it.

Equation of the tangent to the given conic at the point $\theta = \alpha$ is

$$\frac{l}{r} = e \cos \theta + \cos(\theta - \alpha) \tag{1}$$

$$\text{or,} \quad l = r \cos \theta [e + \cos \alpha] + r \sin \theta \sin \alpha.$$

Now, equation of the perpendicular from focus (i.e. origin) to the tangent is

$$r \cos \theta \sin \alpha - r \sin \theta [e + \cos \alpha] = 0$$

$$\text{or,} \quad \sin(\theta - \alpha) = -e \sin \theta. \tag{2}$$

To get the locus of the point of intersection of the lines given by Eqns. (1) and (2), we eliminate θ as follows:

$$\sin^2(\theta - \alpha) + \cos^2(\theta - \alpha) = 1$$

$$\text{or,} \quad \left[\frac{l}{r} - e\cos\theta\right]^2 + e^2\sin^2\theta = 1$$

$$\text{or,} \quad l^2 - 2elr\cos\theta + r^2e^2 = r^2$$

$$\text{or,} \quad l^2 - 2elr\cos\theta + r^2(e^2 - 1) = 0.$$

▶▶ **18**. *Find the polar equation of the director circle of the conic* $\frac{l}{r} = 1 + e\cos\theta$.

Solution: Director circle of a conic is the locus of the point of intersection of two perpendicular tangents.

Equation of the tangent to the given conic at the point $\theta = \alpha$ is

$$\frac{l}{r} = e\cos\theta + \cos(\theta - \alpha)$$

$$\text{or,} \quad l = r\cos\theta[e + \cos\alpha] + r\sin\theta\sin\alpha.$$

Slope of the tangent is

$$m_1 = -\frac{e + \cos\alpha}{\sin\alpha}.$$

Similarly, slope of the tangent at the point $\theta = \beta$ is

$$m_2 = -\frac{e + \cos\beta}{\sin\beta}.$$

If the tangents are perpendicular, then we have

$$\frac{e + \cos\alpha}{\sin\alpha} \times \frac{e + \cos\beta}{\sin\beta} = -1$$

$$\text{or,} \quad e^2 + e(\cos\alpha + \cos\beta) + \cos(\alpha - \beta) = 0$$

$$\text{or,} \quad e^2 + 2e\cos\left(\frac{\alpha + \beta}{2}\right)\cos\left(\frac{\alpha - \beta}{2}\right) + 2\cos^2\left(\frac{\alpha - \beta}{2}\right) - 1 = 0. \quad (1)$$

Now, if the tangents meet at the point (ρ, ϕ), then we have

$$(\phi - \alpha) = \pm(\phi - \beta)$$

$$\text{or,} \quad \phi = \left(\frac{\alpha + \beta}{2}\right).$$

Using this in the equation of tangent we get

$$\frac{l}{\rho} = e\cos\phi + \cos\left(\frac{\alpha - \beta}{2}\right). \quad (2)$$

From Eqns. (1) and (2) we get

$$e^2 + 2e\cos\phi\cos\left(\frac{\alpha - \beta}{2}\right) + 2\cos^2\left(\frac{\alpha - \beta}{2}\right) - 1 = 0$$

$$\text{or,} \quad e^2 + 2e\cos\phi\left[\frac{l}{\rho} - e\cos\phi\right] + 2\left[\frac{l}{\rho} - e\cos\phi\right]^2 - 1 = 0$$

$$\text{or,} \quad \rho^2(e^2 - 1) + 2el\rho\cos\phi - 2\rho^2e^2\cos^2\phi + 2[l^2 - 2le\rho\cos\phi + \rho^2e^2\cos^2\phi] = 0$$

$$\text{or,} \quad \rho^2(e^2 - 1) - 2el\rho\cos\phi + 2l^2 = 0.$$

Hence, locus of the point (ρ, ϕ) is

$$r^2(e^2-1)-2elr\cos\theta+2l^2=0.$$

▶▷▶ **19.** *Show that the equation of the tangent to the conic* $\frac{l}{r}=1+e\cos\theta$, ***parallel to the*** *tangent at* $\theta=\alpha$, *is given by*

$$l(e^2+2e\cos\alpha+1)=r(e^2-1)[\cos(\theta-\alpha)+e\cos\theta].$$

Solution: Equations of the tangents to the given conic at the point $\theta=\alpha$ and $\theta=\beta$ are

$$\frac{l}{r}=e\cos\theta+\cos(\theta-\alpha) \tag{1}$$

and $$\frac{l}{r}=e\cos\theta+\cos(\theta-\beta). \tag{2}$$

Transforming the equations into Cartesian form and applying the condition of parallelism, we get

$$\frac{e+\cos\alpha}{e+\cos\beta}=\frac{\sin\alpha}{\sin\beta}=\frac{1}{k},\quad \text{where } k\neq 1.$$

From it we get

$$\sin\beta=k\sin\alpha \quad\text{and}\quad \cos\beta=k(e+\cos\alpha)-e.$$

Squaring and adding (to eliminate β) we get

$$[k(e+\cos\alpha)-e]^2+k^2\sin^2\alpha=1$$

or, $k^2(e^2+2e\cos\alpha+\cos^2\alpha)-2ek(e+\cos\alpha)+e^2+k^2\sin^2\alpha=1$

or, $k^2(e^2+2e\cos\alpha+1)-2ek(e+\cos\alpha)+e^2-1=0$

or, $k^2(e^2+2e\cos\alpha+1)-k(e^2+2e\cos\alpha+1)+k-ke^2+e^2-1=0$

or, $k(k-1)(e^2+2e\cos\alpha+1)-(e^2-1)(k-1)=0$

or, $(k-1)[k(e^2+2e\cos\alpha+1)-(e^2-1)]=0$

or, $k=\dfrac{e^2-1}{e^2+2e\cos\alpha+1}$, as $k\neq 1$.

From Eqn. (2), we get

$$\frac{l}{r}=\cos\theta[e+\cos\beta]+\sin\theta\sin\beta$$

or, $\dfrac{l}{r}=k[\cos\theta(e+\cos\alpha)+\sin\theta\sin\alpha]$

or, $l(e^2+2e\cos\alpha+1)=r(e^2-1)[\cos(\theta-\alpha)+e\cos\theta]$.

▶▷▶ **20.** *Find the equation of the circle which passes through the focus of the conic* $\frac{l}{r}=1-e\cos\theta$ *and touches it at the point* $\theta=\alpha$.

Solution: Let the equation of the circle be

$$r=2b\cos(\theta-\beta).$$

At the point $\theta = \alpha$, we have

$$2b\cos(\alpha - \beta) = \frac{l}{1 - e\cos\alpha}. \tag{1}$$

Since, the normal to a circle always passes through its centre, in this case normal to the conic at the point $\theta = \alpha$ will pass through the centre (b, β) of the circle, i.e.

$$\frac{el\sin\alpha}{b(1 - e\cos\alpha)} = e\sin\beta - \sin(\beta - \alpha). \tag{2}$$

From Eqns. (1) and (2), we get

$$\begin{aligned}
& 2e\sin\alpha\cos(\alpha-\beta) = e\sin\beta - \sin(\beta-\alpha)\\
\text{or,}\quad & 2e\sin\alpha\cos(\alpha-\beta) = e\sin[(\beta-\alpha)+\alpha] - \sin(\beta-\alpha)\\
\text{or,}\quad & 2e\sin\alpha\cos(\alpha-\beta) = e\sin(\beta-\alpha)\cos\alpha + e\cos(\beta-\alpha)\sin\alpha - \sin(\beta-\alpha)\\
\text{or,}\quad & e\sin\alpha\cos(\alpha-\beta) = e\sin(\beta-\alpha)\cos\alpha - \sin(\beta-\alpha)\\
\text{or,}\quad & e\sin\alpha\cos(\alpha-\beta) = \sin(\alpha-\beta)[1 - e\cos\alpha]\\
\text{or,}\quad & \tan(\alpha-\beta) = \frac{e\sin\alpha}{1 - e\cos\alpha}.
\end{aligned} \tag{3}$$

Now, the equation of circle is

$$\begin{aligned}
r &= 2b\cos(\theta-\beta)\\
&= 2b\cos[(\theta-\alpha)-(\beta-\alpha)]\\
&= 2b[\cos(\theta-\alpha)\cos(\beta-\alpha) + \sin(\theta-\alpha)\sin(\beta-\alpha)]\\
&= \frac{l}{\cos(\alpha-\beta)(1-e\cos\alpha)}[\cos(\theta-\alpha)\cos(\beta-\alpha) + \sin(\theta-\alpha)\sin(\beta-\alpha)]\\
&= \frac{l}{1-e\cos\alpha}[\cos(\theta-\alpha) + \sin(\theta-\alpha)\tan(\beta-\alpha)]\\
&= \frac{l}{1-e\cos\alpha}\left[\cos(\theta-\alpha) - \sin(\theta-\alpha)\frac{e\sin\alpha}{1-e\cos\alpha}\right]\\
&= \frac{l}{(1-e\cos\alpha)^2}[\cos(\theta-\alpha)(1-e\cos\alpha) - \sin(\theta-\alpha)e\sin\alpha]\\
&= \frac{l}{(1-e\cos\alpha)^2}[\cos(\theta-\alpha) - e\cos(\theta-2\alpha)].
\end{aligned}$$

$$\text{or,}\quad r(1-e\cos\alpha)^2 = l[\cos(\theta-\alpha) - e\cos(\theta-2\alpha)].$$

▶▶ **21.** *P, Q, R are three points on the parabola* $\frac{l}{r} = 1 + \cos\theta$ *with vectorial angles* α, β, γ *respectively. Show that the equation of the circle circumscribing the triangle formed by the tangents at these points to the parabola is*

$$r = \frac{l}{2}\sec\left(\frac{\alpha}{2}\right)\sec\left(\frac{\beta}{2}\right)\sec\left(\frac{\gamma}{2}\right)\cos\left(\theta - \frac{\alpha+\beta+\gamma}{2}\right).$$

Solution: Equations of tangents at P, Q, R to the parabola are

$$\frac{l}{r} = \cos\theta + \cos(\theta - \alpha), \tag{1}$$

$$\frac{l}{r} = \cos\theta + \cos(\theta - \beta) \tag{2}$$

$$\text{and} \qquad \frac{l}{r} = \cos\theta + \cos(\theta - \gamma). \tag{3}$$

Tangents at P, Q meet at a point whose vectorial angle is given by

$$\cos(\theta - \alpha) = \cos(\theta - \beta)$$

$$\text{or,} \quad (\theta - \alpha) = \pm(\theta - \beta)$$

$$\text{or,} \quad \theta = \left(\frac{\alpha + \beta}{2}\right).$$

Putting this in Eqn. (1) we get the radius vector of the point of intersection

$$= \frac{l}{\cos\left(\frac{\alpha+\beta}{2}\right) + \cos\left(\frac{\alpha-\beta}{2}\right)} = \frac{l}{2\cos\left(\frac{\alpha}{2}\right)\cos\left(\frac{\beta}{2}\right)}.$$

Hence, the point of intersection of the tangents (1) and (2) is

$$A: \left(\frac{l}{2}\sec\frac{\alpha}{2}\sec\frac{\beta}{2}, \frac{\alpha+\beta}{2}\right).$$

Similarly, the points of intersection of the tangents (2), (3) and (1), (3) are

$$B: \left(\frac{l}{2}\sec\frac{\gamma}{2}\sec\frac{\beta}{2}, \frac{\gamma+\beta}{2}\right) \quad \text{and} \quad C: \left(\frac{l}{2}\sec\frac{\gamma}{2}\sec\frac{\alpha}{2}, \frac{\gamma+\alpha}{2}\right) \text{ respectively.}$$

Let the equation of circle be $r = d\cos(\theta - \lambda)$. If it passes through the points A, B, C, then we have

$$\sec\left(\frac{\alpha}{2}\right)\sec\left(\frac{\beta}{2}\right)\sec\left(\frac{\alpha+\beta}{2} - \lambda\right) = \sec\left(\frac{\gamma}{2}\right)\sec\left(\frac{\beta}{2}\right)\sec\left(\frac{\gamma+\beta}{2} - \lambda\right)$$

$$= \sec\left(\frac{\alpha}{2}\right)\sec\left(\frac{\gamma}{2}\right)\sec\left(\frac{\alpha+\gamma}{2} - \lambda\right).$$

Taking first two expressions into consideration, we get

$$\frac{\cos\left(\frac{\alpha+\beta}{2} - \lambda\right)}{\cos\left(\frac{\gamma}{2}\right)} = \frac{\cos\left(\frac{\gamma+\beta}{2} - \lambda\right)}{\cos\left(\frac{\alpha}{2}\right)}$$

$$\text{or,} \quad \frac{\cos\left(\frac{\alpha+\beta}{2} - \lambda\right) + \cos\left(\frac{\gamma+\beta}{2} - \lambda\right)}{\cos\left(\frac{\alpha}{2}\right) + \cos\left(\frac{\gamma}{2}\right)} = \frac{\cos\left(\frac{\alpha+\beta}{2} - \lambda\right) - \cos\left(\frac{\gamma+\beta}{2} - \lambda\right)}{\cos\left(\frac{\gamma}{2}\right) - \cos\left(\frac{\alpha}{2}\right)}$$

$$\text{or,} \quad \frac{\cos\left(\frac{\alpha+2\beta+\gamma}{4} - \lambda\right)}{\cos\left(\frac{\alpha+\gamma}{4}\right)} = -\frac{\sin\left(\frac{\alpha+2\beta+\gamma}{4} - \lambda\right)}{\sin\left(\frac{\alpha+\gamma}{4}\right)}$$

or, $\tan\left(\dfrac{\alpha+2\beta+\gamma}{4}-\lambda\right)=-\tan\dfrac{\alpha+\gamma}{4}$

or, $\left(\dfrac{\alpha+2\beta+\gamma}{4}-\lambda\right)=-\dfrac{\alpha+\gamma}{4}$

or, $\lambda=\left(\dfrac{\alpha+\beta+\gamma}{2}\right).$

As the points A, B, C lie on the circle, using any one of these in the equation of circle we get

$$\frac{l}{2}\sec\left(\frac{\alpha}{2}\right)\sec\left(\frac{\beta}{2}\right)=d\cos\left(\frac{\alpha+\beta}{2}-\frac{\alpha+\beta+\gamma}{2}\right)$$

$$\text{or,}\quad d=\frac{l}{2}\sec\left(\frac{\alpha}{2}\right)\sec\left(\frac{\beta}{2}\right)\sec\left(\frac{\gamma}{2}\right).$$

Hence, the equation of circle is

$$r=\frac{l}{2}\sec\left(\frac{\alpha}{2}\right)\sec\left(\frac{\beta}{2}\right)\sec\left(\frac{\gamma}{2}\right)\cos\left(\theta-\frac{\alpha+\beta+\gamma}{2}\right).$$

▶▷ **22.** *If the conics* $\dfrac{l}{r}=1+e\cos\theta$ *and* $\dfrac{l'}{r}=1+e\cos(\theta-\gamma)$ *touch at the point* $\theta=\alpha$, *then show that*

$$l'=\frac{l(1-e^2)}{e^2+2e\cos\alpha+1}.$$

Solution: Since the conics touch at the point $\theta=\alpha$, the tangents to the conics at that point should be identical.

Now, equations of the tangents to the given conics at the point $\theta=\alpha$ are

$$\frac{l}{r}=e\cos\theta+\cos(\theta-\alpha) \tag{1}$$

$$\text{and}\quad \frac{l'}{r}=e\cos(\theta-\gamma)+\cos(\theta-\alpha). \tag{2}$$

Comparing Eqns. (1) and (2) we get

$$\frac{l}{l'}=\frac{e+\cos\alpha}{e\cos\gamma+\cos\alpha}=\frac{\sin\alpha}{e\sin\gamma+\sin\alpha}$$

from which we get

$$el\cos\gamma=l'e+(l'-l)\cos\alpha\quad\text{and}\quad el\sin\gamma=(l'-l)\sin\alpha.$$

Squaring and adding we get

$$e^2l^2=l'^2e^2+2l'(l'-l)e\cos\alpha+(l'-l)^2$$

$$\text{or,}\quad e^2(l+l')=-2l'e\cos\alpha-(l'-l)$$

$$\text{or,}\quad l'(e^2+2e\cos\alpha+1)=l(1-e^2)$$

$$\text{or,}\quad l'=\frac{l(1-e^2)}{e^2+2e\cos\alpha+1}.$$

▶▶ **23.** *If d be the diameter of the circle passing through the pole and the points* (r_1, θ_1), (r_2, θ_2), *show that*

$$d^2 \sin^2(\theta_1 - \theta_2) = r_1^2 + r_2^2 - 2r_1r_2\cos(\theta_1 - \theta_2).$$

Solution: If the circle passes through the pole S and P: (r_1, θ_1), Q: (r_2, θ_2), then from the following figure we see that $\angle PSQ = \theta_1 - \theta_2$, $\angle SPQ = \angle STQ$ and $ST = d$.

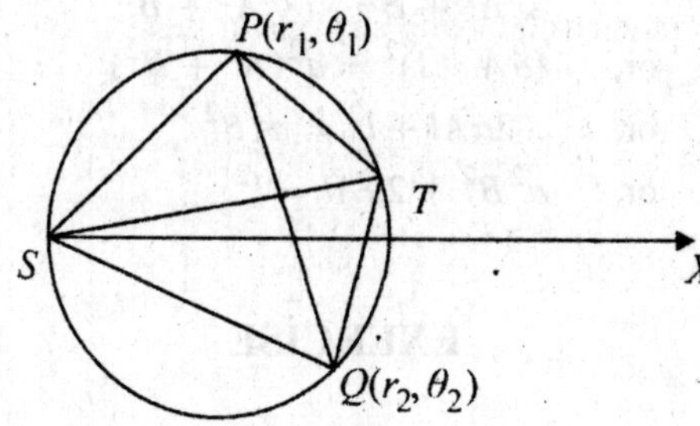

Now, from the triangle ΔPSQ

$$PQ^2 = r_1^2 + r_2^2 - 2r_1r_2\cos(\theta_1 - \theta_2). \tag{1}$$

From the triangle ΔSTQ

$$\frac{ST}{\sin 90^\circ} = \frac{SQ}{\sin \angle STQ} = \frac{SQ}{\sin \angle SPQ}$$

$$\text{or,} \quad d = \frac{SQ}{\sin \angle SPQ}. \tag{2}$$

Again from the triangle ΔSPQ

$$\frac{PQ}{\sin \angle PSQ} = \frac{SQ}{\sin \angle SPQ}. \tag{3}$$

From Eqns. (2) and (3) we get

$$d = \frac{PQ}{\sin \angle PSQ}$$

$$\text{or,} \quad d^2\sin^2(\theta_1 - \theta_2) = PQ^2 = r_1^2 + r_2^2 - 2r_1r_2\cos(\theta_1 - \theta_2).$$

▶▶ **24.** *Show that the line* $\frac{l}{r} = A\cos\theta + B\sin\theta$ *touches the conic* $r = 2a\cos\theta$, *if*

$$a^2B^2 + 2aAl = l^2.$$

Solution: Putting $A = d\cos\alpha$ and $B = d\sin\alpha$ in the equation of straight line we get

$$\frac{l}{d} = r\cos(\theta - \alpha), \quad \text{where } d^2 = A^2 + B^2 \text{ and } \tan\alpha = \frac{B}{A}.$$

Now, we know that the perpendicular distance of the point (ρ, ϕ) from the line $r\cos(\theta - \alpha) = p$ is $\rho\cos(\phi - \alpha) - p$.

If the given line touches the circle $r = 2a\cos\theta$, then the perpendicular distance of the centre $(a, 0)$ from the line $r\cos(\theta - \alpha) = \dfrac{l}{\sqrt{A^2 + B^2}}$ is equal to the radius of the circle, i.e.

$$a\cos\alpha - \frac{l}{\sqrt{A^2 + B^2}} = a$$

$$\text{or,} \quad \frac{aA}{\sqrt{A^2 + B^2}} - \frac{l}{\sqrt{A^2 + B^2}} = a$$

$$\text{or,} \quad (aA - l)^2 = a^2(A^2 + B^2)$$

$$\text{or,} \quad -2aAl + l^2 = a^2B^2$$

$$\text{or,} \quad a^2B^2 + 2aAl = l^2.$$

EXERCISE

1. Transform the following equations into polar form:
   (a) $y^2 = 4ax$;
   (b) $x^2 + y^2 = axy$;
   (c) $(x^2 + y^2)^2 = ay^2$;
   (d) $x^3 = y^2(a - x)$.
2. Transform the following equations into Cartesian form:
   (a) $\sqrt{r}\cos(\theta/2) = \sqrt{a}$;
   (b) $r(\sin 2\theta + \cos 2\theta) = 2\sin 2\theta$;
   (c) $2/r = \cos\theta + \sin\theta$.
3. Find the polar equation of the straight line joining $(2, 30°)$ and $(1, 60°)$.
4. Through a point O, a straight line is drawn, meeting n given straight lines in the points $R_1, R_2, \ldots, R_n$ and on it a point P is taken such that

$$\frac{n}{OP} = \frac{1}{OR_1} + \frac{1}{OR_2} + \frac{1}{OR_3} + \cdots + \frac{1}{OR_n};$$

show that the locus of P is straight line.

[*Hints*: Let O be the pole, and let the equations of n straight lines be

$$\frac{l}{r} = A_k\cos\theta + B_k\sin\theta, \qquad k = 1, 2, 3, \ldots, n.$$

Let P be the point (r', θ'); then equation of the line OP is $\theta = \theta'$. Using this in the above equations we get

$$\frac{l}{OR_k} = A_k\cos\theta' + B_k\sin\theta', \qquad k = 1, 2, 3, \ldots, n.$$

Adding these n equations we get

$$\frac{nl}{r'} = \left(\sum A_k\right)\cos\theta' + \left(\sum B_k\right)\sin\theta'.$$

Hence, locus of P is

$$\frac{nl}{r} = \left(\sum A_k\right)\cos\theta + \left(\sum B_k\right)\sin\theta$$

which is a straight line.]

5. Show that two lines $r\cos(\theta - \alpha) = p$ and $r\sin(\theta - \alpha) = q$ are perpendicular.
6. Find the locus of the feet of the perpendiculars drawn from the pole on the lines passing through a given point (r', θ').

   [*Hints*: If (p, α) be the foot of the perpendicular on one of the lines, then equation of that line is $r\cos(\theta - \alpha) = p$. If this line passes through the point (r', θ'), then $r'\cos(\theta' - \alpha) = p$. So, locus of the point (p, α) is $r = r'\cos(\theta' - \theta)$.]
7. Find the nature of the following conics:

   (a) $8/r = 4 - 5\cos\theta$;
   (b) $5/r = 3 + 7\cos\theta$;
   (c) $3/r = 2 - \sqrt{2}\cos\theta$;
   (d) $15/r = 3 - 3\cos\theta$.
8. Find the point on the conics

   (a) $\frac{8}{r} = 5 - 4\cos\theta$, whose radius vector is 2;

   (b) $\frac{5}{r} = 1 + 3\cos\theta$, whose vectorial angle is $\pi/6$.
9. (a) Find the polar coordinate of the centre of the circle $r = 3\cos\theta + 2\sin\theta$.
   (b) Find the equation of the circle which passes through the pole and the points $(d, 0)$ and $(2d, \pi/3)$.
10. Prove that the length of the focal chord of the conic $\frac{l}{r} = 1 + e\cos\theta$ which is inclined to the axis at an angle α is $\frac{2l}{1 - e^2\cos^2\alpha}$.
11. Show that the semi-latus rectum of a conic is the harmonic mean of the segments of any focal chord of a conic.
12. (a) If PSP' be a focal chord of a conic, prove that the angle between the tangents at P and P' is $\tan^{-1}\left(\frac{2e\sin\alpha}{1 - e^2}\right)$, where α is the angle between the focal chord and major axis.

    [*Hints*: Equation of tangent to the conic $\frac{l}{r} = 1 + e\cos\theta$ at the point $\theta = \alpha$ is

$$\frac{l}{r} = e\cos\theta + \cos(\theta - \alpha)$$

whose slope is $m_1 = \frac{-(e + \cos\alpha)}{\sin\alpha}$; then the slope of the tangent at the point $\theta = \pi + \alpha$, is $m_2 = \frac{e - \cos\alpha}{\sin\alpha}$. So, the angle between the tangents is given by

$$\tan^{-1}\left(\frac{m_1 - m_2}{1 + m_1 m_2}\right) = \tan^{-1}\left(\frac{2e\sin\alpha}{1 - e^2}\right).]$$

(b) Prove that two points on the conic $\frac{l}{r} = 1 + e\cos\theta$, whose vectorial angles are α, β, will be extremities of a diameter if

$$\tan(\alpha/2)\tan(\beta/2) + \frac{1+e}{1-e} = 0.$$

[*Hints*: Equation of tangent to the conic $\frac{l}{r} = 1 + e\cos\theta$ at the point $\theta = \alpha$ is

$$\frac{l}{r} = e\cos\theta + \cos(\theta - \alpha)$$

whose slope is $m_1 = \frac{-(e+\cos\alpha)}{\sin\alpha}$.

Similarly, equation of tangent to the conic $\frac{l}{r} = 1 + e\cos\theta$ at the point $\theta = \beta$ is

$$\frac{l}{r} = e\cos\theta + \cos(\theta - \beta)$$

whose slope is $m_2 = \frac{-(e+\cos\beta)}{\sin\beta}$.

$\therefore$ The tangents will be parallel if the points are extremities of a diameter; in this case we get

$$\frac{e+\cos\alpha}{\sin\alpha} = \frac{e+\cos\beta}{\sin\beta}$$

or, $\quad e(\sin\beta - \sin\alpha) = \sin\alpha\cos\beta - \cos\alpha\sin\beta$

or, $\quad -2e\cos\left(\frac{\alpha+\beta}{2}\right)\sin\left(\frac{\alpha-\beta}{2}\right) = 2\cos\left(\frac{\alpha-\beta}{2}\right)\sin\left(\frac{\alpha-\beta}{2}\right)$

or, $\quad \frac{\cos\left(\frac{\alpha+\beta}{2}\right)}{\cos\left(\frac{\alpha-\beta}{2}\right)} = -\frac{1}{e}.$

By componendo and dividendo

$$\frac{\cos(\alpha/2)\cos(\beta/2)}{\sin(\alpha/2)\sin(\beta/2)} = -\frac{1-e}{1+e}$$

or, $\quad \tan(\alpha/2)\tan(\beta/2) + \frac{1+e}{1-e} = 0.$]

13. A point moves so that sum of its distances from two fixed points S and S' is constant and equal to $2a$. Show that P lies on a conic

$$\frac{a(1-e^2)}{r} = 1 - e\cos\theta$$

with S as focus and SS' as initial line, SS' being equal to $2ae$.

14. Show that the equation $\frac{l}{r} = 1 + e\cos\theta$ and $\frac{l}{r} = -1 + e\cos\theta$ represent the same conic.
[*Hints*: It is already discussed in Section 9.1 that the polar coordinate of any point (r, θ) may also be expressed as $(-r, \pi + \theta)$. Use this in the equation $\frac{l}{r} = 1 + e\cos\theta$.]

15. If the tangent at any point on the ellipse makes an angle α with its major axis and an angle β with the focal radius to the point of contact, show that $e^2\cos^2\alpha = \cos^2\beta$.

[*Hints*: Equation of tangent to the conic $\frac{l}{r} = 1 + e\cos\theta$ at the point $\theta = \delta$ is

$$\frac{l}{r} = e\cos\theta + \cos(\theta - \delta)$$

whose slope is $m_1 = \dfrac{-(e + \cos\delta)}{\sin\delta} = \tan\alpha$ (given), which gives $\cos(\alpha - \delta) = -e\cos\alpha$.
From the figure, $\alpha - \delta = \beta$, so $\cos\beta = -e\cos\alpha$.

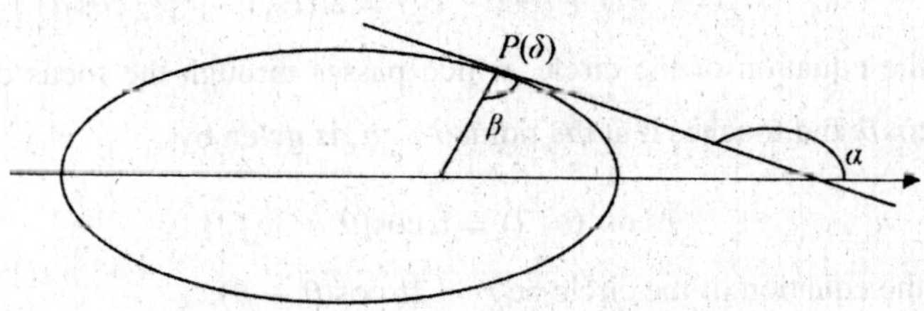

If we take the focal radius through the other focus of the ellipse, then changing e by $-e$ we get $\cos\beta = e\cos\alpha$. Hence, in either case $e^2\cos^2\alpha = \cos^2\beta$.]

16. Find the polar equation of the straight line joining two points on the parabola

$$\frac{2a}{r} = 1 + \cos\theta \text{ with } \alpha, \beta \text{ as their vectorial angles.}$$

17. Find the condition that the line $\frac{l}{r} = a\cos\theta + b\sin\theta$ may touch the conic

$$\frac{l}{r} = 1 + e\cos\theta.$$

18. Prove that the two conics

$$\frac{l_1}{r} = 1 - e_1\cos\theta \quad \text{and} \quad \frac{l_2}{r} = 1 - e_2\cos(\theta - \alpha)$$

will touch one another, if

$$l_2^2(1 - e_1^2) + l_1^2(1 - e_2^2) = 2l_1l_2(1 - e_1e_2\cos\alpha).$$

[*Hints*: If the given conics touch one another at the point $\theta = \beta$, then equations of tangents to both the conics at the point $\theta = \beta$ will be identical.
Now, equation of the tangents at the point $\theta = \beta$ to the given conics are

$$\frac{l_1}{r} = -e_1\cos\theta + \cos(\theta - \beta) \tag{1}$$

and

$$\frac{l_2}{r} = -e_2\cos(\theta - \alpha) + \cos(\theta - \beta) \tag{2}$$

respectively.

Comparing the equations we get

$$\frac{l_1}{l_2} = \frac{\cos\beta - e_1}{\cos\beta - e_2\cos\alpha} = \frac{\sin\beta}{\sin\beta - e_2\sin\alpha}.$$

From the first two ratios, we get

$$\cos\beta = \frac{l_1 e_2\cos\alpha - e_1 l_2}{l_1 - l_2}$$

and from first and third ratios, we get

$$\sin\beta = \frac{l_1 e_2\sin\alpha}{l_1 - l_2}.$$

Squaring and adding we get

$$(l_1 e_2\cos\alpha - e_1 l_2)^2 + (l_1 e_2\sin\alpha)^2 = (l_1 - l_2)^2$$

or, $$l_2^2(1 - e_1^2) + l_1^2(1 - e_2^2) = 2l_1 l_2(1 - e_1 e_2\cos\alpha).]$$

19. Show that the equation of the circle, which passes through the focus of the parabola $\frac{2a}{r} = 1 + \cos\theta$ and touches it at the point $\theta = \alpha$, is given by

$$r\cos^3(\alpha/2) = a\cos(\theta - 3\alpha/2).$$

[*Hints*: Let the equation of the circle be $r = 2b\cos(\theta - \beta)$.
At the point $\theta = \alpha$, we have

$$2b\cos(\alpha - \beta) = \frac{2a}{1 + \cos\alpha}. \tag{1}$$

Equation of the normal at the point $\theta = \alpha$ is

$$\frac{2a\sin\alpha}{r(1 + \cos\alpha)} = \sin\theta + \sin(\theta - \alpha).$$

It passes through the centre of the circle (b, β), so

$$\frac{2a\sin\alpha}{b(1 + \cos\alpha)} = \sin\beta + \sin(\beta - \alpha).$$

From Eqn. (1) we get

$$2b\cos(\alpha - \beta) = \frac{2a}{1 + \cos\alpha}$$

or, $$\frac{2d\sin\alpha}{\sin\beta + \sin(\beta - \alpha)} = \frac{a}{\cos(\alpha - \beta)}$$

or, $$2a\sin\alpha\cos(\alpha - \beta) = a[\sin\beta + \sin(\beta - \alpha)]$$

or, $$\sin(2\alpha - \beta) = \sin(\beta - \alpha)$$

or, $$\beta = 3\alpha/2.$$

Again from Eqn. (1)

$$b = \frac{a}{\cos(\alpha - \beta)(1 + \cos\alpha)} = \frac{a}{2\cos^3(\alpha/2)}.$$

So, the equation of circle is $r\cos^3(\alpha/2) = a\cos(\theta - 3\alpha/2)$.]

20. If P, Q be variable points on a conic $\dfrac{l}{r} = 1 - e\cos\theta$ with vectorial angles α, β such that $\alpha - \beta = 2\gamma =$ constant, then show that the chord PQ touches the conic $\dfrac{l\cos\gamma}{r} = 1 - e\cos\gamma\,\cos\theta$ and this conic has the same directrix as the original one.

21. If PQ be a variable chord of a conic $\dfrac{l}{r} = 1 - e\cos\theta$ subtending a constant angle 2β at the focus S, where S is the pole, then show that the locus of the foot of the perpendicular from S on PQ is

$$r^2(e^2 - \sec^2\beta) + 2elr\cos\theta + l^2 = 0.$$

22. If the normal be drawn at one extremity $(l, \pi/2)$ of the latus rectum PSP' of the conic $\dfrac{l}{r} = 1 + e\cos\theta$, where S is the pole, then show that the distance from the focus S of the other point in which the normal meets the conic is $\dfrac{l(1+3e^2+e^4)}{(1+e^2-e^4)}$.

[*Hints*: Equation of the normal at the point $(l, \pi/2)$ is

$$\frac{el\sin(\pi/2)}{1+e\cos(\pi/2)}\frac{1}{r} = e\sin\theta + \sin\{\theta - (\pi/2)\}$$

$$\text{or,}\qquad \frac{el}{r} = e\sin\theta - \cos\theta.$$

At the common point, we have

$$e(1+e\cos\theta) = e\sin\theta - \cos\theta$$

$$\text{or,}\qquad \frac{\cos\theta}{1-\sin\theta} = -\frac{e}{1+e^2}$$

$$\text{or,}\qquad \frac{\cos^2\theta}{(1-\sin\theta)^2} = \frac{e^2}{(1+e^2)^2}$$

$$\text{or,}\qquad \frac{1+\sin\theta}{1-\sin\theta} = \frac{e^2}{(1+e^2)^2}$$

$$\text{or,}\qquad \sin\theta = \frac{e^2-(1+e^2)^2}{e^2+(1+e^2)^2}.$$

$$\text{So,}\qquad \cos\theta = -\frac{e(1-\sin\theta)}{1+e^2}$$

$$= -\frac{e}{1+e^2}\left(1 - \frac{e^2-(1+e^2)^2}{e^2+(1+e^2)^2}\right) = \frac{-2e(1+e^2)}{e^2+(1+e^2)^2}.$$

So, the required distance is

$$\frac{l}{1+e\cos\theta} = \frac{l}{1 - \dfrac{2e^2(1+e^2)}{e^2+(1+e^2)^2}} = \frac{l(1+3e^2+e^4)}{(1+3e^2+e^4)-2e^2-2e^4} = \frac{l(1+3e^2+e^4)}{(1+e^2-e^4)}.$$

23. If the normals at α, β, γ on the parabola $\dfrac{l}{r} = 1 + \cos\theta$ be concurrent at the point (ρ, ϕ), then show that $\alpha + \beta + \gamma = 2\phi$.

24. If the normals at $\alpha, \beta, \gamma, \delta$ on the conic $\frac{l}{r} = 1 + e\cos\theta$ be concurrent at a point, then show that

$$\tan(\alpha/2)\tan(\beta/2)\tan(\gamma/2)\tan(\delta/2) + \left(\frac{1+e}{1-e}\right)^2 = 0.$$

25. Prove that the equation of the chord joining the points $\theta = \theta_1$ and $\theta = \theta_2$ on the circle $r = 2a\cos\theta$ is $r\cos(\theta - \theta_1 - \theta_2) = 2a\cos\theta_1\cos\theta_2$.

26. A circle of given diameter d passes through the focus of a given conic and cuts it in four points whose distances from the focus are r_1, r_2, r_3, r_4. Prove that

$$\frac{1}{r_1} + \frac{1}{r_2} + \frac{1}{r_3} + \frac{1}{r_4} = \frac{2}{l} \quad \text{and} \quad r_1r_2r_3r_4 = \left(\frac{ld}{e}\right)^2$$

where l is the semi-latus rectum and e is the eccentricity of the conic.

[*Hints*: Let us take the equations of circle and conic as

$$r = d\cos(\theta - \alpha) \tag{1}$$

$$\text{and} \quad \frac{l}{r} = 1 - e\cos\theta. \tag{2}$$

From the equation of circle

$$r = d[\cos\theta\cos\alpha + \sin\theta\sin\alpha]$$

$$\text{or,} \quad (r - d\cos\theta\cos\alpha)^2 = d^2\sin^2\alpha(1 - \cos^2\theta). \tag{3}$$

Eliminating θ from Eqns. (2) and (3) we get

$$\left[er - d\left(1 - \frac{l}{r}\right)\cos\alpha\right]^2 = d^2\sin^2\alpha\left[e^2 - \left(1 - \frac{l}{r}\right)^2\right]$$

$$\text{or,} \quad [er^2 - d(r - l)\cos\alpha]^2 = d^2\sin^2\alpha[r^2e^2 - (r - l)^2]$$

$$\text{or,} \quad e^2r^4 - 2er^2d(r - l)\cos\alpha + d^2(r - l)^2\cos^2\alpha$$

$$= r^2e^2d^2\sin^2\alpha - (r - l)^2d^2\sin^2\alpha$$

$$\text{or,} \quad e^2r^4 - 2er^3d\cos\alpha + dr^2[2el\cos\alpha + d - e^2d\sin^2\alpha] - 2rld^2 + d^2l^2 = 0.$$

If r_1, r_2, r_3, r_4 be its roots, then

$$\sum r_1r_2r_3 = \frac{2ld^2}{e^2} \quad \text{and} \quad r_1r_2r_3r_4 = \left(\frac{ld}{e}\right)^2.$$

Dividing we get

$$\frac{1}{r_1} + \frac{1}{r_2} + \frac{1}{r_3} + \frac{1}{r_4} = \frac{2}{l}.]$$

27. Find the polar equation of the straight line joining two points on the circle $r = 2a\cos\theta$ with α, β as their vectorial angles, and deduce the equation of the tangent at $\theta = \alpha$.

Chapter 10

General Second Degree Equation

10.1 Introduction

In Cartesian coordinate system, a curve is represented by a second degree equation in x, y. The joint equation of two straight lines, whose individual equations are linear in nature, can also be represented by a second degree equation. The general form of a second degree equation in Cartesian system is

$$ax^2 + 2hxy + by^2 + 2gx + 2fy + c = 0 \tag{1}$$

where at least one of the coefficients a, b, h is non-zero.

The well-known forms of parabola, ellipse and hyperbola, which are also known as basic/canonical/normal/standard form are

$$y^2 = 4ax, \quad \frac{x^2}{a^2} + \frac{y^2}{b^2} = 1 \quad \text{and} \quad \frac{x^2}{a^2} - \frac{y^2}{b^2} = 1.$$

So, by transformation of coordinates, i.e. by translation and rotation, it is possible to reduce the above general equation to the basic forms and the following successive transformations are required to do that:

(i) the terms in x and y are to be removed by shifting the origin suitably, i.e. by translation;
(ii) the xy-term is to be removed by rotating the coordinate axes suitably.

The nature of the curve, i.e. whether it is a parabola, ellipse, hyperbola or pair of straight lines, represented by the general equation is determined by the following quantities

$$\Delta = \begin{vmatrix} a & h & g \\ h & b & f \\ g & f & c \end{vmatrix}, \quad D = ab - h^2 \quad \text{and} \quad K = a + b.$$

which are invariant under translation and rotation of coordinate axes.

If $\Delta = 0$, then Eqn. (1) represents a pair of straight lines which are intersecting or parallel (or coincident) according as $D \neq 0$ or $D = 0$. Further, if in the case of intersecting lines, $K = 0$, then the lines will be perpendicular.

If $\Delta \neq 0$, then Eqn. (1) represents a conic.

If we take the line $ax + by + c = 0$ as directrix and the point (h, k) as focus, then for a point $P(x, y)$ on the conic

$$SP = \sqrt{(x-h)^2 + (y-k)^2}$$

and

$$PM = \frac{ax + by + c}{\sqrt{a^2 + b^2}}.$$

Using $SP = ePM$, we get

$$[(x-h)^2 + (y-k)^2][a^2 + b^2] = e^2(ax + by + c)^2$$

or, $$[a^2 + b^2][x^2 + y^2 - 2xh - 2yk + h^2 + k^2]$$
$$= e^2[a^2x^2 + b^2y^2 + 2abxy + 2axc + 2bcy + c^2]$$

or, $$Ax^2 + 2Hxy + By^2 + 2Gx + 2Fy + C = 0. \tag{1a}$$

where

$A = a^2(1 - e^2) + b^2$,
$B = b^2(1 - e^2) + a^2$,
$H = -abe^2$,
$G = -(a^2 + b^2)h - e^2ac$,
$F = -(a^2 + b^2)k - e^2bc$ and
$C = (a^2 + b^2)(h^2 + k^2) - e^2c^2$.

In this case

$$D = AB - H^2$$
$$= [a^2(1 - e^2) + b^2][b^2(1 - e^2) + a^2] - [-abe^2]^2$$
$$= (a^2 + b^2)^2(1 - e^2).$$

So, if $H = 0$, then $e^2 = 0 \Rightarrow e = 0$ and in that case $A = a^2 + b^2 = B$.
So, the Eqn. (1a) represents a circle if $A = B$ and $H = 0$, and the same represents hyperbola, parabola or ellipse according as $e >, =, < 1$, i.e. according as $AB <, =, > H^2$. Again, if in the case of hyperbola, $K = 0$, then the hyperbola will be rectangular. The natures of the conics are given in the following table:

D	Δ	**Standard form**	**Nature**
$D < 0$	$\Delta = 0$	$y^2 = k^2x^2$	Pair of intersecting straight lines
$D = 0$	$\Delta = 0$	$y^2 = k^2$	Pair of parallel/ coincident straight lines
$D > 0$	$\Delta = 0$	$A^2x^2 + B^2y^2 = 0$	Point ellipse/pair of imaginary straight lines
$D > 0$	$\Delta < 0$	$\frac{x^2}{A^2} + \frac{y^2}{B^2} = 1$ or, $x^2 + y^2 = A^2$	Ellipse/circle
$D > 0$	$\Delta > 0$	$\frac{x^2}{A^2} + \frac{y^2}{B^2} = -1$	Imaginary ellipse
$D < 0$	$\Delta \neq 0$	$\frac{x^2}{A^2} - \frac{y^2}{B^2} = \pm 1$	Hyperbola
$D = 0$	$\Delta \neq 0$	$y^2 = 4ax$	Parabola

10.2 Conics Having Origin as the Centre

Point of intersection of any two diameters of a conic is called its centre, i.e. at the centre, any chord of a conic through it is bisected there. Conics having centre are called *central conics*, otherwise it is called a *non-central conic*. Circle, ellipse, hyperbola are central conics and the parabola is a non-central conic.

Let POQ be a chord of conic given by Eqn. (1), whose centre is origin, O. If the coordinates of P be (x_1, y_1), then the coordinates of Q will be $(-x_1, -y_1)$. As both the points lie on the conic given by Eqn. (1), then we have

$$ax_1^2 + 2hx_1y_1 + by_1^2 + 2gx_1 + 2fy_1 + c = 0$$

and
$$ax_1^2 + 2hx_1y_1 + by_1^2 - 2gx_1 - 2fy_1 + c = 0.$$

Subtracting we get

$$gx_1 + fy_1 = 0.$$

As the above equation is true for all points P and Q, we must have

$$g = 0 \quad \text{and} \quad f = 0$$

which implies that equation of a conic, whose centre is the origin, will not contain terms in x and y only.

10.3 Centre of a Conic

If (α, β) be the centre of the conic given by Eqn. (1), then shifting the origin to the point (α, β) without rotating the coordinate axes, i.e. putting $x = X + \alpha$ and $y = Y + \beta$ in the Eqn. (1) we get

$$a(X+\alpha)^2 + 2h(X+\alpha)(Y+\beta) + b(Y+\beta)^2 + 2g(X+\alpha) + 2f(Y+\beta) + c = 0$$

or,
$$aX^2 + 2hXY + bY^2 + 2X(a\alpha + h\beta + g) + 2Y(h\alpha + b\beta + f) + a\alpha^2 + 2h\alpha\beta + b\beta^2 + 2g\alpha + 2f\beta + c = 0.$$

As the centre of the above conic is origin, there would be no term in X and Y only, which leads that

$$a\alpha + h\beta + g = 0 \qquad \text{(2a)}$$

$$h\alpha + b\beta + f = 0 \qquad \text{(2b)}$$

The above system of equations is solvable if $ab \neq h^2$, i.e. the conic given by Eqn. (1) has unique centre, if $ab \neq h^2$. Now, solving the above equations we get

$$\alpha = \frac{hf - hg}{ab - h^2} \quad \text{and} \quad \beta = \frac{hg - af}{ab - h^2}.$$

So, the centre of the conic given by Eqn. (1) is

$$\left(\frac{hf - bg}{ab - h^2}, \frac{hg - af}{ab - h^2}\right), \quad \text{if } ab \neq h^2.$$

If $ab = h^2$ and $hf - bg \neq 0$, $hg - af \neq 0$, then the conic has a centre at infinity, i.e. the conic is non-central.

Again, if $ab = h^2$ and $hf - bg = 0$, $hg - af = 0$, i.e. when the Eqns. (2a) and (2b) are identical, then any point on the line can be taken as centre of the conic and hence, in this case, the conic has infinitely many centres.

10.4 Reduction of a Central Conic to Standard Form

If C: (α, β) be the centre of the conic given by Eqn. (1), then shifting the origin to the point (α, β) without rotating the coordinate axes, i.e. putting $x = X + \alpha$ and $y = Y + \beta$ in the Eqn. (1) we get

$$aX^2 + 2hXY + bY^2 + 2X(a\alpha + h\beta + g) + 2Y(h\alpha + b\beta + f) \\ + a\alpha^2 + 2h\alpha\beta + b\beta^2 + 2g\alpha + 2f\beta + c = 0.$$

As the centre of the above conic is origin, there would be no term in X and Y only, which leads that

$$a\alpha + h\beta + g = 0$$
$$h\alpha + b\beta + f = 0.$$

Now, solving the above equations we get the centre of the conic given by Eqn. (1) is

$$\left(\frac{hf - bg}{ab - h^2}, \frac{hg - af}{ab - h^2}\right).$$

We further note that

$$\begin{aligned} & a\alpha^2 + 2h\alpha\beta + b\beta^2 + 2g\alpha + 2f\beta + c \\ &= \alpha(a\alpha + h\beta + g) + \beta(h\alpha + b\beta + f) + g\alpha + f\beta + c \\ &= g\alpha + f\beta + c \\ &= g\frac{hf - bg}{ab - h^2} + f\frac{hg - af}{ab - h^2} + c \\ &= \frac{abc + 2fgh - af^2 - bg^2 - ch^2}{ab - h^2} = \frac{\Delta}{D} = -d. \quad \text{(say)} \end{aligned}$$

So, the equation is transformed to

$$aX^2 + 2hXY + bY^2 = d. \tag{3}$$

If the coordinate axes be rotated through an angle θ, then the new coordinates x', y' are related to old coordinates X, Y as

$$X = x'\cos\theta - y'\sin\theta \quad \text{and} \quad Y = x'\sin\theta + y'\cos\theta.$$

Putting these in the Eqn. (3) we get

$$a(x'\cos\theta - y'\sin\theta)^2 + 2h(x'\cos\theta - y'\sin\theta)(x'\sin\theta + y'\cos\theta) \\ + b(x'\sin\theta + y'\cos\theta)^2 = d$$

i.e., $$(a\cos^2\theta + 2h\sin\theta\cos\theta + b\sin^2\theta)x'^2 + 2[h(\cos^2\theta - \sin^2\theta) \\ - (a - b)\sin\theta\cos\theta]x'y' + (a\sin^2\theta - 2h\sin\theta\cos\theta + b\cos^2\theta)y'^2 = d.$$

Next we choose θ in such a way that the $x'y'$ term removes which leads that θ must satisfy the equation

$$h(\cos^2\theta - \sin^2\theta) - (a-b)\sin\theta\cos\theta = 0$$

$$\Rightarrow \quad \tan 2\theta = \frac{2h}{a-b}$$

$$\Rightarrow \quad \theta = \frac{1}{2}\tan^{-1}\left(\frac{2h}{a-b}\right).$$

For this value of θ, the Eqn. (3) becomes

$$a'x'^2 + b'y'^2 = d \tag{4}$$

which is the normal form of the central conic given by Eqn. (1), where a', b' satisfy the equations (by the properties of invariance)

$$a' + b' = a + b \quad \text{and} \quad a'b' = ab - h^2.$$

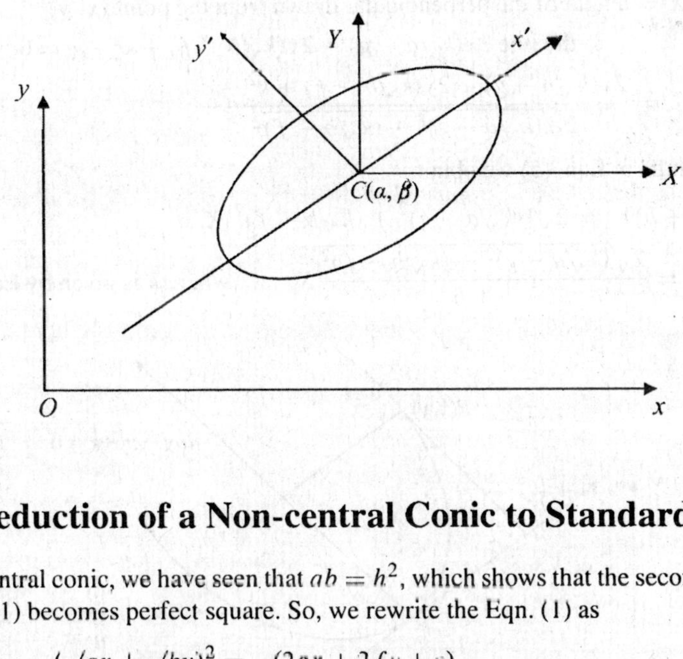

10.5 Reduction of a Non-central Conic to Standard Form

For a non-central conic, we have seen that $ab = h^2$, which shows that the second degree term of the Eqn. (1) becomes perfect square. So, we rewrite the Eqn. (1) as

$$(\sqrt{a}x + \sqrt{b}y)^2 = -(2gx + 2fy + c)$$

$$\Rightarrow \quad (\sqrt{a}x + \sqrt{b}y)^2 + 2k(\sqrt{a}x + \sqrt{b}y) + k^2$$

$$= -(2gx + 2fy + c) + 2k(\sqrt{a}x + \sqrt{b}y) + k^2$$

$$\Rightarrow \quad (\sqrt{a}x + \sqrt{b}y + k)^2 = 2x(k\sqrt{a} - g) + 2y(k\sqrt{b} - f) + k^2 - c \tag{5}$$

which is of the form $Y^2 = 4AX$, i.e. of the form of a parabola and in case of a parabola the axis and the tangent at the vertex are perpendicular. So, we choose k in such a way that the lines

$$\sqrt{a}x + \sqrt{b}y + k = 0 \tag{6}$$

and

$$2x(k\sqrt{a} - g) + 2y(k\sqrt{b} - f) + k^2 - c = 0 \tag{7}$$

become perpendicular, which gives

$$\frac{\sqrt{a}(k\sqrt{a}-g)}{\sqrt{b}(k\sqrt{b}-f)} = -1$$

$$\Rightarrow \quad k = \frac{g\sqrt{a}+f\sqrt{b}}{a+b}. \tag{8}$$

The lines given by the Eqns. (6) and (7) are the axis and the tangent at the vertex respectively which intersect at the vertex of the parabola.

If we choose these lines as coordinate axes, i.e. as $Y = 0$ and $X = 0$ respectively, then we get

$$\begin{aligned} Y &= \text{length of the perpendicular drawn from the point } (x, y) \\ &\qquad \text{to the line } \sqrt{a}x + \sqrt{b}y + k = 0 \\ &= \frac{\sqrt{a}x+\sqrt{b}y+k}{\sqrt{a+b}} \end{aligned}$$

and

$$\begin{aligned} X &= \text{length of the perpendicular drawn from the point } (x, y) \\ &\qquad \text{to the line } 2x(k\sqrt{a}-g)+2y(k\sqrt{b}-f)+k^2-c=0 \\ &= \frac{2x(k\sqrt{a}-g)+2y(k\sqrt{b}-f)+k^2-c}{2\sqrt{(k\sqrt{a}-g)^2+(k\sqrt{b}-f)^2}}. \end{aligned}$$

Using these results in Eqn. (5) we obtain

$$(a+b)Y^2 = 2\sqrt{[(k\sqrt{a}-g)^2+(k\sqrt{b}-f)^2]}X$$

$$\text{or,} \quad Y^2 = \frac{2\sqrt{(k\sqrt{a}-g)^2+(k\sqrt{b}-f)^2}}{a+b}X, \qquad \text{where } k \text{ is given by Eqn. (8).}$$

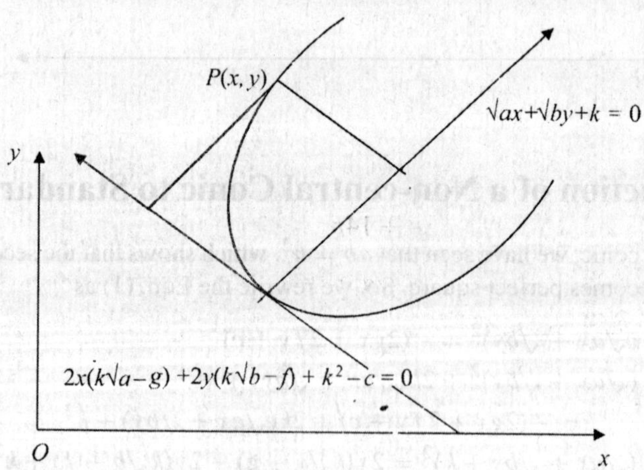

10.6 Contact of Two Conics

We have seen that the general equation of a conic consists of six constants, viz. a, h, b, g, f, c, of which five are independent as one can obtain an equation of a conic with five constants dividing the general equation by any one of those constants. To determine five constants we need five equations which can be formed using five conditions.

If we eliminate any one of x, y from two general second degree equations, we would get a fourth degree equation in x or y, which implies that two conics, in general, intersect at four points two or more of which may coincide and two or all of them may be imaginary.

If the conics intersect in four real points all of which are distinct, then the contact is said to be of zero order and if two, three or four points of intersection coincide, then the contact is said to be first, second or third order respectively.

The equation of a conic which passes through the point of intersection of the conics $S_1 = 0$ and $S_2 = 0$ is given by $S_1 + \lambda S_2 = 0$ as it is satisfied by the common points of the conics $S_1 = 0$ and $S_2 = 0$. Similarly, equation of a conic which passes through the point of intersection of the conic $S = 0$ and the lines $u = 0$ and $v = 0$ is $S + \lambda uv = 0$. If the straight lines $u = 0$ and $v = 0$ coincide, then the above equation becomes $S + \lambda u^2 = 0$. In all the cases the unknown constant λ is to be determined using some given conditions.

Examples

▶▷▶ **1**. *Reduce the following equation to its canonical form:*

$$11x^2 - 4xy + 14y^2 - 58x - 44y + 71 = 0.$$

Solution: Here $a = 11, b = 14, c = 71, h = -2, g = -29$ and $f = -22$.

So, $abc + 2fgh - af^2 - bg^2 - ch^2 = -9000 \neq 0$, hence the given equation is not of pair of straight lines. Also, $ab - h^2 = 150 \neq 0$, so the given equation is of central conic.

Let (α, β) be the centre of the conic. Shifting origin to the centre of the conic, i.e. putting $x = X + \alpha$ and $y = Y + \beta$ in the given equation we get

$$\begin{aligned} &11X^2 - 4XY + 14Y^2 + 2X(11\alpha - 2\beta - 29) + 2Y(-2\alpha + 14\beta - 22) \\ &\quad + 11\alpha^2 - 4\alpha\beta + 14\beta^2 - 58\alpha - 44\beta + 71 = 0. \end{aligned} \tag{1}$$

As the centre of the above conic is origin, there would be no term in X and Y only, which leads that

$$\begin{aligned} 11\alpha - 2\beta - 29 &= 0 \\ -2\alpha + 14\beta - 22 &= 0. \end{aligned}$$

Solving, we get the centre of the conic is (3, 2). Hence, the Eqn. (1) becomes

$$11X^2 - 4XY + 14Y^2 - 60 = 0. \tag{2}$$

Now, we rotate the X-Y axes in such a way that XY-term of the Eqn. (2) removes and the Eqn. (2) reduces to

$$ax'^2 + by'^2 - 60. \tag{3}$$

Then by properties of invariance (discussed in Chapter 1) we have

$$a + b = 11 + 14 = 25$$

and $$ab - 0^2 = 11 \times 14 - (-2)^2 = 150.$$

Solving we get $a = 10, 15$ and $b = 15, 10$. Hence, the canonical form of the given equation is

$$10x'^2 + 15y'^2 = 60 \quad \text{or,} \quad 15x'^2 + 10y'^2 = 60$$

$$\text{i.e.,} \quad \frac{x'^2}{6} + \frac{y'^2}{4} = 1 \quad \text{or,} \quad \frac{x'^2}{4} + \frac{y'^2}{6} = 1.$$

▶▷ **2.** *Reduce the following equation to its canonical form:*

$$x^2 + 4xy + y^2 - 2x + 2y + 6 = 0.$$

Solution: Here $a = 1, b = 1, c = 6, h = 2, g = -1$ and $f = 1$.

So, $abc + 2fgh - af^2 - bg^2 - ch^2 = -24 \neq 0$; hence the given equation is not of pair of straight lines. Also, $ab - h^2 = -3 \neq 0$, so the given equation is of central conic.

Let (α, β) be the centre of the conic. Shifting origin to the centre of the conic, i.e. putting $x = X + \alpha$ and $y = Y + \beta$ in the given equation we get

$$X^2 + 4XY + Y^2 + 2X(\alpha + 2\beta - 1) + 2Y(2\alpha + \beta + 1)$$
$$+ \alpha^2 + 4\alpha\beta + \beta^2 - 2\alpha + 2\beta + 6 = 0. \qquad (1)$$

As the centre of the above conic is origin, there would be no term in X and Y only, which leads that

$$\alpha + 2\beta - 1 = 0$$
$$2\alpha + \beta + 1 = 0.$$

Solving, we get the centre of the conic is $(-1, 1)$. Hence, the Eqn. (1) becomes

$$X^2 + 4XY + Y^2 + 8 = 0. \qquad (2)$$

Now, we rotate the X-Y axes in such a way that XY-term of the Eqn. (2) removes and the Eqn. (2) reduces to

$$ax'^2 + by'^2 + 8 = 0. \qquad (3)$$

Then by properties of invariance, we have

$$a + b = 1 + 1 = 2$$

and $$ab - 0^2 = 1 \times 1 - (2)^2 = -3.$$

Solving we get $a = -1, 3$ and $b = 3, -1$. Hence, the canonical form of the given equation is

$$-x'^2 + 3y'^2 + 8 = 0 \quad \text{or,} \quad 3x'^2 - y'^2 + 8 = 0$$

$$\text{i.e.,} \quad \frac{x'^2}{8} - \frac{y'^2}{8/3} = 1 \quad \text{or,} \quad \frac{y'^2}{8} - \frac{x'^2}{8/3} = 1.$$

▶▷ **3.** *Reduce the following equation to its canonical form:*

$$2x^2 - 4xy - y^2 + 20x - 2y + 17 = 0.$$

Solution: Here $a = 2, b = -1, c = 17, h = -2, g = 10$ and $f = -1$.

So, $abc + 2fgh - af^2 - bg^2 - ch^2 = -34 + 40 - 2 + 100 - 68 = 36 \neq 0$; hence the given equation is not of pair of straight lines. Also, $ab - h^2 = -2 - 4 = -6 \neq 0$, so the given equation is of central conic.

Let (α, β) be the centre of the conic. Shifting origin to the centre of the conic, i.e. putting $x = X + \alpha$ and $y = Y + \beta$ in the given equation we get

$$2X^2 - 4XY - Y^2 + 2X(2\alpha - 2\beta + 10) + 2Y(-2\alpha - \beta - 1) \\ + 2\alpha^2 - 4\alpha\beta - \beta^2 + 20\alpha - 2\beta + 17 = 0. \tag{1}$$

As the centre of the above conic is origin, there would be no term in X and Y only, which leads that

$$2\alpha - 2\beta + 10 = 0$$
$$2\alpha + \beta + 1 = 0.$$

Solving, we get the centre of the conic is $(-2, 3)$. Hence, the Eqn. (1) becomes

$$2X^2 - 4XY - Y^2 - 6. \tag{2}$$

Now, we rotate the X-Y axes through an angle θ. So the Eqn. (2) becomes

$$2(x'\cos\theta - y'\sin\theta)^2 - 4(x'\cos\theta - y'\sin\theta)(x'\sin\theta + y'\cos\theta) \\ - (x'\sin\theta + y'\cos\theta)^2 = 6$$

$$\Rightarrow \quad (2\cos^2\theta - 4\cos\theta\sin\theta - \sin^2\theta)x'^2 + [-4\cos 2\theta - 3\sin 2\theta]x'y' \\ + (2\sin^2\theta + 4\cos\theta\sin\theta - \cos^2\theta)y'^2 = 6.$$

To remove $x'y'$ we need that

$$-4\cos 2\theta - 3\sin 2\theta = 0$$

$$\text{i.e.,} \quad \frac{\cos 2\theta}{3} = \frac{\sin 2\theta}{-4} = \frac{1}{5}.$$

Hence, the above equation becomes

$$[(1 + 3/5) + 8/5 - (1 - 3/5)/2]x'^2 + [(1 - 3/5) - 8/5 - (1 + 3/5)/2]y'^2 = 6$$

$$\text{or,} \quad 3x'^2 - 2y'^2 = 6$$

$$\text{or,} \quad \frac{x'^2}{2} - \frac{y'^2}{3} - 1.$$

▶▶ 4. *Reduce the following equation to its canonical form:*

$$7x^2 - 2xy + 7y^2 - 16x + 16y - 8 = 0.$$

Solution: Here $a = 7, b = 7, c = -8, h = -1, g = -8$ and $f = 8$.

So, $abc + 2fgh - af^2 - bg^2 - ch^2 = -256 \neq 0$, hence the given equation is not of pair of straight lines. Also, $ab - h^2 = 48 \neq 0$, so the given equation is of central conic.

Let (α, β) be the centre of the conic. Shifting origin to the centre of the conic, i.e. putting $x = X + \alpha$ and $y = Y + \beta$ in the given equation we get

$$7X^2 - 2XY + 7Y^2 + 2X(7\alpha - \beta - 8) + 2Y(-\alpha + 7\beta + 8) \\ + 7\alpha^2 - 2\alpha\beta + 7\beta^2 - 16\alpha + 16\beta - 8 = 0. \tag{1}$$

As the centre of the above conic is origin, there would be no terms in X and Y only, which leads that

$$7\alpha - \beta - 8 = 0$$
$$\alpha - 7\beta - 8 = 0.$$

Solving, we get the centre of the conic is $(1, -1)$. Hence the Eqn. (1) becomes

$$7X^2 - 2XY + 7Y^2 = 24. \tag{2}$$

Now, we rotate the X-Y axes through an angle θ. So the Eqn. (2) becomes

$$7(x'\cos\theta - y'\sin\theta)^2 - 2(x'\cos\theta - y'\sin\theta)(x'\sin\theta + y'\cos\theta)$$
$$+ 7(x'\sin\theta + y'\cos\theta)^2 = 24$$
$$\Rightarrow \quad (7 - \sin 2\theta)x'^2 - 2\cos 2\theta x'y' + (7 + \sin 2\theta)y'^2 = 24.$$

To remove $x'y'$ we need that

$$\cos 2\theta = 0$$
$$\text{i.e.,} \quad \theta = \pi/4.$$

Hence the above equation becomes

$$6x'^2 + 8y'^2 = 24$$
$$\text{or,} \quad \frac{x'^2}{4} + \frac{y'^2}{3} = 1.$$

▶▷▶ 5. *Reduce the following equation to its canonical form:*

$$4x^2 - 4xy + y^2 + 2x - 26y + 9 = 0.$$

Solution: Here $a = 4, b = 1, c = 9, h = -2, g = 1$ and $f = -13$.

So, $abc + 2fgh - af^2 - bg^2 - ch^2 = -625 \neq 0$, hence the given equation is not of pair of straight lines. Also, $ab - h^2 = 0$, so the given equation is of non-central conic.

Now rotating the x-y axes through an angle θ, the given equation reduces to

$$4(x'\cos\theta - y'\sin\theta)^2 - 4(x'\cos\theta - y'\sin\theta)(x'\sin\theta + y'\cos\theta)$$
$$+ (x'\sin\theta + y'\cos\theta)^2 + 2(x'\cos\theta - y'\sin\theta) - 26(x'\sin\theta + y'\cos\theta) + 9 = 0$$
$$\text{or,} \quad (4\cos^2\theta - 4\cos\theta\sin\theta + \sin^2\theta)x'^2 + (4\sin^2\theta + 4\cos\theta\sin\theta + \cos^2\theta)y'^2$$
$$+ (-8\cos\theta\sin\theta + 2\cos\theta\sin\theta + 4\sin^2\theta - 4\cos^2\theta)x'y'$$
$$+ (2\cos\theta - 26\sin\theta)x' - (2\sin\theta + 26\cos\theta)y' + 9 = 0.$$

If we choose θ in such a way that

$$(-8\cos\theta\sin\theta + 2\cos\theta\sin\theta + 4\sin^2\theta - 4\cos^2\theta) = 0$$
$$\text{i.e.,} \quad \tan 2\theta = -4/3$$
$$\text{i.e.,} \quad \tan\theta = -1/2 \text{ or } 2, \text{ then the } x'y' \text{ will be removed.}$$

Taking $\tan\theta = 2$, we get $\sin\theta = \dfrac{2}{\sqrt{5}}$ and $\cos\theta = \dfrac{1}{\sqrt{5}}$, and using these values we get

$$
\begin{aligned}
& 5y'^2 - 6\sqrt{5}y' - 10\sqrt{5}x' + 9 = 0 \\
\text{or,}\quad & (\sqrt{5}y' - 3)^2 = 10\sqrt{5}x' \\
\text{or,}\quad & \left(y' - \frac{3}{\sqrt{5}}\right)^2 = \frac{10}{\sqrt{5}}x'.
\end{aligned}
$$

Now, taking $X = x'$ and $Y = y' - \dfrac{3}{\sqrt{5}}$, we get

$$Y^2 = \frac{10}{\sqrt{5}}X, \quad \text{which is a parabola.}$$

▶▷▶ **6**. *Reduce the following equation to its canonical form:*

$$16x^2 - 24xy + 9y^2 - 104x - 172y + 44 = 0.$$

Solution: Here $a = 16, b = 9, c = 44, h = -12, g = -52$ and $f = -86$.

So, $abc + 2fgh - af^2 - bg^2 - ch^2 \neq 0$, hence the given equation is not of pair of straight lines. Also, $ab - h^2 = 0$, so the given equation is of non-central conic. Now, we rewrite the equation as

$$
\begin{aligned}
(4x - 3y)^2 &= 104x + 172y - 44 \\
\text{or,}\quad (4x - 3y + k)^2 &= (4x - 3y)^2 + 2k(4x - 3y) + k^2 \\
&= 104x + 172y - 44 + 8kx - 6ky + k^2 \\
&= 4x(26 + 2k) + 2y(86 - 3k) + k^2 - 44.
\end{aligned}
$$

Now, we choose k in such a way that the lines

$$4x - 3y + k = 0 \quad \text{and} \quad 4x(26 + 2k) + 2y(86 - 3k) + k^2 - 44 = 0$$

become perpendicular to each other.

$$
\begin{aligned}
\text{i.e.,}\quad & \frac{4}{3} \times \frac{2(26 + 2k)}{86 - 3k} = 1 \\
\text{or,}\quad & 208 + 16k = 258 - 9k \\
\text{or,}\quad & k = 2.
\end{aligned}
$$

Thus the above equation becomes

$$(4x - 3y + 2)^2 = 120x + 160y - 40 = 40(3x + 4y - 1).$$

Now, if we take $4x - 3y + 2 = 0$ as X-axis and $3x + 4y - 1 = 0$ as Y-axis, then from any point (x, y) length of the perpendicular drawn to the X and Y axes are given by

$$Y = \frac{4x - 3y + 2}{5} \quad \text{and} \quad X = \frac{3x + 4y - 1}{5}.$$

Using these results in the above equation, we get

$$
\begin{aligned}
& 25Y^2 = 200X \\
\text{or,}\quad & Y^2 = 8X.
\end{aligned}
$$

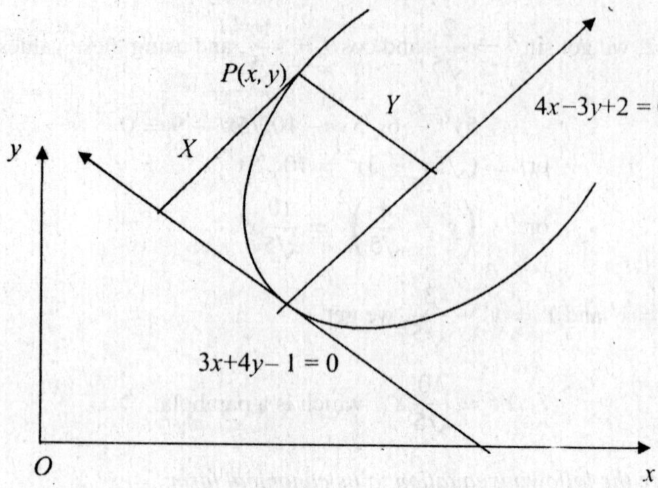

▶▷▶ 7. *Reduce the following equation to its canonical form:*

$$4x^2 + 4xy + y^2 - 12x - 6y + 5 = 0.$$

Solution: Here $a = 4$, $b = 1$, $c = 5$, $h = 2$, $g = -6$ and $f = -3$.

So, $abc + 2fgh - af^2 - bg^2 - ch^2 = 0$, hence the given equation is of pair of straight lines. Also, $ab - h^2 = 0$, so the given equation represents a pair of parallel straight lines.

Now, rotating the x-y axes through an angle θ, the given equation reduces to

$$4(x'\cos\theta - y'\sin\theta)^2 + 4(x'\cos\theta - y'\sin\theta)(x'\sin\theta + y'\cos\theta)$$
$$+ (x'\sin\theta + y'\cos\theta)^2 - 12(x'\cos\theta - y'\sin\theta) - 6(x'\sin\theta + y'\cos\theta) + 5 = 0$$

or, $$(4\cos^2\theta + 4\cos\theta\sin\theta + \sin^2\theta)x'^2 + (4\sin^2\theta - 4\cos\theta\sin\theta + \cos^2\theta)y'^2$$
$$+ (-8\cos\theta\sin\theta + 2\cos\theta\sin\theta - 4\sin^2\theta + 4\cos^2\theta)x'y'$$
$$- (12\cos\theta + 6\sin\theta)x' + (12\sin\theta - 6\cos\theta)y' + 5 = 0$$

or, $$\left[\frac{5}{2} + \frac{3}{2}\cos 2\theta + 2\sin 2\theta\right]x'^2 + \left[\frac{5}{2} - \frac{3}{2}\cos 2\theta - 2\sin 2\theta\right]y'^2$$
$$+ (-3\sin 2\theta + 4\cos 2\theta)x'y' - 2(6\cos\theta + 3\sin\theta)x'$$
$$+ 2(6\sin\theta - 3\cos\theta)y' + 5 = 0.$$

If we choose θ in such a way that

$$(-3\sin 2\theta + 4\cos 2\theta) = 0$$

i.e., $$\frac{\cos 2\theta}{3} = \frac{\sin 2\theta}{4} = \frac{1}{5},$$ then the $x'y'$ will be removed.

Now, $\tan 2\theta = 4/3$ gives $\sin\theta = 1/\sqrt{5}$ and $\cos\theta = 2/\sqrt{5}$, and using these values we get

$$5x'^2 - 6\sqrt{5}x' + 5 = 0$$

or, $$(x' - \sqrt{5})(x' - 1/\sqrt{5}) = 0$$

which represents two parallel straight lines $x' - \sqrt{5} = 0$ and $x' - 1/\sqrt{5} = 0$.

▶▷▶ **8**. *Reduce the following equation to its canonical form:*

$$6x^2 - 5xy - 6y^2 + 14x + 5y + 4 = 0.$$

Solution: Here $a = 6$, $b = -6$, $c = 4$, $h = -5/2$, $g = 7$ and $f = 5/2$.

So, $abc + 2fgh - af^2 - bg^2 - ch^2 = 0$, hence the given equation is of pair of straight lines. Also, $ab - h^2 \neq 0$, so the given equation represents a pair of intersecting straight lines.

Let (α, β) be the point of intersection of the given pair of straight lines. Shifting origin to the point (α, β), i.e. putting $x = X + \alpha$ and $y = Y + \beta$ in the given equation we get

$$\begin{aligned} &6X^2 - 5XY - 6Y^2 + 2X(6\alpha - 5\beta/2 + 7) + 2Y(-5\alpha/2 - 6\beta + 5/2) \\ &\quad + 6\alpha^2 - 5\alpha\beta - 6\beta^2 + 14\alpha + 5\beta + 4 = 0. \end{aligned} \tag{1}$$

As the pair of straight lines passing through the origin is a second degree homogeneous equation in X, Y, there would be no constant term and the terms in X and Y only, which leads that

$$\begin{aligned} &12\alpha - 5\beta + 14 = 0, \\ &5\alpha + 12\beta - 5 = 0 \\ \text{and} \quad &6\alpha^2 - 5\alpha\beta - 6\beta^2 + 14\alpha + 5\beta + 4 = 0. \end{aligned}$$

Solving, the first two equations, we get the $\alpha = -11/13$ and $\beta = 10/13$ which also satisfy the third equation. Hence, the Eqn. (1) becomes

$$6X^2 - 5XY - 6Y^2 = 0. \tag{2}$$

Now, we rotate the X-Y axes through an angle θ. So the Eqn. (2) becomes

$$\begin{aligned} &6(x'\cos\theta - y'\sin\theta)^2 - 5(x'\cos\theta - y'\sin\theta)(x'\sin\theta + y'\cos\theta) \\ &\quad - 6(x'\sin\theta + y'\cos\theta)^2 = 0 \\ \Rightarrow \quad &[6\cos 2\theta - (5/2)\sin 2\theta]x'^2 + [-5\cos 2\theta - 6\sin 2\theta]x'y' \\ &\quad - [6\cos 2\theta - (5/2)\sin 2\theta]y'^2 = 0. \end{aligned}$$

To remove $x'y'$ we need that

$$\begin{aligned} &-5\cos 2\theta - 6\sin 2\theta = 0 \\ \text{i.e.,} \quad &\frac{\cos 2\theta}{6} = \frac{\sin 2\theta}{-5} = \frac{1}{\sqrt{61}}. \end{aligned}$$

Hence, the above equation becomes

$$x'^2 - y'^2 = 0$$

which represents two intersecting straight lines $x' + y' = 0$ and $x' - y' = 0$ which are also perpendicular to each other.

▶▷▶ **9**. *Reduce the following equation to its canonical form*

$$x^2 - 4xy + 4y^2 + 2x - 4y + c = 0$$

and determine the type of the conic represented by it for different values of c.

F-14

Solution: Here $a = 1, b = 4, c = c, h = -2, g = 1$ and $f = -2$.

So, $abc+2fgh-af^2-bg^2-ch^2 = 4c+8-4-4-4c = 0$, hence the given equation is of pair of straight lines. Also, $ab-h^2 = 0$, so the given equation represents a pair of parallel straight lines.

We write the given equation as

$$(x-2y)^2+2(x-2y)+1+c-1=0$$

$$\text{or,}\quad (x-2y+1)^2+c-1=0.$$

Taking $x-2y+1=0$ as Y-axis and any line perpendicular to it as X-axis, we see that the perpendicular distance from any point (x, y) to the line $x-2y+1=0$ is X, i.e.

$$X = \frac{x-2y+1}{\sqrt{5}}.$$

So, the above equation can be written as

$$X^2 + \frac{c-1}{5} = 0$$

which represents a pair of parallel straight lines if $c < 1$ and a pair of coincident straight lines if $c = 1$.

▶▷ **10**. *Find the values of a and g for which the curve $ax^2+8xy+4y^2+2gx+4y+1=0$ represents a conic having infinitely many centres.*

Solution: The centre of the conic lies on the lines

$$ax+4y+g=0$$

and

$$2x+2y+1=0.$$

Solving we get

$$\frac{x}{4-2g} = \frac{y}{2g-a} = \frac{1}{2a-8}.$$

So, the conic will have infinitely many centres if $2a-8=0$; $2g=4$ and $a=2g$, i.e. $a=4$ and $g=2$.

▶▷ **11**. *Show that the equation $(a^2+1)x^2+2(a+b)xy+(b^2+1)y^2=c$ represents an ellipse of area $\pi c/(ab-1)$.*

Solution: Here $abc+2fgh-af^2-bg^2-ch^2 = -c(ab-1)^2 \neq 0$ and $ab-h^2 = (ab-1)^2 > 0$. So, the given equation is of ellipse. Now, we rotate the x-y axes in such a way that the above equation becomes $AX^2+BY^2=c$, i.e. $\dfrac{X^2}{c/A}+\dfrac{Y^2}{c/B}=1$. By properties of invariance we have

$$AB-0^2 = (a^2+1)(b^2+1)-(a+b)^2 = (ab-1)^2.$$

$$\text{Area of the ellipse} = \pi\sqrt{\frac{c}{A}}\sqrt{\frac{c}{B}} = \frac{\pi c}{\sqrt{AB}} = \frac{\pi c}{ab-1}.$$

▶▷▶ **12**. *If two conics* $ax^2+2hxy+by^2+2gx+2fy+c=0$ *and* $a'x^2+2h'xy+b'y^2+2g'x+2f'y+c'=0$ *intersect in four concyclic points, then show that* $\dfrac{a-b}{h}=\dfrac{a'-b'}{h'}$.

Solution: Equation of the conic which passes through the point of intersections of the conics $ax^2+2hxy+by^2+2gx+2fy+c=0$ and $a'x^2+2h'xy+b'y^2+2g'x+2f'y+c'=0$ can be written as

$$\begin{aligned} &(ax^2+2hxy+by^2+2gx+2fy+c)\\ &\quad+\lambda(a'x^2+2h'xy+b'y^2+2g'x+2f'y+c')=0\\ \text{or,}\quad &(a+\lambda a')x^2+2(h+\lambda h')xy+(b+\lambda b')y^2\\ &\quad+2(g+\lambda g')x+2(f+\lambda f')y+c+\lambda c'=0. \end{aligned}$$

This will be a circle if

$$\begin{aligned} &a+\lambda a'=b+\lambda b'\\ \text{and}\quad &h+\lambda h'=0\\ \text{i.e. if}\quad &-\frac{a-b}{a'-b'}=\lambda=-\frac{h}{h'}\\ \text{or,}\quad &\frac{a-b}{h}=\frac{a'-b'}{h'}. \end{aligned}$$

▶▷▶ **13**. *Obtain the focus of the parabola* $\sqrt{ax}+\sqrt{by}=1$. *If its axis passes through a fixed point, show that the locus of the focus is a rectangular hyperbola.*

Solution: Squaring $\sqrt{ax}+\sqrt{by}=1$, we get

$$ax+by-1=-2\sqrt{abxy}.$$

Again, squaring we obtain

$$\begin{aligned} &a^2x^2+b^2y^2-2abxy=2ax+2by-1\\ \text{or,}\quad &(ax-by)^2=2ax+2by-1\\ \text{or,}\quad &(ax-by+k)^2=2ax(1+k)+2by(1-k)+k^2-1. \qquad (1) \end{aligned}$$

Next we choose the value of k in such a way that the lines $ax-by+k=0$ and $2ax(1+k)+2by(1-k)+k^2-1=0$ become perpendicular. Applying the condition of perpendicularity, we get

$$\begin{aligned} &\frac{a}{b}\left[\frac{a(1+k)}{b(1-k)}\right]=1\\ \text{or,}\quad &a^2(1+k)=b^2(1-k)\\ \text{or,}\quad &k=\frac{b^2-a^2}{b^2+a^2}. \end{aligned}$$

So, the Eqn. (1) becomes

$$\begin{aligned} &\left(ax-by+\frac{b^2-a^2}{b^2+a^2}\right)^2=\frac{4ab^2}{a^2+b^2}x+\frac{4ba^2}{a^2+b^2}y-\frac{4a^2b^2}{(a^2+b^2)^2}\\ \text{or,}\quad &\left(ax-by+\frac{b^2-a^2}{b^2+a^2}\right)^2=\frac{4ab}{a^2+b^2}\left(bx+ay-\frac{ba}{a^2+b^2}\right) \end{aligned}$$

$$\text{or,}\quad \left(\frac{ax - by + \dfrac{b^2 - a^2}{b^2 + a^2}}{\sqrt{a^2 + b^2}}\right)^2 = \frac{4ab}{(a^2 + b^2)^{3/2}}\left(\frac{bx + ay - \dfrac{ab}{b^2 + a^2}}{\sqrt{a^2 + b^2}}\right)$$

which is of the form $Y^2 = 4AX$, whose latus ractum $= \dfrac{4ab}{(a^2 + b^2)^{3/2}}$.

Equation of the axis $ax - by + \dfrac{b^2 - a^2}{b^2 + a^2} = 0$, and equation of the tangent at the vertex is $bx + ay - \dfrac{ab}{b^2 + a^2} = 0$. Focus of the parabola is given by the equations $X = A$. $Y = 0$, i.e.

$$bx + ay - \frac{ab}{b^2 + a^2} = \frac{ab}{b^2 + a^2}$$

and

$$ax - by + \frac{b^2 - a^2}{b^2 + a^2} = 0.$$

Solving we get the focus as $(x, y) = \left(\dfrac{a}{a^2 + b^2}, \dfrac{b}{a^2 + b^2}\right)$. If the axis passes through the point (h, k), then we get

$$ah - bk + \frac{b^2 - a^2}{b^2 + a^2} = 0. \tag{2}$$

Now, $x^2 + y^2 = \dfrac{1}{a^2 + b^2}$, so $a = \dfrac{x}{x^2 + y^2}$ and $b = \dfrac{y}{x^2 + y^2}$. Eliminating a, b from Eqn. (2) we get $x^2 - y^2 - hx + ky = 0$, which is a rectangular hyperbola.

EXERCISE

1. Reduce the following equations to standard form:
   (a) $5x^2 - 20xy - 5y^2 - 16x + 8y - 7 = 0$
   (b) $8x^2 - 12xy + 17y^2 + 16x - 12y + 3 = 0$
   (c) $4x^2 - 12xy + 9y^2 - 52x + 26y + 21 = 0$
   (d) $2x^2 - 4xy - y^2 + 20x - 2y + 17 = 0$
   (e) $5x^2 - 20xy - 5y^2 - 16x + 8y - 7 = 0$
   (f) $9x^2 + 24xy + 16y^2 - 126x + 82y - 59 = 0$
   (g) $x^2 - 2xy + 2y^2 - 4x - 6y + 3 = 0$
   (h) $x^2 + 4xy + 4y^2 + 4x + y - 15 = 0$
   (i) $x^2 + 6xy + 9y^2 + 4x + 12y - 5 = 0$
   (j) $3x^2 - 8xy - 3y^2 + 10x - 13y + 8 = 0$
   (k) $3x^2 - 5xy + 6y^2 + 11x - 17y + 13 = 0$
   (l) $20x^2 + 15xy + 9x + 3y + 1 = 0$
   (m) $14x^2 - 4xy + 11y^2 - 44x - 58y + 71 = 0$
   (n) $x^2 - 3xy + y^2 + 10x - 10y + 21 = 0$
   (o) $x^2 - 5xy + y^2 + 8x - 20y + 15 = 0$
   (p) $17x^2 - 12xy + 8y^2 + 46x + 28y + 17 = 0$
   (q) $9x^2 - 24xy + 16y^2 - 18x - 101y + 19 = 0$
   (r) $4x^2 - 4xy + y^2 - 8x - 6y + 5 = 0$
   (s) $7x^2 - 2xy + 7y^2 + 22x - 10y + 7 = 0$
   (t) $x^2 - xy + 2y^2 - 2x - 6y + 7 = 0$

2. Show that the equation $(a^2+b^2)(x^2+y^2)=(ax+by-ab)^2$ represents a parabola of latus rectum $\frac{2ab}{\sqrt{a^2+b^2}}$.

[*Hints*: $(a^2+b^2)(x^2+y^2)=(ax+by-ab)^2 \Rightarrow (bx-ay)^2=-2ab\left(ax+by-\frac{ab}{2}\right)$ and the lines $bx-ay=0$ and $ax+by-\frac{ab}{2}=0$ are perpendicular. Taking these lines as X and Y axes we see that

$$Y=\frac{bx-ay}{\sqrt{a^2+b^2}} \quad \text{and} \quad X=\frac{ax+by-ab/2}{\sqrt{a^2+b^2}}.$$

So, the above equation becomes $Y^2=-\frac{2ab}{\sqrt{a^2+b^2}}X$, etc.]

3. Find the equation of axis and latus rectum of the parabola whose equation is given by $9x^2+24xy+16y^2+8x-156y-95=0$.
4. Show that the curve $3x^2-2xy+3y^2-4x-4y-12=0$ has a single centre.
5. Show that the curve $x^2+4xy+4y^2-20x+10y-50=0$ has no centre.
6. Show that the curve $x^2+4xy+4y^2+4x+8y+4=0$ has infinitely many centres.
7. Find the values of a and f for which the curve $ax^2-20xy+25y^2-14x+2fy-15=0$ (i) has a single centre; (ii) has infinitely many centres.
8. Reduce the equation $x^2+2xy+y^2+2x+2y+k=0$ to its canonical form and determine the type of the conic represented by it for different values of k.
9. Show that the conic passing through the point of intersection of two rectangular hyperbola is also a rectangular hyperbola.

[*Hints*: If $ax^2+2hxy+by^2+2gx+2fy+c=0$ and $a'x^2+2h'xy+b'y^2+2g'x+2f'y+c'=0$ represent two rectangular hyperbola, then we have $a+b=0$ and $a'+b'=0$. Equation of the conic passing through the point of intersection of the above conics can be written as

$$\begin{aligned} &ax^2+2hxy+by^2+2gx+2fy+c+k(a'x^2+2h'xy+b'y^2 \\ &\quad +2g'x+2f'y+c')=0 \\ \text{or,}\quad &(a+ka')x^2+2(h+kh')xy+(b+kb')y^2+2(g+kg')x \\ &\quad ++2(f+kf')y+c+kc'=0 \end{aligned}$$

in which we see that Coeff. of x^2+Coeff. of $y^2=(a+ka')+(b+kb')=a+b+k(a'+b')=0$.]

10. Prove that in general two parabolas can be drawn through the point of intersection of two conics $ax^2+2hxy+by^2+2gx+2fy+c=0$ and $a'x^2+2h'xy+b'y^2+2g'x+2f'y+c'=0$ and their axes are at right angles if $h'(a-b)=h(a'-b')$.

[*Hints*: Equation of the conic passes through the point of intersection of the above conics can be written as

$$\begin{aligned} &ax^2+2hxy+by^2+2gx+2fy+c+k(a'x^2+2h'xy+b'y^2 \\ &\quad +2g'x+2f'y+c')=0 \\ \text{or,}\quad &(a+ka')x^2+2(h+kh')xy+(b+kb')y^2 \\ &\quad +2(g+kg')x+2(f+kf')y+c+kc'=0. \end{aligned}$$

This will be a parabola if

$$(a + ka')(b + kb') = (h + kh')^2$$

$$\text{i.e. if} \quad (a'b' - h'^2)k^2 + (ab' + ba' - 2hh')k + ab - h^2 = 0 \tag{1}$$

which is a second degree equation on k (if both the conics are not parabolas) giving two values of k. Hence, two parabolas can pass through the points of intersection of the given conics.

If k_1 and k_2 be two roots of the above equation in k, then axes of the parabolas can be taken as

$$\sqrt{a + k_1a'}x + \sqrt{b + k_1b'}y + k_1 = 0 \quad \text{and} \quad \sqrt{a + k_2a'}x + \sqrt{b + k_2b'}y + k_2 = 0.$$

These will be at right angle if

$$\sqrt{a + k_1a'}\sqrt{a + k_2a'} + \sqrt{b + k_1b'}\sqrt{b + k_2b'} = 0$$

$$\text{or,} \quad a^2 + (k_1 + k_2)aa' + k_1k_2a'^2 = b^2 + (k_1 + k_2)bb' + k_1k_2b'^2$$

$$\text{or,} \quad (a'b' - h'^2)a^2 - (ab' + ba' - 2hh')aa' + (ab - h^2)a'^2$$

$$= (a'b' - h'^2)b^2 - (ab' + ba' - 2hh')bb' + (ab - h^2)b'^2$$

$$\text{or,} \quad (ah' - a'h)^2 = (bh' - b'h)^2$$

$$\text{or,} \quad h'(a - b) = h(a' - b').]$$

11. Find the locus of the focus of the parabola $\left(\frac{x}{a} + \frac{y}{b} - 1\right)^2 = \frac{4xy}{ab}$, a and b being variables such that $ab = c^2$.

[*Hints*: Proceed as Example 13.]

12. Show that $\frac{1}{x + y - a} + \frac{1}{x - y + a} + \frac{1}{x + y + a} = 0$ represents a parabola whose focus is $\left(\frac{a}{2}, \frac{a}{2}\right)$.

ANALYTICAL GEOMETRY

OF

THREE DIMENSIONS

Chapter 1

Coordinates

1.1 Introduction

In rectangular coordinate system, the position of a point in space is defined by referring it to three fixed concurrent, non-coplanar straight lines which are perpendicular to each other. The common point is called *origin* and the lines are called *coordinate axes*. The planes through lines taken in pairs are called *coordinate planes* which divide the space in eight parts known as *octants*. If we take XOX', YOY' and ZOZ' meeting at $O(0, 0, 0)$ as the coordinate axes, then the position of any point P is determined by the triad of numbers (x, y, z) as shown in the figure.

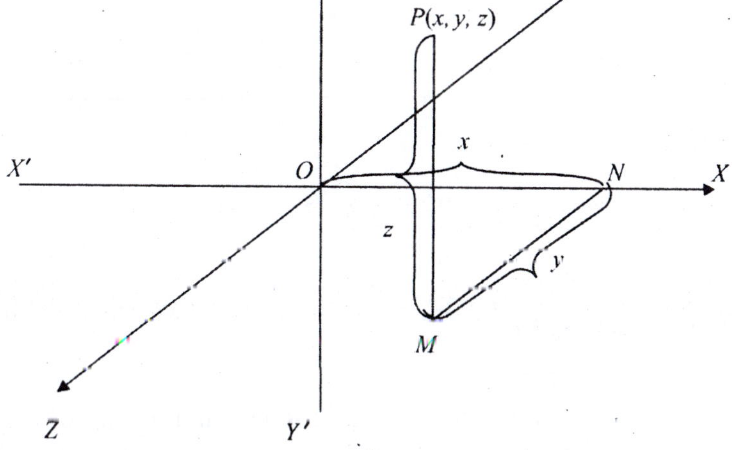

The points which satisfy the equations

(1) $y = 0, z = 0$ lie on x-axis
(2) $x = 0, z = 0$ lie on y-axis
(3) $x = 0, y = 0$ lie on z-axis.

Similarly, the points which satisfy the equations

(4) $x = 0$ lie on yz-plane
(5) $y = 0$ lie on xz-plane
(6) $z = 0$ lie on xy-plane.

The whole space is divided by the coordinate planes in eight octants, namely, $OXYZ$, $OX'YZ$, $OXY'Z$, $OXYZ'$, $OX'Y'Z$, $OX'YZ'$, $OXY'Z'$ and $OX'Y'Z'$. The signs of the coordinates of a point determine the octant in which it lies, i.e. the sign depends on octant.

The following table shows the signs for the octants:

Octant→ Coordinates↓	$OXYZ$	$OX'YZ$	$OXY'Z$	$OXYZ'$	$OX'Y'Z$	$OX'YZ'$	$OXY'Z'$	$OX'Y'Z'$
x	+	−	+	+	−	−	+	−
y	+	+	−	+	−	+	−	−
z	+	+	+	−	+	−	−	−

1.2 Distance between Two Points

Let the coordinates of two points P and Q be (x, y, z) and (x', y', z'). Draw PL parallel to OX, LM parallel to OY and MQ parallel to OZ respectively.

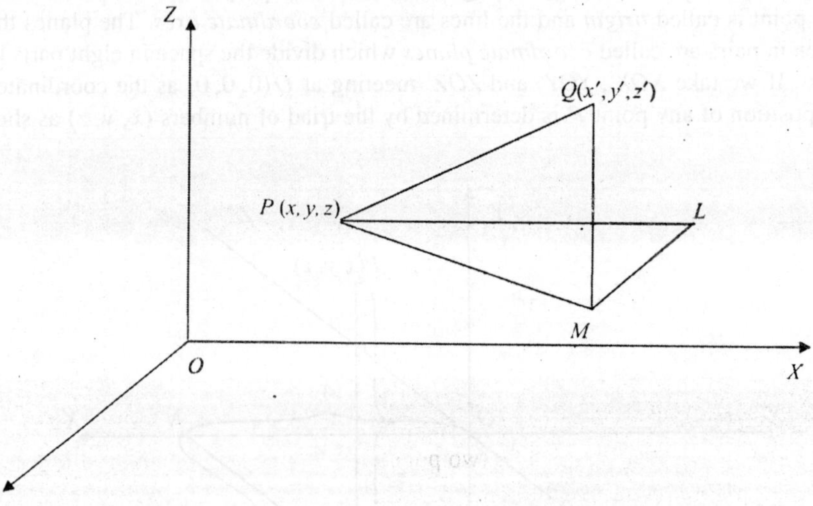

Then from the above figure we have

$$PM^2 = PL^2 + LM^2$$

$= (\text{Difference of the distances of } P, Q \text{ from } YOZ\text{-plane})^2 + (\text{Difference of the distances of } P, Q \text{ from } XOZ\text{-plane})^2$

$= (x' - x)^2 + (y' - y)^2$

and $PQ^2 = PM^2 + QM^2$

$= PM^2 + (\text{Difference of the distances of } P, Q \text{ from } YOX\text{-plane})^2$

$= (x' - x)^2 + (y' - y)^2 + (z' - z)^2.$

So,

$$PQ = \sqrt{(x' - x)^2 + (y' - y)^2 + (z' - z)^2}.$$

If P coincides with O, then distance of the point $Q(x', y', z')$ from origin is given by

$$OQ = \sqrt{x'^2 + y'^2 + z'^2}.$$

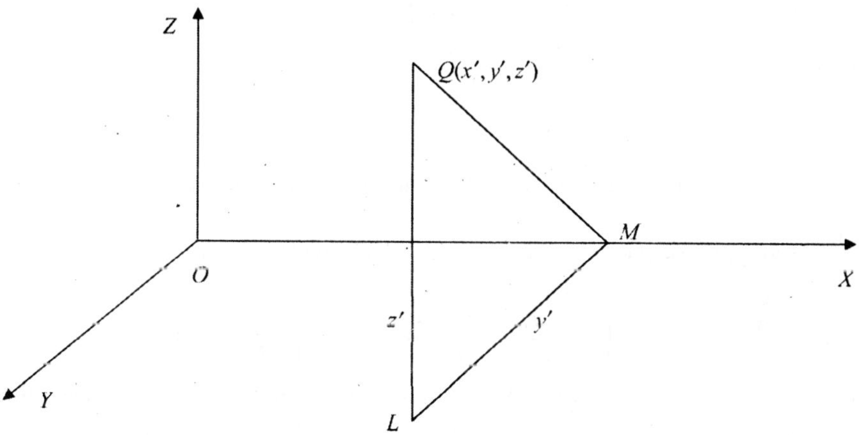

From the above figure, distance of the point Q from X-axis is QM, where

$$QM^2 = LM^2 + LQ^2 = y'^2 + z'^2.$$

Distance of the point $Q(x', y', z')$ from X-axis is $\sqrt{y'^2 + z'^2}$. Similarly, distance of Q from Y and Z axes are $\sqrt{x'^2 + z'^2}$ and $\sqrt{x'^2 + y'^2}$ respectively.

1.3 Coordinate of a Point Which Divides the Line Joining Two Points in a Given Ratio

Let $P(x_1, y_1, z_1)$ and $Q(x_2, y_2, z_2)$ be two points and $R(x, y, z)$ be a point on PQ such that $PR : RQ = \lambda : \mu$.

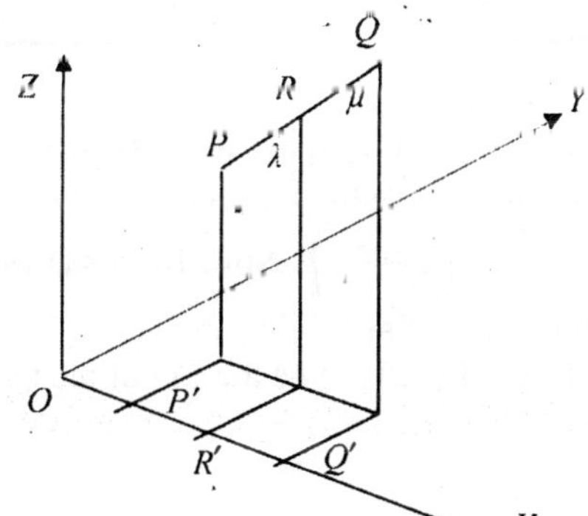

Let the planes through P, Q, R and parallel to the YOZ-plane meet the OX-axis at P', Q', R'. Then from the above figure

$$\frac{PR}{RQ} = \frac{P'R'}{R'Q'} = \frac{x - x_1}{x_2 - x}$$

$$\Rightarrow \quad \frac{\lambda}{\mu} = \frac{x - x_1}{x_2 - x}$$

$$\Rightarrow \quad x = \frac{\lambda x_2 + \mu x_1}{\lambda + \mu}.$$

Similarly,

$$y = \frac{\lambda y_2 + \mu y_1}{\lambda + \mu} \quad \text{and} \quad z = \frac{\lambda z_2 + \mu z_1}{\lambda + \mu}.$$

If $\lambda = \mu$, i.e. R is the middle point of PQ, then its coordinates are $\left(\frac{x_1 + x_2}{2}, \frac{y_1 + y_2}{2}, \frac{z_1 + z_2}{2}\right)$. Again, if we put $\lambda = k\mu = k$, i.e. if R divides PQ in the ratio $k : 1$, then its coordinates are

$$\left(\frac{x_1 + kx_2}{k + 1}, \frac{y_1 + ky_2}{k + 1}, \frac{z_1 + kz_2}{k + 1}\right).$$

If $A(x_1, y_1, z_1)$, $B(x_2, y_2, z_2)$, $C(x_3, y_3, z_3)$ be the vertices of the triangle ABC, then the middle point of CB is, i.e. of D is given by $\left(\frac{x_3 + x_2}{2}, \frac{y_3 + y_2}{2}, \frac{z_3 + z_2}{2}\right)$. From statics, it is well known that the CG of the triangle is a point on the median which divides it in the ratio $2 : 1$. So, if (x, y, z) be the CG of the triangle ABC, then

$$x = \frac{2\dfrac{x_2 + x_3}{2} + 1 \cdot x_1}{2 + 1} = \frac{x_1 + x_2 + x_3}{3}.$$

Similarly,

$$y = \frac{y_1 + y_2 + y_3}{3} \quad \text{and} \quad z = \frac{z_1 + z_2 + z_3}{3}.$$

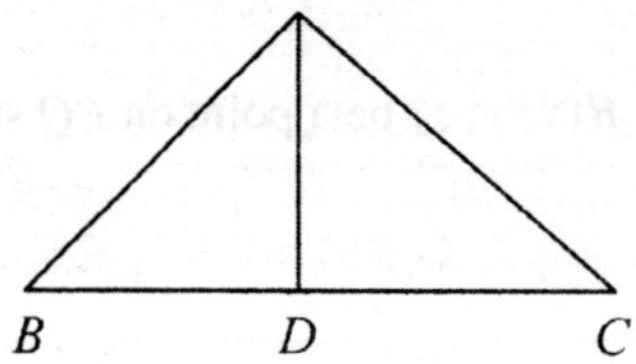

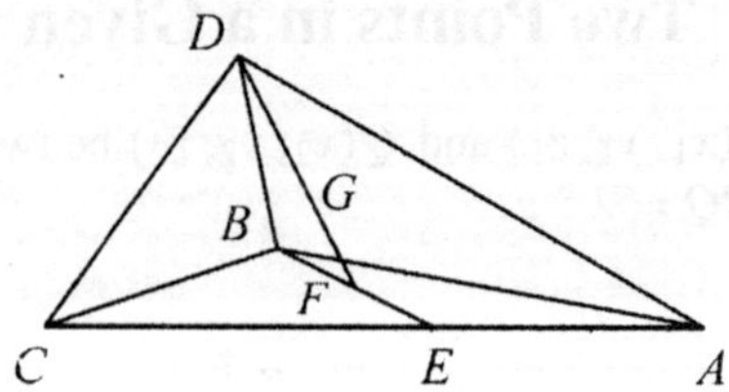

Let $A(x_1, y_1, z_1)$, $B(x_2, y_2, z_2)$, $C(x_3, y_3, z_3)$ and $D(x_4, y_4, z_4)$ be the vertices of the tetrahedron $ABCD$. Using the above result, we get the CG of the triangle ABC as F: $\left(\frac{x_1 + x_2 + x_3}{3}, \frac{y_1 + y_2 + y_3}{3}, \frac{z_1 + z_2 + z_3}{3}\right)$. Now, from statics, we know that the CG of the tetrahedron $ABCD$ is a point on the line joining any vertex to the CG of the opposite face and divides it in the ratio $3 : 1$. So, if (x, y, z) be the CG of the tetrahedron $ABCD$, then

$$x = \frac{3\dfrac{x_1 + x_2 + x_3}{3} + 1 \cdot x_4}{3 + 1} = \frac{x_1 + x_2 + x_3 + x_4}{4}.$$

Similarly,

$$y = \frac{y_1 + y_2 + y_3 + y_4}{4} \quad \text{and} \quad z = \frac{z_1 + z_2 + z_3 + z_4}{4}.$$

1.4 Projection of a Segment

If A' be the foot of the perpendicular drawn from A to a plane, then A' is called the *projection* of A on the plane. Similarly, if B' be the projection of the point B on the same plane, then we say $A'B'$ is the projection of the line segment AB on the plane.

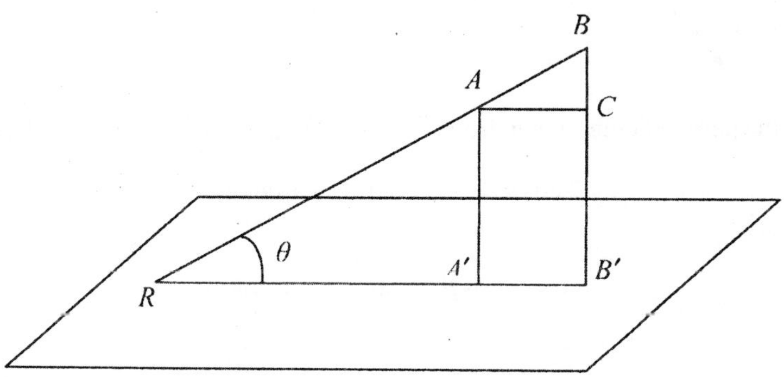

If θ be the angle between AB and $A'B'$, then the length of $A'B'$ is $AB\cos\theta$. If A, B, $C, \ldots, M, N$ be any number of points in a plane and $A', B', C', \ldots, M', N'$ be their projections on a line PQ and these points being on the same straight line, we have

$$A'B' + B'C' + C'D' + \cdots + M'N' = A'N'$$

where $A'B'$ is the projection of AB on PQ. Similarly, $B'C'$ is the projection of BC, $C'D'$ is so of CD and so on. Hence sum of the projections of AB, BC, $CD, \ldots$ on PQ, i.e. projection of AN on PQ is $A'N'$.

It is to be noted further that projection of BA is $B'A'$, i.e. projections of AB and BA differ in sign only.

1.5 Direction Cosines

In three-dimensional coordinate geometry, the angle between two non-intersecting lines AB and CD is measured as the angle between CE and CD where CE is parallel to AB, i.e. if a straight line AB is parallel to OP, where O is the origin, and OP makes α, β, γ with the coordinate axes, then AB is assumed to make the same angles with coordinate axes. Cosines of the angles which a straight line makes with the positive direction of axes are called *direction cosines* of the straight line. If α, β, γ be the angles which the line OP makes with x, y, z axes, then its direction cosines are $\cos\alpha$, $\cos\beta$, $\cos\gamma$ and those of PO are $-\cos\alpha$, $-\cos\beta$, $-\cos\gamma$ as it makes $\pi - \alpha$, $\pi - \beta$, $\pi - \gamma$ with OX, OY and OZ respectively.

The x-axis makes 0°, 90°, 90° with x, y, z axes, so its direction cosines are $(1, 0, 0)$. Similarly, direction cosines of y and z axes are $(0, 1, 0)$ and $(0, 0, 1)$ respectively.

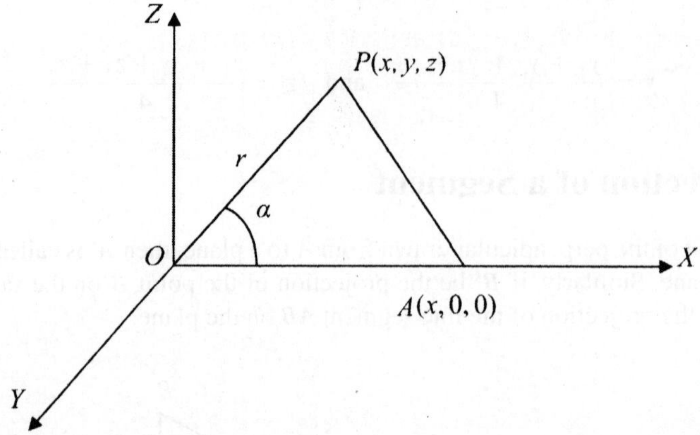

If PA be the perpendicular drawn from $P(x, y, z)$ to the x-axis, then from the above figure

$$\cos\alpha = \frac{x}{r}, \quad \text{i.e. } x = r\cos\alpha.$$

Similarly,

$$y = r\cos\beta \quad \text{and} \quad z = r\cos\gamma.$$

Now,

$$OP^2 = r^2 = x^2 + y^2 + z^2$$

$$\text{i.e.} \quad r^2 = r^2[\cos^2\alpha + \cos^2\beta + \cos^2\gamma]$$

$$\text{i.e.} \quad \cos^2\alpha + \cos^2\beta + \cos^2\gamma = 1.$$

1.6 Direction Ratios

Any three numbers, which are proportional to the direction cosines $\cos\alpha$, $\cos\beta$, $\cos\gamma$ of a straight line OP, are called *direction ratios/numbers* of OP. Let a, b, c be the direction ratios of OP, then we have

$$\frac{a}{\cos\alpha} = \frac{b}{\cos\beta} = \frac{c}{\cos\gamma} = \frac{\pm\sqrt{a^2+b^2+c^2}}{\sqrt{\cos^2\alpha + \cos^2\beta + \cos^2\gamma}} = \pm\sqrt{a^2+b^2+c^2}$$

$$\text{i.e.} \quad \cos\alpha = \frac{\pm a}{\sqrt{a^2+b^2+c^2}}, \ \cos\beta = \frac{\pm b}{\sqrt{a^2+b^2+c^2}} \ \text{and} \ \cos\gamma = \frac{\pm c}{\sqrt{a^2+b^2+c^2}}.$$

This shows that if direction ratios of a straight line are given, then we can determine the direction cosines of that line in terms of direction ratios.

If P', Q' be the projections of $P(x_1, y_1, z_1)$ and $Q(x_2, y_2, z_2)$ on x-axis, then $OP' = x_1$ and $OQ' = x_2$, so that

$$P'Q' = x_2 - x_1.$$

Let the direction cosines of the line PQ be $\cos\alpha$, $\cos\beta$, $\cos\gamma$. Since $P'Q'$ is the projection of PQ on x-axis, we have $P'Q' = PQ\cos\alpha$, i.e. $\cos\alpha = \dfrac{x_2 - x_1}{PQ}$. Similarly, $\cos\beta = \dfrac{y_2 - y_1}{PQ}$ and $\cos\gamma = \dfrac{z_2 - z_1}{PQ}$, which shows that the direction ratios of the line joining $P(x_1, y_1, z_1)$ and $Q(x_2, y_2, z_2)$ are proportional to $(x_2 - x_1)$, $(y_2 - y_1)$, $(z_2 - z_1)$

1.7 Projection of the Straight Line Joining $P(x_1, y_1, z_1)$ and $Q(x_2, y_2, z_2)$ on Another Line

Let AB be a line whose direction cosines are $\cos\alpha$, $\cos\beta$, $\cos\gamma$. If O be the origin, then we have $P_1Q_1 = (x_2 - x_1)$, $P_2Q_2 = (y_2 - y_1)$ and $P_3Q_3 = (z_2 - z_1)$ where P_i's and Q_i's are the projection of PQ on the coordinate axes. So projections of $(x_2 - x_1)$, $(y_2 - y_1)$, $(z_2 - z_1)$ on AB are $(x_2 - x_1)\cos\alpha$, $(y_2 - y_1)\cos\beta$, $(z_2 - z_1)\cos\gamma$ respectively. Using the result of Section 1.4 we get that the projection of PQ on AB is

$$(x_2 - x_1)\cos\alpha + (y_2 - y_1)\cos\beta + (z_2 - z_1)\cos\gamma .$$

1.8 Angle between Two Straight Lines

Let θ be the angle between the straight lines OP and OQ where O is the origin and coordinates of P and Q are given by (x_1, y_1, z_1) and (x_2, y_2, z_2) respectively. If l_1, m_1, n_1 and l_2, m_2, n_2 be the direction cosines of OP and OQ respectively, then we have

$$x_1 = l_1 OP,\ y_1 = m_1 OP,\ z_1 = n_1 OP \quad \text{and} \quad x_2 = l_2 OQ,\ y_2 = m_2 OQ,\ z_2 = n_2 OQ.$$

Now, projection of OP on OQ is given by

$$(x_1 - 0)l_2 + (y_1 - 0)m_2 + (z_1 - 0)n_2 = OP\cos\theta$$

$$\text{or,}\quad OP[l_1l_2 + m_1m_2 + n_1n_2] = OP\cos\theta$$

$$\text{or,}\quad \cos\theta = l_1l_2 + m_1m_2 + n_1n_2 = \frac{a_1a_2 + b_1b_2 + c_1c_2}{\sqrt{a_1^2 + b_1^2 + c_1^2}\sqrt{a_2^2 + b_2^2 + c_2^2}}$$

where (a_1, b_1, c_1), (a_2, b_2, c_2) are direction ratios of OP and OQ respectively.

Now,

$$\begin{aligned}\sin\theta &= \sqrt{1 - \cos^2\theta}\\ &= \sqrt{(l_1^2 + m_1^2 + n_1^2)(l_2^2 + m_2^2 + n_2^2) - (l_1l_2 + m_1m_2 + n_1n_2)^2}\\ &= \sqrt{(l_1m_2 - l_2m_1)^2 + (l_1n_2 - l_2n_1)^2 + (n_1m_2 - n_2m_1)^2}\\ &= \frac{\sqrt{(a_1b_2 - a_2b_1)^2 + (a_1c_2 - a_2c_1)^2 + (c_1b_2 - c_2b_1)^2}}{\sqrt{a_1^2 + b_1^2 + c_1^2}\sqrt{a_2^2 + b_2^2 + c_2^2}}.\end{aligned}$$

So,

$$\begin{aligned}\tan\theta &= \frac{\sqrt{(l_1m_2 - l_2m_1)^2 + (l_1n_2 - l_2n_1)^2 + (n_1m_2 - n_2m_1)^2}}{l_1l_2 + m_1m_2 + n_1n_2}\\ &= \frac{\sqrt{(a_1b_2 - a_2b_1)^2 + (a_1c_2 - a_2c_1)^2 + (c_1b_2 - c_2b_1)^2}}{a_1a_2 + b_1b_2 + c_1c_2}\end{aligned}$$

If the lines are perpendicular, then we have $\cos\theta = 0$

$$l_1l_2 + m_1m_2 + n_1n_2 = 0$$

$$\text{or,}\quad a_1a_2 + b_1b_2 + c_1c_2 = 0.$$

Again, if the lines are parallel, then $\sin\theta = 0$, i.e.

$$\sqrt{(l_1m_2 - l_2m_1)^2 + (l_1n_2 - l_2n_1)^2 + (n_1m_2 - n_2m_1)^2} = 0$$

$$\Rightarrow \quad l_1m_2 - l_2m_1 = 0, \quad l_1n_2 - l_2n_1 = 0, \quad n_1m_2 - n_2m_1 = 0$$

$$\Rightarrow \quad \frac{l_1}{l_2} = \frac{m_1}{m_2} = \frac{n_1}{n_2}$$

$$\Rightarrow \quad \frac{a_1}{a_2} = \frac{b_1}{b_2} = \frac{c_1}{c_2}.$$

Examples

▶▷▶ 1. *If A, B, C be three points on the coordinate axes such that $OA = a$, $OB = b$ and $OC = c$ where O is the origin, then find the point which is equidistant from A, B, C and origin.*

☐ Let $P(x, y, z)$ be the point such that $OP = AP = BP = CP$, then

$$x^2 + y^2 + z^2 = (x-a)^2 + y^2 + z^2 = x^2 + (y-b)^2 + z^2 = x^2 + y^2 + (z-c)^2.$$

From $x^2 + y^2 + z^2 = (x-a)^2 + y^2 + z^2$ we get $x = a/2$; similarly, we get $y = b/2$ and $z = c/2$.

So, the point which is equidistant from A, B, C and origin is $(a/2, b/2, c/2)$.

▶▷▶ 2. *The projections of a line segment on the coordinate axes are 3, 4, 12. Find the length and the direction cosines of the line.*

☐ Let the segment be of length 'r'. If it make angles α, β, γ with the coordinate axes, then we have $r\cos\alpha = 3$, $r\cos\beta = 4$ and $r\cos\gamma = 12$. Squaring and adding we get

$$r^2[\cos^2\alpha + \cos^2\beta + \cos^2\gamma] = 169$$

$$\Rightarrow \quad r = 13.$$

The direction cosines of the line segment are $3/r, 4/r, 12/r$, i.e. $3/13, 4/13, 12/13$.

▶▷▶ 3. *A line makes angles 60° and 45° with y-axis and z-axis. Find the direction cosines of the line.*

☐ If the line makes an angle α with x-axis, then we have,

$$\cos^2\alpha + 1/4 + 1/2 = 1$$

$$\Rightarrow \quad \cos^2\alpha = 1/4$$

$$\Rightarrow \quad \cos\alpha = \pm(1/2)$$

$$\Rightarrow \quad \alpha = 60° \text{ or } 120°.$$

So, the direction cosines of the line are $\pm\cos 60°, \cos 60°, \cos 45°$, i.e. $\pm(1/2), 1/2, 1/\sqrt{2}$.

▶▷▶ 4. *A line in the XOY-plane makes an angle 30° with the x-axis. Find the direction cosines of the line.*

□ Since, the line is on the XOY-plane, it makes 90° with z-axis. Again, as it makes 30° with x-axis, it will make 60° with y-axis. Hence, the direction cosines of the line is cos 30°, cos 60°, cos 90°, i.e. $\sqrt{3}/2, 1/2, 0$.

▶▷▶ 5. *Find the projection of line segment joining the points* (2, −1, 4) *and* (0, 1, 5) *on the straight line joining* (3, 3, 5) *and* (5, 4, 3).

□ Direction ratios of the line segment joining (3, 3, 5) and (5, 4, 3) are −2, −1, 2. So, the direction cosines of that line are −2/3, −1/3, 2/3. So, the projection of the line segment joining the points (2, −1, 4) and (0, 1, 5) on the line whose direction cosines are −2/3, −1/3, 2/3 is

$$-2(2-0)/3 - 1(-1-1)/3 + 2(4-5)/3 = -4/3.$$

▶▷▶ 6. *If* l_1, m_1, n_1; l_2, m_2, n_2 *are two mutually perpendicular straight lines, show that the direction cosines of the line which is perpendicular to both of them are*

$$m_1n_2 - m_2n_1,\ n_1l_2 - l_1n_2,\ m_2l_1 - m_1l_2.$$

□ Let l, m, n be the direction cosines of the line which is perpendicular to both the given lines, then

$$ll_1 + mm_1 + nn_1 = 0$$
$$ll_2 + mm_2 + nn_2 = 0.$$

Solving we get

$$\frac{l}{m_1n_2 - m_2n_1} = \frac{m}{n_1l_2 - n_2l_1} = \frac{n}{m_2l_1 - m_1l_2} = \frac{\sqrt{l^2+m^2+n^2}}{\sqrt{\sum(m_1n_2 - m_2n_1)^2}} = \operatorname{cosec} 90° = 1.$$

Hence, the direction cosines of the line which is perpendicular to both the given lines are

$$m_1n_2 - m_2n_1, n_1l_2 - l_1n_2, m_2l_1 - m_1l_2.$$

▶▷▶ 7. *Show that the straight lines whose direction cosines are given by* $al + bm + cn = 0$ *and* $ul^2 + vm^2 + wn^2 = 0$ *are perpendicular or parallel according as*

$$a^2(w+v) + b^2(u+w) + c^2(u+v) = 0 \quad or, \quad \frac{a^2}{u} + \frac{b^2}{v} + \frac{c^2}{w} = 0.$$

□ Eliminating l from $al + bm + cn = 0$ and $ul^2 + vm^2 + wn^2 = 0$ we get

$$\frac{u(bm+cn)^2}{a^2} + vm^2 + wn^2 = 0$$

$$\text{or,} \quad (b^2u + a^2v)\left(\frac{m}{n}\right)^2 + 2ubc\frac{m}{n} + (c^2u + a^2w) = 0 \tag{1}$$

which is second degree equation in $\frac{m}{n}$. If $\frac{m_1}{n_1}$ and $\frac{m_2}{n_2}$ be its two roots, then we have

$$\frac{m_1m_2}{n_1n_2} = \frac{c^2u + a^2w}{b^2u + a^2v}$$

$$\Rightarrow \frac{m_1m_2}{c^2u + a^2w} = \frac{n_1n_2}{b^2u + a^2v} = \frac{l_1l_2}{c^2v + b^2w} \quad \text{[By symmetry]}$$

Now, the lines will be perpendicular if

$$\begin{aligned}
& l_1 l_2 + m_1 m_2 + n_1 n_2 = 0 \\
\Rightarrow \quad & (b^2 w + c^2 v) + (c^2 u + a^2 w) + (b^2 u + a^2 v) = 0 \\
\Rightarrow \quad & a^2(w + v) + b^2(u + w) + c^2(u + v) = 0.
\end{aligned}$$

Now, the lines will be parallel if

$$\frac{m_1}{m_2} = \frac{n_1}{n_2}$$

$$\Rightarrow \quad \frac{m_1}{n_1} = \frac{m_2}{n_2},$$

i.e., the roots of the Eqn. (1) are equal, i.e. if

$$\begin{aligned}
& 4u^2b^2c^2 = 4(a^2w + c^2u)(b^2u + a^2v) \\
\Rightarrow \quad & u^2b^2c^2 = a^4wv + a^2c^2uv + b^2a^2wu + b^2c^2u^2 \\
\Rightarrow \quad & a^4wv + a^2c^2uv + b^2a^2wu = 0 \\
\Rightarrow \quad & \frac{a^2}{u} + \frac{b^2}{v} + \frac{c^2}{w} = 0.
\end{aligned}$$

▶▷ **8**. *Show that the angle between the straight lines whose direction cosines are given by $l + m + n = 0$ and $fmn + gnl + hlm = 0$ is $\frac{\pi}{3}$ if $\frac{1}{f} + \frac{1}{g} + \frac{1}{h} = 0$.*

□ Eliminating n from $l + m + n = 0$ and $fmn + gnl + hlm = 0$, we obtain

$$g\left(\frac{l}{m}\right)^2 + (f + g - h)\left(\frac{l}{m}\right) + f = 0$$

which is a second degree equation in $\left(\frac{l}{m}\right)$. If $\frac{l_1}{m_1}$ and $\frac{l_2}{m_2}$ be its two roots, then we have

$$\frac{l_1 l_2}{m_1 m_2} = \frac{f}{g} \tag{1}$$

$$\text{and} \quad \frac{l_1}{m_1} + \frac{l_2}{m_2} = -\frac{f + g - h}{g}. \tag{2}$$

From Eqn. (1) we see that

$$\frac{l_1 l_2}{f} = \frac{m_1 m_2}{g} = \frac{n_1 n_2}{h} = k \text{ (say)}.$$

From Eqns. (1) and (2) we get

$$\frac{l_1}{m_1} - \frac{l_2}{m_2} = \frac{\sqrt{(f + g - h)^2 - 4fg}}{g}$$

$$\Rightarrow \quad (l_1 m_2 - l_2 m_1)^2 = k^2[(f + g - h)^2 - 4fg].$$

Similarly,

$$(l_1n_2 - l_2n_1)^2 = k^2[(f+h-g)^2 - 4fh]$$
$$\text{and} \quad (n_1m_2 - n_2m_1)^2 = k^2[(h+g-f)^2 - 4hg].$$

Now, $\tan^2\left(\frac{\pi}{3}\right) = \frac{\sum(l_1m_2 - l_2m_1)^2}{(\sum l_1l_2)^2} = \frac{k^2\sum[(f+g-h)^2 - 4fg]}{k^2(f+g+h)^2}$ gives

$$3(f+g+h)^2 = \sum[(f+g-h)^2 - 4fg]$$
$$= [(f+g-h)^2 - 4fg] + [(f+h-g)^2 - 4fh] + [(h+g-f)^2 - 4hg]$$
$$\Rightarrow \quad 3[f^2+g^2+h^2+2(fg+gh+fh)] = 3[f^2+g^2+h^2] - 6[fg+fh+gh]$$
$$\Rightarrow \quad 12[fg+fh+gh] = 0$$
$$\Rightarrow \quad fg+fh+gh = 0$$
$$\Rightarrow \quad \frac{1}{f}+\frac{1}{g}+\frac{1}{h} = 0.$$

▶▷▶ **9.** *If l_1, m_1, n_1; l_2, m_2, n_2 are the direction cosines of two concurrent lines; show that the direction cosines of the bisectors of the angle between them are proportional to $(l_1 \pm l_2, m_1 \pm m_2, n_1 \pm n_2)$.*

□ Let O be the origin and A, B be two points on the given lines such that $OA = OB = r$, then the coordinates of A and B are (l_1r, m_1r, n_1r) and (l_2r, m_2r, n_2r) respectively. We take a point A' on the extension of OA such that O is the middle point of AA'. Then coordinates of A' will be $(-l_1r, -m_1r, -n_1r)$.

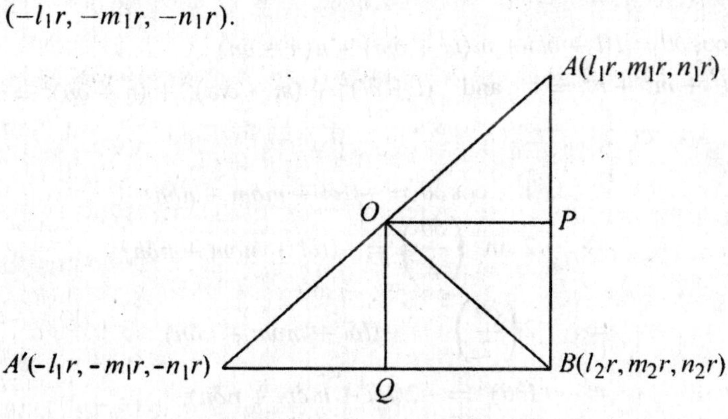

Now, if P, Q be the midpoints of AB, and $A'B$ then OP and OQ are the internal and external bisectors of the angle AOB. Hence, coordinates of P, i.e. direction ratios of OP are

$$\left(\frac{r(l_1+l_2)}{2}, \frac{r(m_1+m_2)}{2}, \frac{r(n_1+n_2)}{2}\right).$$

Similarly, direction ratios of OQ are

$$\left(\frac{r(l_2-l_1)}{2}, \frac{r(m_2-m_1)}{2}, \frac{r(n_2-n_1)}{2}\right).$$

Hence, the direction cosines of the bisectors of the angle between OA and OB are proportional to

$$(l_1 \pm l_2, m_1 \pm m_2, n_1 \pm n_2).$$

▶▷ **10**. *If l_1, m_1, n_1; l_2, m_2, n_2 are the direction cosines of two concurrent lines; show that the line with direction cosines* $\left(\frac{l_1+l_2}{2\cos(\theta/2)}, \frac{m_1+m_2}{2\cos(\theta/2)}, \frac{n_1+n_2}{2\cos(\theta/2)}\right)$ *bisects the angle θ, between the given lines.*

☐ In the Ex. 9 it is shown that the direction cosines of the internal bisector of the angle between the given lines are proportional to $(l_1+l_2, m_1+m_2, n_1+n_2)$.

Now,

$$\begin{aligned}&\sqrt{(l_1+l_2)^2+(m_1+m_2)^2+(n_1+n_2)^2}\\ &= \sqrt{l_1^2+m_1^2+n_1^2+l_2^2+m_2^2+n_2^2+2(l_1l_2+m_1m_2+n_1n_2)}\\ &= \sqrt{1+1+2\cos\theta}, \text{ where } \theta \text{ is the angle between the given lines}\\ &= 2\cos(\theta/2).\end{aligned}$$

So, the direction cosines of the internal bisector are

$$\frac{l_1+l_2}{2\cos(\theta/2)}, \frac{m_1+m_2}{2\cos(\theta/2)}, \frac{n_1+n_2}{2\cos(\theta/2)}.$$

▶▷ **11**. *The direction cosines of a moving line in two adjacent positions are l, m, n and $l+\delta l$, $m+\delta m$, $n+\delta n$. Show that the small angle $\delta\theta$ between the positions is given by*

$$(\delta\theta)^2 = (\delta l)^2 + (\delta m)^2 + (\delta n)^2.$$

☐ We have,

$$\begin{aligned}&\cos\delta\theta = l(l+\delta l) + m(m+\delta m) + n(n+\delta n)\\ &l^2+m^2+n^2 = 1 \quad \text{and} \quad (l+\delta l)^2+(m+\delta m)^2+(n+\delta n)^2 = 1.\end{aligned}$$

Now,

$$\begin{aligned}1-\cos\delta\theta &= -(l\delta l + m\delta m + n\delta n)\\ \Rightarrow \quad 2\sin^2\left(\frac{\delta\theta}{2}\right) &= -(l\delta l + m\delta m + n\delta n)\\ \Rightarrow \quad 2\left(\frac{\delta\theta}{2}\right)^2 &= -(l\delta l + m\delta m + n\delta n)\\ \Rightarrow \quad (\delta\theta)^2 &= -2(l\delta l + m\delta m + n\delta n). \qquad (1)\end{aligned}$$

Again,

$$\begin{aligned}&(l+\delta l)^2+(m+\delta m)^2+(n+\delta n)^2 = l^2+m^2+n^2\\ \Rightarrow \quad &(\delta l)^2+(\delta m)^2+(\delta n)^2 = -2(l\delta l + m\delta m + n\delta n). \qquad (2)\end{aligned}$$

From Eqns. (1) and (2) we get the result.

▶▷ **12**. *If l_k, m_k, n_k; $k = 1, 2, 3$ be the direction cosines of these mutually perpendicular lines then show that*

$$\begin{vmatrix} l_1 & m_1 & n_1\\ l_2 & m_2 & n_2\\ l_3 & m_3 & n_3\end{vmatrix} = \pm 1.$$

□ Given that

$$\begin{aligned}
l_1l_2 + m_1m_2 + n_1n_2 &= 0\\
l_1l_3 + m_1m_3 + n_1n_3 &= 0\\
l_3l_2 + m_3m_2 + n_3n_2 &= 0\\
l_1^2 + m_1^2 + n_1^2 &= 1\\
l_2^2 + m_2^2 + n_2^2 &= 1\\
l_3^2 + m_3^2 + n_3^2 &= 1.
\end{aligned}$$

Solving the first two equations for l_1, m_1, n_1 we get

$$\begin{aligned}
&\frac{l_1}{m_2n_3 - m_3n_2} = \frac{m_1}{n_2l_3 - n_3l_2} = \frac{n_1}{l_2m_3 - l_3m_2}\\
&= \frac{l_1l_1 + m_1m_1 + n_1n_1}{l_1(m_2n_3 - m_3n_2) + m_1(n_2l_3 - n_3l_2) + n_1(l_2m_3 - l_3m_2)}\\
&= \frac{1}{\Delta}
\end{aligned}$$

where Δ is the determinant whose value is to be determined.

From the above result we obtain

$$\begin{aligned}
\Delta l_1 &= m_2n_3 - m_3n_2,\\
\Delta m_1 &= n_2l_3 - n_3l_2,\\
\Delta n_1 &= l_2m_3 - l_3m_2.
\end{aligned}$$

Squaring and adding we get

$$\begin{aligned}
&\Delta^2(l_1^2 + m_1^2 + n_1^2) = \sum(m_2n_3 - m_3n_2)^2 = \sin^2 90^\circ = 1\\
\Rightarrow\quad &\Delta = \pm 1.
\end{aligned}$$

EXERCISE

1. Find the direction cosines of the line joining the points (2, 1, 5) and (−1, −2, 2).
2. Can the numbers $1/\sqrt{2}, 1/\sqrt{2}, 1/\sqrt{2}$ be direction cosines of a line?
3. Find the direction cosines of line which is equally inclined to the coordinate axes.
4. A line in the plane YOZ makes 45° with y-axis. Find the direction cosines of the line.
5. Find the projection of the line joining the points (2, 3, 4) and (4, 1, 5) on the line joining (5, 8, 0) and (−1, 2, 6)
6. Prove that the angle between the diagonals of a cube is $\cos^{-1}(1/3)$.
7. Prove that the angle between the lines whose direction cosines satisfy $l + m + n = 0$ and $l^2 + m^2 - n^2 = 0$ is $\pi/3$.
8. Find the angle between the lines whose direction cosines satisfy $2l + 2m - n = 0$ and $mn + lm + ln = 0$.
9. Prove that the angle between the lines whose direction cosines satisfy $l + m + n = 0$ and $mn - 2lm - 2ln = 0$ is $\cos^{-1}(1/2)$.
10. Show that the straight lines whose direction cosines are given by $a^2l + b^2m + c^2n = 0$ and $mn + lm + ln = 0$ are parallel if $a + b + c = 0$.

11. Show that the straight lines whose direction cosines are given by $al + bm + cn = 0$ and $fmn + gnl + hlm = 0$ are perpendicular or parallel according as

$$\frac{f}{a} + \frac{g}{b} + \frac{h}{c} = 0 \quad \text{or} \quad \sqrt{af} + \sqrt{bg} + \sqrt{ch} = 0.$$

12. If l_1, m_1, n_1; l_2, m_2, n_2 and l_3, m_3, n_3 be direction cosines of three mutually perpendicular straight lines, then find the direction cosines of a line whose direction ratios are $l_1 + l_2 + l_3$, $m_1 + m_2 + m_3$, $n_1 + n_2 + n_3$ and show that the line is equally inclined to the given lines.

13. If the edges of a rectangular parallelopiped be a, b, c, show that the angles between the four diagonals are given by $\cos^{-1}\left[\dfrac{\pm a^2 \pm b^2 \pm c^2}{a^2 + b^2 + c^2}\right]$.

14. A line makes angles $\alpha, \beta, \gamma, \delta$ with four diagonals of a cube; show that

$$\cos^2\alpha + \cos^2\beta + \cos^2\gamma + \cos^2\delta = 4/3.$$

Chapter 2

The Plane

2.1 Introduction

A plane is a surface such that any point on the line segment joining any two points on it is also a point on it, i.e. the line joining any two points on it lies wholly on the surface.

We shall now show that the surface represented by the linear equation in x, y, z,

$$ax + by + cz + d = 0 \tag{1}$$

where a, b, c, d are constants and not all of a, b, c are zero, is a plane.

Let (x_1, y_1, z_1) and (x_2, y_2, z_2) be two points on the surface (1), then we have

$$ax_1 + by_1 + cz_1 + d = 0 \tag{2}$$

$$\text{and} \quad ax_2 + by_2 + cz_2 + d = 0. \tag{3}$$

Multiplying Eqn. (3) by $k(\neq -1)$ and adding it to Eqn. (2) we get

$$a(x_1 + kx_2) + b(y_1 + ky_2) + c(z_1 + kz_2) + d(1 + k) = 0.$$

From this we see that the points $\left(\frac{x_1 + kx_2}{k+1}, \frac{y_1 + ky_2}{k+1}, \frac{z_1 + kx_2}{k+1}\right)$ which lie on the line segment joining (x_1, y_1, z_1) and (x_2, y_2, z_2) satisfy the Eqn. (1), i.e. the line segment joining (x_1, y_1, z_1) and (x_2, y_2, z_2) lie wholly on the surface (1). So, Eqn. (1) represents a plane.

Subtracting Eqn. (2) from Eqn. (1) we get

$$a(x - x_1) + b(y - y_1) + c(z - z_1) = 0$$

which represents the equation of the plane through (x_1, y_1, z_1). From this we get the equation of the plane through origin as

$$ax + by + cz = 0.$$

Further we note that the Eqn. (1) involves four unknown constants but dividing the equation by d we see that the equation of a plane consists of three unknown constants. To determine three constants we need three equations which can be constructed using three conditions given in the problem like

(i) The plane passes through three given non-collinear points,
(ii) The plane passes through two given points and perpendicular to a given plane,
(iii) The plane passes through a given point and perpendicular to two given planes.

2.2 Equation of a Plane through Three Given Points

Let the equation of the plane be

$$ax + by + cz + d = 0. \tag{4}$$

If it passes through the points (x_1, y_1, z_1), (x_2, y_2, z_2) and (x_3, y_3, z_3), then we have

$$ax_1 + by_1 + cz_1 + d = 0, \tag{5}$$

$$ax_2 + by_2 + cz_2 + d = 0 \tag{6}$$

and $$ax_3 + by_3 + cz_3 + d = 0. \tag{7}$$

Eliminating a, b, c, d from Eqns. (4)–(7), we get

$$\begin{vmatrix} x & y & z & 1 \\ x_1 & y_1 & z_1 & 1 \\ x_2 & y_2 & z_2 & 1 \\ x_3 & y_3 & z_3 & 1 \end{vmatrix} = 0$$

which is a first degree equation in x, y, z and represents the plane through the points (x_1, y_1, z_1), (x_2, y_2, z_2) and (x_3, y_3, z_3).

If a plane intercepts the lengths a, b, c from the coordinate axes respectively, then it passes through the points $(a, 0, 0)$, $(0, b, 0)$ and $(0, 0, c)$ and its equation is

$$\begin{vmatrix} x & y & z & 1 \\ a & 0 & 0 & 1 \\ 0 & b & 0 & 1 \\ 0 & 0 & c & 1 \end{vmatrix} = 0$$

or, $$\frac{x}{a} + \frac{y}{b} + \frac{z}{c} = 1.$$

This is the intercepting form of a plane.

2.3 Normal Form of a Plane

Let ON be the perpendicular drawn from origin O to the plane ABC whose direction cosines are l, m, n. If $P(x, y, z)$ be any point on the plane, then ON is perpendicular to PN, i.e. ON is the projection of OP on ON.

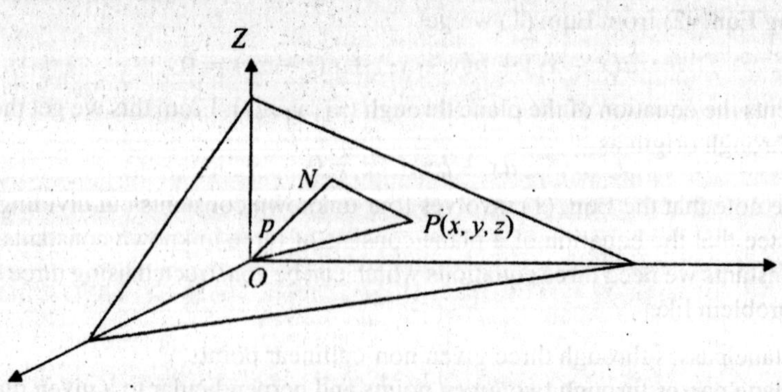

If $ON = p$, then p = projection of OP on $ON = l(x - 0) + m(y - 0) + n(z - 0)$, i.e. $lx + my + nz = p$, which is the equation of the plane in normal form.

Comparing this equation with the general equation of the plane, i.e. $Ax+By+Cz+D=0$ we see that A, B, C are proportional to l, m, n, i.e. proportional to the direction cosines of the normal to the plane. Hence, in the general equation of the plane A, B, C are the direction ratios of the normal to the plane.

2.4 Angle between Two Planes

As the angle between two planes is equal to the angle between the normal to the planes and the coefficients A, B, C of x, y, z in the general equation of a plane are the direction ratios of the normal to the plane, the angle θ between two planes

$$a_1x+b_1y+c_1z+d_1=0$$
$$a_2x+b_2y+c_2z+d_2=0$$

is given by

$$\cos\theta=\frac{a_1a_2+b_1b_2+c_1c_2}{\sqrt{a_1^2+b_1^2+c_1^2}\sqrt{a_2^2+b_2^2+c_2^2}}.$$

So, if the planes are perpendicular, then

$$a_1a_2+b_1b_2+c_1c_2=0$$

and if the planes are parallel, then

$$\frac{a_1}{a_2}=\frac{b_1}{b_2}=\frac{c_1}{c_2}.$$

2.5 Equation of Bisectors of the Angle between Two Planes

Let the equations of the planes be

$$a_1x+b_1y+c_1z+d_1=0$$
$$a_2x+b_2y+c_2z+d_2=0$$

where the constant terms are of same sign.

If (x, y, z) be any point on the bisectors of the angles between the given planes, then the lengths of the perpendiculars from it to the bisectors are same, i.e.

$$\frac{a_1x+b_1y+c_1z+d_1}{\sqrt{a_1^2+b_1^2+c_1^2}}=\pm\frac{a_2x+b_2y+c_2z+d_2}{\sqrt{a_2^2+b_2^2+c_2^2}}.$$

The positive sign in the r.h.s. will give the equation of the bisector of the angle which contains the origin and negative sign will give the equation of the bisector of the other angle.

2.6 Plane through the Line of Intersection of Two Planes

Let the equation of the planes be

$$a_1x+b_1y+c_1z+d_1=0$$
$$a_2x+b_2y+c_2z+d_2=0. \quad (8)$$

As any point, whose coordinates satisfy the above planes also satisfy the equation

$$a_1x + b_1y + c_1z + d_1 + k(a_2x + b_2y + c_2z + d_2) = 0 \quad \textbf{(9)}$$

for any value of k, the Eqn. (9) represents the plane which passes through the line of intersection of the planes given by Eqn. (8). In fact, Eqn. (9) represents a system of planes which passes through the line of intersection of the planes given by Eqn. (8). The constant k is determined using some prescribed condition, viz. "The plane passes through a given point" or "The plane is perpendicular/parallel to a given plane," etc.

2.7 Condition That the Three Planes May Have Common Line of Intersection

Let the equations of the plane be

$$a_1x + b_1y + c_1z + d_1 = 0$$
$$a_2x + b_2y + c_2z + d_2 = 0$$
$$a_3x + b_3y + c_3z + d_3 = 0.$$

Any plane through first two planes is

$$a_1x + b_1y + c_1z + d_1 + k(a_2x + b_2y + c_2z + d_2) = 0.$$

If it is identical to the third plane, then we have

$$\frac{a_1 + ka_2}{a_3} = \frac{b_1 + kb_2}{b_3} = \frac{c_1 + kc_2}{c_3} = \frac{d_1 + kd_2}{d_3} = \lambda \text{ (say)}.$$

Then

$$a_1 + ka_2 - \lambda a_3 = 0$$
$$b_1 + kb_2 - \lambda b_3 = 0$$
$$c_1 + kc_2 - \lambda c_3 = 0$$
$$d_1 + kd_2 - \lambda d_3 = 0.$$

Eliminating k, λ from the above set of equations taking three at a time we get

$$\begin{vmatrix} a_1 & a_2 & a_3 \\ b_1 & b_2 & b_3 \\ c_1 & c_2 & c_3 \end{vmatrix} = 0, \quad \begin{vmatrix} a_1 & a_2 & a_3 \\ b_1 & b_2 & b_3 \\ d_1 & d_2 & d_3 \end{vmatrix} = 0,$$

$$\begin{vmatrix} a_1 & a_2 & a_3 \\ c_1 & c_2 & c_3 \\ d_1 & d_2 & d_3 \end{vmatrix} = 0 \quad \text{and} \quad \begin{vmatrix} b_1 & b_2 & b_3 \\ c_1 & c_2 & c_3 \\ d_1 & d_2 & d_3 \end{vmatrix} = 0.$$

2.8 Pair of Planes

Consider the following equations of planes:

$$ax + by + cz + d = 0 \quad (10)$$
$$a'x + b'y + c'z + d' = 0. \quad (11)$$

Next consider the equation

$$(ax+by+cz+d)(a'x+b'y+c'z+d')=0. \tag{12}$$

Obviously, the points on the planes (10) and/or (11) will satisfy the Eqn. (3) which is second degree equation in x, y, z of the form

$$Ax^2+By^2+Cz^2+2Fyz+2Gxz+2Hxy+2Ux+2Vy+2Wz+D=0. \tag{13}$$

Hence, equation of the form (4) may represent a pair of planes, where it is assumed that at least one of A, B, C, F, G, H is non-zero.

Comparing Eqns. (12) and (13) we get

$$\begin{aligned}&A=aa',\ \ B=bb',\ \ C=cc',\ \ D=dd',\ \ 2F=bc'+cb',\ \ 2G=ac'+ca',\\ &2H=ba'+ab',\ \ 2U=ad'+da',\ \ 2V=bd'+db',\ \ 2W=cd'+dc'.\end{aligned} \tag{14}$$

Putting $d=0=d'$ in Eqn. (14) we see that if the planes pass through origin, then the equation of pair of planes, i.e. Eqn. (13) becomes

$$Ax^2+By^2+Cz^2+2Fyz+2Gxz+2Hxy=0 \tag{15}$$

which is a second degree homogeneous equation in x, y, z.

If θ be the angle between the planes given by Eqn. (15), then

$$\begin{aligned}\tan\theta &= \frac{\sqrt{(bc'-cb')^2+(ba'-ab')^2+(ac'-ca')^2}}{aa'+bb'+cc'}\\ &= \frac{\sqrt{4F^2-4BC+4G^2-4AC+4H^2-4AB}}{A+B+C}\\ &= \frac{2\sqrt{F^2-BC+G^2-AC+H^2-AB}}{A+B+C}.\end{aligned}$$

So, the plane will be right-angled if $A+B+C=0$, and the planes will be parallel if $F^2+G^2+H^2=AB+BC+AC$.

2.9 To Find the Necessary Condition that the Equation $ax^2+by^2+cz^2+2fyz+2gxz+2hxy=0$, May Represent a Pair of Planes

Multiplying the given equation by a and rearranging we get

$$\begin{aligned}&a^2x^2+2a(hy+gz)x=-aby^2-acz^2-2afyz\\ \Rightarrow\ \ &a^2x^2+2a(hy+gz)x+(hy+gz)^2=-aby^2-acz^2-2afyz+(hy+gz)^2\\ \Rightarrow\ \ &(ax+hy+gz)^2=(h^2-ab)y^2+2(gh-af)yz+(g^2-ac)z^2\\ \Rightarrow\ \ &(ax+hy+gz)=\pm\sqrt{(h^2-ab)y^2+2(gh-af)yz+(g^2-ac)z^2}.\end{aligned}$$

So, the given equation will of a pair of planes if

$$(h^2-ab)y^2+2(gh-af)yz+(g^2-ac)z^2$$

be a perfect square for which we have

$$4(gh - af)^2 = 4(h^2 - ab)(g^2 - ac)$$

$$\text{or,} \quad abc + 2fgh - af^2 - bg^2 - ch^2 = 0.$$

This is the required condition.

Examples

▶▷ 1. *Find the equation of the plane passing through the points* (1, 2, 3), (2, 3, 4) *and* (7, 5, 3).

□ Equation of the plane through the point (1, 2, 3) is

$$A(x-1) + B(y-2) + C(z-3) = 0. \tag{1}$$

If it passes through the points (2, 3, 4) and (7, 5, 3), then we have

$$A + B + C = 0 \tag{2}$$

$$\text{and} \quad 6A + 3B = 0. \tag{3}$$

Eliminating A, B, C from the Eqns. (1)–(3), we get

$$\begin{vmatrix} x-1 & y-2 & z-3 \\ 1 & 1 & 1 \\ 6 & 3 & 0 \end{vmatrix} = 0.$$

Expanding we get

$$-3(x-1) + 6(y-2) - 3(z-3) = 0$$

$$\text{or,} \quad x - 1 - 2(y-2) + z - 3 = 0$$

$$\text{or,} \quad x - 2y + z = 0.$$

▶▷ 2. *Find the equation of the plane passing through the points* (1, 2, 3), (2, 3, 4) *and perpendicular to the plane* $2x + 3y - 4z = 9$.

□ Equation of the plane through the point (1, 2, 3) is

$$A(x-1) + B(y-2) + C(z-3) = 0. \tag{1}$$

If it passes through the point (2, 3, 4), then we have

$$A + B + C = 0. \tag{2}$$

Again, the plane (1) is perpendicular to the plane $2x + 3y - 4z = 9$, so we have

$$2A + 3B - 4C = 0. \tag{3}$$

Eliminating A, B, C from the Eqns. (1)–(3), we get

$$\begin{vmatrix} x-1 & y-2 & z-3 \\ 1 & 1 & 1 \\ 2 & 3 & -4 \end{vmatrix} = 0.$$

Expanding we get

$$-7(x-1)+6(y-2)+(z-3)=0$$
$$\text{or,} \quad 7x-6y-z=-8.$$

▶▷▶ 3. *Find the equation of the plane passing through the point* (1, 2, 3), *and perpendicular to the planes* $2x+3y-4z=9$ *and* $x+2y+2z=5$.

☐ Equation of the plane through the point (1, 2, 3) is

$$A(x-1)+B(y-2)+C(z-3)=0. \tag{1}$$

Again, the plane (1) is perpendicular to the planes $2x+3y-4z=9$ and $x+2y+2z=5$, so we have

$$2A+3B-4C=0 \tag{2}$$
$$\text{and} \quad A+2B+2C=0. \tag{3}$$

Eliminating A, B, C from the Eqns. (1)–(3), we get

$$\begin{vmatrix} x-1 & y-2 & z-3 \\ 1 & 2 & 2 \\ 2 & 3 & -4 \end{vmatrix}=0.$$

Expanding we get

$$-14(x-1)+8(y-2)-(z-3)=0$$
$$\text{or,} \quad 14x-8y+z-1=0.$$

▶▷▶ 4. *Find the equation of the plane which passes through the line of intersection of the planes* $2x-3y+4z=0$ *and* $4x+y-2z=8$, *and through the point* (1, 1, 1).

☐ Equation of the plane which passes through the line of intersection of the planes $2x-3y+4z=0$ and $4x+y-2z=8$ can be written as

$$2x-3y+4z+k(4x+y-2z-8)=0, \quad \text{where } k \text{ is any arbitrary constant}$$
$$\text{or,} \quad 2x(1+2k)+(k-3)y+2z(2-k)-8k=0.$$

If it passes through the point (1, 1, 1), then we have

$$2(1+2k)+k-3+2(2-k)-8k=0$$
$$\text{or,} \quad -5k+3=0$$
$$\text{or,} \quad k=3/5.$$

Putting the value of k in the equation of the plane, we get

$$2x-3y+4z+(3/5)(4x+y-2z-8)=0$$
$$\text{or,} \quad 10x-15y+20z+12x+3y-6z-24=0$$
$$\text{or,} \quad 22x-12y+14z=24$$
$$\text{or,} \quad 11x-6y+7z=12.$$

▶▷▶ 5. *Find the equation of the plane which passes through the line of intersection of the planes* $2x-3y+4z=0$ *and* $4x+y-2z=8$, *and perpendicular to the plane* $x+y+z=9$.

□ Equation of the plane which passes through the line of intersection of the planes $2x-3y+4z=0$ and $4x+y-2z=8$ can be written as

$$2x-3y+4z+k(4x+y-2z-8)=0, \quad \text{where } k \text{ is any arbitrary constant}$$

$$\text{or,} \quad 2x(1+2k)+(k-3)y+2z(2-k)-8k=0.$$

If it is perpendicular to the plane $x+y+z=9$, then we have

$$2(1+2k)+k-3+2(2-k)=0$$

$$\text{or,} \quad 3+3k=0$$

$$\text{or,} \quad k=-1.$$

Putting the value of k in the equation of the plane, we get

$$2x-3y+4z-(4x+y-2z-8)=0$$

$$\text{or,} \quad -2x-4y+6z+8=0$$

$$\text{or,} \quad x+2y-3z-4=0.$$

▶▷▶ 6. *If the plane* $x-2y+3z=4$ *is rotated through a right angle about the line of intersection with the plane* $2x+3y-4z=5$, *show that the equation of the plane in its new position is* $22x+5y-4z=67$.

□ Equation of the plane which passes through the line of intersection of the planes $x-2y+3z=4$ and $2x+3y-4z=5$ can be written as

$$x-2y+3z-4+k(2x+3y-4z-5)=0, \quad \text{where } k \text{ is any arbitrary constant}$$

$$\text{or,} \quad (1+2k)x+(3k-2)y-(4k-3)z=4+5k.$$

If it is perpendicular to the plane $x-2y+3z=4$, then we have

$$(1+2k)-2(3k-2)-3(4k-3)=0$$

$$\text{or,} \quad -16k+14=0$$

$$\text{or,} \quad k=7/8.$$

Putting this value in the equation of the plane, we get

$$x-2y+3z-4+(7/8)(2x+3y-4z-5)=0$$

$$\text{or,} \quad 8x-16y+24z-32+14x+21y-28z-35=0$$

$$\text{or,} \quad 22x+5y-4z=67.$$

▶▷▶ 7. *The plane* $lx+my=0$ *is rotated about the line of intersection with plane* $z=0$ *through an angle* α. *Prove that its equation in its new position is*

$$lx+my\pm z\sqrt{l^2+m^2}\tan\alpha=0.$$

☐ Equation of the plane which passes through the line of intersection of the planes $lx + my = 0$ and $z = 0$ can be written as

$$lx + my + kz = 0, \text{ where } k \text{ is any arbitrary constant.}$$

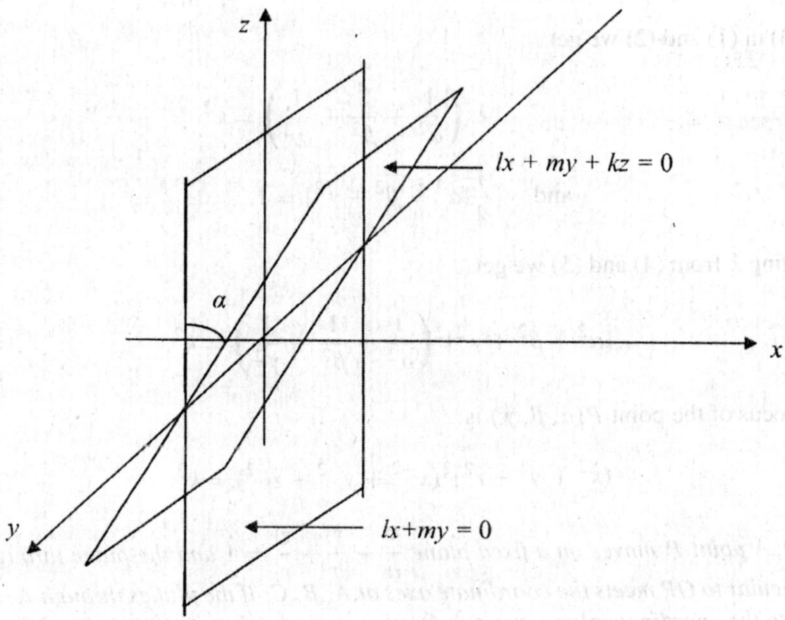

Given that the angle between the planes $lx + my = 0$ and $lx + my + kz = 0$ is α, so we have

$$\cos\alpha = \frac{l \times l + m \times m + k \times 0}{\sqrt{l^2 + m^2 + k^2}\sqrt{l^2 + m^2 + 0^2}}$$

$$\text{or,} \quad \sec^2\alpha = 1 + \frac{k^2}{l^2 + m^2}$$

$$\text{or,} \quad \tan^2\alpha = \frac{k^2}{l^2 + m^2}.$$

Hence, the equation of the required plane is

$$lx + my \pm z\sqrt{l^2 + m^2}\tan\alpha = 0.$$

▸▸ **8.** *A variable plane has intercepts on coordinate axes, the sum of whose squares is a constant k^2. Show that locus of the foot of the perpendicular from origin to the plane is*

$$(x^2 + y^2 + z^2)^2(x^{-2} + y^{-2} + z^{-2}) = k^2.$$

☐ Let the equation of the plane be

$$\frac{x}{a} + \frac{y}{b} + \frac{z}{c} = 1$$

where

$$a^2 + b^2 + c^2 = k^2. \qquad (1)$$

If $P(\alpha, \beta, \gamma)$ be the foot of the perpendicular from origin 'O' to the plane (1), then we have

$$\frac{\alpha}{a} + \frac{\beta}{b} + \frac{\gamma}{c} = 1 \tag{2}$$

$$\text{and} \quad a\alpha = b\beta = c\gamma = \lambda \text{ (say).} \tag{3}$$

Using (3) in (1) and (2) we get

$$\lambda^2\left(\frac{1}{\alpha^2} + \frac{1}{\beta^2} + \frac{1}{\gamma^2}\right) = k^2 \tag{4}$$

$$\text{and} \quad \frac{1}{\lambda}[\alpha^2 + \beta^2 + \gamma^2] = 1. \tag{5}$$

Eliminating λ from (4) and (5) we get

$$[\alpha^2 + \beta^2 + \gamma^2]^2\left(\frac{1}{\alpha^2} + \frac{1}{\beta^2} + \frac{1}{\gamma^2}\right) = k^2.$$

Hence, locus of the point $P(\alpha, \beta, \gamma)$ is

$$(x^2 + y^2 + z^2)^2(x^{-2} + y^{-2} + z^{-2}) = k^2.$$

▶▷▶ **9.** *A point P moves on a fixed plane* $\frac{x}{a} + \frac{y}{b} + \frac{z}{c} = 1$ *and the plane through P and perpendicular to OP meets the coordinate axes at A, B, C. If the planes through A, B, C and parallel to the coordinate planes meet at Q, then show that locus of the point Q is*

$$\frac{1}{x^2} + \frac{1}{y^2} + \frac{1}{z^2} = \frac{1}{ax} + \frac{1}{by} + \frac{1}{cz}.$$

☐ Let $P(\alpha, \beta, \gamma)$ be the moving point on the fixed plane

$$\frac{x}{a} + \frac{y}{b} + \frac{z}{c} = 1;$$

then we have

$$\frac{\alpha}{a} + \frac{\beta}{b} + \frac{\gamma}{c} = 1. \tag{1}$$

Equation of the plane through P and perpendicular to OP is

$$\alpha(x - \alpha) + \beta(y - \beta) + \gamma(z - \gamma) = 0$$

$$\text{or,} \quad \frac{\alpha x}{\sum \alpha^2} + \frac{\beta y}{\sum \alpha^2} + \frac{\gamma z}{\sum \alpha^2} = 1.$$

It intersects the coordinate axes at $A\left(\frac{\sum \alpha^2}{\alpha}, 0, 0\right)$, $B\left(0, \frac{\sum \alpha^2}{\beta}, 0\right)$ and $C\left(0, 0, \frac{\sum \alpha^2}{\gamma}\right)$. If the planes through A, B, C and parallel to the coordinate planes meet at Q, then coordinates of Q will be $\left(\frac{\sum \alpha^2}{\alpha}, \frac{\sum \alpha^2}{\beta}, \frac{\sum \alpha^2}{\gamma}\right) = (x', y', z')$, say.

Now,

$$\frac{1}{x'^2}+\frac{1}{y'^2}+\frac{1}{z'^2}=\frac{\sum\alpha^2}{(\sum\alpha^2)^2}=\frac{1}{\sum\alpha^2} \quad (2)$$

and $$\frac{1}{ax'}+\frac{1}{by'}+\frac{1}{cz'}=\frac{1}{\sum\alpha^2}\left(\frac{\alpha}{a}+\frac{\beta}{b}+\frac{\gamma}{c}\right)=\frac{1}{\sum\alpha^2}. \quad (3)$$

From (2) and (3) we obtain

$$\frac{1}{x'^2}+\frac{1}{y'^2}+\frac{1}{z'^2}=\frac{1}{ax'}+\frac{1}{by'}+\frac{1}{cz'}.$$

Hence, locus of the point $Q(x', y', z')$ is

$$\frac{1}{x^2}+\frac{1}{y^2}+\frac{1}{z^2}=\frac{1}{ax}+\frac{1}{by}+\frac{1}{cz}.$$

EXERCISE

1. Find the equation of the planes which passes through the points (1, 2, 3); (3, 4, −2) and parallel to coordinate axes.
2. If the sum of reciprocals of the intercepts on the coordinate axes be constant, show that the plane passes through a fixed point.
3. Find the equation of the plane through the point $P(a, b, c)$ and perpendicular to OP, O being the origin.
4. If P be the point (−3, 1, 1) and Q be the point (3, 4, 2), then find the equation of the plane through R and perpendicular to PQ where R divides PQ internally in the ratio 2 : 1.
5. Find the equations of the planes passing through the points (0, 4, 3); (6, −4, 3) and cutting off intercepts from axes whose sum is zero.
6. Perpendiculars PL, PM, PN are drawn from $P(a, b, c)$ to the coordinate planes. Show that equation of the plane LMN is $\frac{x}{a}+\frac{y}{b}+\frac{z}{c}=2$.

   [*Hints*: Coordinates of L, M, N are $(a, b, 0)$; $(0, b, c)$ and $(a, 0, c)$ respectively. Let the equation of the plane be $\frac{x}{\alpha}+\frac{y}{\beta}+\frac{z}{\gamma}=1$. It passes through L, M, N, so we have

$$\frac{a}{\alpha}+\frac{b}{\beta}=1; \quad \frac{b}{\beta}+\frac{c}{\gamma}=1 \quad \text{and} \quad \frac{a}{\alpha}+\frac{c}{\gamma}=1.$$

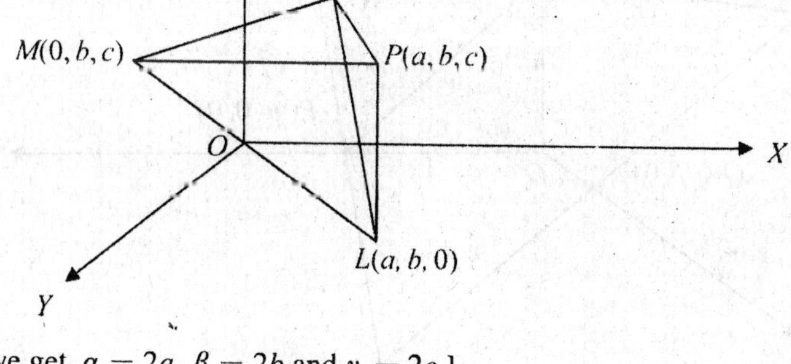

Solving we get, $\alpha = 2a$, $\beta = 2b$ and $\gamma = 2c$.]

7. A plane cuts the axes in A, B, C and if the centroid of the triangle ABC is (a, b, c), show that the equation of the plane is $\frac{x}{a}+\frac{y}{b}+\frac{z}{c}=3$.
8. A variable plane which is at a constant distance $3p$ from the origin O cuts axes in A, B, C; show that the locus of the centroid of the triangle ABC is

$$x^{-2}+y^{-2}+z^{-2}=p^{-2}.$$

9. A variable plane which is at a constant distance $3p$ from the origin O cuts axes in A, B, C; show that the locus of the centroid of the tetrahedron $OABC$ is

$$9(x^{-2}+y^{-2}+z^{-2})=16p^{-2}.$$

10. A variable plane which is at a constant distance $3p$ from the origin O cuts axes in A, B, C; show that the locus of the point of intersection of the planes through A, B, C drawn parallel to the coordinate planes is

$$9(x^{-2}+y^{-2}+z^{-2})=p^{-2}.$$

11. Find the equation of the plane passing through the line of intersection of the planes $ax+by+cz+d=0$ and $a'x+b'y+c'z+d'=0$, and perpendicular to the xy-plane.
12. If the plane $x-2y+3z=0$ is rotated through a right angle about the line of intersection with the plane $x+3y-4z=5$, find the equation of the plane in its new position.
13. Show that the equation $\frac{a}{y-z}+\frac{b}{z-x}+\frac{c}{x-y}=0$ represents a pair of planes.
14. A triangle of sides a, b, c is placed so that the middle points of the sides are on the axes. Show that the equation of the plane is

$$\frac{x}{\alpha}+\frac{y}{\beta}+\frac{z}{\gamma}=1$$

where $8\alpha^2=b^2+c^2-a^2$, $8\beta^2=a^2+c^2-b^2$ and $8\gamma^2=b^2+a^2-c^2$. Find also the coordinates of vertices.

[*Hints*: From the following figure we get $\alpha^2+\beta^2=c^2/4$; $\gamma^2+\beta^2=a^2/4$ and $\alpha^2+\gamma^2=b^2/4$.

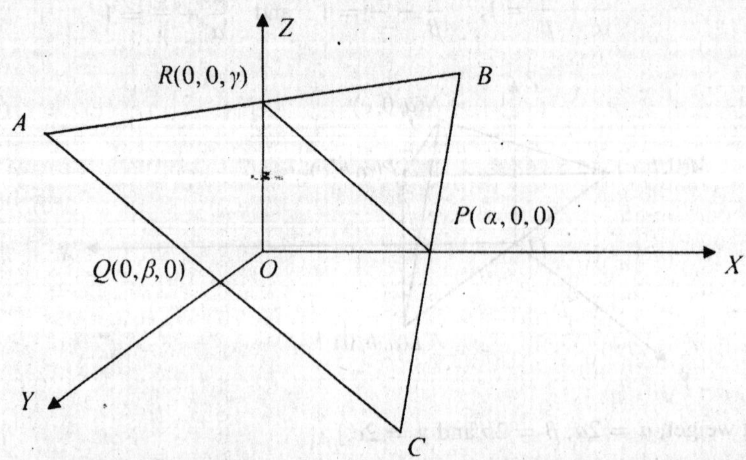

Solving the above equations for α, β, γ we get the results. Again, taking the coordinates of A, B, C as (x_1, y_1, z_1), (x_2, y_2, z_2), (x_3, y_3, z_3) and using the results

$$\begin{aligned} x_1 + x_2 = 0; \quad & y_1 + y_2 = 0; \quad & z_1 + z_2 = 2\gamma; \\ x_1 + x_3 = 0; \quad & y_1 + y_3 = 2\beta; \quad & z_1 + z_3 = 0; \\ x_2 + x_3 = 2\alpha; \quad & y_2 + y_3 = 0; \quad & z_2 + z_3 = 0, \end{aligned}$$

we get the coordinates of the vertices.]

15. The plane $ax + by + cz + d = 0$ bisects the angle between a pair of planes of which one is $lx + my + nz + p = 0$. Find the equation of the other plane.

[*Hints*: Equation of any plane passing through the line of intersection of the planes $ax + by + cz + d = 0$ and $lx + my + nz + p = 0$ can be written as

$$(ax + by + cz + d) + k(lx + my + nz + p) = 0. \tag{1}$$

Since, the plane $ax+by+cz+d = 0$ bisects the angle between planes (1) and $lx+my+nz+p = 0$, the length of perpendiculars from any point on the plane $ax+by+cz+d = 0$ drawn of the planes will be of equal length. This gives the value of k.]

Chapter 3

The Straight Line

3.1 Introduction

A straight line is represented by the intersection of two non-parallel planes and hence can be represented by two equations of the first degree in x, y, z together. Thus equation of a straight line can be written as

$$ax + by + cz + d = 0 = a'x + b'y + c'z + d'.$$

So, the equation of a straight line passing through the origin can be written as

$$ax + by + cz = 0 = a'x + b'y + c'z.$$

The x-axis is the intersection of xy-plane and xz-plane, whose equations are $z = 0$ and $y = 0$. So, equation of x-axis is $y = 0$, $z = 0$. Similarly, equation of y-axis is $x = 0$, $z = 0$ and that of z-axis is $x = 0$, $y = 0$.

3.2 Symmetrical Form

Let $P(x_1, y_1, z_1)$ be a fixed point on a straight line whose direction cosines are l, m, n. If $Q(x, y, z)$ be any point on that line such that $PQ = r$, then projecting PQ on coordinate axes, we obtain

$$x - x_1 = lr; \quad y - y_1 = mr \quad \text{and} \quad z - z_1 = nr$$

from which we get

$$\frac{x - x_1}{l} = \frac{y - y_1}{m} = \frac{z - z_1}{n} = r.$$

Any point on this line may be taken as $(x_1 + lr, y_1 + mr, z_1 + nr)$. For different values of r we get different points on the line whose distances from the point P are values of r.

Since, direction ratios of a straight line are proportional to its direction cosines, equation of a straight line having direction ratios a, b, c and passing through the point $P(x_1, y_1, z_1)$ can be put in the form

$$\frac{x - x_1}{a} = \frac{y - y_1}{b} = \frac{z - z_1}{c} = r.$$

If $P(x_1, y_1, z_1)$ and $R(x_2, y_2, z_2)$ be two points on a straight line, then its direction ratios are proportional to $x_2 - x_1$, $y_2 - y_1$, $z_2 - z_1$. So, equation of the line PR can be put in the form

$$\frac{x - x_1}{x_2 - x_1} = \frac{y - y_1}{y_2 - y_1} = \frac{z - z_1}{z_2 - z_1} = r.$$

3.3 Distance of a Point from a Line

Let $P(\alpha, \beta, \gamma)$ be a point whose distance from a straight line through $A(a, b, c)$ having direction cosines l, m, n, is to be measured.

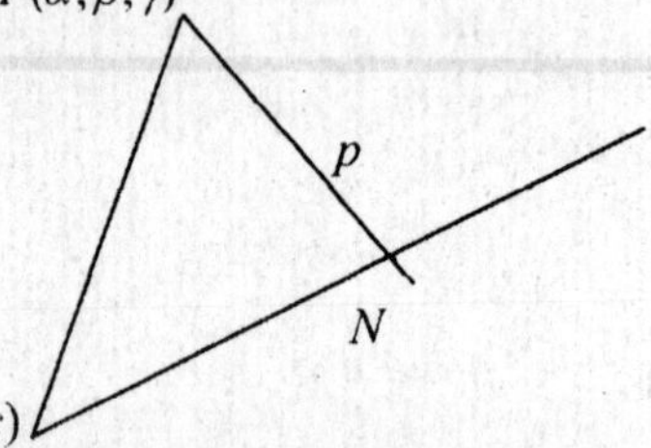

Then AN = projection of AP on the straight line $= l(\alpha - a) + m(\beta - b) + n(\gamma - c)$. So length of the perpendicular from P on the line

$$\begin{aligned} PN &= \sqrt{AP^2 - AN^2} = \sqrt{(\alpha - a)^2 + (\beta - b)^2 + (\gamma - c)^2 - [l(\alpha - a) + m(\beta - b) + n(\gamma - c)]^2} \\ &= \sqrt{[(\alpha - a)^2 + (\beta - b)^2 + (\gamma - c)^2](l^2 + m^2 + n^2) - [l(\alpha - a) + m(\beta - b) + n(\gamma - c)]^2} \\ &= \sqrt{[(\gamma - c)l - (\alpha - a)n]^2 + [(\gamma - c)m - (\beta - b)n]^2 + [(\beta - b)l - (\alpha - a)m]^2}. \end{aligned}$$

3.4 Expression of Line of Intersection of Two Planes in Symmetrical Form

If A, B, C be the direction ratios of the line of intersection of the planes

$$ax + by + cz = d \quad \text{and} \quad a'x + b'y + c'z = d'$$

then we have, $Aa + Bb + Cc = 0$ and $Aa' + Bb' + Cc' = 0$. Solving we get

$$\frac{A}{bc' - cb'} = \frac{B}{ca' - ac'} = \frac{C}{ab' - ba'}.$$

Now putting $z = 0$ in the given equations of the planes we get

$$ax + by = d \quad \text{and} \quad a'x + b'y = d'.$$

Solving we get, $x = \dfrac{b'd - bd'}{ab' - a'b}$ and $y = \dfrac{ad' - da'}{ab' - a'b}$. Hence the line of intersection of the given planes intersects xy-plane at $\left(\dfrac{b'd - bd'}{ab' - a'b}, \dfrac{ad' - da'}{ab' - a'b}, 0\right)$. Thus the symmetrical form of the required line is

$$\frac{x - \dfrac{b'd - bd'}{ab' - a'b}}{bc' - cb'} = \frac{y - \dfrac{ad' - da'}{ab' - a'b}}{ca' - ac'} = \frac{z - 0}{ab' - ba'}.$$

3.5 Intersection of a Straight Line and a Plane

If we take the plane and the straight line as

$$ax + by + cz + d = 0 \quad \text{and} \quad \frac{x-\alpha}{l} = \frac{y-\beta}{m} = \frac{z-\gamma}{n} = r$$

respectively, then any point on the straight line can be taken as $(\alpha + lr, \beta + mr, \gamma + nr)$. If this point lies on the plane, then we have

$$a(\alpha + lr) + b(\beta + mr) + c(\gamma + nr) + d = 0$$

$$\text{or,} \quad a\alpha + b\beta + c\gamma + d + r(al + bm + cn) = 0 \tag{1}$$

$$\text{or,} \quad r = -\frac{a\alpha + b\beta + c\gamma + d}{al + bm + cn}$$

provided $al + bm + cn \neq 0$, i.e. the straight line is not parallel to the plane.

If the point (α, β, γ) through which the line passes, lies on the plane, then $a\alpha + b\beta + c\gamma + d = 0$, which gives $r = 0$.

If the plane contains the line, then Eqn. (1) is satisfied for all values of r implying that all the points on the line are also the points of the plane. So the equation of the plane through the given line can be written as

$$a(x-\alpha) + b(y-\beta) + c(z-\gamma) = 0$$

$$\text{with} \quad al + bm + cn = 0.$$

3.6 Necessary and Sufficient Conditions of Coplanarity of Two Straight Lines

If we take two straight lines as

$$\frac{x-\alpha}{l} = \frac{y-\beta}{m} = \frac{z-\gamma}{n} \tag{2}$$

$$\text{and} \quad \frac{x-\alpha'}{l'} = \frac{y-\beta'}{m'} = \frac{z-\gamma'}{n'}. \tag{3}$$

Then equation of the plane through the first line can be written as

$$a(x-\alpha) + b(y-\beta) + c(z-\gamma) = 0 \tag{4}$$

$$\text{with} \quad al + bm + cn = 0. \tag{5}$$

If the plane given by Eqn. (4) contains the second line, then we have

$$a(\alpha'-\alpha) + b(\beta'-\beta) + c(\gamma'-\gamma) = 0 \tag{6}$$

$$\text{and} \quad al' + bm' + cn' = 0. \tag{7}$$

Eliminating a, b, c from Eqns. (6), (5) and (7) we get

$$\begin{vmatrix} \alpha'-\alpha & \beta'-\beta & \gamma'-\gamma \\ l & m & n \\ l' & m' & n' \end{vmatrix} = 0. \tag{8}$$

This is the necessary condition for the given lines to be coplanar.

Further, if the condition (8) holds, then we can say that the following system of equations have non-trivial solutions, say (u, v, w):

$$x(\alpha' - \alpha) + y(\beta' - \beta) + z(\gamma' - \gamma) = 0 \tag{9}$$

$$xl + ym + zn = 0 \tag{10}$$

$$xl' + ym' + zn' = 0. \tag{11}$$

Then we have

$$u(\alpha' - \alpha) + v(\beta' - \beta) + w(\gamma' - \gamma) = 0 \tag{12}$$

$$ul + vm + wn = 0 \tag{13}$$

$$ul' + vm' + wn' = 0. \tag{14}$$

From Eqn. (12) we see that

$$u\alpha' + v\beta' + w\gamma' = d = u\alpha + v\beta + w\gamma$$

which with conditions (12) and (13) show that the given lines lie on the same plane whose equation is

$$ux + vy + wz = d.$$

Again, eliminating a, b, c from (4), (5) and (7) we get

$$\begin{vmatrix} x-\alpha & y-\beta & z-\gamma \\ l & m & n \\ l' & m' & n' \end{vmatrix} = 0. \tag{15}$$

This is the equation of the plane containing the given coplanar lines.
If instead of (2) and (3) the lines are given in general form as

$$a_1x + b_1y + c_1z + d_1 = 0 = a_2x + b_2y + c_2z + d_2 \tag{16}$$

$$\text{and} \quad a_3x + b_3y + c_3z + d_3 = 0 = a_4x + b_4y + c_4z + d_4 \tag{17}$$

then any plane through the lines given by Eqns. (16) and (17) can be taken as

$$(a_1 + \lambda a_2)x + (b_1 + \lambda b_2)y + (c_1 + \lambda c_2)z + d_1 + \lambda d_2 = 0 \tag{18}$$

$$\text{and} \quad (a_3 + \mu a_4)x + (b_3 + \mu b_4)y + (c_3 + \mu c_4)z + d_3 + \mu d_4 = 0. \tag{19}$$

These equations will be identical if the given lines are coplanar. Comparing Eqns. (18) and (19) we get

$$\frac{a_1 + \lambda a_2}{a_3 + \mu a_4} = \frac{b_1 + \lambda b_2}{b_3 + \mu b_4} = \frac{c_1 + \lambda c_2}{c_3 + \mu c_4} = \frac{d_1 + \lambda d_2}{d_3 + \mu d_4} = k \text{ (say)}.$$

From this we get

$$a_1 + \lambda a_2 - k(a_3 + \mu a_4) = 0$$

$$b_1 + \lambda b_2 - k(b_3 + \mu b_4) = 0$$

$$c_1 + \lambda c_2 - k(c_3 + \mu c_4) = 0$$

$$d_1 + \lambda d_2 - k(d_3 + \mu d_4) = 0.$$

Eliminating λ, $-k$, $-k\mu$ from above equations we get

$$\begin{vmatrix} a_1 & a_2 & a_3 & a_4 \\ b_1 & b_2 & b_3 & b_4 \\ c_1 & c_2 & c_3 & c_4 \\ d_1 & d_2 & d_3 & d_4 \end{vmatrix} = 0. \tag{20}$$

This is the required condition of coplanarity.

3.7 Skew Lines and the Shortest Distance between Them

Two non-parallel lines are said to be skew lines if they do not intersect and the length of the line, which is perpendicular to both the skew lines, intercepted between them is called *shortest distance* between the skew lines.

If we take two skew lines as

$$\frac{x-\alpha}{l} = \frac{y-\beta}{m} = \frac{z-\gamma}{n} \tag{21}$$

$$\text{and} \quad \frac{x-\alpha'}{l'} = \frac{y-\beta'}{m'} = \frac{z-\gamma'}{n'} \tag{22}$$

and if the common perpendicular to the above lines, AB has direction cosines L, M, N, then we have

$$Ll + Mm + Nn = 0$$

$$\text{and} \quad Ll' + Mm' + Nn' = 0.$$

Solving we get

$$\frac{L}{mn' - nm'} = \frac{M}{nl' - n'l} = \frac{N}{lm' - ml'} = \frac{1}{\sqrt{\sum (mn' - nm')^2}}. \tag{23}$$

Now, the shortest distance between the lines GH and $ST = AB =$ projection of PQ on $AB =$ $L(\alpha' - \alpha) + M(\beta' - \beta) + N(\gamma' - \gamma)$.

Using (23) we get the shortest distance

$$= \frac{(\alpha' - \alpha)(mn' - nm') + (\beta' - \beta)(nl' - n'l) + (\gamma' - \gamma)(lm' - ml')}{\sqrt{(mn' \quad nm')^2 + (nl' - n'l)^2 + (lm' - ml')^2}}$$

$$= \begin{vmatrix} \alpha' - \alpha & \beta' - \beta & \gamma' - \gamma \\ l & m & n \\ l' & m' & n' \end{vmatrix} \div \sqrt{\left[\sum (mn' - m'n)^2\right]}. \tag{24}$$

From the results (8) and (24) it is clear that if the two lines are coplanar, then the shortest distance between them is zero. The lines AB and GH are intersecting and hence coplanar. Equation of the plane containing them is

$$\begin{vmatrix} x-\alpha & y-\beta & z-\gamma \\ l & m & n \\ L & M & N \end{vmatrix} = 0. \tag{25}$$

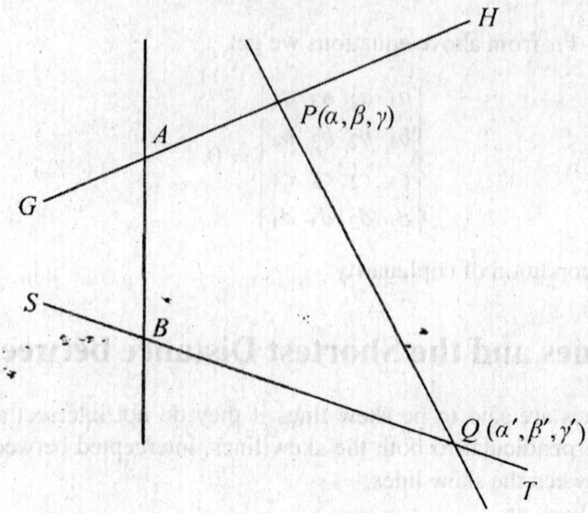

Similarly, equation of the plane containing AB and ST is

$$\begin{vmatrix} x-\alpha' & y-\beta' & z-\gamma' \\ l' & m' & n' \\ L & M & N \end{vmatrix} = 0. \tag{26}$$

Obviously, the above planes intersect along the line AB. Hence, equation of the line of shortest distance is given by the Eqns. (25) and (26).

Alternative Method

Any point on the given lines can be taken as $(\alpha + lr, \beta + mr, \gamma + nr)$ and $(\alpha' + l'R, \beta' + m'R, \gamma' + n'R)$ respectively. If the skew lines intersect the line of shortest distance at these points, then direction ratios of the line of shortest distance are

$$(\alpha' + l'R - \alpha - lr), \;\; (\beta' + m'R - \beta - mr), \;\; (\gamma' + n'R - \gamma - nr).$$

Since AB is perpendicular to both GH and ST, we have

$$(\alpha' + l'R - \alpha - lr)l + (\beta' + m'R - \beta - mr)m + (\gamma' + n'R - \gamma - nr)n = 0$$

and $$(\alpha' + l'R - \alpha - lr)l' + (\beta' + m'R - \beta - mr)m' + (\gamma' + n'R - \gamma - nr)n' = 0$$

or, $$(l^2 + m^2 + n^2)r - (l'l + m'm + n'n)R = (\alpha' - \alpha)l + (\beta' - \beta)m + (\gamma' - \gamma)n \tag{27}$$

and $$(l'l + m'm + n'n)r - (l'^2 + m'^2 + n'^2)R = (\alpha' - \alpha)l' + (\beta' - \beta)m' + (\gamma' - \gamma)n'. \tag{28}$$

For known values of $l, m, n, l', m', n', \alpha, \beta, \gamma, \alpha', \beta', \gamma'$, the above equations can be solved for r, R. Thus the coordinates of A, B are now known to us and hence the shortest distance between given lines, i.e. AB and equation of the line of the shortest distance can be determined easily.

If one of the lines is given in general form as

$$a_1x + b_1y + c_1z + d_1 = 0 = a_2x + b_2y + c_2z + d_2$$

and the other line is in the symmetrical form, i.e.

$$\frac{x-\alpha}{l}=\frac{y-\beta}{m}=\frac{z-\gamma}{n}$$

then we consider a plane through the first line as

$$(a_1+\lambda a_2)x+(b_1+\lambda b_2)y+(c_1+\lambda c_2)z+d_1+\lambda d_2=0 \tag{29}$$

where the unknown constant λ is to be determined in such a way that the plane becomes parallel to the second line, i.e.

$$(a_1+\lambda a_2)l+(b_1+\lambda b_2)m+(c_1+\lambda c_2)n=0.$$

In this case shortest distance is the perpendicular distance of any point on the line from the plane given by Eqn. (29).
Similarly, if both the lines are given in general form as

$$a_1x+b_1y+c_1z+d_1=0=a_2x+b_2y+c_2z+d_2$$
$$\text{and}\quad a_3x+b_3y+c_3z+d_3=0=a_4x+b_4y+c_4z+d_4$$

then any plane through the given lines can be taken as

$$(a_1+\lambda a_2)x+(b_1+\lambda b_2)y+(c_1+\lambda c_2)z+d_1+\lambda d_2=0 \tag{30}$$
$$\text{and}\quad (a_3+\mu a_4)x+(b_3+\mu b_4)y+(c_3+\mu c_4)z+d_3+\mu d_4=0 \tag{31}$$

where the unknowns λ, μ are determined in such a way that the above planes are parallel. In this case the shortest distance between the lines is the distance between the parallel planes.

3.8 Equations of Two Skew Lines

Let AB and CD be two skew lines, and the line of shortest distance between them meet them at L and M where $LM = 2c$. Let the middle point of LM, say O be the origin and the direction of ML be z-axis. Through O, draw two lines OP and OQ parallel to AB and CD respectively such that angle between them be 2α.

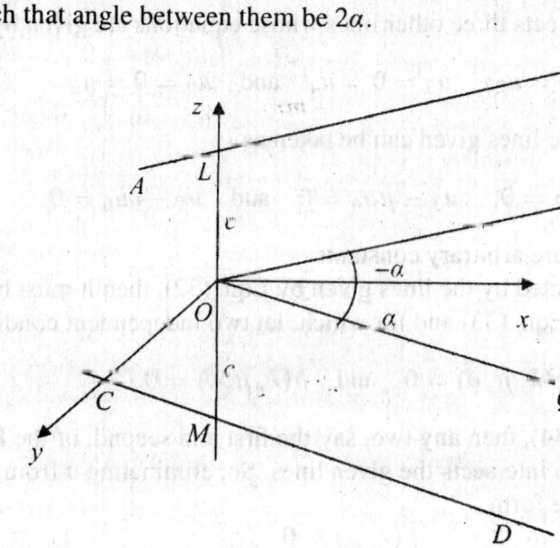

The internal and external bisectors of $\angle POQ$ are taken as x-axis and y-axis respectively. The direction cosines of the line OQ are $\cos\alpha$, $\cos(90° - \alpha)$, $\cos 90°$, i.e. $\cos\alpha$, $\sin\alpha$, 0. Similarly, direction cosines of the line OP are $\cos(-\alpha)$, $\cos(90° + \alpha)$, $\cos 90°$, i.e. $\cos\alpha$, $-\sin\alpha$, 0. Since $OL = c = OM$, coordinates of L, M are $(0, 0, c)$ and $(0, 0, -c)$. Hence equations of AB and CD are respectively

$$\frac{x}{\cos\alpha} = \frac{y}{\sin\alpha} = \frac{z-c}{0} \quad \text{and} \quad \frac{x}{\cos\alpha} = \frac{y}{-\sin\alpha} = \frac{z+c}{0}$$

$$\text{or,} \quad y = mx,\ z = c \quad \text{and} \quad y = -mx,\ z = -c, \quad \text{where } m = \tan\alpha.$$

3.9 A Straight Line Intersecting Other Given Lines

Let a straight line intersects two other lines whose equations are given by

$$a_1x + b_1y + c_1z + d_1 = 0 = a_2x + b_2y + c_2z + d_2 \quad \text{or} \quad u_1 = 0 = u_2$$

$$\text{and} \quad a_3x + b_3y + c_3z + d_3 = 0 = a_4x + b_4y + c_4z + d_4 \quad \text{or} \quad u_3 = 0 = u_4$$

then any plane through the lines given can be taken as

$$(a_1 + \lambda a_2)x + (b_1 + \lambda b_2)y + (c_1 + \lambda c_2)z + d_1 + \lambda d_2 = 0 \quad \text{or} \quad u_1 + \lambda u_2 = 0$$

$$\text{and} \quad (a_3 + \mu a_4)x + (b_3 + \mu b_4)y + (c_3 + \mu c_4)z + d_3 + \mu d_4 = 0 \quad \text{or} \quad u_3 + \mu u_4 = 0.$$

Now, the line of intersection of the above planes will be the required line where the constants λ and μ are to be determined using given condition, say the line passes through the point (α, β, γ). Then

$$(a_1 + \lambda a_2)\alpha + (b_1 + \lambda b_2)\beta + (c_1 + \lambda c_2)\gamma + d_1 + \lambda d_2 = 0 \quad \text{or} \quad u_1' + \lambda u_2' = 0$$

$$\text{and} \quad (a_3 + \mu a_4)\alpha + (b_3 + \mu b_4)\beta + (c_3 + \mu c_4)\gamma + d_3 + \mu d_4 = 0 \quad \text{or} \quad u_3' + \mu u_4' = 0.$$

Eliminating λ and μ we get the equation of required line is

$$u_1u_2' - u_1'u_2 = 0 = u_3u_4' - u_3'u_4.$$

Now, if a straight line intersects three other lines whose equations are given by

$$u_1 = 0 = u_2; \quad u_3 = 0 = u_4 \quad \text{and} \quad u_5 = 0 = u_6 \tag{32}$$

then any plane through these lines given can be taken as

$$u_1 - \lambda u_2 = 0; \quad u_3 - \mu u_4 = 0 \quad \text{and} \quad u_5 - \delta u_6 = 0 \tag{33}$$

respectively, where λ, μ, δ are arbitrary constants.

Now, if a line be intersected by the lines given by Eqn. (32), then it must be the common line of the planes given by Eqn. (33) and for which, let two independent conditions be

$$g(\lambda, \mu, \delta) = 0 \quad \text{and} \quad h(\lambda, \mu, \delta) = 0. \tag{34}$$

If we can solve the Eqns. (34), then any two, say the first and second, of the Eqns. (33) will give the required line which intersects the given lines. So, eliminating δ from the Eqns. (34) we obtain an equation of the form

$$f(\lambda, \mu) = 0 \tag{35}$$

which is satisfied by infinite number of values of λ, μ implying that there exists infinite number of lines which intersect the given three lines and these lines generate a surface whose equation will be λ, μ eliminant of Eqn. (35), i.e.

$$f\left(\frac{u_1}{u_2}, \frac{u_3}{u_4}\right) = 0.$$

Examples

▶▶ 1. (a) *Find the equation of the straight line passing through the points* $(3, -9, 4)$ *and* $(-9, 6, -4)$.

(b) *Find the equation of the median AD of the triangle whose vertices are* $A(3, 4, 8)$, $B(1, -6, 2)$ *and* $C(1, 4, -2)$.

□ (a) Direction ratios of the line joining $A(3, -9, 4)$ and $B(-9, 6, -4)$ are $3 - (-9)$, $-9 - 6, 4 - (-4)$, i.e. $12, -15, 8$. So, the equation of the required line is

$$\frac{x-3}{12} = \frac{y+9}{-15} = \frac{z-4}{8}.$$

□ (b) Coordinates of D are $\left(\frac{1+1}{2}, \frac{-6+4}{2}, \frac{2-2}{2}\right)$, i.e. $(1, -1, 0)$. So, the direction ratios of AD are $(3-1)$, $(4+1)$, $(8-0)$, i.e. $2, 5, 8$. Hence, equation of AD is

$$\frac{x-1}{2} = \frac{y+1}{5} = \frac{z}{8}.$$

▶▶ 2. *Show that the straight line through the points* (a, b, c) *and* (A, B, C) *passes through the origin, if* $Aa + Bb + Cc = Pp$ *where* p, P *are the distances of the points from origin.*

□ Direction ratios of the line joining (a, b, c) and (A, B, C) are $a - A, b - B, c - C$. So equation of the line joining them is

$$\frac{x-a}{a-A} = \frac{y-b}{b-B} = \frac{z-c}{c-C}$$

$$\text{or,} \quad \frac{x-A}{a-A} = \frac{y-B}{b-B} = \frac{z-C}{c-C}.$$

If the line passes through origin, we have

$$\frac{a}{a-A} = \frac{b}{b-B} = \frac{c}{c-C}$$

$$\text{or,} \quad \frac{A}{a-A} = \frac{B}{b-B} = \frac{C}{c-C}.$$

This gives $\frac{a}{A} = \frac{b}{B} = \frac{c}{C} = k$ (say).

Now,

$$\begin{aligned}
Pp &= \sqrt{A^2 + B^2 + C^2}\sqrt{a^2 + b^2 + c^2} \\
&= \sqrt{A^2 + B^2 + C^2}\, k\sqrt{A^2 + B^2 + C^2} \\
&= k(A^2 + B^2 + C^2) \\
&= A(kA) + B(kB) + C(kC) \\
&= Aa + Bb + Cc.
\end{aligned}$$

▶▷ 3. *Put the equations in symmetrical form:* $x + 5y - z - 7 = 0 = 2x - 5y + 3z + 1$.

□ Let a, b, c be the direction ratios of the line of intersection of the given planes, then we have

$$a + 5b - c = 0$$
$$2a - 5b + 3c = 0.$$

Solving we get

$$\frac{a}{15-5} = \frac{b}{-2-3} = \frac{c}{-5-10}$$

$$\text{or,} \quad \frac{a}{-2} = \frac{b}{1} = \frac{c}{3}.$$

To get a point on the line we put $z = 0$ in the given equations, i.e. the resulting equations are

$$x + 5y = 7 \quad \text{and} \quad 2x - 5y + 1 = 0.$$

Solving we get

$$x = 2 \quad \text{and} \quad y = 1.$$

Thus the given line intersects the xy-plane at (2, 1, 0). So, the equation of the given line in symmetrical form is

$$\frac{x-2}{-2} = \frac{y-1}{1} = \frac{z}{3}.$$

▶▷ 4. *Find the point where the line joining* (−2, 3, −1) *and* (−3, 4, 5) *intersects the plane* $3x + y + z = 8$.

□ Direction ratios of the line joining (−2, 3, −1) and (−3, 4, 5) are 1, −1, −6, and hence its equation is $\frac{x+2}{1} = \frac{y-3}{-1} = \frac{z+1}{-6} = r$. Any point on this line can be taken as $(-2+r, 3-r, -1-6r)$ and if this line intersect the given plane at this point, then we have

$$3(-2+r) + 3 - r - 1 - 6r = 8$$
$$\text{or,} \quad 4r = -12, \ \text{i.e. } r = -3.$$

Hence, the coordinate of the point of intersection of the given line and plane is (−5, 6, 17).

▶▷ 5. *Find the distance of the point* (1, −2, 3) *from the plane* $x - y + z = 5$ *measured parallel to the line* $\frac{x-2}{2} = \frac{y-1}{3} = \frac{z}{-6}$.

□ Equation of the straight line through the point (1, −2, 3) and parallel to the line

$$\frac{x-2}{2} = \frac{y-1}{3} = \frac{z}{-6}$$

is

$$\frac{x-1}{2} = \frac{y+2}{3} = \frac{z-3}{-6} = r.$$

Any point on this line can be taken as $(1+2r, -2+3r, 3-6r)$ and if this line intersect the given plane at this point, then we have

$$(1+2r) - (-2+3r) + (3-6r) = 5$$
$$\text{or,} \quad 7r = 1, \ \text{i.e. } r = 1/7.$$

Hence, the coordinate of the point of intersection of the given line and plane is $(9/7, -11/7, 15/7)$. So the required distance $= \sqrt{4r^2 + 9r^2 + 36r^2} = 7r = 1$.

▶▷▶ 6. *Find the image of the point* $(-3, 8, 4)$ *in the plane* $6x - 3y - 2z + 1 = 0$.

□ If B be the foot of the perpendicular drawn from the point $A(-3, 8, 4)$ on the plane

$$6x - 3y - 2z + 1 = 0$$

then B will be the middle point of AC, where C is the image of A in the given plane. Now equation of AC is

$$\frac{x+3}{6} = \frac{y-8}{-3} = \frac{z-4}{-2} = r.$$

Any point on AC can be taken as $(-3+6r, 8-3r, 4-2r)$. If it be the point C, then coordinates of B are $(-3+3r, 8-3r/2, 4-r)$. Since B lies on the plane, we have

$$6(-3+3r) - 3(8-3r/2) - 2(4-r) + 1 = 0 \quad \text{or,} \quad r = 2.$$

Hence, the image of the point $(-3, 8, 4)$ in the plane $6x - 3y - 2z + 1 = 0$ is $(9, 2, 0)$.

▶▷▶ 7. *Find the equation of the image line of the line* $\frac{x-1}{3} = \frac{y-3}{5} = \frac{z-4}{2}$ *in the plane* $2x - y + z + 6 = 0$.

□ Equation of the perpendicular drawn from the point $(1, 3, 4)$ to the plane $2x - y + z + 6 = 0$ is

$$\frac{x-1}{2} = \frac{y-3}{-1} = \frac{z-4}{1} = r.$$

Any point on this line can be taken as $(1+2r, 3-r, 4+r)$. If this be the image of the point $(1, 3, 4)$ in the plane $2x - y + z + 6 = 0$, then middle point of $(1+2r, 3-r, 4+r)$ and $(1, 3, 4)$ must lie in the plane, i.e.

$$2(1+r) - (3-r/2) + (4+r/2) + 6 = 0 \quad \text{or,} \quad r = -3.$$

So, the image of the point $(1, 3, 4)$ in the plane $2x - y + z + 6 = 0$ is $(-5, 6, 1)$.

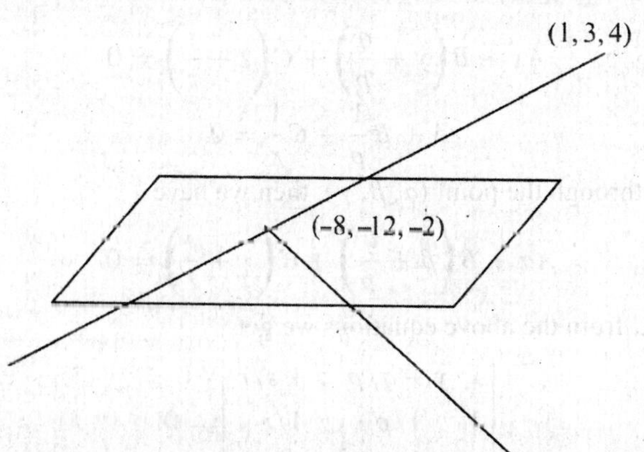

Any point on the given line can be taken as $(1+3r, 3+5r, 4+2r)$. If this point lies on the given plane, then

$$2(1+3r) - (3+5r) + (4+2r) + 6 = 0$$

$$\text{or,} \quad 3r = -9, \quad \text{i.e. } r = -3.$$

So, the given line intersects the plane $2x-y+z+6=0$ at $(-8,-12,-2)$. Hence, the image of the given line in the plane $2x-y+z+6=0$ is

$$\frac{x+8}{1}=\frac{y+12}{6}=\frac{z+2}{1}.$$

▶▷ 8. *Find the equation of the straight line passing through the point* $(1,-1,3)$ *and perpendicular to the straight line* $\frac{x-1}{2}=\frac{y-3}{-1}=\frac{z-4}{1}$.

□ Any point on this line can be taken as $B(1+2r, 3-r, 4+r)$. If we take $(1,-1,3)$ as the point A, then direction ratios of the line AB are $2r, 4-r, 1+r$. If AB is perpendicular to the given line, then we have

$$2(2r)-(4-r)+1+r=0,$$

which gives $r=1/2$.

Thus the coordinate of the foot of the perpendicular from the point $(1,-1,3)$ on the straight line $\frac{x-1}{2}=\frac{y-3}{-1}=\frac{z-4}{1}$ is $(2, 5/2, 9/2)$. So equation of AB is

$$\frac{x-1}{2}=\frac{y+1}{7}=\frac{z-3}{3}.$$

▶▷ 9. *Prove that the equation of the plane through the point* (α,β,γ) *and the line* $x=py+q=rz+s$ *is*

$$\begin{vmatrix} x & py+q & rz+s \\ 1 & 1 & 1 \\ \alpha & p\beta+q & r\gamma+s \end{vmatrix}=0.$$

□ We express the given straight line in symmetrical form as

$$\frac{x-0}{1}=\frac{y+q/p}{1/p}=\frac{z+s/r}{1/r}.$$

Equation of the plane through the above line can be written as

$$Ax+B\left(y+\frac{q}{p}\right)+C\left(z+\frac{s}{r}\right)=0 \tag{1}$$

where

$$A+B\frac{1}{p}+C\frac{1}{r}=0. \tag{2}$$

If this plane passes through the point (α,β,γ), then we have

$$A\alpha+B\left(\beta+\frac{q}{p}\right)+C\left(\gamma+\frac{s}{r}\right)=0. \tag{3}$$

Eliminating A, B, C from the above equations we get

$$\begin{vmatrix} x & y+q/p & z+s/r \\ 1 & 1/p & 1/r \\ \alpha & \beta+q/p & \gamma+s/r \end{vmatrix}=0.$$

Multiplying the 2nd and 3rd columns respectively by p and r we obtain

$$\begin{vmatrix} x & py+q & rz+s \\ 1 & 1 & 1 \\ \alpha & p\beta+q & r\gamma+s \end{vmatrix}=0.$$

▶▷▶ **10**. *Show that the equation of the plane through the line* $\frac{x}{l} = \frac{y}{m} = \frac{z}{n}$ *and perpendicular to the plane containing* $\frac{x}{m} = \frac{y}{n} = \frac{z}{l}$ *and* $\frac{x}{n} = \frac{y}{l} = \frac{z}{m}$ *is* $(m-n)x+(n-l)y+(l-m)z = 0$.

□ Let the equation of the plane through the line $\frac{x}{m} = \frac{y}{n} = \frac{z}{l}$ be

$$Ax + By + Cz = 0$$
$$\text{where} \quad Am + Bn + Cl = 0.$$

If it contains the line $\frac{x}{n} = \frac{y}{l} = \frac{z}{m}$, then we have

$$An + Bl + Cm = 0.$$

Eliminating A, B, C from the above equations we get

$$\begin{vmatrix} x & y & z \\ m & n & l \\ n & l & m \end{vmatrix} = 0.$$

Thus the equation of the plane containing $\frac{x}{m} = \frac{y}{n} = \frac{z}{l}$ and $\frac{x}{n} = \frac{y}{l} = \frac{z}{m}$ is

$$(mn - l^2)x + (ln - m^2)y + (ml - n^2)z = 0. \tag{1}$$

Now, equation of the plane through the line $\frac{x}{l} = \frac{y}{m} = \frac{z}{n}$ is

$$Ax + By + Cz = 0$$
$$\text{where} \quad Al + Bm + Cn = 0.$$

If this plane is perpendicular to the plane (1), then

$$A(mn - l^2) + B(ln - m^2) + C(ml - n^2) = 0.$$

Eliminating A, B, C from the above equations we get

$$\begin{vmatrix} x & y & z \\ l & m & n \\ mn - l^2 & ln - m^2 & ml - n^2 \end{vmatrix} = 0.$$

Simplifying it and cancelling the factor $(mn + ml + nl)$ we get the equation of the required plane is

$$(m - n)x + (n - l)y + (l - m)z = 0.$$

▶▷▶ **11**. *Prove that the straight lines* $\frac{x-3}{3} = \frac{y-2}{-4} = \frac{z+1}{1}$ *and* $x + 2y + 3z = 0 = 2x + 4y + 3z + 3$ *are coplanar. Find also the equation of the plane containing them.*

□ Any point on the line $\frac{x-3}{3}=\frac{y-2}{-4}=\frac{z+1}{1}$ can be taken as $(3+3r, 2-4r, -1+r)$. If this line intersects the plane $x+2y+3z=0$, then we have

$$3+3r+2(2-4r)+3(-1+r)=0 \text{ which gives } r=2.$$

Similarly, if the given line intersect the plane $2x+4y+3z+3=0$, then we have

$$2(3+3r)+4(2-4r)+3(-1+r)+3=0 \text{ which gives } r=2.$$

Thus the given lines intersect both the planes at the same point whose coordinates are (9, –6, 1), i.e. the given line intersect the planes at their line of intersection and so the given lines are coplanar. Now, equation of the plane through the line $x+2y+3z=0=2x+4y+3z+3$ is

$$x+2y+3z+k(2x+4y+3z+3)=0.$$

If it contains the first line, then the point $(3, 2, -1)$ lies on the above plane. This gives

$$3+4-3+k(6+8-3+3)=0, \quad \text{i.e. } k=-2/7.$$

So, equation of the plane containing given lines is

$$7(x+2y+3z)-2(2x+4y+3z+3)=0$$

or, $\quad 3x+6y+15z=6$

or, $\quad x+2y+5z=2.$

▶▶ **12.** *Show that the straight lines*

$$\frac{x-a+d}{\alpha-\delta}=\frac{y-a}{\alpha}=\frac{z-a-d}{\alpha+\delta}$$

and $$\frac{x-b+c}{\beta-\gamma}=\frac{y-b}{\beta}=\frac{z-b-c}{\beta+\gamma}$$

are coplanar and they lie on the plane $x+z-2y=0$.

□ The given lines will be coplanar if the determinant

$$\begin{vmatrix} b-c-a+d & b-a & b+c-a-d \\ \alpha-\delta & \alpha & \alpha+\delta \\ \beta-\gamma & \beta & \beta+\gamma \end{vmatrix}$$

vanishes.

Now, performing C_1+C_3 we get

$$\begin{vmatrix} 2(b-a) & b-a & b+c-a-d \\ 2\alpha & \alpha & \alpha+\delta \\ 2\beta & \beta & \beta+\gamma \end{vmatrix}=0.$$

So, the condition of coplanarity is satisfied.

Equation of the plane through the first line is

$$A(x-a+d)+B(y-a)+C(z-a-d)=0 \tag{1}$$

where $$A(\alpha-\delta)+B\alpha+C(\alpha+\delta)=0. \tag{2}$$

If the plane (1) contains the second line we have

$$A(\beta-\gamma)+B\beta+C(\beta+\gamma)=0. \tag{3}$$

Eliminating A, B, C from (1), (2) and (3) we get

$$\begin{vmatrix} x-a+d & y-a & z-a-d \\ \alpha-\delta & \alpha & \alpha+\delta \\ \beta-\gamma & \beta & \beta+\gamma \end{vmatrix}=0.$$

Performing $C_1+C_3-2C_2$ we obtain

$$\begin{vmatrix} x+z-2y & y-a & z-a-d \\ 0 & \alpha & \alpha+\delta \\ 0 & \beta & \beta+\gamma \end{vmatrix}=0$$

or, $x+z-2y=0$

which is the equation of the plane containing given lines.

▶▶ **13**. *Show that the straight lines*

$$\frac{px}{l}=\frac{qy}{m}=\frac{rz}{n};\quad \frac{x}{l}=\frac{y}{m}=\frac{z}{n};\quad \frac{x}{pl}=\frac{y}{qm}=\frac{z}{rn}$$

are coplanar if $p=q$ *or* $q=r$ *or* $r=p$.

□ Let A, B, C be the direction ratios of the normal to the plane containing the given lines, then we have

$$\frac{Al}{p}+\frac{Bm}{q}+\frac{Cn}{r}=0,$$
$$Al+Bm+Cn=0$$

and $$Alp+Bmq+Cnr=0.$$

Eliminating A, B, C from the above equations we get

$$\begin{vmatrix} l/p & m/q & n/r \\ l & m & n \\ lp & mq & nr \end{vmatrix}=0.$$

Taking common l, m, n from C_1, C_2, C_3 respectively we get

$$\begin{vmatrix} 1/p & 1/q & 1/r \\ 1 & 1 & 1 \\ p & q & r \end{vmatrix}=0.$$

Performing $C_3 - C_2$ and $C_2 - C_1$ we get

$$\begin{vmatrix} 1/p & 1/q - 1/p & 1/r - 1/q \\ 1 & 0 & 0 \\ p & q - p & r - q \end{vmatrix} = 0$$

or, $\begin{vmatrix} 1/q - 1/p & 1/r - 1/q \\ q - p & r - q \end{vmatrix} = 0$

or, $(q-p)(r-q)\left(\dfrac{1}{rq} - \dfrac{1}{pq}\right) = 0$

i.e., $(q-p)(r-q)(p-r) = 0$

i.e., $p = q$ or $q = r$ or $r = p$.

▶▷ **14**. *Prove that the square of the distance of the point (x_0, y_0, z_0) from the line $u \equiv ax + by + cz + d = 0$ and $v \equiv a'x + b'y + c'z + d' = 0$ is given by*

$$\frac{(a'u_0 - av_0)^2 + (b'u_0 - bv_0)^2 + (c'u_0 - cv_0)^2}{(bc' - b'c)^2 + (ca' - c'a)^2 + (ab' - a'b)^2}$$

where u_0 and v_0 are the values of u and v at (x_0, y_0, z_0).

□ Any plane through the line of intersection of $u = 0$ and $v = 0$ can be taken as $u + kv = 0$. If (x_0, y_0, z_0) lies on it, then we have $u_0 + kv_0 = 0$. So the equation of the plane through the given line and the point (x_0, y_0, z_0) is $vu_0 - uv_0 = 0$, i.e.

$$(av_0 - a'u_0)x + (bv_0 - b'u_0)y + (cv_0 - c'u_0)z + \text{constant term} = 0.$$

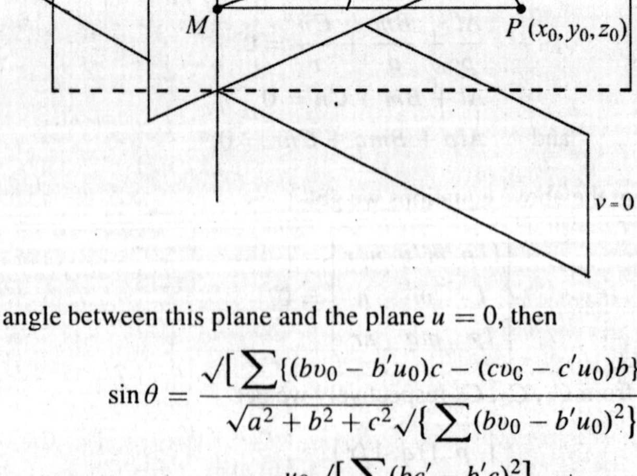

If θ be the angle between this plane and the plane $u = 0$, then

$$\sin\theta = \frac{\sqrt{\left[\sum\{(bv_0 - b'u_0)c - (cv_0 - c'u_0)b\}^2\right]}}{\sqrt{a^2+b^2+c^2}\sqrt{\{\sum(bv_0 - b'u_0)^2\}}}$$

$$= \frac{u_0\sqrt{\left[\sum(bc' - b'c)^2\right]}}{\sqrt{a^2+b^2+c^2}\sqrt{\{\sum(bv_0 - b'u_0)^2\}}}.$$

If PN be the perpendicular to the plane $u = 0$ drawn from P and if PM be perpendicular from P on the line of intersection of $u = 0$ and $v = 0$, then $PN = PM \sin\theta$.
So, the required distance

$$= PM = \frac{PN}{\sin\theta} = \frac{u_0}{\sin\theta\sqrt{a^2+b^2+c^2}} = \frac{\sqrt{\{\sum(bv_0 - b'u_0)^2\}}}{\sqrt{[\sum(bc' - b'c)^2]}}.$$

Hence, the result.

▶▷ **15.** *A line through the origin makes angles α, β, γ with its projection on coordinate planes. The distances of any point (x, y, z) from the line and its projections are d, a, b, c respectively. Prove that*

$$d^2 = (a^2 - x^2)\cos^2\alpha + (b^2 - y^2)\cos^2\beta + (c^2 - z^2)\cos^2\gamma.$$

□ Given that the line makes angles α, β, γ with its projection on coordinate planes, i.e. it makes angles $90 - \alpha, 90 - \beta, 90 - \gamma$ with coordinate axes. Hence, direction cosines of the line are $\sin\alpha, \sin\beta, \sin\gamma$ and $\sin^2\alpha + \sin^2\beta + \sin^2\gamma = 1$. So the equations of the line and its projections are respectively

$$\frac{x}{\sin\alpha} = \frac{y}{\sin\beta} = \frac{z}{\sin\gamma};\quad \frac{x}{0} = \frac{y}{\sin\beta} = \frac{z}{\sin\gamma};\quad \frac{x}{\sin\alpha} = \frac{y}{0} = \frac{z}{\sin\gamma};\quad \frac{x}{\sin\alpha} = \frac{y}{\sin\beta} = \frac{z}{0}.$$

Now,

$$d^2 = x^2 + y^2 + z^2 - \frac{(x\sin\alpha + y\sin\beta + z\sin\gamma)^2}{\sin^2\alpha + \sin^2\beta + \sin^2\gamma}$$
$$= x^2 + y^2 + z^2 - (x\sin\alpha + y\sin\beta + z\sin\gamma)^2$$
$$a^2 = x^2 + y^2 + z^2 - \frac{(y\sin\beta + z\sin\gamma)^2}{\sin^2\beta + \sin^2\gamma} = x^2 + y^2 + z^2 - \frac{(y\sin\beta + z\sin\gamma)^2}{\cos^2\alpha}$$
$$b^2 = x^2 + y^2 + z^2 - \frac{(x\sin\alpha + z\sin\gamma)^2}{\sin^2\alpha + \sin^2\gamma} = x^2 + y^2 + z^2 - \frac{(x\sin\alpha + z\sin\gamma)^2}{\cos^2\beta}$$
$$c^2 = x^2 + y^2 + z^2 - \frac{(x\sin\alpha + y\sin\beta)^2}{\sin^2\alpha + \sin^2\beta} = x^2 + y^2 + z^2 - \frac{(x\sin\alpha + y\sin\beta)^2}{\cos^2\gamma}.$$

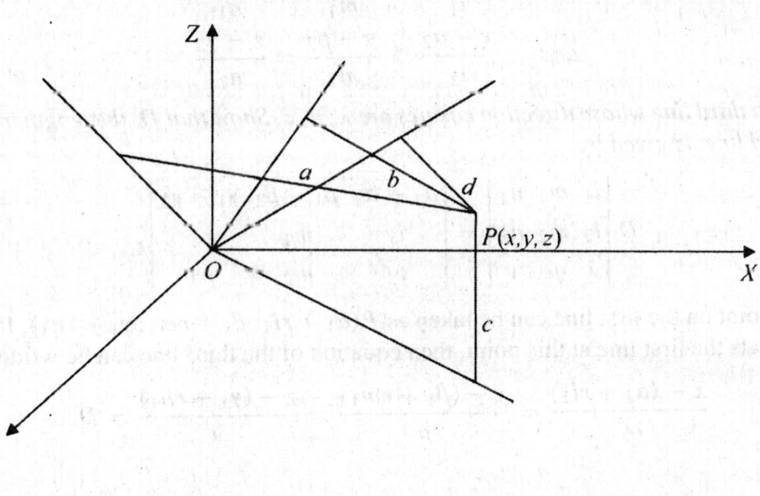

Now,

$$\begin{aligned}
&(a^2-x^2)\cos^2\alpha+(b^2-y^2)\cos^2\beta+(c^2-z^2)\cos^2\gamma \\
&=(y^2+z^2)\cos^2\alpha-(y\sin\beta+z\sin\gamma)^2+(x^2+z^2)\cos^2\beta \\
&\quad -(x\sin\alpha+z\sin\gamma)^2+(y^2+x^2)\cos^2\gamma-(y\sin\beta+x\sin\alpha)^2 \\
&=2(x^2+y^2+z^2)-x^2\cos^2\alpha-y^2\cos^2\beta-z^2\cos^2\gamma-x^2\sin^2\alpha \\
&\quad -y^2\sin^2\beta-z^2\sin^2\gamma-(x\sin\alpha+y\sin\beta+z\sin\gamma)^2 \\
&=x^2+y^2+z^2-(x\sin\alpha+y\sin\beta+z\sin\gamma)^2=d^2.
\end{aligned}$$

▶▶ **16.** *A square ABCD of diagonal 2a is folded about the diagonal AC so that the planes ABC and ADC are at right angles. Show that the distance between AB and DC is $2a/\sqrt{3}$.*

☐ Let us take OB, AC and OD as x, y and z axes respectively as shown in the figure, where O is the point of intersection of the diagonals AC and BD.

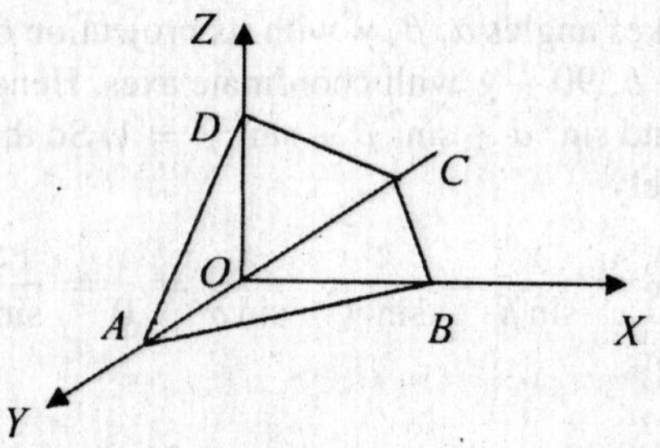

Then the coordinates of A, B, C and D are $(0, a, 0)$; $(a, 0, 0)$; $(0, -a, 0)$ and $(0, 0, a)$ respectively. So direction ratios of AB and CD are $(a, -a, 0)$ and $(0, a, a)$. Hence, the shortest distance between AB and CD

$$=\begin{vmatrix} a & a & 0 \\ a & -a & 0 \\ 0 & a & a \end{vmatrix} \div \sqrt{[(-a^2-0)^2+(0-a^2)^2+(a^2-0)^2]}=2a/\sqrt{3}.$$

▶▶ **17.** *Two straight lines*

$$\frac{x-\alpha_1}{l_1}=\frac{y-\beta_1}{m_1}=\frac{z-\gamma_1}{n_1}$$

$$\text{and}\quad \frac{x-\alpha_2}{l_2}=\frac{y-\beta_2}{m_2}=\frac{z-\gamma_2}{n_2}$$

are cut by a third line whose direction cosines are λ, μ, ν. Show that D, the length intercepted on the third line, is given by

$$D\begin{vmatrix} l_1 & m_1 & n_1 \\ l_2 & m_2 & n_2 \\ \lambda & \mu & \nu \end{vmatrix}=\begin{vmatrix} \alpha_1-\alpha_2 & \beta_1-\beta_2 & \gamma_1-\gamma_2 \\ l_1 & m_1 & n_1 \\ l_2 & m_2 & n_2 \end{vmatrix}.$$

☐ Any point on the first line can be taken as $P(\alpha_1+rl_1, \beta_1+rm_1, \gamma_1+rn_1)$. If the third line intersects the first line at this point, then equation of the third line can be written as

$$\frac{x-(\alpha_1+rl_1)}{\lambda}=\frac{y-(\beta_1+rm_1)}{\mu}=\frac{z-(\gamma_1+rn_1)}{\nu}=D.$$

So, the coordinate of a point on the third line at a distance D from P is

$$Q(\lambda D + \alpha_1 + rl_1, \mu D + \beta_1 + rm_1, \nu D + \gamma_1 + rn_1).$$

Comparing these coordinates with any point on the second line, i.e. with $P(\alpha_2 + Rl_2, \beta_2 + Rm_2, \gamma_2 + Rn_2)$ we get

$$\alpha_2 - \alpha_1 - \lambda D - rl_1 + Rl_2 = 0$$
$$\beta_2 - \beta_1 - \mu D - rm_1 + Rm_2 = 0$$
$$\gamma_2 - \gamma_1 - \nu D - rn_1 + Rn_2 = 0.$$

Eliminating r, R, we get

$$\begin{vmatrix} \alpha_2 - \alpha_1 - \lambda D & l_1 & l_2 \\ \beta_2 - \beta_1 - \mu D & m_1 & m_2 \\ \gamma_2 - \gamma_1 - \nu D & n_1 & n_2 \end{vmatrix} = 0$$

$$\text{or,} \quad \begin{vmatrix} \alpha_2 - \alpha_1 & l_1 & l_2 \\ \beta_2 - \beta_1 & m_1 & m_2 \\ \gamma_2 - \gamma_1 & n_1 & n_2 \end{vmatrix} = D \begin{vmatrix} \lambda & l_1 & l_2 \\ \mu & m_1 & m_2 \\ \nu & n_1 & n_2 \end{vmatrix}.$$

Interchanging rows and columns we get the required result.

▶▷▶ **18.** *Prove that the shortest distance between any two opposite edges of the tetrahedron formed by the planes $y + z = 0$; $x + z = 0$; $x + y = 0$ and $x + y + z = a$ is $2a/\sqrt{6}$ and these lines of shortest distances meet at the point $(-a, -a, -a)$.*

☐ The equation of the edge determined by the planes $y + z = 0$; $x + z = 0$ is

$$\frac{x}{1} = \frac{y}{1} = \frac{z}{-1}$$

and the planes $x + y = 0$; $x + y + z = a$ intersect along the line

$$\frac{x}{1} = \frac{y}{-1} = \frac{z - a}{0}.$$

Now, shortest distance between them

$$= \begin{vmatrix} 0 & 0 & a \\ 1 & 1 & -1 \\ 1 & -1 & 0 \end{vmatrix} \div \sqrt{[1 + 1 + 4]} = 2a/\sqrt{6}.$$

If L, M, N be the direction ratios of the line of shortest distance, then we have

$$L + M - N = 0 \quad \text{and} \quad L - M = 0.$$

Solving we get

$$\frac{L}{1} = \frac{M}{1} = \frac{N}{2}.$$

Equation of the line of shortest distance is the line of intersection of the planes

$$\begin{vmatrix} x & y & z \\ 1 & 1 & -1 \\ 1 & 1 & 2 \end{vmatrix} = 0 \quad \text{and} \quad \begin{vmatrix} x & y & z-a \\ 1 & -1 & 0 \\ 1 & 1 & 2 \end{vmatrix} = 0$$

i.e., $x = y$ and $x + y - z = -a$, which is satisfied by $(-a, -a, -a)$.

Due to symmetry in the equations of the edges we can say that the shortest distance between other edges is also $2a/\sqrt{6}$ and lines of shortest distances pass through the point $(-a, -a, -a)$.

▶▷ **19**. *Prove that the shortest distances between the diagonals of a rectangular parallelopiped whose sides are a, b, c and the edges not meeting it, are*

$$\frac{bc}{\sqrt{(b^2+c^2)}}; \quad \frac{ac}{\sqrt{(a^2+c^2)}}; \quad \frac{ba}{\sqrt{(b^2+a^2)}}.$$

☐ Let us take the sides OA, OB, OC as x, y, z axes. We first calculate the shortest distance between the diagonal AL and the edge OB.

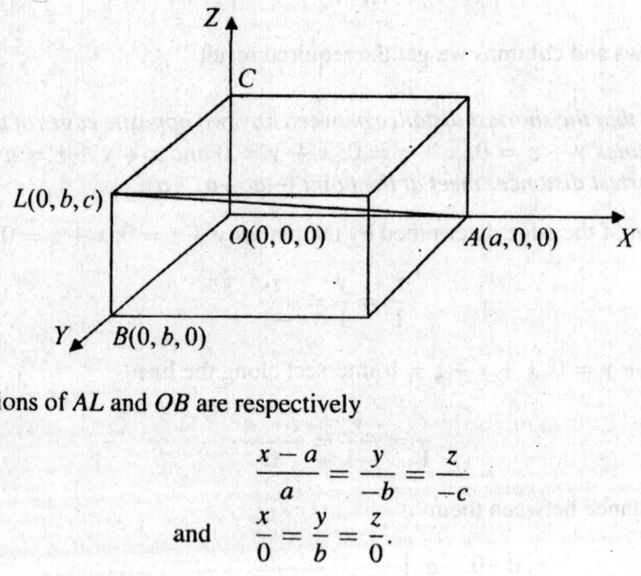

Now, equations of AL and OB are respectively

$$\frac{x-a}{a} = \frac{y}{-b} = \frac{z}{-c}$$

$$\text{and} \quad \frac{x}{0} = \frac{y}{b} = \frac{z}{0}.$$

Shortest distance between them

$$= \begin{vmatrix} a & 0 & 0 \\ a & -b & -c \\ 0 & b & 0 \end{vmatrix} \div \sqrt{(b^2c^2 + a^2b^2)} = \frac{ac}{\sqrt{a^2+c^2}}.$$

Similarly, other results can also be easily established.

▶▷ **20**. *Show that the length of the shortest distance between the lines $z = x \tan\alpha$, $y = 0$ and any tangent to the ellipse $x^2 \sin^2\alpha + y^2 = a^2$, $z = 0$ is constant.*

☐ Equation of the tangent at the point $(a \operatorname{cosec}\alpha\cos\theta, a\sin\theta, 0)$ on the ellipse $x^2\sin^2\alpha + y^2 = a^2, z = 0$ is

$$\frac{x\cos\theta}{a\operatorname{cosec}\alpha} + \frac{y\sin\theta}{a} = 1, \quad z = 0$$

$$\text{or,} \quad \frac{x - a\operatorname{cosec}\alpha\sec\theta}{a\operatorname{cosec}\alpha\sec\theta} = \frac{y}{-a\operatorname{cosec}\theta} = \frac{z}{0}.$$

Writing the given straight line $z = x\tan\alpha$, $y = 0$ in symmetrical form as

$$\frac{x}{\cos\alpha} = \frac{y}{0} = \frac{z}{\sin\alpha}$$

we see that the shortest distance between them

$$= \begin{vmatrix} a\operatorname{cosec}\alpha\sec\theta & 0 & 0 \\ \cos\alpha & 0 & \sin\alpha \\ a\operatorname{cosec}\alpha\sec\theta & -a\operatorname{cosec}\theta & 0 \end{vmatrix} \div \sqrt{a^2\operatorname{cosec}^2\theta\sin^2\alpha + a^2\sec^2\theta + a^2\operatorname{cosec}^2\theta\cos^2\alpha}$$

$$= a\operatorname{cosec}\alpha\sec\theta\, a\operatorname{cosec}\theta\sin\alpha \div \sqrt{a^2\operatorname{cosec}^2\theta + a^2\sec^2\theta}$$

$$= a^2\sec\theta\operatorname{cosec}\theta \div a\sqrt{\operatorname{cosec}^2\theta + \sec^2\theta}$$

$$= a = \text{constant.}$$

▶▷▶ **21.** *Prove that the planes $x = cy + bz$; $y = az + cx$ and $z = ay + bx$ pass through one line if*

$$a^2 + b^2 + c^2 + 2abc = 1$$

and also show that the equation of their line of intersection is

$$\frac{x}{\sqrt{1-a^2}} = \frac{y}{\sqrt{1-b^2}} = \frac{z}{\sqrt{1-c^2}}.$$

☐ Writing the planes as

$$x - cy - bz = 0$$
$$cx - y + az = 0$$
$$bx + ay - z = 0$$

we see that the planes will pass through one line

$$\text{if} \quad \begin{vmatrix} 1 & -c & -b \\ c & -1 & a \\ b & a & -1 \end{vmatrix} = 0 \quad \text{or, if} \quad a^2 + b^2 + c^2 + 2abc = 1.$$

Next, using the third equation in first and second equations we get

$$x - cy = b(bx + ay) \quad \text{or,} \quad x(1 - b^2) = (c + ab)y$$

$$\text{and} \quad y - cx = a(bx + ay) \quad \text{or,} \quad y(1 - a^2) = (c + ab)x.$$

Dividing we get

$$\frac{x}{\sqrt{1-a^2}} = \frac{y}{\sqrt{1-b^2}}.$$

Similarly, using the second equation in first and third equations, and dividing the results we get

$$\frac{x}{\sqrt{1-a^2}} = \frac{z}{\sqrt{1-c^2}}.$$

Combining these two we get the required result.

▶▷▶ **22.** ***A line of constant length has its extremities on two given straight lines. Prove that the locus of the middle point is an ellipse.***

☐ Let us take the fixed straight lines as

$$\frac{x}{1} = \frac{y}{m} = \frac{z-c}{0} = r; \quad \frac{x}{1} = \frac{y}{-m} = \frac{z+c}{0} = R.$$

Then any point on the lines can be taken as $A(r, mr, c)$ and $B(R, -mR, -c)$. If $C(\alpha, \beta, \gamma)$ be the middle point of the segment, then

$$\alpha = \frac{r+R}{2}, \quad \beta = \frac{m(r-R)}{2} \quad \text{and} \quad \gamma = 0.$$

Now, given that $AB = 2k =$ constant,

$$\text{i.e.,} \quad (r-R)^2 + m^2(r+R)^2 + 4c^2 = 4k^2$$

$$\text{or,} \quad \frac{4\beta^2}{m^2} + 4m^2\alpha^2 = 4(k^2 - c^2).$$

Hence, locus of the middle point is

$$\frac{y^2}{m^2} + m^2x^2 = k^2 - c^2; \quad z = 0.$$

▶▷▶ **23.** *Show that the surface generated by a straight line which intersects the lines* $y = 0$, $z = c$; $x = 0$, $z = -c$ *and the curve* $z = 0$, $xy = -c^2$ *is* $z^2 - c^2 = xy$.

☐ Equation of the plane through the line $y = 0$, $z = c$ is

$$y + \lambda(z - c) = 0$$

and equation of the plane through the line $x = 0$, $z = -c$ is

$$x + \mu(z + c) = 0.$$

If the line of intersection of the above planes intersect the curve $z = 0$, $xy = -c^2$, then putting $z = 0$ in the above equations we get $y = \lambda c$ and $x = -\mu c$ which satisfy the equation $xy = -c^2$, so we get

$$\lambda\mu = 1.$$

Eliminating λ, μ, we get

$$xy = z^2 - c^2.$$

▶▷▶ **24.** *Find the surface generated by a straight line which meets two straight lines* $y = \pm mx$; $z = \pm c$ *at the same angle.*

☐ In symmetrical form the given equations are

$$\frac{x}{1} = \frac{y}{m} = \frac{z-c}{0} = r; \quad \frac{x}{1} = \frac{y}{-m} = \frac{z+c}{0} = R.$$

Then any point on the lines can be taken as $A(r, mr, c)$ and $B(R, -mR, -c)$. So equation of AB is

$$\frac{x-r}{r-R} = \frac{y-mr}{m(r+R)} = \frac{z-c}{2c}.$$

Since the above lines meet the given lines at same angle, we have

$$(r-R) + m^2(r+R) + 0 = \pm[(r-R) - m^2(r+R) + 0]$$

i.e. either $r = R$ or $r = -R$.
When $r = -R$, we get

$$\frac{x-r}{2r} = \frac{y-mr}{0} = \frac{z-c}{2c}.$$

To get the locus we eliminate r and get

$$c\left(x - \frac{y}{m}\right) = \frac{y(z-c)}{m}$$

$$\text{or,} \quad mcx = yz.$$

When $r = R$, we get

$$\frac{x-r}{0} = \frac{y-mr}{2mr} = \frac{z-c}{2c}.$$

Eliminating r we get

$$c(y - mx) = mx(z-c)$$

$$\text{or,} \quad mxz = cy.$$

Hence, the equation of the surface is $mcx = yz$ or $mxz = cy$.

▶▷▶ **25.** *A line moves so as to meet the lines*

$$\frac{x}{\cos\alpha} = \frac{y}{\pm\sin\alpha} = \frac{z \mp c}{0}$$

in A and B, and passes through the curve $yz = k^2$, $x = 0$. Prove that the locus of the middle point is a third degree curve, two of whose asymptotes are parallel to the given lines.

☐ Any points on the given lines, i.e. on

$$\frac{x}{\cos\alpha} = \frac{y}{\sin\alpha} = \frac{z-c}{0} = r; \quad \frac{x}{\cos\alpha} = \frac{y}{-\sin\alpha} = \frac{z+c}{0} = R$$

can be taken as $A(r\cos\alpha, r\sin\alpha, c)$ and $B(R\cos\alpha, -R\sin\alpha, -c)$. If (x, y, z) be the middle point of AB, then

$$x = \frac{(r+R)\cos\alpha}{2}, \quad y = \frac{(r-R)\sin\alpha}{2}, \quad z = 0.$$

Now, the equation of the line AB is

$$\frac{x - r\cos\alpha}{(r-R)\cos\alpha} = \frac{y - r\sin\alpha}{(r+R)\sin\alpha} = \frac{z-c}{2c}.$$

If it intersects the curve $yz = k^2$, $x = 0$, then putting $x = 0$ we get

$$y = r\sin\alpha - \frac{r(r+R)\sin\alpha\cos\alpha}{(r-R)\cos\alpha} = -\frac{2rR\sin\alpha}{r-R}$$

$$z = c - \frac{2cr}{r-R} = -c\frac{r+R}{r-R}.$$

Using these values of y and z in $yz = k^2$, we get

$$\frac{2rcR(r+R)\sin\alpha}{(r-R)^2} = k^2$$

or, $$\frac{c[(r+R)^2-(r-R)^2](r+R)\sin\alpha}{(r-R)^2} = 2k^2.$$

Eliminating r, R, we get the locus of the line is

$$\frac{c[(2x\sec\alpha)^2-(2y\operatorname{cosec}\alpha)^2](2x\sec\alpha)\sin\alpha}{(2y\operatorname{cosec}\alpha)^2} = 2k^2, \quad z = 0$$

or, $$cx(x^2\sec^2\alpha - y^2\operatorname{cosec}^2\alpha)\tan\alpha - k^2y^2\operatorname{cosec}^2\alpha = 0, \quad z = 0$$

which is a third degree curve in xy-plane, two of whose asymptotes are parallel to

$$x^2\sec^2\alpha - y^2\operatorname{cosec}^2\alpha = 0, \quad z = 0$$

or, $$\frac{x}{\cos\alpha} = \frac{y}{\pm\sin\alpha} = \frac{z}{0}$$

i.e. parallel to the given lines.

▶▷▶ **26.** *A variable line intersects the x-axis and the curve $x = y$, $y^2 = cz$ and is parallel to the plane $x = 0$. Prove that it generates the surface $xy = cz$.*

☐ Equation of the plane through x-axis is $y - kz = 0$ and equation of the plane parallel to $x = 0$ is $x = k'$. So the variable line is the line of intersection of the planes $y - kz = 0$ and $x = k'$. If this line intersects the curve $x = y$, $y^2 = cz$, then we have

$$x = k' = y \quad \text{and} \quad k = y/z = k'/z, \quad \text{i.e.} \quad z = k'/k,$$

so $y^2 = cz$ gives $k'^2 = ck'/k$ or $kk' = c$. Eliminating k, k' we get

$$yx = cz.$$

▶▷▶ **27.** *P, P' are two variable points on two non-intersecting lines such that $PP' = 2k$. Find the surface generated by the line PP'.*

☐ Let us take the fixed straight lines as

$$\frac{x}{1} = \frac{y}{m} = \frac{z-c}{0} = r; \quad \frac{x}{1} = \frac{y}{-m} = \frac{z+c}{0} = R.$$

Then any point on the lines can be taken as $P(r, mr, c)$ and $P'(R, -mR, -c)$. Given that $PP' = 2k$. So

$$(r-R)^2 + m^2(r+R)^2 + 4c^2 = 4k^2 \tag{1}$$

and equation of PP' is

$$\frac{x-r}{r-R}=\frac{y-mr}{m(r+R)}=\frac{z-c}{2c}.$$

From which we get

$$x-r=\frac{(r-R)(z-c)}{2c} \quad \text{or,} \quad r(z+c)-R(z-c)=2cx$$

$$\text{and} \quad y-mr=\frac{m(r+R)(z-c)}{2c} \quad \text{or,} \quad mr(z+c)+mR(z-c)=2cy.$$

Solving these equations for r, R, we get

$$r=\frac{c(y+mx)}{m(z+c)}, \quad R=\frac{c(y-mx)}{m(z-c)}.$$

Using these values of r, R in Eqn. (1) we get

$$\frac{(cy-mzx)^2}{m^2}+(yz-cmx)^2=\frac{k^2-c^2}{c^2}(z^2-c^2)^2.$$

▶▷▶ **28**. *AP and A'P' are two given lines such that A and A' are fixed, and P and P' are two variable points on those, such that AP · A'P' = constant. Find the surface generated by the line PP'.*•

☐ Let us take the fixed straight lines as

$$\frac{x}{1}=\frac{y}{m}=\frac{z-c}{0}=r; \quad \frac{x}{1}=\frac{y}{-m}=\frac{z+c}{0}=R.$$

Then taking the fixed points A, A' as $(0, 0, c)$, and $(0, 0, -c)$ and variable points as $P(r, mr, c)$ and $P'(R, -mR, -c)$, we get

$$AP=\sqrt{r^2+m^2r^2} \quad \text{and} \quad A'P'=\sqrt{R^2+m^2R^2}.$$

Using the given condition we obtain

$$rR=k^2 \text{ (say)}.$$

Now, equation of PP' is

$$\frac{x-r}{r-R}=\frac{y-mr}{m(r+R)}=\frac{z-c}{2c}$$

from which we get

$$x-r=\frac{(r-R)(z-c)}{2c} \quad \text{or,} \quad r(z+c)-R(z-c)=2cx$$

$$\text{and} \quad y-mr=\frac{m(r+R)(z-c)}{2c} \quad \text{or,} \quad mr(z+c)+mR(z-c)=2cy.$$

Solving these equations for r, R, we get

$$r=\frac{c(y+mx)}{m(z+c)}, \quad R=\frac{c(y-mx)}{m(z-c)}.$$

Using these values of r, R in Eqn. (1) we get

$$\frac{c(y+mx)}{m(z+c)} \times \frac{c(y-mx)}{m(z-c)} = k^2$$

$$\text{or,} \quad c^2(y^2 - m^2x^2) = k^2m^2(z^2 - c^2).$$

▶▶ **29.** *Find the equation of the surface generated by a straight line which is parallel to the line $y = mx$ and $z = nx$, and intersects the ellipse*

$$\frac{x^2}{a^2} + \frac{y^2}{b^2} = 1; \quad z = 0.$$

☐ Equation of the line parallel to the given line can be taken as

$$\frac{x-\alpha}{1} = \frac{y-\beta}{m} = \frac{z-\gamma}{n}.$$

It meets the given ellipse at $\left(\alpha - \frac{\gamma}{n}, \beta - \frac{m\gamma}{n}, 0\right)$ which satisfy the equation

$$\frac{x^2}{a^2} + \frac{y^2}{b^2} = 1.$$

Then

$$\frac{(\alpha - \gamma/n)^2}{a^2} + \frac{(\beta - m\gamma/n)^2}{b^2} = 1.$$

Hence, locus of the variable point (α, β, γ) is

$$b^2(nx - z)^2 + a^2(ny - mz)^2 = a^2b^2n^2.$$

▶▶ **30.** *A line is parallel to the plane $y + z = 0$ and intersects the curves $x^2 + y^2 = a^2$, $z = 0$ and $x^2 + z^2 = a^2$, $y = 0$. Find the surface generated by it.*

☐ If L, M, N be the direction ratios of the line which is parallel to the plane $y + z = 0$, then $M + N = 0$, so its equation can be written as

$$\frac{x-\alpha}{L} = \frac{y-\beta}{M} = \frac{z-\gamma}{-M}.$$

It meets the given circle $x^2 + y^2 = a^2$, $z = 0$ at $\left(\alpha + \frac{L\gamma}{M}, \beta + \gamma, 0\right)$, so we have

$$\left(\alpha + \frac{L\gamma}{M}\right)^2 + (\beta + \gamma)^2 = a^2$$

$$\text{or,} \quad \frac{L}{M} = \frac{\sqrt{[a^2 - (\beta + \gamma)^2]} - \alpha}{\gamma}.$$

Again, it meets the given circle $x^2 + z^2 = a^2$, $y = 0$ at $\left(\alpha - \frac{L\beta}{M}, 0, \beta + \gamma\right)$, so we have

$$\left(\alpha - \frac{L\beta}{M}\right)^2 + (\beta + \gamma)^2 = a^2$$

$$\text{or,} \quad \frac{L}{M} = -\frac{\sqrt{[a^2 - (\beta + \gamma)^2]} - \alpha}{\beta}.$$

Equating them we get

$$\frac{\sqrt{[a^2-(\beta+\gamma)^2]}-\alpha}{\gamma}=-\frac{\sqrt{[a^2-(\beta+\gamma)^2]}-\alpha}{\beta}$$

or, $\sqrt{[a^2-(\beta+\gamma)^2]}(\beta+\gamma)=\alpha(\beta+\gamma)$

or, $\alpha^2+(\beta+\gamma)^2=a^2.$

Hence, locus of the variable point (α,β,γ) is

$$x^2+(y+z)^2=a^2.$$

▶▷ 31. *Find the locus of the line which moves parallel to the xz-plane and meets the curves $xy=c^2$, $z=0$ and $y^2=4cz$, $x=0$.*

□ If L, M, N be the direction ratios of the line which parallel to the xz-plane, i.e. $y=0$, then $M=0$, so its equation can be written as

$$\frac{x-\alpha}{L}=\frac{y-\beta}{0}=\frac{z-\gamma}{N}.$$

It meets the curve $xy=c^2$, $z=0$ at $\left(\alpha-\frac{L\gamma}{N},\beta,0\right)$, so we have

$$\left(\alpha-\frac{L\gamma}{N}\right)\beta=c^2 \quad \text{or,} \quad \frac{L}{N}=\frac{1}{\gamma}\left(\alpha-\frac{c^2}{\beta}\right).$$

Similarly, it meets the curve $y^2=4cz$, $x=0$ at $\left(0,\beta,\gamma-\frac{N\alpha}{L}\right)$, so we have

$$\beta^2=4c\left(\gamma-\frac{N\alpha}{L}\right) \quad \text{or,} \quad \frac{N}{L}=\frac{1}{\alpha}\left(\gamma-\frac{\beta^2}{4c}\right).$$

Eliminating L/N we obtain

$$\frac{\alpha-\frac{c^2}{\beta}}{\gamma}=\frac{\alpha}{\gamma-\frac{\beta^2}{4c}}$$

or, $\left(\gamma-\frac{\beta^2}{4c}\right)\left(\alpha-\frac{c^2}{\beta}\right)=\alpha\gamma$

or, $(4c\gamma-\beta^2)(\alpha\beta-c^2)=4c\alpha\beta\gamma$

or, $\beta^2(\alpha\beta-c^2)+4c^3\gamma=0.$

Hence, locus of (α,β,γ) is $y^2(xy-c^2)+4c^3z=0.$

▶▷ 32. *A line meets two straight lines $y=\pm x\tan\alpha$; $z=\pm c$, so that the length intercepted on it is constant. Show that its equation may be written as*

$$\frac{x-k\sin\theta\cot\alpha}{k\cos\theta}=\frac{y-k\cos\theta\tan\alpha}{k\sin\theta}=\frac{z}{c}$$

where k is a constant and θ is a parameter. Find also the equation of surface generated by the line.

☐ In symmetrical form the given equations are

$$\frac{x}{1}=\frac{y}{m}=\frac{z-c}{0}=r;\quad \frac{x}{1}=\frac{y}{-m}=\frac{z+c}{0}=R,\quad \text{where } m=\tan\alpha$$

Then any point on the lines can be taken as $A(r, mr, c)$ and $B(R, -mR, -c)$. So, equation of AB is

$$\frac{x-r}{r-R}=\frac{y-mr}{m(r+R)}=\frac{z-c}{2c}$$

$$\text{or,}\qquad \frac{x-\dfrac{r+R}{2}}{(r-R)/2}=\frac{y-\dfrac{m(r-R)}{2}}{m(r+R)/2}=\frac{z}{c}.$$

Given that $PP'=2d$. So

$$(r-R)^2+m^2(r+R)^2+4c^2=4d^2$$

$$\text{or,}\qquad \left(\frac{r-R}{2}\right)^2+\left(\frac{m(r+R)}{2}\right)^2=k^2,\quad \text{where } k^2=d^2-c^2.$$

This is satisfied by $\dfrac{r-R}{2}=k\cos\theta$ and $\dfrac{m(r+R)}{2}=k\sin\theta$. So, the above equation of the line becomes

$$\frac{x-k\sin\theta\cot\alpha}{k\cos\theta}=\frac{y-k\cos\theta\tan\alpha}{k\sin\theta}=\frac{z}{c}.$$

From this we get

$$k\sin\theta(c\cot\alpha)+k\cos\theta(z)-cx=0$$
$$k\sin\theta(z)+k\cos\theta(c\tan\alpha)-cy=0.$$

Solving for $\cos\theta$ and $\sin\theta$ we get

$$\frac{k\sin\theta}{c^2x\tan\alpha-cyz}=\frac{k\cos\theta}{c^2y\cot\alpha-cxz}=\frac{1}{c^2-z^2}.$$

Eliminating θ we get

$$k^2(c^2-z^2)^2=c^2[(cx\tan\alpha-yz)^2+(cy\cot\alpha-xz)^2].$$

EXERCISE

1. Find the points where the line joining the points $(2, -3, 1)$ and $(3, -4, -5)$ intersects the xy-plane.
2. Find the points where the line joining the points $(2, 3, 1)$ and $(3, 4, 5)$ intersects the plane $3x+2y+z=10$.
3. Express the equation of the straight line $x-2y+3z=4$, $2x-3y+4z=5$ in symmetrical form.
4. Find the image of the point $(1, 2, 4)$ in the plane $2x-y+z+4=0$.
5. Find the image of the straight line $\dfrac{x-2}{1}=\dfrac{y-3}{2}=\dfrac{z-1}{3}$ in the plane $x-2y+z+3=0$.
6. Find the projection of the line $3x-y+2z-1=0=x+2y-z-2$ in the plane $3x+2y+z=0$ in the symmetrical form.

   [*Hints*: Equation of the plane through the line $3x-y+2z-1=0=x+2y-z-2$ is

   $$3x-y+2z-1-k(x+2y-z-2)=0.$$

   This will be perpendicular to the plane $3x+2y+z=0$ if $k=3/2$.

Hence, equation of the plane through the line $3x - y + 2z - 1 = 0 = x + 2y - z - 2$ and perpendicular to the plane $3x + 2y + z = 0$ is $3x - 8y + 7z + 4 = 0$. Now express the line of the intersection of the planes $3x + 2y + z = 0$ and $3x - 8y + 7z + 4 = 0$ in symmetrical form.]

7. Show that the lines $x + 2y - 5z + 9 = 0 = 3x - y + 2z - 5$ and $2x + 3y - z - 3 = 0 = 4x - 5y + z + 3$ are coplanar.
8. Show that the shortest distance between the lines $y = az+b, z = \alpha x+\beta$ and $y = a'z+b'$, $z = \alpha' x + \beta'$ is

$$\frac{(\alpha - \alpha')(b - b') - (\alpha'\beta - \alpha\beta')(a - a')}{\sqrt{\alpha^2\alpha'^2(a - a')^2 + (\alpha - \alpha')^2 - (a\alpha - a'\alpha')^2}}.$$

[*Hints*: Equation of the lines in symmetrical form is

$$\frac{x + \frac{\beta}{\alpha}}{\frac{1}{\alpha}} = \frac{y - b}{a} = \frac{z - 0}{1}, \quad \frac{x + \frac{\beta'}{\alpha'}}{\frac{1}{\alpha'}} = \frac{y - b'}{a'} = \frac{z - 0}{1}.$$

Now use the formula of shortest distance.]

9. Prove that the lines $\frac{x - \alpha}{l} = \frac{y - \beta}{m} = \frac{z - \gamma}{n}$, $ax+by+cz+d = 0 = a'x+b'y+c'z+d'$ are coplanar if

$$\frac{a\alpha + b\beta + c\gamma + d}{al + bm + cn} = \frac{a'\alpha + b'\beta + c'\gamma + d'}{a'l + b'm + c'n}.$$

[*Hints*: Any point on the line

$$\frac{x - \alpha}{l} = \frac{y - \beta}{m} = \frac{z - \gamma}{n}$$

can be taken as $(\alpha + lr, \beta + mr, \gamma + nr)$. Since the lines are coplanar, the given line will intersect the planes $ax + by + cz + d = 0$ and $a'x + b'y + c'z + d' = 0$ at the same point. Putting the point in two planes and equating the value of r we get the required result.]

10. Show that the straight lines

$$\frac{x}{\alpha} = \frac{y}{\beta} = \frac{z}{\gamma}; \quad \frac{x}{a\alpha} = \frac{y}{b\beta} = \frac{z}{c\gamma}; \quad \frac{x}{l} = \frac{y}{m} = \frac{z}{n}$$

are coplanar if

$$\frac{l}{\alpha}(b - c) + \frac{m}{\beta}(c - a) + \frac{n}{\gamma}(a - b) = 0.$$

[*Hints*: Proceed as Example 13.]

11. Show that the lines $x = ay + b = cz + d$ and $x = \alpha y + \beta = \gamma z + \delta$ are coplanar if

$$(\gamma - c)(a\beta - b\alpha) = (\alpha - a)(c\delta - d\gamma).$$

[*Hints*: Equation of the straight lines in symmetrical form is

$$\frac{x - 0}{1} = \frac{y + \frac{b}{a}}{\frac{1}{a}} = \frac{z + \frac{d}{c}}{\frac{1}{c}}, \quad \frac{x - 0}{1} = \frac{y + \frac{\beta}{\alpha}}{\frac{1}{\alpha}} = \frac{z + \frac{\delta}{\gamma}}{\frac{1}{\gamma}}.$$

Now use the condition of coplanarity.]

12. Find the shortest distance between the line $ax + by + cz + d = 0 = a'x + b'y + c'z + d'$ and x-axis.

13. Show that the shortest distance between the lines

$$\frac{x-x_1}{\cos\alpha_1}=\frac{y-y_1}{\cos\beta_1}=\frac{z-z_1}{\cos\gamma_1},\quad \frac{x-x_2}{\cos\alpha_2}=\frac{y-y_2}{\cos\beta_2}=\frac{z-z_2}{\cos\gamma_2}$$

meets the first line at a point whose distance from the point (x_1, y_1, z_1) is

$$\frac{\sum(x_1-x_2)(\cos\alpha_1-\cos\theta\cos\alpha_2)}{\sin^2\theta}$$

where θ is the angle between the lines.

14. Show that the equation of the plane containing the line $\frac{y}{b}+\frac{z}{c}=1, x=0$ and parallel to the line $\frac{x}{a}-\frac{z}{c}=1, y=0$ is $\frac{x}{a}-\frac{y}{b}-\frac{z}{c}+1=0$ and if $2d$ is the shortest distance, then prove that

$$\frac{1}{a^2}+\frac{1}{b^2}+\frac{1}{c^2}=\frac{1}{d^2}.$$

15. A line moves so as to intersect the line $z=0, x=y$; and intersects the circles $x^2+y^2=a^2, z=0$ and $x^2+z^2=a^2, y=0$. Find the surface generated by it.

16. Find the locus of the line which moves parallel to the yz-plane and meets the curves $x^2+y^2=a^2, z=0$ and $x^2=az, y=0$.

[*Hints*: Proceed as Example 31.]

17. Show that all the lines which cut the z-axis and the lines

$$\frac{x+3}{2}=\frac{y-6}{3}=\frac{z-3}{-2};\quad \frac{x}{2}=\frac{y-6}{2}=\frac{z}{-1}$$

lie on the surface $7(x-y+6)(x+z)=(3x-2y+21)(x+2z)$.

18. Show that the locus of the point which is equidistant from the lines $y=\pm mx; z=\pm c$ is $mxy+(1+m^2)cz=0$.

19. Show that the locus of the lines which intersects the lines $y=\pm mx, z=\pm c$ and $y=z$, $mx=-c$ is the surface $y^2-m^2x^2=z^2-c^2$.

20. Find the locus of the line which meets the lines $y=\pm mx, z=\pm c$ and

(i) the circle $x^2+y^2=a^2, z=0$;

(ii) the ellipse $\frac{x^2}{a^2}+\frac{y^2}{b^2}=1, z=0$;

(iii) the hyperbola $xy=c^2, z=0$.

[*Hints*: (i) Equation of the planes through the given lines is

$$(y-mx)+\lambda(z-c)=0 \quad\text{and}\quad (y+mx)+\mu(z+c)=0.$$

If it intersects the curve $x^2+y^2=a^2, z=0$, then putting $z=0$ in the equation of the planes we get $y-mx=\lambda c$ and $y+mx=-\mu c$. On solving for x, y, we get $x=-\frac{c(\lambda+\mu)}{2m}$ and $y=\frac{c(\lambda-\mu)}{2}$. Using these values in $x^2+y^2=a^2$ we get

$$\frac{c^2(\lambda+\mu)^2}{4m^2}+\frac{c^2(\lambda-\mu)^2}{4}=a^2.$$

Now,

$$\lambda+\mu=-\left[\frac{y-mx}{z-c}+\frac{y+mx}{z+c}\right]=\frac{2(cmx-yz)}{z^2-c^2};$$

$$\lambda-\mu=-\left[\frac{y-mx}{z-c}-\frac{y+mx}{z+c}\right]=\frac{2(zmx-yc)}{z^2-c^2}.$$

Putting these values we get the required result.
For (ii) and (iii) proceed in the same way.]

21. A line is parallel to the plane $x = 0$ and intersects the curves $x^2 + y^2 = a^2$, $z = 0$ and $x^2 = az$, $y = 0$. Find the surface generated by it.

    [*Hints*: Proceed as Example 31.]

22. A variable line intersects the z-axis and the curve $z = y$, $y^2 = ax$ and is parallel to the plane $z = 0$. Find the surface generated by it.

    [*Hints*: Proceed as Example 26.]

23. Find the locus of the middle points of lines whose extremities are on two skew lines and which are parallel to a given plane.

    [*Hints*: Let us take the skew lines as

$$\frac{x}{1} = \frac{y}{m} = \frac{z-c}{0} = r; \quad \frac{x}{1} = \frac{y}{-m} = \frac{z+c}{0} = R.$$

Then direction ratios of the line joining $P(r, mr, c)$ and $P'(R, -mR, -c)$ are $r - R$, $m(r + R)$, $2c$. If it is parallel to $Ax + By + Cz + D = 0$, then $(r - R)A + m(r + R)B + 2cC = 0$. If the middle point of PP' is (f, g, h), then $Ag/m + Bmf + cC = 0$; $h = 0$. Thus the locus of (f, g, h) is $Bm^2x + Ay + cmC = 0$, $z = 0$.]

24. A variable line intersects the lines

$$x = b,\ y + c = 0; \quad y = c,\ z + a = 0; \quad z = a,\ x + b = 0.$$

Show that the locus of the line is $axy + byz + cxz + abc = 0$.

25. Prove that the line of s.d. between the z-axis and the variable line

$$\frac{x}{a} + \frac{z}{c} = \lambda\left(1 + \frac{y}{b}\right), \quad \frac{x}{a} - \frac{z}{c} = \frac{1}{\lambda}\left(1 - \frac{y}{b}\right)$$

generates the surface $abz(x^2 + y^2) = (a^2 - b^2)cxy$ as λ varies.

[*Hints*: If l, m, n are the direction ratios of the variable line, then we have

$$\frac{l}{a} - \frac{\lambda m}{b} + \frac{n}{c} = 0 \quad \text{and} \quad \frac{l}{a} + \frac{m}{\lambda b} - \frac{n}{c} = 0.$$

Solving we get

$$\frac{l}{\frac{1}{bc}\left(\lambda - \frac{1}{\lambda}\right)} = \frac{m}{\frac{2}{ac}} = \frac{n}{\frac{1}{ab}\left(\lambda + \frac{1}{\lambda}\right)}.$$

Putting $z = 0$ in the equation of variable line and solving for x, y we get the symmetrical form of the variable line is

$$\frac{x - \dfrac{2a}{\left(\lambda + \frac{1}{\lambda}\right)}}{\frac{1}{bc}\left(\lambda - \frac{1}{\lambda}\right)} = \frac{y + \dfrac{b\left(\lambda - \frac{1}{\lambda}\right)}{\left(\lambda + \frac{1}{\lambda}\right)}}{\frac{2}{ac}} = \frac{z - 0}{\frac{1}{ab}\left(\lambda + \frac{1}{\lambda}\right)}$$

or, $$\frac{x - \alpha}{l} = \frac{y - \beta}{m} = \frac{z}{n}.$$

If L, M, N be the direction ratios of the line of s.d. between this line and z-axis, then we have

$$N = 0 \quad \text{and} \quad La(\lambda^2 - 1) + 2Mb\lambda = 0.$$

Equation of the shortest distance is

$$\begin{vmatrix} x & y & z \\ L & M & 0 \\ 0 & 0 & 1 \end{vmatrix} = 0 \quad \text{and} \quad \begin{vmatrix} x-\alpha & y-\beta & z \\ L & M & 0 \\ l & m & n \end{vmatrix} = 0$$

or, $$xM - yL = 0 \quad \text{and} \quad \begin{vmatrix} -\alpha & -\beta & z \\ x & y & 0 \\ l & m & n \end{vmatrix} = 0, \quad \text{i.e. } z(mx - ly) = n(\alpha y - \beta x) \quad (1)$$

Again, from $xM - yL = 0$ and $La(\lambda^2-1)+2Mb\lambda = 0$, we get $xa(\lambda^2-1)+2yb\lambda = 0$, from which we get $\lambda - \frac{1}{\lambda} = -2\frac{by}{ax}$ and $\lambda + \frac{1}{\lambda} = 2\frac{\sqrt{b^2y^2 + a^2x^2}}{ax}$. Hence α, β, l, n can be determined and using these values in Eqn. (1) we get the result.]

26. Find the surface generated by a straight line which meets two straight lines

$$\frac{x \pm a}{0} = \frac{y}{\sin\alpha} = \frac{z}{\pm\cos\alpha}$$

at the same angle.

[*Hints*: Proceed as Example 24.]

27. Find the locus of the lines which are parallel to the plane $x + y = 0$ and which intersect the straight line $x - y = 0 = z$ and the curve $x^2 = 2az, y = 0$.

28. Obtain the locus of the straight line which intersects the circle $x^2 + y^2 = r^2, z = 0$ and the lines $x = 0 = z + a; y = 0 = x - a$.

29. The ends of the diameters of the ellipse

$$\frac{x^2}{a^2} + \frac{y^2}{b^2} = 1, \quad z = c$$

are joined to the corresponding ends of the conjugates of the parallel diameters of the ellipse

$$\frac{x^2}{a^2} + \frac{y^2}{b^2} = 1, \quad z = -c.$$

Show that the surface generated by the joining lines is

$$\frac{2x^2}{a^2} + \frac{2y^2}{b^2} - \frac{z^2}{c^2} = 1.$$

Chapter 4

Transformation of Axes

4.1 Transformation of Axes

To express the coordinates of a point in the space referred to a set of given axes, with respect to another set of axes often simplifies a problem. Here we discuss few transformations of axes.

(a) Transformation of Axes by Change of Origin Only

Let OX, OY, OZ be the axes of coordinates with O as origin. Let $C(f, g, h)$ be the new origin through which the axes Cx, Cy, Cz are drawn parallel to OX, OY, OZ; then if the point $P(X, Y, Z)$ becomes $P(x, y, z)$ with respect to new set of axes, we have

$$X = x + f; \quad Y = y + g \quad \text{and} \quad Z = z + h.$$

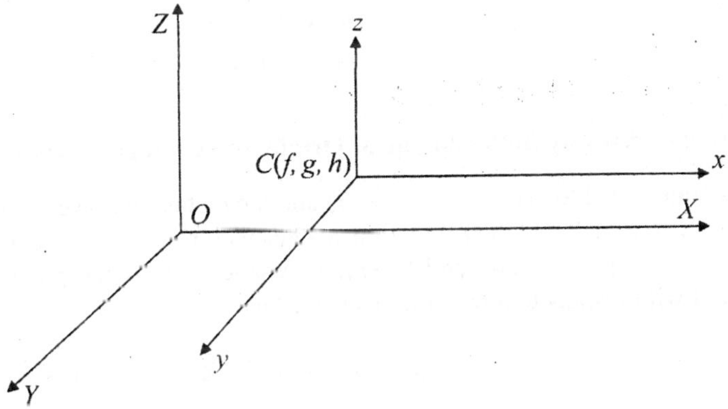

(b) Transformation of Axes by Change of Axes Only

Let the rectangular coordinate axes OX, OY, OZ with O as origin be rotated about the origin so that they become Ox, Oy, Oz whose direction cosines with respect to the coordinate axes OX, OY, OZ are l_1, m_1, n_1; l_2, m_2, n_2 and l_3, m_3, n_3 respectively. Let the coordinates of P be (X, Y, Z) and (x, y, z) with respect to old and new set of axes respectively.

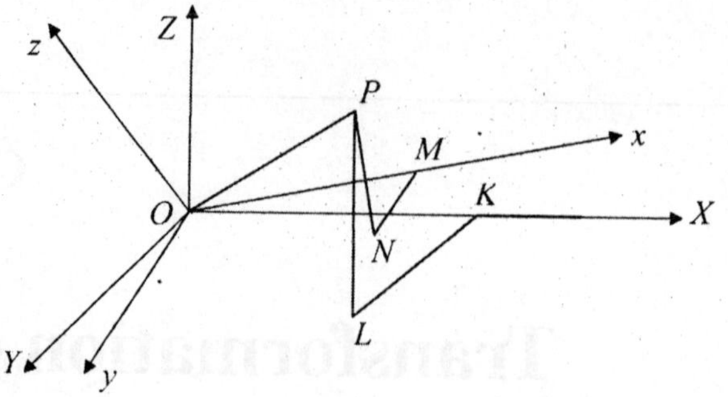

Now,

$$\text{proj. of } OP \text{ on } OX = \text{proj. of } OM \text{ on } OX + \text{proj. of } MN \text{ on } OX + \text{proj. of } NP \text{ on } OX$$

i.e., $X = l_1x + l_2y + l_3z.$

Similarly,

$$Y = m_1x + m_2y + m_3z \quad \text{and} \quad Z = n_1x + n_2y + n_3z.$$

The above results can be written as

	x	y	z
X	l_1	l_2	l_3
Y	m_1	m_2	m_3
Z	n_1	n_2	n_3

From the above table we can also express x, y, z in terms of X, Y, Z as

$$x = l_1X + m_1Y + n_1Z,$$
$$y = l_2X + m_2Y + n_2Z$$
$$\text{and} \quad z = l_3X + m_3Y + n_3Z.$$

(c) Transformation of Axes by Both Change of Origin and Change of Axes

If the origin O be transferred to the point $C(f, g, h)$ and the rectangular axes be rotated to its new position Cx, Cy, Cz such that their direction cosines with respect to the coordinate axes OX, OY, OZ are l_1, m_1, n_1; l_2, m_2, n_2 and l_3, m_3, n_3 respectively. If the point $P(X, Y, Z)$ becomes $P(x, y, z)$ with respect to new set of axes, we have

$$X = f + l_1x + l_2y + l_3z; \; Y = g + m_1x + m_2y + m_3z; \; Z = h + n_1x + n_2y + n_3z.$$

In matrix notation the above orthogonal transformations can be expressed as

$$\mathbf{X} = A\mathbf{x} + F$$

where $\mathbf{X} = \begin{bmatrix} X \\ Y \\ Z \end{bmatrix}$; $\mathbf{x} = \begin{bmatrix} x \\ y \\ z \end{bmatrix}$; $F = \begin{bmatrix} f \\ g \\ h \end{bmatrix}$ and $A = \begin{bmatrix} l_1 & l_2 & l_3 \\ m_1 & m_2 & m_3 \\ n_1 & n_2 & n_3 \end{bmatrix}$.

4.2 Invariants

(i) *The square of the distance between two points is an invariant*: Let the coordinates of $P(x_1, y_1, z_1)$ and $Q(x_2, y_2, z_2)$ become $P(x_1', y_1', z_1')$ and $Q(x_2', y_2', z_2')$ due to orthogonal transformation. Then we have

$$x_1 = f + l_1x_1' + l_2y_1' + l_3z_1'; \quad y_1 = g + m_1x_1' + m_2y_1' + m_3z_1';$$
$$z_1 = h + n_1x_1' + n_2y_1' + n_3z_1'$$
and
$$x_2 = f + l_1x_2' + l_2y_2' + l_3z_2'; \quad y_2 = g + m_1x_2' + m_2y_2' + m_3z_2';$$
$$z_2 = h + n_1x_2' + n_2y_2' + n_3z_2'.$$

Now,

$$\begin{aligned} PQ^2 &= (x_1 - x_2)^2 + (y_1 - y_2)^2 + (z_1 - z_2)^2 \\ &= [l_1(x_1' - x_2') + l_2(y_1' - y_2') + l_3(z_1' - z_2')]^2 + [m_1(x_1' - x_2') + m_2(y_1' - y_2') \\ &\quad + m_3(z_1' - z_2')]^2 + [n_1(x_1' - x_2') + n_2(y_1' - y_2') + n_3(z_1' - z_2')]^2 \\ &= (x_1' - x_2')^2[l_1^2 + m_1^2 + n_1^2] + (y_1' - y_2')^2[l_2^2 + m_2^2 + n_2^2] \\ &\quad + (z_1' - z_2')^2[l_3^2 + m_3^2 + n_3^2] + 2(l_1l_2 + m_1m_2 + n_1n_2)(x_1' - x_2')(y_1' - y_2') \\ &\quad + 2(l_1l_3 + m_1m_3 + n_1n_3)(x_1' - x_2')(z_1' - z_2') + 2(l_3l_2 + m_3m_2 + n_3n_2) \\ &\quad \times (z_1' - z_2')(y_1' - y_2') \\ &= (x_1' - x_2')^2 + (y_1' - y_2')^2 + (z_1' - z_2')^2 = P'Q'^2. \end{aligned}$$

Since, for orthogonal coordinate system $l_k^2 + m_k^2 + n_k^2 = 1$ and $l_il_j + m_im_j + n_in_j = 0$ for $i, j, k = 1, 2, 3$.

(ii) If the expression $ax^2 + by^2 + cz^2 + 2fyz + 2gxz + 2hxy$ be transformed to $AX^2 + BY^2 + CZ^2 + 2FYZ + 2GXZ + 2HXY$ by the transformation of rotation of axes about the origin, then show that the following expressions are invariant:

(a) $a + b + c$;
(b) $ab + bc + ac - f^2 - g^2 - h^2$;
(c) $abc + 2fgh - af^2 - bg^2 - ch^2$.

Since, distance between two points is an invariant and the expression

$$ax^2 + by^2 + cz^2 + 2fyz + 2gxz + 2hxy \quad \text{becomes}$$
$$AX^2 + BY^2 + CZ^2 + 2FYZ + 2GXZ + 2HXY$$

after transformation, the expression

$$ax^2 + by^2 + cz^2 + 2fyz + 2gxz + 2hxy + k(x^2 + y^2 + z^2)$$
i.e.,
$$(a + k)x^2 + (b + k)y^2 + (c + k)z^2 + 2fyz + 2gxz + 2hxy \tag{1}$$

will also be transformed to

$$(A + k)X^2 + (B + k)Y^2 + (C + k)Z^2 + 2FYZ + 2GXZ + 2HXY \tag{2}$$

for any value of k.

If for some value of k, Eqn. (1) represents two linear factors in x, y, z, then for the same value of k Eqn. (2) can be factorized to two linear factors. This implies that the equations in k

$$\begin{vmatrix} a+k & h & g \\ h & b+k & f \\ g & f & c+k \end{vmatrix} = 0 \quad \text{and} \quad \begin{vmatrix} A+k & H & G \\ H & B+k & F \\ G & F & C+k \end{vmatrix} = 0$$

will be same. Comparing the equations

$$k^3 - (a+b+c)k^2 + (ab+bc+ac-f^2-g^2-h^2)k + (abc+2fgh-af^2-bg^2-ch^2) = 0$$

and
$$k^3 - (A+B+C)k^2 + (AB+BC+AC-F^2-G^2-H^2)k + (ABC+2FGH-AF^2-BG^2-CH^2) = 0$$

we get

$$\frac{a+b+c}{A+B+C} = \frac{ab+bc+ac-f^2-g^2-h^2}{AB+BC+AC-F^2-G^2-H^2} = \frac{abc+2fgh-af^2-bg^2-ch^2}{ABC+2FGH-AF^2-BG^2-CH^2} = 1.$$

This gives the required result.

4.3 Area of a Triangle and Volume of a Tetrahedron

If Δ denotes the area of the triangle ABC whose projection on a plane which makes an angle θ with the plane of the triangle ABC be Δ', then it can be shown that

$$\Delta' = \Delta\cos\theta.$$

If l, m, n be the direction cosines of the normal to the plane ABC, then its projection on the yz, zx and xy planes are given by

$$\Delta_x = l\Delta, \quad \Delta_y = m\Delta, \quad \Delta_z = n\Delta.$$

Squaring and adding we get

$$\Delta = \sqrt{\Delta_x^2 + \Delta_y^2 + \Delta_z^2}.$$

If the vertices of the triangle ABC are $A(x_1, y_1, z_1)$; $B(x_2, y_2, z_2)$ and $C(x_3, y_3, z_3)$, then their projections on yz-plane are $A'(0, y_1, z_1)$; $B'(0, y_2, z_2)$ and $C'(0, y_3, z_3)$ respectively. So,

$$\Delta_x = (1/2)\begin{vmatrix} y_1 & z_1 & 1 \\ y_2 & z_2 & 1 \\ y_3 & z_3 & 1 \end{vmatrix}.$$

Similarly,

$$\Delta_y = (1/2)\begin{vmatrix} x_1 & z_1 & 1 \\ x_2 & z_2 & 1 \\ x_3 & z_3 & 1 \end{vmatrix} \quad \text{and} \quad \Delta_z = (1/2)\begin{vmatrix} x_1 & y_1 & 1 \\ x_2 & y_2 & 1 \\ x_3 & y_3 & 1 \end{vmatrix}.$$

If $A(x_1, y_1, z_1)$; $B(x_2, y_2, z_2)$; $C(x_3, y_3, z_3)$ and $D(x_4, y_4, z_4)$ be the vertices of a tetrahedron of volume V, then $V = \frac{1}{3}p\Delta$, where Δ is the area of any face, say BCD, of the tetrahedron and p is the perpendicular drawn from the fourth vertex, i.e. from A to that face BCD. Now equation of the plane BCD is

$$\begin{vmatrix} x & y & z & 1 \\ x_2 & y_2 & z_2 & 1 \\ x_3 & y_3 & z_3 & 1 \\ x_4 & y_4 & z_4 & 1 \end{vmatrix} = 0.$$

If Δ_x, Δ_y, Δ_z be the projections of ΔBCD on the yz, zx and xy planes, then we have

$$\Delta_x = (1/2)\begin{vmatrix} y_2 & z_2 & 1 \\ y_3 & z_3 & 1 \\ y_4 & z_4 & 1 \end{vmatrix}; \quad \Delta_y = (1/2)\begin{vmatrix} x_2 & z_2 & 1 \\ x_3 & z_3 & 1 \\ x_4 & z_4 & 1 \end{vmatrix} \quad \text{and} \quad \Delta_z = (1/2)\begin{vmatrix} x_2 & y_2 & 1 \\ x_3 & y_3 & 1 \\ x_4 & y_4 & 1 \end{vmatrix}$$

and

$$p = \text{length of the perpendicular drawn from } A \text{ to the plane } BCD$$

$$= \frac{1}{2\sqrt{\Delta_x^2 + \Delta_y^2 + \Delta_z^2}} \begin{vmatrix} x_1 & y_1 & z_1 & 1 \\ x_2 & y_2 & z_2 & 1 \\ x_3 & y_3 & z_3 & 1 \\ x_4 & y_4 & z_4 & 1 \end{vmatrix}.$$

$$\text{So, volume of the tetrahedron} = (1/6)\begin{vmatrix} x_1 & y_1 & z_1 & 1 \\ x_2 & y_2 & z_2 & 1 \\ x_3 & y_3 & z_3 & 1 \\ x_4 & y_4 & z_4 & 1 \end{vmatrix}.$$

4.4 Tetrahedron Formed by Four Given Planes

Let the equations of the given planes be

$$a_k x + b_k y + c_k z + d_k = 0, \quad k = 1, 2, 3, 4.$$

Solving any three of these equations, say for $k = 2, 3, 4$, we obtain

$$\frac{x}{A_1} = \frac{y}{B_1} = \frac{z}{C_1} = \frac{1}{D_1}$$

where A_1, B_1, C_1, D_1 are the cofactors of a_1, b_1, c_1, d_1 in

$$\Delta = \begin{vmatrix} a_1 & b_1 & c_1 & d_1 \\ a_2 & b_2 & c_2 & d_2 \\ a_3 & b_3 & c_3 & d_3 \\ a_4 & b_4 & c_4 & d_4 \end{vmatrix}.$$

Proceeding in this way we get the vertices of the tetrahedron are

$$\left(\frac{A_1}{D_1}, \frac{B_1}{D_1}, \frac{C_1}{D_1}\right); \left(\frac{A_2}{D_2}, \frac{B_2}{D_2}, \frac{C_2}{D_2}\right); \left(\frac{A_3}{D_3}, \frac{B_3}{D_3}, \frac{C_3}{D_3}\right) \text{ and } \left(\frac{A_4}{D_4}, \frac{B_4}{D_4}, \frac{C_4}{D_4}\right).$$

So, volume of the tetrahedron

$$= (1/6)\begin{vmatrix} A_1/D_1 & B_1/D_1 & C_1/D_1 & 1 \\ A_2/D_2 & B_2/D_2 & C_2/D_2 & 1 \\ A_3/D_3 & B_3/D_3 & C_3/D_3 & 1 \\ A_4/D_4 & B_4/D_4 & C_4/D_4 & 1 \end{vmatrix} = \begin{vmatrix} A_1 & B_1 & C_1 & D_1 \\ A_2 & B_2 & C_2 & D_2 \\ A_3 & B_3 & C_3 & D_3 \\ A_4 & B_4 & C_4 & D_4 \end{vmatrix} \div 6D_1D_2D_3D_4$$

$$= \text{Adj of a fourth order determinant } \Delta \div 6D_1D_2D_3D_4$$

$$= \frac{\Delta^3}{6D_1D_2D_3D_4}.$$

Examples

▶▶ 1. *Show that the equation* $x^2 + y^2 + z^2 - 6x - 10y - 8z + 49 = 0$ *is transformed to* $X^2 + Y^2 + Z^2 = 1$ *when the origin is transferred to the point* (3, 5, 4) *without changing the direction of axes.*

□ Putting $X = x - 3$, $Y = y - 5$ and $Z = z - 4$ in the given equation we get

$$(X+3)^2 + (Y+5)^2 + (Z+4)^2 - 6(X+3) - 10(Y+5) - 8(Z+4) + 49 = 0$$

or, $$X^2 + Y^2 + Z^2 = 1.$$

▶▶ 2. *The coordinate axes are rotated about the origin in such a way that the direction ratios of the new axes with respect to the old axes are* $-b, a, 0$; $-ac, -bc, a^2 + b^2$; a, b, c. *Show that due to this trcnsformation the equation* $ax + by + cz = 0$, *where* $a^2 + b^2 + c^2 = 1$, *is reduced to* $Z = 0$.

□ The direction cosines of the new axis are

$$\frac{-b}{\sqrt{a^2+b^2}}, \frac{a}{\sqrt{a^2+b^2}}, 0; \quad \frac{-ac}{\sqrt{a^2+b^2}}, \frac{-bc}{\sqrt{a^2+b^2}}, \sqrt{a^2+b^2}; \quad a,\ b,\ c.$$

Now, if $P(x, y, z)$ becomes $P(X, Y, Z)$ after transformation, then we have

$$x = \frac{-b}{\sqrt{a^2+b^2}}X - \frac{ac}{\sqrt{a^2+b^2}}Y + aZ; \quad y = \frac{a}{\sqrt{a^2+b^2}}X - \frac{bc}{\sqrt{a^2+b^2}}Y + bZ;$$

$$z = \sqrt{a^2+b^2}Y + cZ.$$

Hence, $ax + by + cz = 0$ is transformed to

$$a\left(\frac{-b}{\sqrt{a^2+b^2}}X - \frac{ac}{\sqrt{a^2+b^2}}Y + aZ\right) + b\left(\frac{a}{\sqrt{a^2+b^2}}X - \frac{bc}{\sqrt{a^2+b^2}}Y + bZ\right)$$
$$+ c(\sqrt{a^2+b^2}Y + cZ) = 0$$

i.e., $(a^2 + b^2 + c^2)Z = 0$ or $Z = 0$.

▶▷▶ **3.** *If the point (α, β, γ) be equidistant from three mutually perpendicular planes $l_k x + m_k y + n_k z = p_k$, $k = 1, 2, 3$ referred to a rectangular axes, then show that*

$$\frac{\alpha - (l_1 p_1 + l_2 p_2 + l_3 p_3)}{l_1 + l_2 + l_3} = \frac{\beta - (m_1 p_1 + m_2 p_2 + m_3 p_3)}{m_1 + m_2 + m_3} = \frac{\gamma - (n_1 p_1 + n_2 p_2 + n_3 p_3)}{n_1 + n_2 + n_3} = d.$$

☐ Given that

$$d = \frac{\alpha l_1 + \beta m_1 + \gamma n_1 - p_1}{\sqrt{l_1^2 + m_1^2 + n_1^2}} = \alpha l_1 + \beta m_1 + \gamma n_1 - p_1.$$

Similarly,

$$d = \alpha l_2 + \beta m_2 + \gamma n_2 - p_2;$$
$$d = \alpha l_3 + \beta m_3 + \gamma n_3 - p_3.$$

Now, multiplying the first, second and third equations by l_1, l_2, l_3 and adding we get

$$d(l_1 + l_2 + l_3) = \alpha \sum l_i^2 + \beta \sum l_i m_i + \gamma \sum l_i n_i - \sum l_i p_i$$

or, $$\frac{\alpha - (l_1 p_1 + l_2 p_2 + l_3 p_3)}{l_1 + l_2 + l_3} = d.$$

Again, multiplying the first, second and third equations by m_1, m_2, m_3 and n_1, n_2, n_3 respectively and adding we get

$$d = \frac{\beta - (m_1 p_1 + m_2 p_2 + m_3 p_3)}{m_1 + m_2 + m_3} \quad \text{and} \quad d = \frac{\gamma - (n_1 p_1 + n_2 p_2 + n_3 p_3)}{n_1 + n_2 + n_3}.$$

Hence the result.

▶▷▶ **4.** *Show that the volume of the tetrahedron formed by the four planes $lx + my + nz = p$; $lx + my = 0$; $my + nz = 0$ and $lx + nz = 0$ is $\frac{2p^3}{3lmn}$.*

☐ It is easy to see that the vertices of the tetrahedron, i.e. the point of intersection of the planes taking three at a time are

$$\left(\frac{p}{l}, \frac{-p}{m}, \frac{p}{n}\right); \quad \left(\frac{p}{l}, \frac{p}{m}, \frac{-p}{n}\right); \quad \left(\frac{-p}{l}, \frac{p}{m}, \frac{p}{n}\right); \quad \text{and} \quad (0, 0, 0).$$

So, the volume of the tetrahedron is

$$(1/6)\begin{vmatrix} 0 & 0 & 0 & 1 \\ p/l & -p/m & p/n & 1 \\ p/l & p/m & -p/n & 1 \\ -p/l & p/m & p/n & 1 \end{vmatrix} = \frac{2p^3}{3lmn}.$$

▶▶ **5.** *Two systems of rectangular axes have same origin. If a plane cuts them at distances a, b, c and a', b', c' from origin, then show that*

$$\frac{1}{a^2}+\frac{1}{b^2}+\frac{1}{c^2}=\frac{1}{a'^2}+\frac{1}{b'^2}+\frac{1}{c'^2}.$$

□ Let the equation of the plane with respect to the first coordinate system be

$$\frac{x}{a}+\frac{y}{b}+\frac{z}{c}=1, \quad \text{where the axes are } Ox,\ Oy \text{ and } Oz$$

and equation of the same plane with respect to the coordinate system with OX, OY and OZ as axes be

$$\frac{X}{a'}+\frac{Y}{b'}+\frac{Z}{c'}=1.$$

Since the length of a line segment is an invariant, length of the perpendicular to the plane from origin in both coordinate systems must be same. So,

$$\frac{1}{\sqrt{\frac{1}{a^2}+\frac{1}{b^2}+\frac{1}{c^2}}}=\frac{1}{\sqrt{\frac{1}{a'^2}+\frac{1}{b'^2}+\frac{1}{c'^2}}}$$

$$\text{i.e.,}\quad \frac{1}{a^2}+\frac{1}{b^2}+\frac{1}{c^2}=\frac{1}{a'^2}+\frac{1}{b'^2}+\frac{1}{c'^2}.$$

▶▶ **6.** *Through the point $P(\alpha, \beta, \gamma)$ a plane is drawn at right angles to OP to meet the coordinate axes in A, B, C. Prove that the area of the triangle ABC is $\dfrac{r^5}{2\alpha\beta\gamma}$, where $OP = r$.*

□ Equation of plane through the point $P(\alpha, \beta, \gamma)$ and right angles to OP is

$$(\alpha-0)(x-\alpha)+(\beta-0)(y-\beta)+(\gamma-0)(z-\gamma)=0$$

$$\text{or,}\quad \alpha x+\beta y+\gamma z=\alpha^2+\beta^2+\gamma^2$$

$$\text{or,}\quad \frac{x}{a}+\frac{y}{b}+\frac{z}{c}=1 \tag{1}$$

where $a=\dfrac{\sum\alpha^2}{\alpha}$, $b=\dfrac{\sum\alpha^2}{\beta}$, $c=\dfrac{\sum\alpha^2}{\gamma}$.

If the plane (1) meets the coordinate axes at A, B, C, then areas of the triangles OAB, OBC and OAC are respectively $\frac{1}{2}ab$, $\frac{1}{2}bc$ and $\frac{1}{2}ac$. Hence, area of the triangle ABC is

$$\frac{1}{2}\sqrt{b^2c^2+a^2c^2+b^2a^2}$$

$$=\frac{1}{2}\sqrt{\left(\frac{\sum\alpha^2}{\alpha}\right)^2\left(\frac{\sum\alpha^2}{\beta}\right)^2+\left(\frac{\sum\alpha^2}{\beta}\right)^2\left(\frac{\sum\alpha^2}{\gamma}\right)^2+\left(\frac{\sum\alpha^2}{\gamma}\right)^2\left(\frac{\sum\alpha^2}{\alpha}\right)^2}$$

$$=\frac{1}{2}\left(\sum\alpha^2\right)^2\sqrt{\frac{1}{\alpha^2\beta^2}+\frac{1}{\gamma^2\beta^2}+\frac{1}{\alpha^2\gamma^2}}$$

$$=\frac{1}{2\alpha\beta\gamma}\left(\sum\alpha^2\right)^2\sqrt{\sum\alpha^2}$$

$$=\frac{r^5}{2\alpha\beta\gamma}, \quad \text{where } OP=r=\sqrt{\sum\alpha^2}.$$

EXERCISE

1. What does the equation $x^2 + 7y^2 + z^2 + 8yz + 16xz - 8xy = 9$ reduce to, when the straight lines with direction ratios (1, 2, 2); (2, −2, 1) and (2, 1, −2) are taken as axes?
2. If $ax^2 + by^2 + cz^2 + 2fyz + 2gxz + 2hxy + 2ux + 2vy + 2wz + d$ be transformed by changing from one set of coordinate axes to other with same origin, then show that the expressions $a + b + c$, $u^2 + v^2 + w^2$ remain unaltered.
3. Show that in an orthogonal transformation of coordinate axes, the expression

$$\begin{vmatrix} x_1 & y_1 & z_1 & 1 \\ x_2 & y_2 & z_2 & 1 \\ x_3 & y_3 & z_3 & 1 \\ x_4 & y_4 & z_4 & 1 \end{vmatrix}$$

is an invariant, where (x_k, y_k, z_k), $k = 1, 2, 3, 4$ are four non-coplanar points.

[*Hints*: We know that the distance between two points is an invariant under an orthogonal transformation of axes. So, the distance between any two of the non-coplanar points (x_k, y_k, z_k), $k = 1, 2, 3, 4$, i.e. the edges of the tetrahedron formed by above points and hence the volume of the tetrahedron is invariant, etc.]

4. A variable plane makes with the coordinate planes a tetrahedron of constant volume $32p^3$. Show that the locus of the centroid of the tetrahedron is $xyz = 3p^3$.

[*Hints*: Let the variable plane be $\frac{x}{a} + \frac{y}{b} + \frac{z}{c} = 1$, then the vertices of the tetrahedron are (0, 0, 0); $(a, 0, 0)$; $(0, b, 0)$ and $(0, 0, c)$. Hence, its volume $= abc/6 = 32p^3$. So $abc = 192p^3$. If (f, g, h) be the centroid of the tetrahedron, then $a = 4f$, $b = 4g$, $c = 4h$ and thus $64fgh = 192p^3$ or, $fgh = 3p^3$. So, the locus of the centroid of the tetrahedron is $xyz = 3p^3$.]

5. A variable plane makes with coordinate planes a tetrahedron of constant volume k^3. Show that the locus of the foot of the perpendicular from origin to the plane is

$$(x^2 + y^2 + z^2)^3 = 6k^3xyz.$$

[*Hints*: Let the variable plane be $\frac{x}{a} + \frac{y}{b} + \frac{z}{c} = 1$; then the vertices of the tetrahedron are (0, 0, 0); $(a, 0, 0)$; $(0, b, 0)$ and $(0, 0, c)$. Hence, its volume $= abc/6 = k^3$. So, $abc = 6k^3$. If (f, g, h) be the foot of the perpendicular from origin to the plane, then we have

$$\frac{f}{a} + \frac{g}{b} + \frac{h}{c} = 1 \quad \text{and} \quad af = bg = ch = \lambda$$

which gives $\lambda = f^2 + g^2 + h^2$.
So,

$$\frac{\lambda^3}{fgh} = 6k^3, \quad \text{i.e.} \quad \frac{(f^2 + g^2 + h^2)^3}{fgh} = 6k^3 \quad \text{or,} \quad (f^2 + g^2 + h^2)^3 = 6k^3 fgh.$$

Hence, locus of the foot of the perpendicular from origin to the plane, i.e. of (f, g, h) is

$$(x^2 + y^2 + z^2)^3 = 6k^3xyz.]$$

6. Three mutually perpendicular lines PA, PB, PC meet the axes OX, OY, OZ at A, B, C respectively. If the volume of the tetrahedron $OABC$ is equal to $k^3/6$, show that the locus of the point P is $(x^2+y^2+z^2)^3 = 8k^3xyz$.

   [*Hints*: If coordinates of P, A, B, C are taken as (α, β, γ); $(a, 0, 0)$; $(0, b, 0)$ and $(0, 0, c)$ respectively, then we have $abc = k^3$. Now direction cosines of the lines PA, PB, PC are $\alpha - a, \beta, \gamma$; $\alpha, \beta - b, \gamma$ and $\alpha, \beta, \gamma - c$. Using conditions of perpendicularity of the lines PA, PB, PC we obtain

$$\alpha^2+\beta^2+\gamma^2 = a\alpha + b\beta$$
$$\alpha^2+\beta^2+\gamma^2 = b\beta + c\gamma$$
$$\alpha^2+\beta^2+\gamma^2 = a\alpha + c\gamma.$$

   Subtracting the third equation from the sum of first two we get

$$\alpha^2+\beta^2+\gamma^2 = 2b\beta.$$

   In a similar way we get $\alpha^2+\beta^2+\gamma^2 = 2a\alpha$ and $\alpha^2+\beta^2+\gamma^2 = 2c\gamma$. From these results we get

$$(\alpha^2+\beta^2+\gamma^2)^3 = 8abc\alpha\beta\gamma = 8k^3\alpha\beta\gamma.$$

   Generalizing we get the result.]

7. A variable plane is a constant distance d from the origin and meets the coordinate axes at A, B, C. Show that the locus of the centroid of the tetrahedron $OABC$ is

$$\frac{1}{x^2}+\frac{1}{y^2}+\frac{1}{z^2} = \frac{16}{d^2}.$$

   [*Hints*: Let the variable plane be $\frac{x}{a}+\frac{y}{b}+\frac{z}{c} = 1$, then the vertices of the tetrahedron are $(0, 0, 0)$; $(a, 0, 0)$; $(0, b, 0)$ and $(0, 0, c)$. Given that

$$\frac{1}{a^2}+\frac{1}{b^2}+\frac{1}{c^2} = \frac{1}{d^2}.$$

   If (f, g, h) be the centroid of the tetrahedron, then $a = 4f$, $b = 4g$, $c = 4h$, etc.]

8. Find the area of the triangle whose vertices are $(2, -3, 5)$, $(2, -5, 3)$ and $(1, -7, 4)$.

9. Prove that the area of the triangle OAB where O is the origin and A, B are given by (x_1, y_1, z_1) and (x_2, y_2, z_2) respectively is

$$\frac{1}{2}\sqrt{(y_1z_2 - y_2z_1)^2 + (z_1x_2 - z_2x_1)^2 + (y_1x_2 - y_2x_1)^2}.$$

10. The lengths of two opposite edges of a tetrahedron are a, b and shortest distance between them is d, and angle between them is θ. Prove that the volume of the tetrahedron is

$$\frac{abd\sin\theta}{6}.$$

   [*Hints*: If the direction cosines of the opposite edges AB and CD be l, m, n and l', m', n' and A, C be taken as origin and the point (f, g, h), then coordinates of B and D will be (al, am, an) and $(f + bl', g + bm', h + bn')$ respectively.

Now, shortest distance between the lines, i.e.

$$d = \begin{vmatrix} f & g & h \\ l & m & n \\ l' & m' & n' \end{vmatrix} \div \sqrt{\sum (mn' - nm')^2}.$$

Again we have, $\sin\theta = \sqrt{\sum (mn' - nm')^2}$.
So, the volume of the tetrahedron is

$$(1/6)\begin{vmatrix} 0 & 0 & 0 & 1 \\ f & g & h & 1 \\ al & am & an & 1 \\ f+bl' & g+bm' & h+bn' & 1 \end{vmatrix} = (ab/6)\begin{vmatrix} f & g & h \\ l & m & n \\ l' & m' & n' \end{vmatrix} = \frac{abd\sin\theta}{6}.]$$

11. Show that the volume of a tetrahedron of which a pair of opposite edges is formed by lengths r and r' on the straight lines

$$\frac{x-a}{l} = \frac{y-b}{m} = \frac{z-c}{n} \quad \text{and} \quad \frac{x-a'}{l'} = \frac{y-b'}{m'} = \frac{z-c'}{n'}$$

is

$$\frac{rr'}{6}\begin{vmatrix} a-a' & b-b' & c-c' \\ l & m & n \\ l' & m' & n' \end{vmatrix}.$$

[*Hints;* Let the given lines be AB and CD. If we take $A(a, b, c)$ as one vertex, then the other vertex will be $(a + lr, b + mr, c + nr)$. Similarly, if we take $C(a', b', c')$ as one vertex, then other vertex will be $(a' + l'r', b' + m'r', c' + n'r')$.

So, the volume of the tetrahedron $= (1/6)\begin{vmatrix} a & b & c & 1 \\ a+lr & b+mr & c+nr & 1 \\ a' & b' & c' & 1 \\ a'+l'r' & b'+m'r' & c'+n'r' & 1 \end{vmatrix}$, etc.]

Chapter 5

The Sphere

5.1 Definition and Equation of a Sphere

The sphere is a quadric surface (defined by a second degree equation in x, y, z) generated by a moving point, which is always at a constant distance, called its *radius*, from a fixed point, called its *centre*.

If $C(f, g, h)$ be the centre of the sphere and $P(x, y, z)$ be a point on the sphere, then we have

$$(x-f)^2+(y-g)^2+(z-h)^2=a^2 \quad \text{(constant)}.$$

This is the equation of sphere of radius 'a' and whose centre is at (f, g, h). If the centre of the sphere be the origin, then its equation reduces to

$$x^2+y^2+z^2=a^2.$$

The above equation of sphere can be written as

$$x^2+y^2+z^2-2fx-2gy-2hz+d=0, \quad \text{where } d=f^2+g^2+h^2-a^2$$

which is the general form of a sphere whose centre is at (f, g, h) and radius

$$\sqrt{f^2+g^2+h^2-d}.$$

5.2 Equation of a Sphere When the End Points of a Diameter are Given

If $P(x_1, y_1, z_1)$ and $Q(x_2, y_2, z_2)$ be the end points of a diameter and $V(x, y, z)$ be any point on the sphere, then direction ratios of VP and VQ are $x-x_1, y-y_1, z-z_1$ and $x-x_2, y-y_2, z-z_2$. Since the lines VP and VQ are perpendicular, we have

$$(x-x_1)(x-x_2)+(y-y_1)(y-y_2)+(z-z_1)(z-z_2)=0$$

which is the equation of the sphere having PQ as diameter and its centre is at the point

$$\left(\frac{x_1+x_2}{2}, \frac{y_1+y_2}{2}, \frac{z_1+z_2}{2}\right).$$

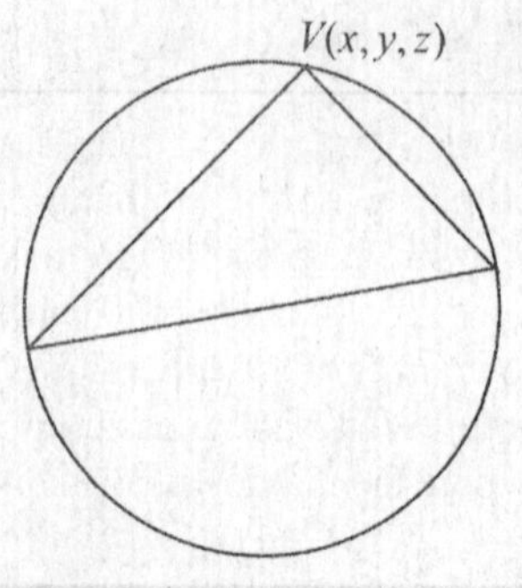

5.3 Plane Section of a Sphere

Since the plane section of a sphere is a circle, the equation

$$S(x, y, z) = x^2 + y^2 + z^2 - 2fx - 2gy - 2hz + d = 0$$

along with the equation of plane

$$L(x, y, z) = ax + by + cz - p = 0$$

represents a circle. If 'a' and 'r' be the radii of the sphere and the circle, then we have

$$r = \sqrt{a^2 - q^2}, \quad \text{where } q = \frac{af + bg + ch - p}{\sqrt{a^2 + b^2 + c^2}}.$$

If $q = 0$, i.e. when the plane passes through the centre of the sphere, the circle of intersection is called *great circle*.

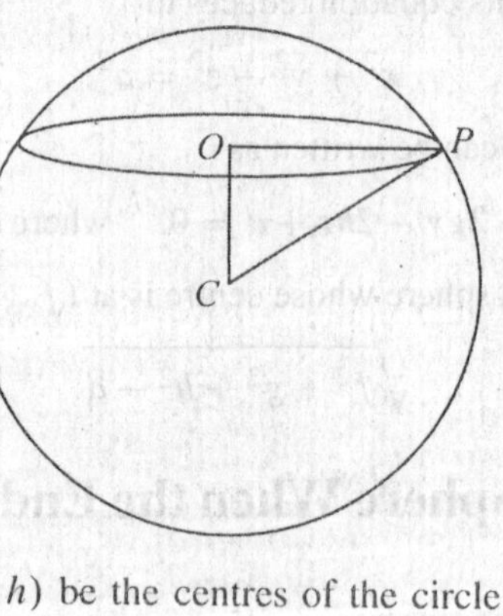

If $O(a, b, c)$ and $C(f, g, h)$ be the centres of the circle and sphere respectively and if $P(x, y, z)$ be any point on the plane section, then direction ratios of OC and OP are $a - f$, $b - g$, $c - h$ and $x - a$, $y - b$, $z - c$ respectively. Since they are perpendicular

$$(x - a)(a - f) + (y - b)(b - g) + (z - c)(c - h) = 0$$

which is the equation of the plane section as it is satisfied by the points on the plane.

If $S'(x, y, z) = x^2 + y^2 + z^2 - 2f'x - 2g'y - 2h'z + d' = 0$ be another sphere, the common points to the spheres $S = 0$ and $S' = 0$ satisfy the equation $S - S' = 0$, i.e.

$$(f' - f)x + (g' - g)y + (h' - h)z + d - d' = 0.$$

This is an equation of a plane. This plane cuts either sphere in a circle through which the other sphere passes.

5.4 Spheres through a Given Circle

From the above article we have seen that a plane cuts a sphere in a circle through which a sphere can pass. The points common to both $S = 0$ and $L = 0$ satisfy the equation $S + kL = 0$, so this equation represents sphere passes through the circle of intersection of the sphere $S = 0$ and the line $L = 0$, where the unknown constant k is to be determined using given condition. Similarly, the common points of the spheres $S = 0$ and $S' = 0$ satisfy the equation $S + \lambda S' = 0$, so this equation represents sphere passes through the circle of intersection of the spheres $S = 0$ and $S' = 0$, where the unknown constant λ is to be determined using given condition.

5.5 Intersection of a Sphere and a Line

Let the equations of the sphere and the line be

$$x^2 + y^2 + z^2 - 2fx - 2gy - 2hz + d = 0$$

$$\text{and} \quad \frac{x-\alpha}{l} = \frac{y-\beta}{m} = \frac{z-\gamma}{n} \quad \text{respectively.}$$

Any point on the line can be taken as $(\alpha + lr, \beta + mr, \gamma + nr)$. If it lies on the sphere, then we have

$$(\alpha + lr)^2 + (\beta + mr)^2 + (\gamma + nr)^2$$
$$- 2f(\alpha + lr) - 2g(\beta + mr) - 2h(\gamma + nr) + d = 0$$

$$\text{or,} \quad (l^2 + m^2 + n^2)r^2 + 2r[(\alpha - f)l$$
$$+ (\beta - g)m + (\gamma - h)n] + S(\alpha, \beta, \gamma) = 0. \tag{1}$$

The above second degree equation in 'r' shows that the line intersects the sphere in two points and the nature of the points depend on the nature of the roots of the equation. For example, if the roots of the equation be equal, then the lines intersects the sphere in two coincident points and hence it becomes a tangent line to the sphere.

If l, m, n be the direction cosines of the line, then the Eqn. (1) reduces to

$$r^2 + 2r[(\alpha - f)l + (\beta - g)m + (\gamma - h)n] + S(\alpha, \beta, \gamma) = 0. \tag{2}$$

If r_1, r_2 be its two roots, then

$$r_1 r_2 = S(\alpha, \beta, \gamma)$$

which is independent of l, m, n, implying that the quantity $r_1 r_2$ is independent of the line and dependent on (α, β, γ). This is called *power of the point* (α, β, γ) with respect to the sphere.

If the line intersects the sphere at two coincident points, i.e. $r_1 = r_2 = r$, then the length of the tangent drawn from (α, β, γ) is $r = \sqrt{S(\alpha, \beta, \gamma)}$.

5.6 Length of the Chord and Tangency of a Line

From Eqn. (2) we get

$$r_1 + r_2 = -2[(\alpha - f)l + (\beta - g)m + (\gamma - h)n].$$

So, length of a chord $= r_1 - r_2 = \sqrt{(r_1+r_2)^2 - 4r_1r_2}$. If the length of the chord $= 0$, i.e. when $r_1 = r_2$ we get

$$[(\alpha - f)l + (\beta - g)m + (\gamma - h)n]^2 = \alpha^2 + \beta^2 + \gamma^2 - 2f\alpha - 2g\beta - 2h\gamma + d.$$

This is the condition of tangency of a line.

5.7 Tangent Plane

From the above article we see that if the point (α, β, γ) lies on the sphere $S = 0$, then $S(\alpha, \beta, \gamma) = 0$ and the Eqn. (1), in this case, reduces to

$$r[(l^2 + m^2 + n^2)r + 2\{(\alpha - f)l + (\beta - g)m + (\gamma - h)n\}] = 0. \qquad (3)$$

Hence, the line $\dfrac{x-\alpha}{l} = \dfrac{y-\beta}{m} = \dfrac{z-\gamma}{n}$ will be a tangent line to the sphere at (α, β, γ) if the Eqn. (3) has two equal roots, i.e. if

$$(\alpha - f)l + (\beta - g)m + (\gamma - h)n = 0.$$

As the tangent plane at (α, β, γ) is the locus of all tangent lines passing through (α, β, γ), eliminating l, m, n we get

$$(x - \alpha)(\alpha - f) + (y - \beta)(\beta - g) + (z - \gamma)(\gamma - h) = 0$$

or, $$x(\alpha - f) + y(\beta - g) + z(\gamma - h) - (\alpha^2 + \beta^2 + \gamma^2 - \alpha f - \beta g - \gamma h) = 0$$

or, $$x(\alpha - f) + y(\beta - g) + z(\gamma - h) - \alpha f - \beta g - \gamma h + d = 0$$

or, $$x\alpha + y\beta + z\gamma - f(x + \alpha) - g(y + \beta) - h(z + \gamma) + d = 0.$$

This is the equation of tangent plane at (α, β, γ) to the sphere

$$x^2 + y^2 + z^2 - 2fx - 2gy - 2hz + d = 0.$$

The direction ratios of the normal to the tangent plane are $\alpha - f, \beta - g, \gamma - h$; also, the direction ratios of the joining (f, g, h) and (α, β, γ) are $\alpha - f, \beta - g, \gamma - h$, which shows that the normal to the tangent plane at the point of contact passes through the centre of the sphere.

5.8 Condition of Tangency of a Plane

Let the equation of a tangent plane to the sphere

$$x^2 + y^2 + z^2 - 2fx - 2gy - 2hz + d = 0 \quad \text{at } (\alpha, \beta, \gamma) \text{ be}$$

$$lx + my + nz = p.$$

Then it must be identical with $x(\alpha - f) + y(\beta - g) + z(\gamma - h) - \alpha f - \beta g - \gamma h + d = 0$.

Comparing we get

$$\frac{\alpha - f}{l} = \frac{\beta - g}{m} = \frac{\gamma - h}{n} = \frac{\alpha f + \beta g + \gamma h - d}{p}$$

$$= \frac{f(\alpha - f) + g(\beta - g) + h(\gamma - h) - (\alpha f + \beta g + \gamma h - d)}{lf + gm + hn - p}$$

$$= \frac{d - f^2 - g^2 - h^2}{fl + gm + hn - p}.$$

Again,

$$\frac{\alpha - f}{l} = \frac{\beta - g}{m} = \frac{\gamma - h}{n}$$
$$= \frac{\sqrt{(\alpha - f)^2 + (\beta - g)^2 + (\gamma - h)^2}}{\sqrt{l^2 + m^2 + n^2}} = \frac{\sqrt{f^2 + g^2 + h^2 - d}}{\sqrt{l^2 + m^2 + n^2}}.$$

From the above results we obtain

$$\frac{\sqrt{f^2 + g^2 + h^2 - d}}{\sqrt{l^2 + m^2 + n^2}} = \frac{d - f^2 - g^2 - h^2}{fl + gm + hn - p}$$

or, $$(fl + gm + hn - p)^2 = (l^2 + m^2 + n^2)(f^2 + g^2 + h^2 - d).$$

This is the required condition of tangency.

Alternative Method

A plane will be a tangent plane to the sphere if length of the perpendicular from centre to the plane is equal to the radius of the sphere. So,

$$\frac{fl + gm + hn - p}{\sqrt{l^2 + m^2 + n^2}} = \sqrt{f^2 + g^2 + h^2 - d}$$

or, $$(fl + gm + hn - p)^2 = (l^2 + m^2 + n^2)(f^2 + g^2 + h^2 - d). \quad (4)$$

5.9 Plane of Contact of the Tangent Planes to a Sphere

The plane containing the point of contacts of the tangent planes drawn through a point to a sphere is called *plane of contact*. If we take the point as (α, β, γ) and the equation of sphere as

$$x^2 + y^2 + z^2 - 2fx - 2gy - 2hz + d = 0$$

then equation of the tangent plane to the sphere at (x', y', z') is

$$xx' + yy' + zz' - f(x + x') - g(y + y') - h(z + z') + d = 0.$$

If it passes through (α, β, γ), then

$$\alpha x' + \beta y' + \gamma z' - f(\alpha + x') - g(\beta + y') - h(\gamma + z') + d = 0$$

or, $$(\alpha - f)x' + (\beta - g)y' + (\gamma - h)z' - f\alpha - g\beta - h\gamma + d = 0.$$

Hence, locus of the point (x', y', z') is

$$(\alpha - f)x + (\beta - g)y + (\gamma - h)z - f\alpha - g\beta - h\gamma + d = 0$$

which is the equation of plane of contact of the tangent planes drawn through the point (α, β, γ) to the sphere

$$x^2 + y^2 + z^2 - 2fx - 2gy - 2hz + d = 0.$$

5.10 Polar Plane of a Point with Respect to a Sphere

The locus of the point of intersection of the tangent planes drawn at the extremities of chords passing through a point to the sphere is called the *polar plane* of the point with respect to the sphere. Let the equation of the sphere be

$$x^2 + y^2 + z^2 - 2fx - 2gy - 2hz + d = 0$$

then the equation of the plane of contact of the tangent planes drawn through the point (x', y', z') is

$$xx' + yy' + zz' - f(x + x') - g(y + y') - h(z + z') + d = 0.$$

If it passes through (α, β, γ), then

$$\alpha x' + \beta y' + \gamma z' - f(\alpha + x') - g(\beta + y') - h(\gamma + z') + d = 0$$

or, $$(\alpha - f)x' + (\beta - g)y' + (\gamma - h)z' - f\alpha - g\beta - h\gamma + d = 0.$$

Hence, locus of the point (x', y', z') is

$$(\alpha - f)x + (\beta - g)y + (\gamma - h)z - f\alpha - g\beta - h\gamma + d = 0.$$

This is the polar plane of the point (α, β, γ) with respect to the given sphere and the point (α, β, γ) is called *pole of the polar plane.*

5.11 Pole of a Plane

Let (α, β, γ) be the pole of the plane $lx + my + nz = p$ with respect to the sphere $x^2 + y^2 + z^2 - 2fx - 2gy - 2hz + d = 0$; then the plane $lx + my + nz = p$ should be identical with

$$(\alpha - f)x + (\beta - g)y + (\gamma - h)z - f\alpha - g\beta - h\gamma + d = 0.$$

Comparing we get

$$\frac{\alpha - f}{l} = \frac{\beta - g}{m} = \frac{\gamma - h}{n} = \frac{f\alpha + g\beta + h\gamma - d}{p} = \frac{f^2 + g^2 + h^2 - d}{p - (lf + mg + nh)}$$

from which we get

$$\alpha = \frac{l(g^2 + h^2 - d) - f(mg + nh - p)}{p - (lf + mg + nh)}$$

$$\beta = \frac{m(f^2 + h^2 - d) - g(lf + nh - p)}{p - (lf + mg + nh)}$$

$$\gamma = \frac{n(g^2 + f^2 - d) - h(mg + lf - p)}{p - (lf + mg + nh)}.$$

5.12 Properties of Pole and Polar Planes

(i) The polar plane is perpendicular to the line joining the centre of the sphere to the pole.
(ii) The pole and the point of intersection of the polar plane with the join of the centre and the pole are inverse points with respect to the sphere.
(iii) If the polar plane of a point P passes through a point Q, then the polar plane of Q passes through the point P and such points are called *conjugate points*.
(iv) If the pole of a plane L_1 lies on another plane L_2, then the pole of L_2 lies on the plane L_1 and such planes are called *conjugate planes*.
(v) The polar planes of all points on a line pass through another line. Two lines such that polar plane of each point of one line passes through the other line are called *polar lines*.

5.13 Orthogonal Intersection of Two Spheres

The angle of intersection of two spheres at a point is the angle between the tangent planes at that point drawn to the spheres and two spheres are said to cut orthogonally if their angle of intersection is a right angle. Let us take two spheres as

$$x^2 + y^2 + z^2 - 2f_1x - 2g_1y - 2h_1z + d_1 = 0$$
$$\text{and} \quad x^2 + y^2 + z^2 - 2f_2x - 2g_2y - 2h_2z + d_2 = 0$$

then their centres are $A(f_1, g_1, h_1)$ and $B(f_2, g_2, h_2)$ and radii are

$$r_1 = \sqrt{f_1^2 + g_1^2 + h_1^2 - d_1} \quad \text{and} \quad r_2 = \sqrt{f_2^2 + g_2^2 + h_2^2 - d_2} \quad \text{respectively.}$$

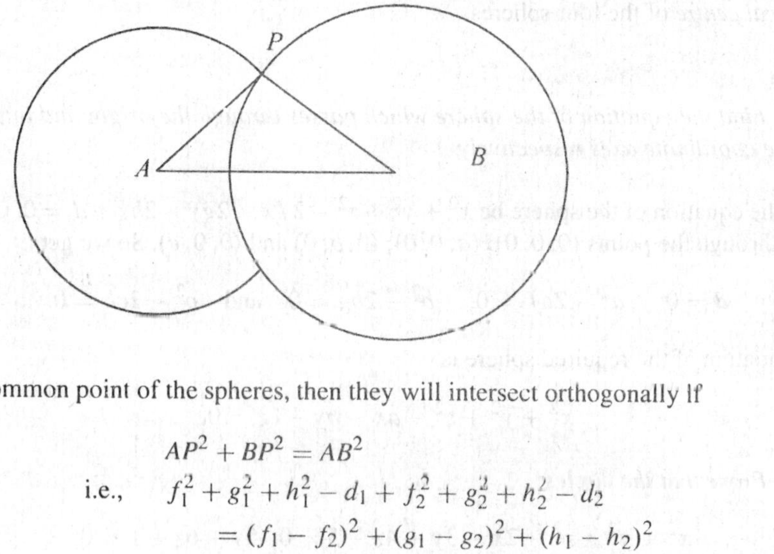

If P is a common point of the spheres, then they will intersect orthogonally if

$$AP^2 + BP^2 = AB^2$$
$$\text{i.e.,} \quad f_1^2 + g_1^2 + h_1^2 - d_1 + f_2^2 + g_2^2 + h_2^2 - d_2$$
$$= (f_1 - f_2)^2 + (g_1 - g_2)^2 + (h_1 - h_2)^2$$
$$\Rightarrow \quad 2f_1f_2 + 2g_1g_2 + 2h_1h_2 = d_1 + d_2.$$

This is the condition of orthogonal intersection of the given spheres.

5.14 Radical Plane

The locus of a moving point whose powers with respect to two spheres are equal is called the *radical plane* of the spheres. Let us take two spheres as

$$x^2+y^2+z^2-2f_1x-2g_1y-2h_1z+d_1=0$$
$$\text{and}\quad x^2+y^2+z^2-2f_2x-2g_2y-2h_2z+d_2=0.$$

If $P(x', y', z')$ be the moving point, such that its powers with respect to the given spheres are of equal magnitude, then

$$x'^2+y'^2+z'^2-2f_1x'-2g_1y'-2h_1z'+d_1$$
$$=x'^2+y'^2+z'^2-2f_2x'-2g_2y'-2h_2z'+d_2$$
$$\text{or,}\quad 2(f_1-f_2)x'+2(g_1-g_2)y'+2(h_1-h_2)z'=d_1-d_2.$$

Hence, the locus of the point $P(x', y', z')$ is

$$2(f_1-f_2)x+2(g_1-g_2)y+2(h_1-h_2)z=d_1-d_2.$$

This is the equation of radical plane of the given spheres.

5.15 Properties of Radical Planes

(i) The radical plane of two spheres is perpendicular to the line joining their centres.
(ii) The radical planes of three spheres taken in pairs pass through a line. This line is called *radical line* of the three spheres.
(iii) The radical lines of four spheres taken three at a time meet in a point. This point is called *radical centre* of the four spheres.

Examples

▶▷ 1. *Find the equation of the sphere which passes through the origin and intercepts a, b, c on the coordinate axes respectively.*

□ Let the equation of the sphere be $x^2+y^2+z^2-2fx-2gy-2hz+d=0$. Given that it passes through the points $(0, 0, 0)$; $(a, 0, 0)$; $(0, b, 0)$ and $(0, 0, c)$. So we get

$$d=0;\quad a^2-2af=0;\quad b^2-2bg=0\ \text{ and }\ c^2-2ch=0.$$

So, the equation of the required sphere is

$$x^2+y^2+z^2-ax-by-cz=0.$$

▶▷ 2. *Prove that the circles*

$$x^2+y^2+z^2-2x+3y+4z-5=0,\ 5y+6z+1=0;$$
$$x^2+y^2+z^2-3x-4y+5z-6=0,\ x+2y-7z=0$$

lie on the same sphere and find the equation of the sphere.

☐ Equation of the sphere through the first circle is

$$x^2+y^2+z^2-2x+3y+4z-5+k_1(5y+6z+1)=0$$

and equation of the sphere through the second circle is

$$x^2+y^2+z^2-3x-4y+5z-6+k_2(x+2y-7z)=0.$$

The given circles will lie on same sphere if the above equations are identical. Comparing we get

$$-3+k_2=-2;\quad 3+5k_1=-4+2k_2;\quad 4+6k_1=5-7k_2 \quad \text{and} \quad k_1-5=-6.$$

Solving the first and second equations we get

$k_2=1$ and $k_1=-1$, which also satisfy the third and fourth equations.

So, the given circles lie on same sphere whose equation is

$$x^2+y^2+z^2-2x-2y-2z-6=0$$

$$\text{or,}\quad (x-1)^2+(y-1)^2+(z-1)^2=3^2.$$

▶▷▶ 3. *Find the centre and radius of the circle $x^2+y^2+z^2=49,\ 2x-y+3z=14$.*

☐ Length of the perpendicular from the centre (0, 0, 0) of the sphere $x^2+y^2+z^2=49$ to the plane $2x-y+3z=14$ is

$$p=\frac{14}{\sqrt{14}}=\sqrt{14}.$$

So, the radius of the circle $=\sqrt{(49-14)}=\sqrt{35}$.

Equation of the perpendicular from origin to the given plane is

$$\frac{x}{2}=\frac{y}{-1}=\frac{z}{3}=r.$$

Any point on the line is $(2r,-r,3r)$. If this point lies on the plane $2x-y+3z=14$, then

$$4r+r+9r=14$$

$$\text{or,}\quad r=1.$$

So, the centre of the circle is $(2,-1,3)$.

▶▷▶ 4. *Find the equation of the sphere which passes through the circle $x^2+y^2+z^2=4$, $z=0$ and is cut by the plane $x+2y+2z=0$ in a circle of radius 3.*

☐ Equation of the sphere through the given circle is

$$x^2+y^2+z^2-4+kz=0.$$

Its centre is $(0,0,-k/2)$ and radius is $\sqrt{4+\dfrac{k^2}{4}}$. Now the length of the perpendicular from the centre of the sphere to the plane is $-k/3$. Given that

$$4+\frac{k^2}{4}=9+\frac{k^2}{9} \Rightarrow \frac{5k^2}{36}=5 \Rightarrow k=\pm 6.$$

So, equation of the required sphere is $x^2+y^2+z^2\pm 6z-4=0$.

▶▶ 5. *Find the equation of the sphere which passes through the point* (α, β, γ) *and the circle*

$$x^2 + y^2 = a^2, z = 0.$$

□ Let the equation of the sphere be $x^2 + y^2 + z^2 - 2fx - 2gy - 2hz + d = 0$. Putting $z = 0$ we get

$$x^2 + y^2 - 2fx - 2gy + d = 0.$$

Comparing this with $x^2 + y^2 = a^2$ we get

$$f = 0, \quad g = 0 \quad \text{and} \quad d = -a^2.$$

So, the equation of the sphere becomes

$$x^2 + y^2 + z^2 - 2hz - a^2 = 0.$$

Given that the sphere passes through the point (α, β, γ), so we get

$$\alpha^2 + \beta^2 + \gamma^2 - 2h\gamma - a^2 = 0.$$

Eliminating h, we get

$$\gamma(x^2 + y^2 + z^2 - a^2) = z(\alpha^2 + \beta^2 + \gamma^2 - a^2).$$

▶▶ 6. *A plane passing through a fixed point* (a, b, c) *cuts the axes at A, B, C. Show that the locus of the centre of the sphere OABC is*

$$\frac{a}{x} + \frac{b}{y} + \frac{c}{z} = 2.$$

□ Let the equation of the plane be

$$\frac{x}{\alpha} + \frac{y}{\beta} + \frac{z}{\gamma} = 1.$$

As it passes through the point (a, b, c) we get

$$\frac{a}{\alpha} + \frac{b}{\beta} + \frac{c}{\gamma} = 1. \tag{1}$$

Now equation of the sphere $OABC$ is

$$x^2 + y^2 + z^2 - \alpha x - \beta y - \gamma z = 0$$

whose centre is $\left(\frac{\alpha}{2}, \frac{\beta}{2}, \frac{\gamma}{2}\right)$. Hence, from (1) we get

$$\frac{a}{2\left(\frac{\alpha}{2}\right)} + \frac{b}{2\left(\frac{\beta}{2}\right)} + \frac{c}{2\left(\frac{\gamma}{2}\right)} = 1.$$

So, locus of the centre of the sphere is $\frac{a}{x} + \frac{b}{y} + \frac{c}{z} = 2.$

▶▶ 7. *Find the equation of the sphere passing through four non-coplanar points* (x_k, y_k, z_k), $k = 1, 2, 3, 4$.

□ Let the equation of the sphere be

$$x^2 + y^2 + z^2 - 2fx - 2gy - 2hz + d = 0. \qquad (1)$$

If it passes through the points (x_k, y_k, z_k), $k = 1, 2, 3, 4$, we have

$$x_1^2 + y_1^2 + z_1^2 - 2fx_1 - 2gy_1 - 2hz_1 + d = 0 \qquad (2)$$
$$x_2^2 + y_2^2 + z_2^2 - 2fx_2 - 2gy_2 - 2hz_2 + d = 0 \qquad (3)$$
$$x_3^2 + y_3^2 + z_3^2 - 2fx_3 - 2gy_3 - 2hz_3 + d = 0 \qquad (4)$$
$$x_4^2 + y_4^2 + z_4^2 - 2fx_4 - 2gy_4 - 2hz_4 + d = 0. \qquad (5)$$

Eliminating f, g, h, d from (1)–(5) we get the equation of sphere as

$$\begin{vmatrix} x^2+y^2+z^2 & x & y & z & 1 \\ x_1^2+y_1^2+z_1^2 & x_1 & y_1 & z_1 & 1 \\ x_2^2+y_2^2+z_2^2 & x_2 & y_2 & z_2 & 1 \\ x_3^2+y_3^2+z_3^2 & x_3 & y_3 & z_3 & 1 \\ x_4^2+y_4^2+z_4^2 & x_4 & y_4 & z_4 & 1 \end{vmatrix} = 0.$$

▶▷▶ **8.** *Find the equation of the sphere circumscribing the tetrahedron formed by the planes*

$$\frac{x}{a} + \frac{y}{b} + \frac{z}{c} = 1; \quad \frac{y}{b} + \frac{z}{c} = 0; \quad \frac{x}{a} + \frac{z}{c} = 0 \ \textit{and} \ \frac{x}{a} + \frac{y}{b} = 0.$$

□ To get the vertices of the tetrahedron we solve the above equations taking three at a time and these are $O(0, 0, 0)$; $A(a, b, -c)$; $B(a, -b, c)$ and $C(-a, b, c)$. Let the equation of the sphere be

$$x^2 + y^2 + z^2 - 2fx - 2gy - 2hz + d = 0.$$

Since it passes through origin, $d = 0$. Again, it passes through the points A, B and C, so we get

$$a^2 + b^2 + c^2 - 2fa - 2gb + 2hc = 0,$$
$$a^2 + b^2 + c^2 - 2fa + 2gb - 2hc = 0$$
$$\text{and} \quad a^2 + b^2 + c^2 + 2fa - 2gb - 2hc = 0.$$

Solving these equations taking in pairs we obtain

$$f = \frac{a^2 + b^2 + c^2}{2a}; \quad g = \frac{a^2 + b^2 + c^2}{2b} \quad \text{and} \quad h = \frac{a^2 + b^2 + c^2}{2c}.$$

So, the equation of the sphere is

$$x^2 + y^2 + z^2 - (a^2 + b^2 + c^2)\left(\frac{x}{a} + \frac{y}{b} + \frac{z}{c}\right) = 0$$

$$\text{or,} \quad \frac{x^2 + y^2 + z^2}{a^2 + b^2 + c^2} - \left(\frac{x}{a} + \frac{y}{b} + \frac{z}{c}\right) = 0.$$

▶▷ **9.** *A sphere of radius k passes through O and meets the axes at A, B, C. Prove that the locus of the foot of the perpendicular to the plane ABC from O is given by*

$$(x^2+y^2+z^2)^2(x^{-2}+y^{-2}+z^{-2})=4k^2.$$

□ If the sphere cuts the axes at $A(a,0,0)$; $B(0,b,0)$ and $C(0,0,c)$, then its equation is

$$x^2+y^2+z^2-ax-by-cz=0, \quad \text{where } a^2+b^2+c^2=4k^2$$

and equation of the plane ABC is

$$\frac{x}{a}+\frac{y}{b}+\frac{z}{c}=1.$$

If P be the foot of the perpendicular drawn from O to the plane ABC, then equation of OP is

$$\frac{x}{\frac{1}{a}}=\frac{y}{\frac{1}{b}}=\frac{z}{\frac{1}{c}}=r.$$

Any point on it is $\left(\frac{r}{a},\frac{r}{b},\frac{r}{c}\right)$. If it lies on the plane ABC, then

$$r\left(\frac{1}{a^2}+\frac{1}{b^2}+\frac{1}{c^2}\right)=1 \quad\Rightarrow\quad r=\frac{1}{a^{-2}+b^{-2}+c^{-2}}.$$

Now, if (f,g,h) be the foot of the perpendicular, then

$$f=\frac{r}{a} \quad \text{or,} \quad a=\frac{r}{f} \quad \text{and similarly, } b=\frac{r}{g}, \quad c=\frac{r}{h}.$$

Therefore,

$$r^2\left(\frac{1}{f^2}+\frac{1}{g^2}+\frac{1}{h^2}\right)=4k^2. \tag{1}$$

Again,

$$r^2=\left(\frac{1}{a^{-2}+b^{-2}+c^{-2}}\right)^2=\frac{r^4}{(f^2+g^2+h^2)^2}. \tag{2}$$

Eliminating r from Eqns. (1) and (2), we get

$$(f^2+g^2+h^2)^2(f^{-2}+g^{-2}+h^{-2})=4k^2.$$

Generalizing, we get

$$(x^2+y^2+z^2)^2(x^{-2}+y^{-2}+z^{-2})=4k^2.$$

▶▷ **10.** *A variable plane passes through a fixed point. Show that the locus of the foot of the perpendicular to the plane is a sphere.*

□ Let the variable plane be $\frac{x}{a}+\frac{y}{b}+\frac{z}{c}=1$. If it passes through a fixed point (f,g,h), then we have

$$\frac{f}{a}+\frac{g}{b}+\frac{h}{c}=1.$$

If P be the foot of the perpendicular drawn from O to the given plane, then equation of OP is

$$\frac{x}{\frac{1}{a}} = \frac{y}{\frac{1}{b}} = \frac{z}{\frac{1}{c}} = r.$$

Any point on it is $\left(\frac{r}{a}, \frac{r}{b}, \frac{r}{c}\right)$. If it lies on the given plane, then

$$r\left(\frac{1}{a^2} + \frac{1}{b^2} + \frac{1}{c^2}\right) = 1 \quad \Rightarrow \quad r = \frac{1}{a^{-2} + b^{-2} + c^{-2}}.$$

Now, if (x', y', z') be the foot of the perpendicular, then

$$x' = \frac{r}{a} \quad \text{or,} \quad a = \frac{r}{x'} \quad \text{and} \quad \text{similarly, } b = \frac{r}{y'}, \quad c = \frac{r}{z'}.$$

Therefore,

$$r = fx' + gy' + hz' \quad \text{and} \quad r = \frac{r^2}{x'^2 + y'^2 + z'^2}.$$

Eliminating r we get

$$x'^2 + y'^2 + z'^2 = fx' + gy' + hz'.$$

Generalizing, we get the locus of the foot of the perpendicular is

$$x^2 + y^2 + z^2 = fx + gy + hz.$$

This is a sphere.

Alternative Method

Let the fixed point be (f, g, h). Equation of any plane through it can be taken as

$$l(x - f) + m(y - g) + n(z - h) = 0.$$

Equation of the perpendicular from O to the plane is

$$\frac{x}{l} = \frac{y}{m} = \frac{z}{n}.$$

To get the locus of the foot of the perpendicular, we eliminate l, m, n from the above equation and get

$$x(x - f) + y(y - g) + z(z - h) = 0$$

$$\text{or,} \quad x^2 + y^2 + z^2 = fx + gy + hz.$$

▶▶ **11**. *The plane* $\frac{x}{a} + \frac{y}{b} + \frac{z}{c} = 1$ *meets the axes in* A, B, C. *Find equation of the circumcircle of the triangle ABC and also find its centre.*

□ The triangle meets the axes at $A(a, 0, 0)$; $B(0, b, 0)$ and $C(0, 0, c)$. So equation of the sphere $OABC$ is

$$x^2 + y^2 + z^2 - ax - by - cz = 0.$$

Circumcircle of the triangle ABC is the plane section of the above sphere by the given plane, so its equation is

$$x^2 + y^2 + z^2 - ax - by - cz = 0; \quad \frac{x}{a} + \frac{y}{b} + \frac{z}{c} = 1.$$

Now, centre of the sphere $OABC$ is $\left(\frac{a}{2}, \frac{b}{2}, \frac{c}{2}\right)$, so equation of the perpendicular to the plane ABC from the centre of the sphere is

$$\frac{x - \frac{a}{2}}{\frac{1}{a}} = \frac{y - \frac{b}{2}}{\frac{1}{b}} = \frac{z - \frac{c}{2}}{\frac{1}{c}} = r.$$

Any point on the above line can be taken as

$$\left(\frac{a}{2} + \frac{r}{a}, \frac{b}{2} + \frac{r}{b}, \frac{c}{2} + \frac{r}{c}\right).$$

This will lie on the plane if

$$\frac{1}{a}\left(\frac{a}{2} + \frac{r}{a}\right) + \frac{1}{b}\left(\frac{b}{2} + \frac{r}{b}\right) + \frac{1}{c}\left(\frac{c}{2} + \frac{r}{c}\right) = 1$$

$$\Rightarrow \quad r\left(\frac{1}{a^2} + \frac{1}{b^2} + \frac{1}{c^2}\right) = -\frac{1}{2}$$

$$\Rightarrow \quad r = -\frac{1}{2(a^{-2} + b^{-2} + c^{-2})}.$$

Since centre of the circle of intersection is the foot of the perpendicular drawn from the centre of the sphere to the plane, centre of the circumcircle is

$$\left(\frac{a}{2} - \frac{1}{2a(a^{-2} + b^{-2} + c^{-2})}, \frac{b}{2} - \frac{1}{2b(a^{-2} + b^{-2} + c^{-2})}, \frac{c}{2} - \frac{1}{2c(a^{-2} + b^{-2} + c^{-2})}\right)$$

or, $$\left(\frac{a(b^{-2} + c^{-2})}{2(a^{-2} + b^{-2} + c^{-2})}, \frac{b(a^{-2} + c^{-2})}{2(a^{-2} + b^{-2} + c^{-2})}, \frac{c(b^{-2} + a^{-2})}{2(a^{-2} + b^{-2} + c^{-2})}\right).$$

▶▶ **12.** *Prove that the centre of all sections of the sphere* $x^2 + y^2 + z^2 = r^2$ *by the planes through the point* (x', y', z') *lie on the sphere* $x(x - x') + y(y - y') + z(z - z') = 0$.

□ Let $P(f, g, h)$ be the centre of the circle of intersection of the sphere $x^2 + y^2 + z^2 = r^2$ by the planes through the point (x', y', z'). Then the direction ratios of OP, i.e. normal to the plane are f, g, h, where $O(0, 0, 0)$ is the centre of the sphere. Hence equation of the plane which cuts the sphere in a circle with centre at (f, g, h) is

$$f(x - f) + g(y - g) + h(z - h) = 0.$$

If it passes through the point (x', y', z'), then

$$f(x' - f) + g(y' - g) + h(z' - h) = 0.$$

Hence, locus of the centre (f, g, h) is

$$x(x - x') + y(y - y') + z(z - z') = 0.$$

▶▷▶ **13.** *Prove that the sphere* $x^2 + y^2 + z^2 + 2ux + 2vy + 2wz + d = 0$ *cuts the sphere*

$$x^2 + y^2 + z^2 + 2u'x + 2v'y + 2w'z + d' = 0$$

in a great circle if

$$2(uu' + vv' + ww') = 2r'^2 + d + d'$$

where r' is the radius of the second sphere.

☐ Equation of the plane through the given spheres is given by $S - S' = 0$, so its equation is

$$2(u - u')x + 2(v - v')y + 2(w - w')z + d - d' = 0.$$

If the sphere $S = 0$ cuts $S' = 0$ in a great circle, the centre of $S' = 0$ will lie on the above plane, i.e.

$$2(u - u')u' + 2(v - v')v' + 2(w - w')w' = d - d'.$$

Now, given that $r'^2 = u'^2 + v'^2 + w'^2 - d'$, so from the above result we get

$$2(uu' + vv' + ww') = 2(r'^2 + d') + d - d'$$

or, $$2(uu' + vv' + ww') = 2r'^2 + d + d'.$$

▶▷▶ **14.** *If P is a variable point on a given line and A, B, C are its projections on coordinate axes, show that the sphere $OABC$ passes through a fixed circle.*

☐ Let the given line be

$$\frac{x - a}{\alpha} = \frac{y - b}{\beta} = \frac{z - c}{\gamma} = r.$$

Any point on the line can be taken as $P(a + \alpha r, b + \beta r, c + \gamma r)$. Its projections on coordinate axes are $A(a + \alpha r, 0, 0)$; $B(0, b + \beta r, 0)$ and $C(0, 0, c + \gamma r)$. So, equation of the sphere $OABC$ is

$$x^2 + y^2 + z^2 - (a + \alpha r)x - (b + \beta r)y - (c + \gamma r)z = 0$$

or, $$x^2 + y^2 + z^2 - ax - by - cz + r(\alpha x + \beta y + \gamma z) = 0.$$

The above equation shows that the sphere $OABC$ passes through the circle

$$x^2 + y^2 + z^2 - ax - by - cz = 0; \quad \alpha x + \beta y + \gamma z = 0$$

which is fixed.

▶▷▶ **15.** *POP' is a variable diameter of the ellipse $\frac{x^2}{a^2} + \frac{y^2}{b^2} = 1$, $z = 0$ and a circle is described in the plane $PP'ZZ'$ on PP' as diameter; prove that as PP' varies, the circle generates the surface*

$$(x^2 + y^2 + z^2)\left(\frac{x^2}{a^2} + \frac{y^2}{b^2}\right) = x^2 + y^2.$$

□ If we take the points P, P' as $(a\cos\theta, b\sin\theta, 0)$ and $(-a\cos\theta, -b\sin\theta, 0)$ respectively, then equation of the sphere describing PP' as diameter is

$$x^2+y^2+z^2=a^2\cos^2\theta+b^2\sin^2\theta. \tag{1}$$

Equation of the plane $PP'ZZ'$ is

$$\begin{vmatrix} x & y & z \\ a\cos\theta & b\sin\theta & 0 \\ 0 & 0 & 1 \end{vmatrix}=0$$

$$\text{or,}\quad \frac{x}{a\cos\theta}=\frac{y}{b\sin\theta}. \tag{2}$$

Now the plane section of the sphere (1) by the plane (2) will be a circle in the plane $PP'ZZ'$ on PP' as diameter. In order to find the surface generated by the circle we eliminate θ from (1) and (2) as follows:

From (2) we get

$$\frac{x}{a\cos\theta}=\frac{y}{b\sin\theta}\quad\Rightarrow\quad\frac{\frac{x}{a}}{\cos\theta}=\frac{\frac{y}{b}}{\sin\theta}=\frac{\sqrt{\frac{x^2}{a^2}+\frac{y^2}{b^2}}}{1}$$

$$\Rightarrow\quad a\cos\theta=\frac{x}{\sqrt{\frac{x^2}{a^2}+\frac{y^2}{b^2}}}\quad\text{and}\quad b\sin\theta=\frac{y}{\sqrt{\frac{x^2}{a^2}+\frac{y^2}{b^2}}}.$$

Using these in (1) we get

$$(x^2+y^2+z^2)\left(\frac{x^2}{a^2}+\frac{y^2}{b^2}\right)=x^2+y^2.$$

▶▷ **16**. *Find the locus of the centre of sphere which passes through a fixed point and intercepts a constant length on a given straight line.*

□ Without loss of generality, we take the fixed straight line as x-axis and the fixed point as the point $(0, k, 0)$, i.e. a point in y-axis. Let the equation of the sphere be

$$x^2+y^2+z^2-2ux-2vy-2wz+d=0.$$

Then we have

$$k^2-2vk+d=0. \tag{1}$$

Putting $y=0=z$ in the equation of sphere we obtain

$$x^2-2ux+d=0$$

which gives

$$(x_2-x_1)^2=(x_2+x_1)^2-4x_1x_2=4u^2-4d=4\lambda\ \text{(constant)}. \tag{2}$$

Eliminating d from (1) and (2) we get

$$k^2-2vk+u^2-\lambda=0.$$

Hence, locus of the centre, i.e. of (u, v, w) is

$$x^2 - 2yk + k^2 - \lambda = 0.$$

▶▷▶ 17. *Find the locus of the centre of sphere of constant radius which passes through a fixed point and touches a given straight line.*

☐ Without loss of generality, we take the fixed straight line as x-axis and the fixed point as the point $(0, k, 0)$, i.e. a point in y-axis. Let the equation of the sphere be

$$x^2 + y^2 + z^2 - 2ux - 2vy - 2wz + d = 0$$
$$\text{where} \quad u^2 + v^2 + w^2 - d = \lambda \text{ (constant).} \tag{1}$$

Then we have

$$k^2 - 2vk + d = 0. \tag{2}$$

Since it touches x-axis, the equation $x^2 - 2ux + d = 0$ will have same roots for which the condition is

$$u^2 = d. \tag{3}$$

Now, eliminating d we get from (1)–(3)

$$v^2 + w^2 = \lambda \quad \text{and} \quad k^2 - 2vk + u^2 = 0.$$

Hence, locus of the centre, i.e. of (u, v, w) is

$$x^2 - 2yk + k^2 = 0 \quad \text{and} \quad y^2 + z^2 = \lambda.$$

▶▷▶ 18. *If any tangent plane to the sphere $x^2 + y^2 + z^2 = r^2$ makes intercepts a, b, c on coordinate axes, then show that $a^{-2} + b^{-2} + c^{-2} = r^{-2}$.*

☐ Equation of tangent plane to the sphere $x^2 + y^2 + z^2 = r^2$ at the point (f, g, h) is

$$xf + yg + zh = r^2$$
$$\text{or,} \quad \frac{x}{r^2/f} + \frac{y}{r^2/g} + \frac{z}{r^2/h} = 1.$$

So,

$$a = \frac{r^2}{f}; \quad b = \frac{r^2}{g}; \quad c = \frac{r^2}{h}$$
$$\text{i.e.,} \quad f = \frac{r^2}{a}; \quad g = \frac{r^2}{b}; \quad h = \frac{r^2}{c}.$$

Since (f, g, h) lies on the sphere, we have

$$f^2 + g^2 + h^2 = r^2$$
$$\text{or,} \quad \frac{r^4}{a^2} + \frac{r^4}{b^2} + \frac{r^4}{c^2} = r^2$$
$$\text{i.e.,} \quad a^{-2} + b^{-2} + c^{-2} = r^{-2}.$$

▶▷ **19.** *Tangent planes at any point on the sphere $x^2+y^2+z^2=r^2$ meet the coordinate axes at A, B, C. Show that the locus of the point of intersection of the planes drawn parallel to the coordinate planes through A, B, C is the surface $x^{-2}+y^{-2}+z^{-2}=r^{-2}$.*

☐ Equation of tangent plane to the sphere $x^2+y^2+z^2=r^2$ at the point (f, g, h) is

$$xf+yg+zh=r^2$$

$$\text{or,}\quad \frac{x}{r^2/f}+\frac{y}{r^2/g}+\frac{z}{r^2/h}=1.$$

It meets the coordinate axes at $A\left(\frac{r^2}{f}, 0, 0\right)$; $B\left(0, \frac{r^2}{g}, 0\right)$ and $C\left(0, 0, \frac{r^2}{h}\right)$. Equations of the planes through A, B, C and parallel to the coordinate planes are

$$x=\frac{r^2}{f};\quad y=\frac{r^2}{g};\quad z=\frac{r^2}{h}.$$

These planes meet at a point whose coordinate is $\left(\frac{r^2}{f}, \frac{r^2}{g}, \frac{r^2}{h}\right)=(\alpha, \beta, \gamma)$, say. Then

$$f=\frac{r^2}{\alpha};\quad g=\frac{r^2}{\beta};\quad h=\frac{r^2}{\gamma}.$$

Since (f, g, h) lies on the sphere, we have

$$f^2+g^2+h^2=r^2$$

$$\text{or,}\quad \frac{r^4}{\alpha^2}+\frac{r^4}{\beta^2}+\frac{r^4}{\gamma^2}=r^2$$

$$\text{i.e.,}\quad \alpha^{-2}+\beta^{-2}+\gamma^{-2}=r^{-2}.$$

Hence, locus of the point (α, β, γ) is

$$x^{-2}+y^{-2}+z^{-2}=r^{-2}.$$

▶▷ **20.** *Prove that the equation of sphere which lies in the octant OXYZ and touches the coordinate planes is of the form $x^2+y^2+z^2-2k(x+y+z)+2k^2=0$.*

☐ Let the equation of the sphere be

$$x^2+y^2+z^2-2ux-2vy-2wz+d=0.$$

Since it touches the $x=0$ plane, length of the perpendicular from the centre (u, v, w) to the $x=0$ plane will be equal to its radius, which gives

$$\frac{u}{1}=\sqrt{u^2+v^2+w^2-d}\quad\Rightarrow\quad v^2+w^2=d.$$

Similarly, from the conditions of tangency to the planes $y=0$ and $z=0$, we get

$$u^2+w^2=d\quad\text{and}\quad v^2+u^2=d.$$

Solving we get

$$u = v = w = k, \quad \text{where } d = 2k^2.$$

So, equation of the required circle can be written in the form

$$x^2 + y^2 + z^2 - 2k(x + y + z) + 2k^2 = 0.$$

▶▷▶ **21.** *Two spheres of radii r_1 and r_2 cut orthogonally. Prove that radius of the common circle is*

$$\frac{r_1 r_2}{\sqrt{r_1^2 + r_2^2}}.$$

□ Let the equation of the common circle be

$$x^2 + y^2 = a^2; \quad z = 0.$$

Equations of spheres through the above circle can be written as

$$x^2 + y^2 + z^2 - 2kz - a^2 = 0$$

$$\text{and} \quad x^2 + y^2 + z^2 - 2k'z - a^2 = 0.$$

The above spheres will cut orthogonally if

$$2kk' = -a^2 - a^2$$

$$\text{or,} \quad k^2 k'^2 = a^4$$

$$\text{or,} \quad (a^2 - r_1^2)(a^2 - r_2^2) = a^4$$

$$\text{or,} \quad a^2(r_1^2 + r_2^2) = r_1^2 r_2^2$$

$$\text{or,} \quad a = \frac{r_1 r_2}{\sqrt{r_1^2 + r_2^2}}.$$

▶▷▶ **22.** *Find the equation of the sphere which passes through the circle $z = 0$; $(x - a)^2 + (y - b)^2 = r^2$ and touches the plane $y = 0$. Show that the area which it cuts off from the plane $x = 0$ is $\pi(b^2 - a^2)$.*

□ Let the equation of the sphere through the circle $z = 0$; $(x - a)^2 + (y - b)^2 = r^2$ be

$$x^2 + y^2 + z^2 + 2ux + 2vy + 2wz + d = 0.$$

Putting $z = 0$ in it we get $x^2 + y^2 + 2ux + 2vy + d = 0$. Comparing it with the given circle we get $u = -a$; $v = -b$ and $d = a^2 + b^2 - r^2$.

Hence, the equation of the sphere through the given circle is

$$x^2 + y^2 + z^2 - 2ax - 2by + 2wz + a^2 + b^2 - r^2 = 0.$$

Putting $y = 0$ in the above equation we get

$$x^2 + z^2 - 2ax + 2wz + a^2 + b^2 - r^2 = 0$$

$$\text{or,} \quad (x - a)^2 + (z + w)^2 = w^2 + r^2 - b^2.$$

The above equation shows that the sphere touches the plane $y = 0$ if $w = \sqrt{b^2 - r^2}$. Hence the equation of the sphere which passes through the circle $z = 0$; $(x - a)^2 + (y - b)^2 = r^2$ and touches the plane $y = 0$ is

$$x^2 + y^2 + z^2 - 2ax - 2by + 2\sqrt{b^2 - r^2}z + a^2 + b^2 - r^2 = 0.$$

Equation of the circle in which the plane $x = 0$ cuts the above sphere is

$$y^2 + z^2 - 2by + 2\sqrt{b^2 - r^2}z + a^2 + b^2 - r^2 = 0$$

$$\text{or,} \quad (y - b)^2 + (z + \sqrt{b^2 - r^2})^2 = b^2 - a^2.$$

whose area is $\pi(b^2 - a^2)$.

EXERCISE

1. Find the equation of sphere describing the line segment joining (1, 2, 0) and (2, 0, 3) as diameter. Discuss the position of the point (0, 2, 1) with respect to this sphere.
2. Find the equation of the sphere for which $x^2+y^2+z^2+7y-2z+2 = 0$; $2x+3y+4z = 8$ is a great circle.
3. Show that if all the plane sections of a surface, represented by a second degree equation, are circles, then the surface is a sphere.

   [*Hints*: Let the surface be

   $$ax^2 + by^2 + cz^2 + 2hxy + 2gxz + 2fyz + 2ux + 2vy + 2wz + d = 0.$$

   Consider the coordinate planes, i.e. $x = 0$; $y = 0$ and $z = 0$. Putting $x = 0$ in the equation of the surface we get

   $$by^2 + cz^2 + 2fyz + 2vy + 2wz + d = 0.$$

   This will be a circle if $b = c$ and $f = 0$. Similarly, we get $a = c$; $g = 0$ and $a = b$; $h = 0$.]
4. Obtain the equation of the circle lying on the sphere

   $$x^2 + y^2 + z^2 - 2x + 2y - 4z + 3 = 0$$

   and having its centre at the point (2, 2, −3).
5. Find the equation of the sphere which passes through the circle $x^2 + y^2 = 4$; $z = 0$ and is cut by the plane $x + 2y + 2z = 0$ in a circle of radius 3.
6. Find the equation of the sphere which passes through the points (3, 1, −3); (−2, 4, 1); (−5, 0, 0) and whose centre lies on the plane $2x + y - z + 3 = 0$.
7. Find the equation of sphere which makes equal intercepts on the axes and has least possible radius.

   [*Hints*: Let the intercepted length on the axes be k. If we take the equation of sphere as

   $$(x - a)^2 + (y - b)^2 + (z - c)^2 = r^2,$$

   then it will pass through the points $(k, 0, 0)$; $(0, k, 0)$ and $(0, 0, k)$. Thus we get

   $$(k - a)^2 + b^2 + c^2 = r^2;$$

   $$a^2 + (k - b)^2 + c^2 = r^2$$

   $$\text{and} \quad a^2 + b^2 + (k - c)^2 = r^2.$$

The above equations give $a = b = c$. So,

$$(k-a)^2 + a^2 + a^2 = r^2$$

$$\text{or,}\quad 3\left[\left(a - \frac{k}{3}\right)^2 + \frac{2k^2}{9}\right] = r^2$$

which gives the minimum value of $r^2 = 2k^2/3$ which occurs when $a = k/3$, etc.]

8. A sphere of constant radius $2a$ passes through the origin O and meets the axes in A, B, C. Show that the locus of the centroid of the tetrahedron $OABC$ is the sphere

$$x^2 + y^2 + z^2 = a^2.$$

[*Hints*: Let the sphere meets the axes at $A(a', 0, 0)$; $B(0, b', 0)$ and $C(0, 0, c')$. So equation of the sphere $OABC$ is

$$x^2 + y^2 + z^2 - a'x - b'y - c'z = 0$$

where $a'^2 + b'^2 + c'^2 = 16a^2$. If (f, g, h) be the centroid of the tetrahedron, then

$$f = \frac{a' + 0 + 0 + 0}{4} = \frac{a'}{4}, \quad \text{i.e. } a' = 4f.$$

Similarly, $b' = 4g$ and $c' = 4h$. Hence, $f'^2 + g'^2 + h'^2 = a^2$. Generalizing we get the result.]

9. A sphere of constant radius k passes through the origin O and meets the axes in A, B, C. Show that the locus of the centroid of the triangle ABC is the sphere

$$9(x^2 + y^2 + z^2) = 4k^2.$$

10. A variable sphere which passes through the origin O and meets the axes at A, B, C, so that the volume of the tetrahedron $OABC$ is equal to k. Find the locus of the centre of the sphere.
11. Show that the diameter of the circumcircle of the triangle whose vertices are $(a, 0, 0)$; $(0, b, 0)$ and $(0, 0, c)$ is

$$\sqrt{\frac{(b^2+c^2)(c^2+a^2)(a^2+b^2)}{b^2c^2 + c^2a^2 + a^2b^2}}.$$

[*Hints*: Proceed as Example 11.]

12. A variable plane is parallel to the plane $\frac{x}{a} + \frac{y}{b} + \frac{z}{c} = 0$ and meets the axes at A, B, C respectively. Prove that the circle ABC lies on the surface

$$yz\left(\frac{b}{c} + \frac{c}{b}\right) + yx\left(\frac{b}{a} + \frac{a}{b}\right) + xz\left(\frac{a}{c} + \frac{c}{a}\right) = 0.$$

[*Hints*: If we take the plane as $\frac{x}{a} + \frac{y}{b} + \frac{z}{c} = k$, then the points A, B, C are $(ak, 0, 0)$; $(0, bk, 0)$ and $(0, 0, ck)$. Hence equation of the sphere $OABC$ is

$$x^2 + y^2 + z^2 - akx - bky - ckz = 0.$$

So, the equation of the circumcircle of the triangle ABC is

$$x^2 + y^2 + z^2 - k(ax + by + cz) = 0; \quad \frac{x}{a} + \frac{y}{b} + \frac{z}{c} = k.$$

To get the equation of the surface generated by the circle, we eliminate k, etc.]

13. A is a point on OX and B is a point on OY, so that $\angle OAB = \lambda$ (constant). On AB as diameter a circle is drawn whose plane is parallel to OZ. Prove that as AB varies, the circle generates the surface $2xy = z^2 \sin 2\lambda$.

[*Hints*: Let us take the point A and B as $(a, 0, 0)$ and $(0, b, 0)$, then $\tan\lambda = \dfrac{b}{a}$. Now equation of the plane through A, B and parallel to z-axis is

$$\frac{x}{a} + \frac{y}{b} = 1$$

and equation of the sphere, describing AB as diameter, is

$$x(x - a) + y(y - b) + z^2 = 0.$$

So, the equation of the circle on AB as diameter and parallel to OZ is

$$x^2 + y^2 + z^2 = ax + by; \quad \frac{x}{a} + \frac{y}{b} = 1.$$

To get the surface generated by the above circle, we eliminate the parameters a, b as follows:

$$x^2 + y^2 + z^2 = (ax + by)\left(\frac{x}{a} + \frac{y}{b}\right)$$

or, $$z^2 = xy\left(\frac{b}{a} + \frac{a}{b}\right) = xy(\tan\lambda + \cot\lambda)$$

or, $$2xy = z^2 \sin(2\lambda).]$$

14. Find the locus of the centre of the sphere which passes through a given point and touch a given plane.
15. Find the equation of the spheres which touch the coordinate planes. How many such spheres can be drawn?
16. Show that through the line $3x - 4y - 8 = 0 = y - 3x + 2$ only one tangent plane can be drawn to the sphere $x^2 + y^2 + z^2 - 2x + 6y + 2z + 8 = 0$.
17. Find the sphere of smallest radius that touches the straight lines

$$\frac{x-2}{1} = \frac{y-1}{-2} = \frac{z-6}{1} \quad \text{and} \quad \frac{x+3}{7} = \frac{y+3}{-6} = \frac{z+3}{1}.$$

[*Hints*: Using the condition of coplanarity, we see that the given lines are skew, so the sphere describing the shortest distance between them as diameter will be the sphere of small radius which touches the given line.]

18. Prove that the centres of the spheres which touch the lines $y = \pm mx$; $z = \pm c$ lie on the surface $mxy + (1 + m^2)cz = 0$.

[*Hints*: Let the equation of the sphere be

$$x^2 + y^2 + z^2 - 2ux - 2vy - 2wz + d = 0.$$

Since it touches the line $y = mx; z = c$, the following equation will have two same root:

$$x^2(1+m^2) - 2(u+vm)x + c^2 - 2wc + d = 0$$

which gives

$$4(u+vm)^2 = 4(1+m^2)(c^2 - 2wc + d)$$

i.e., $$(u+vm)^2 = (1+m^2)(c^2 - 2wc + d). \quad (1)$$

As the sphere also touches the line $y = -mx; z = -c$, we have

$$(u-vm)^2 = (1+m^2)(c^2 + 2wc + d). \quad (2)$$

Subtracting (2) from (1) we get

$$4uvm = (1+m^2)(-4wc)$$

or, $$uvm + (1+m^2)wc = 0.$$

Generalizing, we get the locus of the centre of the sphere, i.e. of (u, v, w) is

$$mxy + (1+m^2)cxz = 0.]$$

19. Find the equation of the sphere which makes equal intercepts on the axes and touches the plane $2x + 2y + z = 9$.
20. Find the locus of the centre of the sphere which passes through the fixed point $(0, 0, \pm c)$ and cuts the lines $y = \pm x \tan\alpha; z = \pm c$ at two points A, B, such that $AB = 2a$.

[*Hints*: Let the equation of the sphere be

$$x^2 + y^2 + z^2 - 2ux - 2vy - 2wz + d = 0.$$

As it passes through $(0, 0, \pm c)$, we have $c^2 \pm 2cw + d = 0$, i.e. $w = 0$ and $d = -c^2$. The given equations of the lines can be written as

$$\frac{x-0}{1} = \frac{y-0}{\tan\alpha} = \frac{z-c}{0} = r \quad \text{and} \quad \frac{x-0}{1} = \frac{y-0}{-\tan\alpha} = \frac{z+c}{0} = r'.$$

Any point on the lines can be taken as $(r, r\tan\alpha, c)$ and $(r', -r'\tan\alpha, -c)$. Given that

$$(r-r')^2 + (r+r')^2 \tan^2\alpha + 4c^2 = 4a^2 \quad (1)$$

Again, since the points $(r, r\tan\alpha, c)$ and $(r', -r'\tan\alpha, -c)$ lie on the sphere, we get

$$r^2 + r^2\tan^2\alpha - 2ur - 2vr\tan\alpha = 0 \quad \Rightarrow \quad r = 2(u + v\tan\alpha)\cos^2\alpha$$

and $$r'^2 + r'^2\tan^2\alpha - 2ur' + 2vr'\tan\alpha = 0 \quad \Rightarrow \quad r' = 2(u - v\tan\alpha)\cos^2\alpha.$$

Using these values of r and r' in (1), we get

$$u^2 + v^2 = (a^2 - c^2)\operatorname{cosec}^2 2\alpha.$$

Hence, locus of the centre of the sphere, i.e. of (u, v, w) is $x^2 + y^2 = (a^2 - c^2)\operatorname{cosec}^2 2\alpha$; $z = 0$.]

21. If d be the distance between the centres of two spheres of radii r_1 and r_2, prove that the angle between them is $\cos^{-1}\left(\frac{r_1^2 + r_2^2 - d^2}{2r_1r_2}\right)$.
22. Prove that the sphere which cuts the spheres $S = 0$ and $S' = 0$ at right angles also cuts the sphere $S + kS' = 0$ at right angle.
23. Find the locus of the centres of the spheres of constant radius which pass through a fixed point and touch a given line.

    [*Hints*: Without loss of generality we take the fixed point as $(0, 0, c)$ and x-axis as the fixed line. Then for the general equation of sphere

$$x^2 + y^2 + z^2 + 2ux + 2vy + 2wz + d = 0$$

    which has constant radius k, we have $c^2 + 2wc + d = 0$; $u^2 + v^2 + w^2 - d = k^2$. Again, it touches x-axis, so the equation $x^2 + 2xu + d = 0$ must have equal roots, which gives $u^2 = d$. Hence $v^2 + w^2 = k^2$ and $u^2 + 2wc + c^2 = 0$. Generalizing we get the locus.]

Chapter 6

Cone

6.1 Definition

A cone is a surface generated by a moving line which passes through a fixed point and intersects a given curve or touches a fixed surface.

The fixed point is called its vertex and the fixed curve is called its guiding curve or base. The line which generates the cone is called its generator.

6.2 Equation of Cone when the Vertex and the Base is Given

(i) Let (α, β, γ) be the vertex and $f(x, y) = 0$; $z = 0$ be the base of the cone. Then its generator's equation is

$$\frac{x-\alpha}{l} = \frac{y-\beta}{m} = \frac{z-\gamma}{n} \tag{1}$$

where l, m, n, the direction ratios of the generator, are variables.

Since the generator intersects the curve $f(x, y) = 0$; $z = 0$, then points $\left(\alpha - \frac{l\gamma}{n}, \beta - \frac{m\gamma}{n}\right)$ will satisfy the equation $f(x, y) = 0$. Thus we get

$$f\left(\alpha - \frac{l\gamma}{n}, \beta - \frac{m\gamma}{n}\right) = 0$$

$$\text{or,} \quad f\left(\alpha - \frac{(x-\alpha)\gamma}{(z-\gamma)}, \beta - \frac{(y-\beta)\gamma}{(z-\gamma)}\right) = 0 \quad \text{by Eqn. (1).}$$

This is the equation of the required cone.

(ii) If we take vertex as origin and the guiding curve as $f(x, y) = 0$; $z = k$, then we take the equation of generator as

$$\frac{x}{l} = \frac{y}{m} = \frac{z}{n} = r.$$

Any point on the line can be taken as (lr, mr, nr). Since the generator intersects the curve $f(x, y) = 0$; $z = k$, then points $\left(\frac{lk}{n}, \frac{mk}{n}\right)$ will satisfy the equation $f(x, y) = 0$. Thus we get

$$f\left(\frac{lk}{n}, \frac{mk}{n}\right) = 0$$

$$\text{or,}\quad f\left(\frac{xk}{z}, \frac{yk}{z}\right) = 0.$$

This is the equation of the required cone.

6.3 Equation of a Cone with Vertex as Origin

Let the general second degree equation in x, y, z i.e.,

$$ax^2 + by^2 + cz^2 + 2hxy + 2gxz + 2fyz + 2ux + 2vy + 2wz + d = 0$$

represent a cone with vertex as origin, O. If $P(\alpha, \beta, \gamma)$ be any point on the cone, then equation of the generator OP is

$$\frac{x}{\alpha} = \frac{y}{\beta} = \frac{z}{\gamma} = r.$$

Which shows that the point $(r\alpha, r\beta, r\gamma)$ lies on the cone. Hence we get

$$r^2(a\alpha^2 + b\beta^2 + c\gamma^2 + 2h\alpha\beta + 2g\alpha\gamma + 2f\beta\gamma) + r(2u\alpha + 2v\beta + 2w\gamma) + d = 0.$$

For any value of r this gives

$$a\alpha^2 + b\beta^2 + c\gamma^2 + 2h\alpha\beta + 2g\alpha\gamma + 2f\beta\gamma = 0; \qquad (2)$$

$$u\alpha + v\beta + w\gamma = 0; \qquad (3)$$

$$\text{and}\quad d = 0. \qquad (4)$$

Equation (3) shows that (α, β, γ) lies in the plane $ux + vy + wz = 0$ unless $u = v = w = 0$, which is a contradiction. So equation of the cone becomes

$$ax^2 + by^2 + cz^2 + 2hxy + 2gxz + 2fyz = 0, \qquad (5)$$

which is a second degree homogeneous equation in x, y, z. Hence the equation of a cone with vertex as origin is a second degree homogeneous equation in x, y, z.

From Eqn. (2) we see that the direction ratios of the generator satisfy the equation of the cone.

Conversely, if $P(x', y', z')$ be any point on the surface whose equation is given by Eqn. (5), then we have

$$ax'^2 + by'^2 + cz'^2 + 2hx'y' + 2gx'z' + 2fy'z' = 0$$

$$\text{or,}\quad a(rx')^2 + b(ry')^2 + c(rz')^2 + 2hrx'ry' + 2grx'rz' + 2fry'rz' = 0.$$

From this we see that the point (rx', ry', rz'), which is the general coordinate of a point on the line OP, satisfies the Eqn. (5). So the line OP lies on the surface whose equation is given by (5). Thus the Eqn. (5) represents a cone whose vertex is O i.e., origin.

It is to be noted further Eqn. (5) represents a pair of planes passing through the origin if

$$\begin{vmatrix} a & h & g \\ h & b & f \\ g & f & c \end{vmatrix} \neq 0.$$

6.4 Condition for the General Second Degree Equation Represents a Cone

The general second degree equation in x, y, z is given by

$$ax^2 + by^2 + cz^2 + 2hxy + 2gxz + 2fyz + 2ux + 2vy + 2wz + d = 0$$

where at least one of a, b, c, f, g, h is non-zero.

If it represents a cone with vertex at (α, β, γ) then shifting origin to the vertex we get

$$\begin{aligned} & a(x+\alpha)^2 + b(y+\beta)^2 + c(z+\gamma)^2 + 2h(x+\alpha)(y+\beta) + 2g(x+\alpha)(z+\gamma) \\ & \quad + 2f(y+\beta)(z+\gamma) + 2u(x+\alpha) + 2v(y+\beta) + 2w(z+\gamma) + d = 0 \\ \text{or,} \quad & ax^2 + by^2 + cz^2 + 2hxy + 2gxz + 2fyz + 2x(a\alpha + h\beta + g\gamma + u) \\ & \quad + 2y(h\alpha + b\beta + f\gamma + v) + 2z(g\alpha + f\beta + c\gamma + w) + a\alpha^2 + b\beta^2 + c\gamma^2 \\ & \quad + 2h\alpha\beta + 2g\alpha\gamma + 2f\beta\gamma + 2u\alpha + 2v\beta + 2w\gamma + d = 0. \end{aligned}$$

This is the equation of a cone whose vertex is origin, so we have

$$a\alpha + h\beta + g\gamma + u = 0, \tag{6}$$

$$h\alpha + b\beta + f\gamma + v = 0, \tag{7}$$

$$g\alpha + f\beta + c\gamma + w = 0 \tag{8}$$

and $$a\alpha^2 + b\beta^2 + c\gamma^2 + 2h\alpha\beta + 2g\alpha\gamma + 2f\beta\gamma + 2u\alpha + 2v\beta + 2w\gamma + d = 0. \tag{9}$$

Multiplying (6), (7), (8) by respectively a, b, c and subtracting their sum from (9) we get

$$u\alpha + v\beta + w\gamma + d = 0. \tag{10}$$

Eliminating α, β, γ from (6), (7), (8) and (10) we get

$$\begin{vmatrix} a & h & g & u \\ h & b & f & v \\ g & f & c & w \\ u & v & w & d \end{vmatrix} = 0.$$

which is the required condition. If this condition is satisfied then solving any three of the (6), (7), (8) and (10) we get the vertex of the cone.

6.5 General Equation of the Cone Passing Through the Coordinate Axes

As the axes meet at the origin, the vertex of such cone will be origin. Also from Section 6.3 we have see that the direction ratios of the generator satisfy the equation of the cone. If we take the equation of the cone as

$$ax^2 + by^2 + cz^2 + 2hxy + 2gxz + 2fyz = 0,$$

then direction cosines of the axes i.e., (1, 0, 0), (0, 1, 0) and (0, 0, 1) will satisfy the above equation. This gives

$$a = b = c = 0.$$

Thus the general equation of the cone passing throu; h the coordinate axes is

$$hxy + gxz + fyz = 0.$$

6.6 Intersection of a Cone with a Plane

Let vertex of the cone be origin. If the plane $ux + vy + wz = 0$ cuts the cone $\phi(x, y, z) = ax^2+by^2+cz^2+2hxy+2gxz+2fyz = 0$ along the lines whose equations are of the form

$$\frac{x}{\alpha} = \frac{y}{\beta} = \frac{z}{\gamma}.$$

Then we have

$$u\alpha + v\beta + w\gamma = 0 \tag{11}$$

$$\text{and} \quad a\alpha^2 + b\beta^2 + c\gamma^2 + 2h\alpha\beta + 2g\alpha\gamma + 2f\beta\gamma = 0. \tag{12}$$

Eliminating γ from the above equations we get

$$a\alpha^2 + b\beta^2 + c\left(\frac{u\alpha + v\beta}{w}\right)^2 + 2h\alpha\beta - 2g\alpha\left(\frac{u\alpha + v\beta}{w}\right) - 2f\beta\left(\frac{u\alpha + v\beta}{w}\right) = 0$$

$$\text{or,} \quad (cu^2 + aw^2 - 2guw)\alpha^2 + 2(hw^2 + cuv - gvw - fuw)\alpha\beta + (cv^2 + bw^2 - 2fvw)\beta^2 = 0$$

$$\text{or,} \quad (cu^2 + aw^2 - 2guw)\left(\frac{\alpha}{\beta}\right)^2 + 2(hw^2 + cuv - gvw - fuw)\left(\frac{\alpha}{\beta}\right) + (cv^2 + bw^2 - 2fvw) = 0. \tag{13}$$

This is a quadratic equation in $\left(\frac{\alpha}{\beta}\right)$, giving two value of $\left(\frac{\alpha}{\beta}\right)$ corresponding to which there are two lines of the intersection of the plane and cone. If the roots of the above equation be $\left(\frac{\alpha_1}{\beta_1}\right)$ and $\left(\frac{\alpha_2}{\beta_2}\right)$, then solving the equation for $\left(\frac{\alpha_1}{\beta_1}\right)$ and $\left(\frac{\alpha_2}{\beta_2}\right)$ and using the result (11) we get the values of $\left(\frac{\gamma_1}{\beta_1}\right)$ and $\left(\frac{\gamma_2}{\beta_2}\right)$. In this way we get the equations of the line of intersection.

From Eqn. (13), we get

$$\frac{\alpha_1\alpha_2}{\beta_1\beta_2} = \frac{cv^2 + bw^2 - 2fvw}{aw^2 + cu^2 - 2guw}$$

and from symmetry

$$\frac{\alpha_1\alpha_2}{\gamma_1\gamma_2} = \frac{cv^2 + bw^2 - 2fvw}{av^2 + bu^2 - 2huv}.$$

So,

$$\frac{\alpha_1\alpha_2}{cv^2+bw^2-2fvw}=\frac{\beta_1\beta_2}{aw^2+cu^2-2guw}=\frac{\gamma_1\gamma_2}{av^2+bu^2-2huv}=k$$

and from this we get

$$\begin{aligned}\alpha_1\alpha_2+\beta_1\beta_2+\gamma_1\gamma_2 &\\ &= k[a(v^2+w^2)+b(u^2+w^2)+c(v^2+u^2)-2(fvw+guw+huv)]\\ &= k[(a+b+c)(u^2+v^2+w^2)-\phi(u,v,w)].\end{aligned}$$

If the plane $ux+vy+wz=0$ cuts the cone $ax^2+by^2+cz^2+2hxy+2gxz+2fyz=0$ along two perpendicular generators then we get

$$(a+b+c)(u^2+v^2+w^2)-\phi(u,v,w)=0. \tag{14}$$

Further equation of normal to the plane through the origin is

$$\frac{x}{u}=\frac{y}{v}=\frac{z}{w} \tag{15}$$

which is also a generator of the cone $\phi(x,y,z)=0$, so we have $\phi(u,v,w)=0$. Hence from Eqn. (14) we get

$$a+b+c=0 \quad \text{as} \quad u^2+v^2+w^2\neq 0.$$

The above condition is independent of u, v, w. So the plane $ux+vy+wz=0$ cuts the cone $ax^2+by^2+cz^2+2hxy+2gxz+2fyz=0$ in two perpendicular generators which are also perpendicular to the generator given by the Eqn. (15). Hence the given cone will have three mutually perpendicular generators if

$$\begin{aligned}&a+b+c=0\\ \text{i.e.,}\quad &\text{Coeff. of } x^2+\text{Coeff. of } y^2+\text{Coeff. of } z^2=0\end{aligned}$$

and there exists infinite number of sets of three mutually perpendicular generators as the condition is independent of u, v, w.

6.7 Right Circular Cone

The surface generated by a moving line which passes through a fixed point and makes a constant angle with a fixed line through that fixed point is called *right circular cone*. The fixed point is called vertex and the fixed line is called its axis. The angle which the moving line makes with the axis of the cone is called semi-vertical angle.

If l, m, n be the direction ratios of the axis and θ be the semi-vertical angle and $O(\alpha,\beta,\gamma)$ be the vertex of the cone. Then for any point $P(x,y,z)$ on the cone, the direction ratios of OP are $x-\alpha, y-\beta, z-\gamma$. So

$$\cos\theta=\frac{l(x-\alpha)+m(y-\beta)+n(z-\gamma)}{\sqrt{l^2+m^2+n^2}\sqrt{(x-\alpha)^2+(y-\beta)^2+(z-\gamma)^2}}$$

Hence locus of (x,y,z) i.e., equation of the right circular cone is

$$\begin{aligned}&\{l(x-\alpha)+m(y-\beta)+n(z-\gamma)\}^2\\ &=(l^2+m^2+n^2)\{(x-\alpha)^2+(y-\beta)^2+(z-\gamma)^2\}\cos^2\theta.\end{aligned}$$

Particular Cases

Case I: If we take the vertex of the cone as origin then the above equation becomes

$$\{lx + my + nz\}^2 = (l^2 + m^2 + n^2)\{x^2 + y^2 + z^2\}\cos^2\theta \tag{16}$$

Case II: If the vertex be the point (α, β, γ) and the axis of the cone is any coordinate axis, say x-axis, then putting $l = 1, m = 0, n = 0$ we get

$$(x-\alpha)^2 = \{(x-\alpha)^2 + (y-\beta)^2 + (z-\gamma)^2\}\cos^2\theta$$

$$\Rightarrow \quad (y-\beta)^2 + (z-\gamma)^2 = (x-\alpha)^2\tan^2\theta$$

Case III: If the vertex be origin and the axis of the cone is any coordinate axis, say x-axis, then putting $\alpha = 0, \beta = 0, \gamma = 0$ we get

$$y^2 + z^2 = x^2\tan^2\theta.$$

6.8 Condition that the General Second Degree Homogeneous Equation Represents a Right Circular Cone

Consider the second degree homogeneous equation

$$ax^2 + by^2 + cz^2 + 2hxy + 2gxz + 2fyz = 0 \tag{17}$$

and the equation of a right circular cone

$$\{lx + my + nz\}^2 = (l^2 + m^2 + n^2)\{x^2 + y^2 + z^2\}\cos^2\theta.$$

Taking the semi-vertical angle as

$$\theta = \cos^{-1}(l^2 + m^2 + n^2)^{-1/2}$$

the above equation can be written as

$$\{lx + my + nz\}^2 = \{x^2 + y^2 + z^2\}. \tag{18}$$

Comparing Eqns. (17) and (18) we get

$$\frac{a}{1-l^2} = \frac{b}{1-m^2} = \frac{c}{1-n^2} = \frac{h}{-lm} = \frac{g}{-ln} = \frac{f}{-mn} = k \text{ (say)}.$$

From this we get

$$f = -kmn; \; g = -kln; \; h = -klm; \; a = k(1-l^2); \; b = k(1-m^2); \; c = k(1-n^2).$$

This gives

$$\frac{gh}{f} = -kl^2, \quad \frac{fh}{g} = -km^2, \quad \frac{gf}{h} = -kn^2$$

and

$$af - gh = k^2[-mn(1-l^2) - mnl^2] = -k^2mn = kf,$$

$$bg - fh = kg, \quad ch - fg = kh$$

$$\text{or,} \quad \frac{F}{f} = \frac{G}{g} = \frac{H}{h}$$

This is the required condition, where F, G, H are the cofactors of f, g, h respectively in the determinant

$$\begin{vmatrix} a & h & g \\ h & b & f \\ g & f & c \end{vmatrix}.$$

Examples

▶▷▶ 1. *Find the equation of the cone whose vertex is origin and the base is the curve $x^2 + z^2 = a^2$; $y = b$. Show that the plane $y = k$ cuts the cone in a circle.*

☐ Making homogeneous $x^2 + z^2 = a^2$ with the aid of $y = b$ as follows, we get

$$x^2 + z^2 = a^2\left(\frac{y}{b}\right)^2$$

$$\text{or,}\quad b^2(x^2 + z^2) - a^2y^2 = 0.$$

This is the equation of the required cone.

Putting $y = k$ in the above equation we get

$$x^2 + z^2 = \left(\frac{ak}{b}\right)^2$$

This is a circle in x-z plane.

▶▷▶ 2. *Find the equation of the right circular cone containing the coordinate axes.*

☐ In this case, vertex of the cone is origin and the axis of the cone makes equal angle with the coordinate axes, so the semi-vertical angle of the cone is

$$\cos^{-1}\left(\frac{1}{\sqrt{3}}\right)$$

and equation of the axis is $x = y = z$.

Using the formula

$$\{lx + my + nz\}^2 = (l^2 + m^2 + n^2)\{x^2 + y^2 + z^2\}\cos^2\theta.$$

we get

$$(x + y + z)^2 = x^2 + y^2 + z^2$$

$$\text{or,}\quad xy + yz + xz = 0.$$

▶▷▶ 3. *Find the equation of the right circular cone whose vertex is origin and whose semi-vertical angle is 45° and the axis is the line*

$$\frac{x}{3} = \frac{y}{2} = \frac{z}{4}.$$

☐ Using the formula

$$\{lx + my + nz\}^2 = (l^2 + m^2 + n^2)\{x^2 + y^2 + z^2\}\cos^2\theta$$

we get

$$\{3x+2y+4z\}^2 = (3^2+2^2+4^2)\{x^2+y^2+z^2\}\cos^2 45^\circ$$

or, $$2(9x^2+4y^2+16z^2+12xy+16yz+24xz) = 29(x^2+y^2+z^2)$$

or, $$11x^2+21y^2-3z^2-24xy-32yz-48xz=0.$$

▶▷ 4. *Find the equation of the right circular cone whose vertex is the point* (1, –2, –1) *and whose semi-vertical angle is* 60° *and the axis is the line*

$$\frac{x-1}{3} = \frac{y+2}{-4} = \frac{z+1}{5}.$$

□ Using the formula

$$\{l(x-\alpha)+m(y-\beta)+n(z-\gamma)\}^2 = (l^2+m^2+n^2)\{(x-\alpha)^2+(y-\beta)^2+(z-\gamma)^2\}\cos^2\theta$$

we get

$$\{3(x-1)-4(y+2)+5(z+1)\}^2 = (3^2+4^2+5^2)\{(x-1)^2+(y+2)^2+(z+1)^2\}\cos^2 60^\circ$$

or, $$2\{3(x-1)-4(y+2)+5(z+1)\}^2 = 25\{(x-1)^2+(y+2)^2+(z+1)^2\}$$

or, $$2\{3x-4y+5z-6\}^2 = 25\{(x-1)^2+(y+2)^2+(z+1)^2\}$$

or, $$7x^2-7y^2-25z^2+48xy+80yz-60xz+22x+4y+170z+78=0.$$

▶▷ 5. *Find the equation of the cone whose vertex is the origin and which passes through the curve of intersection of* $4x^2-5y^2-7z^2=2$; $3x-2y+4z=3$.

□ Any generator of the cone is

$$\frac{x}{l} = \frac{y}{m} = \frac{z}{n} = \frac{3}{3l-2m+4n}$$

From this we get

$$x = \frac{3l}{3l-2m+4n}, \quad y = \frac{3m}{3l-2m+4n}, \quad \text{and} \quad z = \frac{3n}{3l-2m+4n}.$$

If this lies on the curve $4x^2-5y^2-7z^2=2$, we get

$$4\left(\frac{3l}{3l-2m+4n}\right)^2 - 5\left(\frac{3m}{3l-2m+4n}\right)^2 - 7\left(\frac{3n}{3l-2m+4n}\right)^2 = 2$$

or, $$36l^2-45m^2-63n^2 = 2(3l-2m+4n)^2$$

or, $$18l^2-53m^2-95n^2+24lm+32mn-48ln=0.$$

Eliminating l, m, n we get

$$18x^2-53y^2-95z^2+24xy+32yz-48xz=0.$$

▶▷▶ 6. *Find the equation of the cone whose vertex is origin and which contains the curve given by* $x^2 - y^2 + 4ax = 0;\ x + y + z = b$.

☐ Any generator of the cone is

$$\frac{x}{l} = \frac{y}{m} = \frac{z}{n} = \frac{b}{l+m+n}.$$

From this we get

$$x = \frac{bl}{l+m+n}, \quad y = \frac{bm}{l+m+n} \quad \text{and} \quad z = \frac{bn}{l+m+n}.$$

If this lies on the curve $x^2 - y^2 + 4ax = 0$, we get

$$\left(\frac{bl}{l+m+n}\right)^2 - \left(\frac{bm}{l+m+n}\right)^2 + 4a\left(\frac{bl}{l+m+n}\right) = 0$$

$$\text{or,} \quad b(l^2 - m^2) + 4al(l+m+n) = 0.$$

Eliminating l, m, n we get

$$b(x^2 - y^2) + 4ax(x + y + z) = 0.$$

▶▷▶ 7. *A variable plane is parallel to the plane* $\frac{x}{a} + \frac{y}{b} + \frac{z}{c} = 0$ *and meets the axes at A, B, C respectively. Prove that the circle ABC lies on the surface*

$$yz\left(\frac{b}{c} + \frac{c}{b}\right) + yx\left(\frac{b}{a} + \frac{a}{b}\right) + xz\left(\frac{a}{c} + \frac{c}{a}\right) = 0.$$

☐ If we take the plane as $\frac{x}{a} + \frac{y}{b} + \frac{z}{c} = k$, then the points A, B, C are $(ak, 0, 0)$; $(0, bk, 0)$ and $(0, 0, ck)$. Hence equation of the sphere $OABC$ is

$$x^2 + y^2 + z^2 - akx - bky - ckz = 0.$$

So the equation of the circumcircle of the triangle ABC is

$$x^2 + y^2 + z^2 - k(ax + by + cz) = 0; \quad \frac{x}{a} + \frac{y}{b} + \frac{z}{c} = k.$$

To get the equation of the cone, we eliminate k and get

$$x^2 + y^2 + z^2 - \left(\frac{x}{a} + \frac{y}{b} + \frac{z}{c}\right)(ax + by + cz) = 0$$

$$\text{or,} \quad yz\left(\frac{b}{c} + \frac{c}{b}\right) + yx\left(\frac{b}{a} + \frac{a}{b}\right) + xz\left(\frac{a}{c} + \frac{c}{a}\right) = 0.$$

▶▷▶ 8. *Find the equation of the cone whose vertex is origin and base is the curve*

$$ax^2 + by^2 + cz^2 - 1; \quad \alpha x^2 + \beta y^2 - 2z.$$

☐ Making homogeneous $ax^2 + by^2 + cz^2 - 1$ with the help of $\alpha x^2 + \beta y^2 - 2z$ we get

$$ax^2 + by^2 + cz^2 = \left(\frac{\alpha x^2 + \beta y^2}{2z}\right)^2$$

$$\text{or,} \quad 4z^2(ax^2 + by^2 + cz^2) = (\alpha x^2 + \beta y^2)^2$$

which is the equation of the required cone.

▶▷▶ **9**. *A variable plane through the x-axis and a variable plane through y-axis are inclined at a constant angle α. Prove that their line of intersection generates the cone*

$$(x^2 + y^2 + z^2)z^2 = x^2y^2 \tan^2 \alpha.$$

□ Equation of the plane through x-axis is $y + \lambda z = 0$ and equation of the plane through y-axis is $x + \mu z = 0$. If the angle between them be α, then we have

$$\cos \alpha = \frac{\lambda\mu}{\sqrt{1+\lambda^2}\sqrt{1+\mu^2}}$$

$$\text{or,} \quad \sec^2 \alpha = \frac{1+\lambda^2+\mu^2+\lambda^2\mu^2}{\lambda^2\mu^2}$$

$$\text{or,} \quad \tan^2 \alpha = \frac{1+\lambda^2+\mu^2}{\lambda^2\mu^2}$$

Eliminating λ, μ we get

$$\tan^2 \alpha = \frac{1+\dfrac{y^2}{z^2}+\dfrac{x^2}{z^2}}{\dfrac{y^2x^2}{z^4}}$$

$$\text{or,} \quad (x^2 + y^2 + z^2)z^2 = x^2y^2 \tan^2 \alpha.$$

▶▷▶ **10**. *Find the equation of the cone which passes through the coordinate axes as well as the two lines*

$$\frac{x}{1} = \frac{y}{-2} = \frac{z}{3} \quad \textit{and} \quad \frac{x}{3} = \frac{y}{-1} = \frac{z}{1}$$

□ Equation of the cone which passes through the coordinate axes can be taken as

$$fyz + gxz + hxy = 0.$$

If it passes through the given lines then we have

$$f(-2)(3) + g(1)(3) + h(1)(-2) = 0$$

$$f(-1)(1) + g(1)(3) + h(-1)(3) = 0.$$

$$\text{i.e.,} \quad -6f + 3g - 2h = 0$$

$$\text{and} \quad -f + 3g - 3h = 0.$$

Solving we get

$$\frac{f}{-9+6} = \frac{g}{2-18} = \frac{h}{-18+3}$$

$$\text{or,} \quad \frac{f}{3} = \frac{g}{16} = \frac{h}{15}.$$

Hence the equation of the cone is

$$3yz + 16xz + 15xy = 0.$$

▶▷▶ **11**. *Find the equation of the straight lines in which the plane* $2x + y - z = 0$ *cuts the cone* $4x^2 - y^2 + 3z^2 = 0$.

☐ If l, m, n be the direction ratios of the line of intersection of the cone by the plane, then we have

$2l + m - n = 0$ (since 2, 1, −1 are the d.r.^{s} of the normal to the plane)

and $4l^2 - m^2 + 3n^2 = 0$. (since d.r.^{s} of the generator satisfy the equation of the cone)

Eliminating 'n' we get

$$4l^2 - m^2 + 3(2l + m)^2 = 0$$

$$\text{or,} \quad 8l^2 + 6lm + m^2 = 0$$

$$\text{or,} \quad (2l + m)(4l + m) = 0$$

$$\text{i.e.,} \quad \frac{l}{m} = -\frac{1}{2} \text{ or } -\frac{1}{4}.$$

Using these in $2l + m - n = 0$ we get

$$\frac{n}{m} = 0 \text{ or } \frac{1}{2}.$$

Thus the direction ratios of the lines of intersection are −1, 2, 0 and −1, 4, 2. So their equations are

$$\frac{x}{-1} = \frac{y}{2} = \frac{z}{0} \quad \text{and} \quad \frac{x}{-1} = \frac{y}{4} = \frac{z}{2}.$$

▶▷▶ **12**. *If the straight line* $\frac{x}{1} = \frac{y}{2} = \frac{z}{3}$ *represents one of a set of three mutually perpendicular generators of the cone* $5yz - 8xz - 3xy = 0$, *then find the equation of the other two.*

☐ Obviously the other two perpendicular generators will lie in the plane $x + 2y + 3z = 0$. If l, m, n be the direction ratios of the line of intersection of the cone by the above plane, then we have

$$l + 2m + 3n = 0 \quad \text{and} \quad 5mn - 8ln - 3lm = 0.$$

Eliminating l form these expressions we obtain

$$5mn + (2m + 3n)(3m + 8n) = 0$$

$$\text{or,} \quad m^2 + 5mn + 4n^2 = 0$$

$$\text{or,} \quad (m + n)(m + 4n) = 0$$

When $m + n = 0$, from $l + 2m + 3n = 0$, we get $l + n = 0$, so

$$\frac{l}{1} = \frac{m}{1} = \frac{n}{-1}.$$

Again when $m + 4n = 0$, from $l + 2m + 3n = 0$, we get $l - 5n = 0$, so

$$\frac{l}{5} = \frac{m}{-4} = \frac{n}{1}.$$

So the equations of other generators are

$$\frac{x}{1}=\frac{y}{1}=\frac{z}{-1} \quad \text{and} \quad \frac{x}{5}=\frac{y}{-4}=\frac{z}{1}.$$

▶▶ **13.** *Show that the angle between the lines in which the plane $x+y+z=0$ cuts the cone $(c-a)(a-b)yz+(b-c)(a-b)xz+(b-c)(c-a)xy=0$ is 60°.*

□ If l, m, n be the direction ratios of the line of intersection of the cone by the plane, then we have

$$l+m+n=0 \quad \text{and} \quad (c-a)(a-b)mn+(b-c)(a-b)ln+(b-c)(c-a)lm=0.$$

Eliminating n we get

$$[(c-a)(a-b)m+(b-c)(a-b)l](l+m)-(b-c)(c-a)lm=0$$

or, $$(b-c)(a-b)l^2+[(c-a)(a-b)+(b-c)(a-b)-(b-c)(c-a)]lm$$
$$+(c-a)(a-b)m^2=0$$

or, $$(b-c)(a-b)\left(\frac{l}{m}\right)^2-[(a-b)^2+(b-c)(c-a)]\left(\frac{l}{m}\right)+(c-a)(a-b)=0.$$

If the roots of the above equation be $\left(\frac{l_1}{m_1}\right)$ and $\left(\frac{l_2}{m_2}\right)$, then we have

$$\frac{l_1l_2}{m_1m_2}=\frac{c-a}{b-c}$$

or, $$\frac{l_1l_2}{(c-a)(a-b)}=\frac{m_1m_2}{(b-c)(a-b)}=\frac{n_1n_2}{(c-a)(b-c)}=k, \quad \text{by symmetry}$$

and

$$\frac{l_1}{m_1}+\frac{l_2}{m_2}=\frac{(a-b)^2+(b-c)(c-a)}{(b-c)(a-b)}$$

or, $$l_1m_2+l_2m_1=k[(a-b)^2+(b-c)(c-a)]$$

or, $$l_1m_2-l_2m_1=k\sqrt{[(a-b)^2+(b-c)(c-a)]^2-4(a-b)^2(b-c)(c-a)}$$
$$=k[(a-b)^2-(b-c)(c-a)].$$

If θ be the angle between the lines then

$$\tan\theta=\frac{\sqrt{\sum(l_1m_2-l_2m_1)^2}}{l_1l_2+m_1m_2+n_1n_2}$$

$$=\frac{\sqrt{\sum[(a-b)^2-(b-c)(c-a)]^2}}{(c-a)(a-b)+(b-c)(a-b)+(c-a)(b-c)}$$

$$=\frac{\sqrt{\sum[(a-b)^2-(b-c)(c-a)]^2}}{-(a-b)^2+(c-a)(b-c)}=\sqrt{3}.$$

Hence the angle between the lines of intersection is 60°.

Alternative Method

Take $f = (c-a)(a-b)$, $g = (b-c)(a-b)$ and $h = (b-c)(c-a)$, then

$$\frac{1}{f}+\frac{1}{g}+\frac{1}{h}=0.$$

Now proceed as the example 8 of Chapter 1.

▶▷▶ **14.** *Show that the locus of the point from which three mutually perpendicular straight lines can be drawn to intersect the conic*

$$ax^2+by^2=1, \quad z=0$$
is
$$ax^2+by^2+(a+b)z^2=1.$$

☐ Let $P(\alpha, \beta, \gamma)$ be the point from which the lines are drawn. Equation of the line through P is

$$\frac{x-\alpha}{l}=\frac{y-\beta}{m}=\frac{z-\gamma}{n}. \tag{1}$$

It intersects the plane $z=0$ at $\left(\alpha-\frac{l\gamma}{n}, \beta-\frac{m\gamma}{n}\right)$. If this point lies on $ax^2+by^2=1$, then

$$a\left(\alpha-\frac{l\gamma}{n}\right)^2+b\left(\beta-\frac{m\gamma}{n}\right)^2=1.$$

Eliminating l, m, n using (1) we get

$$a\left(\alpha-\frac{(x-\alpha)\gamma}{z-\gamma}\right)^2+b\left(\beta-\frac{(y-\beta)\gamma}{z-\gamma}\right)^2=1$$

or, $\quad a(\alpha z-x\gamma)^2+b(\beta z-y\gamma)^2=(z-\gamma)^2$

or, $\quad a\gamma^2x^2+b\gamma^2y^2+(a\alpha^2+b\beta^2-1)z^2-2a\alpha\gamma xz-2b\beta\gamma yz+2z\gamma-\gamma^2=0.$

If the above cone has three mutually perpendicular generators then,

$$\text{Coeff. of } x^2+\text{Coeff. of } y^2+\text{Coeff. of } z^2=0$$

i.e., $\quad a\alpha^2+b\beta^2+(a+b)\gamma^2-1=0.$

Hence locus of P is

$$ax^2+by^2+(a+b)z^2=1.$$

▶▷▶ **15.** *A cone has for its guiding curve the circle $x^2+y^2+2ax+2by=0$, $z=0$ and passes through a fixed point $(0, 0, c)$. If the section of the cone by the plane $x=0$ is a rectangular hyperbola, prove that the vertex lies on the fixed circle*

$$x^2+y^2+z^2+2ax+2by=0; \quad 2ax+2by+cz=0.$$

☐ If $P(\alpha, \beta, \gamma)$ be the vertex of the cone, then equation of the generator can be written as

$$\frac{x-\alpha}{l}=\frac{y-\beta}{m}=\frac{z-\gamma}{n}. \tag{1}$$

If it intersects the circle $x^2+y^2+2ax+2by=0$, $z=0$, then the point $\left(\alpha-\frac{l\gamma}{n}, \beta-\frac{m\gamma}{n}\right)$ will satisfy the equation $x^2+y^2+2ax+2by=0$. Thus we have

$$\left(\alpha-\frac{l\gamma}{n}\right)^2+\left(\beta-\frac{m\gamma}{n}\right)^2+2a\left(\alpha-\frac{l\gamma}{n}\right)+2b\left(\beta-\frac{m\gamma}{n}\right)=0.$$

Using (1) in the above equation we get

$$\left(\alpha-\frac{\gamma(x-\alpha)}{z-\gamma}\right)^2+\left(\beta-\frac{\gamma(y-\beta)}{z-\gamma}\right)^2$$
$$+2a\left(\alpha-\frac{\gamma(x-\alpha)}{z-\gamma}\right)+2b\left(\beta-\frac{\gamma(y-\beta)}{z-\gamma}\right)=0$$

$$\text{or,}\quad (\alpha z-\gamma x)^2+(\beta z-\gamma y)^2+2[a(\alpha z-\gamma x)+b(\beta z-\gamma y)](z-\gamma)=0. \qquad (2)$$

If the cone passes through $(0, 0, c)$, we get

$$(\alpha c)^2+(\beta c)^2+2(c-\gamma)[a(\alpha c)+b(\beta c)]=0$$
$$\text{or,}\quad c[\alpha^2+\beta^2+2(a\alpha+b\beta)]-2\gamma(a\alpha+b\beta)=0. \qquad (3)$$

To get the section of the cone (2) by the plane $x=0$, we put $x=0$ in (2) and get

$$(\alpha z)^2+(\beta z-\gamma y)^2+2[a(\alpha z)+b(\beta z-\gamma y)](z-\gamma)=0.$$

If this represents a rectangular hyperbola in y-z plane then we have

$$\text{Coeff. of } y^2+\text{Coeff. of } z^2=0$$
$$\text{i.e.,}\quad \alpha^2+\beta^2+2a\alpha+2b\beta+\gamma^2=0. \qquad (4)$$

Using (4) in (3) we get

$$c\gamma^2+2\gamma(a\alpha+b\beta)=0$$
$$\text{or,}\quad c\gamma+2a\alpha+2b\beta=0. \qquad (5)$$

From (4) and (5) we get the locus of the vertex $P(\alpha, \beta, \gamma)$ of the cone is

$$x^2+y^2+z^2+2ax+2by=0; \quad 2ax+2by+cz=0.$$

▶▷ **16.** *Prove that the conditions that the lines of section of the plane $lx+my+nz=0$ and the cones $ax^2+by^2+cz^2=0$, $fyz+gzx+hxy=0$ may be coincident are*

$$\frac{bn^2+cm^2}{fmn}=\frac{cl^2+an^2}{gnl}=\frac{bl^2+am^2}{hml}.$$

□ If the plane $lx+my+nz=0$ cuts the cones $ax^2+by^2+cz^2=0$ and $fyz+gzx+hxy=0$ along the line $\frac{x}{l'}=\frac{y}{m'}=\frac{z}{n'}$, then we have

$$ll'+mm'+nn'=0 \qquad (1)$$
$$al'^2+bm'^2+cn'^2=0 \qquad (2)$$
$$\text{and}\quad fm'n'+gn'l'+hm'l'=0 \qquad (3)$$

Eliminating n' from (1) and (3) we get

$$gll'^2 + (fl + gm - hn)m'l' + fmm'^2 = 0. \tag{4}$$

Equation (4) gives the values of l' and m' which when substituted in (1) gives the values of n'.

Eliminating n' from (1) and (2) we get

$$(an^2 + cl^2)l'^2 + 2clmm'l' + (bn^2 + cm^2)m'^2 = 0. \tag{5}$$

Equation (5) gives the values of l' and m' which when substituted in (1) gives the values of n'.

Since the lines of intersection of the plane and the cones are identical, values of l', m', n' obtained from Eqns. (4) and (5) will be same which in turn implies that the Eqns. (4) and (5) are identical. This gives

$$\frac{bn^2 + cm^2}{fm} = \frac{cl^2 + an^2}{gi}. \tag{6}$$

Similarly, eliminating l' from Eqns. (1), (2) and (3) we get the condition of coincidence of the line of intersection as

$$\frac{cl^2 + an^2}{gn} = \frac{bl^2 + am^2}{hm}. \tag{7}$$

From (6) and (7) we get the required condition is

$$\frac{bn^2 + cm^2}{fmn} = \frac{cl^2 + an^2}{gnl} = \frac{bl^2 + am^2}{hml}.$$

▶▷▶ 17. *Find the locus of the vertices of the right circular cones that pass through the ellipse* $\dfrac{x^2}{a^2} + \dfrac{y^2}{b^2} = 1,\ z = 0.$

☐ Equation of the right circular cone, whose vertex is (α, β, γ), semi-vertical angle is θ and direction cosines of the axis are l, m, n is

$$\{l(x - \alpha) + m(y - \beta) + n(z - \gamma)\}^2 = \{(x - \alpha)^2 + (y - \beta)^2 + (z - \gamma)^2\}\cos^2\theta.$$

The section of this cone by the plane $z = 0$ is

$$\{l(x - \alpha) + m(y - \beta) - n\gamma\}^2 = \{(x - \alpha)^2 + (y - \beta)^2 + \gamma^2\}\cos^2\theta$$

or, $$x^2(\cos^2\theta - l^2) + y^2(\cos^2\theta - m^2) - 2lmxy + 2\{l(l\alpha + m\beta + n\gamma) - \alpha\cos^2\theta\}x + 2\{m(l\alpha + m\beta + n\gamma) - \beta\cos^2\theta\}y + (\alpha^2 + \beta^2 + \gamma^2)\cos^2\theta - (l\alpha + m\beta + n\gamma)^2 = 0.$$

If the above equation is identical with $\dfrac{x^2}{a^2} + \dfrac{y^2}{b^2} = 1$, we get

$$\cos^2\theta - l^2 = \frac{1}{a^2} \tag{1}$$

$$\cos^2\theta - m^2 = \frac{1}{b^2} \tag{2}$$

$$(l\alpha + m\beta + n\gamma)^2 - (\alpha^2 + \beta^2 + \gamma^2)\cos^2\theta = 1 \quad (3)$$

$$lm = 0 \quad (4)$$

$$l(l\alpha + m\beta + n\gamma) = \alpha\cos^2\theta \quad (5)$$

$$m(l\alpha + m\beta + n\gamma) = \beta\cos^2\theta. \quad (6)$$

From (4) we get either $l = 0$ or $m = 0$. If $l = 0$, from (5) we get $\alpha = 0$.

Using this in (1), (2), (3) and (6) we get

$$a^2\cos^2\theta = b^2(\cos^2\theta - m^2) = (m\beta + n\gamma)^2 - (\beta^2 + \gamma^2)\cos^2\theta \quad (7)$$

and $\quad m(m\beta + n\gamma) = \beta\cos^2\theta. \quad (8)$

From (7) we get

$$m^2 = \left(1 - \frac{a^2}{b^2}\right)\cos^2\theta. \quad (9)$$

Using (8) in (7) we get

$$a^2\cos^2\theta = \frac{\beta^2}{m^2}\cos^4\theta - (\beta^2 + \gamma^2)\cos^2\theta$$

or, $\quad m^2(a^2 + \beta^2 + \gamma^2)\cos^2\theta = \beta^2\cos^4\theta$

or, $\quad \left(1 - \dfrac{a^2}{b^2}\right)(a^2 + \beta^2 + \gamma^2) = \beta^2$

or, $\quad \dfrac{b^2 - a^2}{b^2}(a^2 + \gamma^2) = \dfrac{\beta^2 a^2}{b^2}$

or, $\quad 1 + \dfrac{\gamma^2}{a^2} = \dfrac{\beta^2}{b^2 - a^2}.$

Hence locus of the point (α, β, γ) is

$$\frac{y^2}{a^2 - b^2} + \frac{z^2}{a^2} + 1 = 0, \quad x = 0.$$

Similarly, when $m = 0$ we get the locus (α, β, γ) is

$$\frac{x^2}{a^2 - b^2} - \frac{z^2}{b^2} = 1, \quad y = 0.$$

EXERCISE

1. Find the equation of the cone with vertex $(1, 2, 3)$ and the guiding curve is $y^2 = 4ax$, $z = 0$.
2. Find the equation of the cone with vertex $(0, 0, 0)$ and the guiding curve is $x^2 + y^2 = 4$, $z = 2$.
3. Find the equation of the cone with vertex $(1, 2, 3)$ and the guiding curve is the circle

$$x^2 + y^2 + z^2 = 4, \quad x + y + z = 1.$$

4. Find the equation of the cone through the coordinate axes and the lines

$$\frac{x}{1} = \frac{y}{-2} = \frac{z}{3} \quad \text{and} \quad \frac{x}{3} = \frac{y}{2} = \frac{z}{-1}.$$

5. Show that the plane $3x + 4y + z = 0$ cuts the cone $15x^2 - 32y^2 - 7z^2 = 0$ along the lines
$$\frac{x}{2} = \frac{y}{-1} = \frac{z}{-2} \quad \text{and} \quad \frac{x}{3} = \frac{y}{-2} = \frac{z}{-1}$$
6. Show that the plane $x - 5y + 3z = 0$ cuts the cone $7x^2 + 5y^2 - 3z^2 = 0$ along the lines
$$\frac{x}{1} = \frac{y}{2} = \frac{z}{3} \quad \text{and} \quad \frac{x}{-1} = \frac{y}{1} = \frac{z}{2}$$
7. Show that the plane $x + y + z = 0$ cuts the cone $ayz + bxz + cxy = 0$ along two perpendicular lines if $a + b + c = 0$.
8. Show that the angle between the lines of intersection of the plane $x + y + z = 0$ and the cone $ayz + bxz + cxy = 0$ is $\pi/3$ if $a^{-1} + b^{-1} + c^{-1} = 0$.
9. Show that the angle between the lines of intersection of the plane $ax + by + cz = 0$ and the cone $yz + xz + xy = 0$ is $\pi/2$ if $a^{-1} + b^{-1} + c^{-1} = 0$.
10. Find the equation of right circular cone whose vertex is $(0, 0, 0)$ and the axis is line $\frac{x}{3} = \frac{y}{2} = \frac{z}{4}$ and semi-vertical angle is $\pi/4$.
11. Find the equation of right circular cone whose vertex is $(1, -2, 1)$ and the axis is line $\frac{x-1}{3} = \frac{y+2}{-4} = \frac{z+1}{5}$ and semi-vertical angle is $\pi/3$.
12. Show that the equation $x^2 - y^2 - 3z^2 - 6yz - 4zx - 2xy = 0$ represents a cone.
13. Show that $ax^2 + by^2 + cz^2 + 2ux + 2vy + 2wz + d = 0$ represents a cone, if $\frac{u^2}{a} + \frac{v^2}{b} + \frac{w^2}{c} = d$.
14. Two cones are described with guiding curves $z^2 = 4ax$; $y = 0$ and $z^2 = 4by, x = 0$ and with common vertex; if the plane $z = 0$ meets them in two conics that intersect in four concyclic points. Show that the vertex lies on the surface $4(x^2 + y^2) = z^2\left(\frac{x}{a} + \frac{y}{b}\right)$.

[*Hints*: Equation of the cones with vertex $P(\alpha, \beta, \gamma)$ and guiding curves $z^2 = 4ax$; $y = 0$ and $z^2 = 4by, x = 0$ are
$$(\beta z - \gamma y)^2 = 4a(\beta - y)(\beta x - \alpha y)$$
$$\text{and} \quad (\alpha z - \gamma x)^2 = 4b(x - \alpha)(\beta x - \alpha y) \quad \text{respectively.}$$

To get the section of the above cones by the plane $z = 0$, we put $z = 0$ in the above equations and get
$$S_1 = \gamma^2 y^2 - 4a(\beta - y)(\beta x - \alpha y)$$
$$\text{and} \quad S_2 = \gamma^2 x^2 - 4b(x - \alpha)(\beta x - \alpha y).$$

Now equation of any curve which passes through the points of intersection of S_1 and S_2 can be written as $S_1 + kS_2 = 0$. If this represents a circle then we have
$$\text{Coeff. of } x^2 = \text{Coeff. of } y^2 \quad \text{and} \quad \text{Coeff. of } xy = 0.$$

These give two results involving k and k eliminant of those yields the required result.]

15. Two cones are described with guiding curves $zx = a^2$; $y = 0$ and $yz = b^2, x = 0$ and with any vertex. Show that if their four common generators meet the plane $z = 0$ in four concyclic points, the vertex lies on the surface $z(x^2 + y^2) = a^2x + b^2y$.

[*Hints*: Equation of the cones with vertex $P(\alpha, \beta, \gamma)$ and guiding curves $zx = a^2; y = 0$ and $yz = b^2, x = 0$ are

$$a^2(y - \beta)^2 = (\alpha y - \beta x)(\gamma y - \beta z)$$
$$\text{and} \quad b^2(x - \alpha)^2 = (\beta x - \alpha y)(\gamma x - \alpha z) \quad \text{respectively.}$$

The general equation of the cone through the common generators of the above cones is

$$a^2(y - \beta)^2 - (\alpha y - \beta x)(\gamma y - \beta z) + k[b^2(x - \alpha)^2 - (\beta x - \alpha y)(\gamma x - \alpha z)] = 0.$$

To get the section of the above cone by the plane $z = 0$, we put $z = 0$ in the above equation and get

$$a^2(y - \beta)^2 - (\alpha y - \beta x)\gamma y + k[b^2(x - \alpha)^2 - (\beta x - \alpha y)\gamma x] = 0.$$

If this represents a circle then we have

$$\text{Coeff. of } x^2 = \text{Coeff. of } y^2 \quad \text{and} \quad \text{Coeff. of } xy = 0.$$

These give two results involving k and k eliminant of those yields the required result.]

16. Prove that a line which passes through (α, β, γ) and intersects the parabola $z^2 = 4ax$, $y = 0$ lies on the surface $(\beta z - \gamma y)^2 = 4a(\beta - y)(\beta x - \alpha y)$.

17. The section of a cone whose guiding curve is $\dfrac{x^2}{a^2} + \dfrac{y^2}{b^2} = 1, z = 0$ by the plane $x = 0$ is a rectangular hyperbola. Show that locus of the vertex is $\dfrac{x^2}{a^2} + \dfrac{y^2 + z^2}{b^2} = 1$.

[*Hints*: If $P(\alpha, \beta, \gamma)$ be the vertex of the cone, then equation of the generator can be written as

$$\frac{x - \alpha}{l} = \frac{y - \beta}{m} = \frac{z - \gamma}{n}. \tag{1}$$

If it intersects the curve $\dfrac{x^2}{a^2} + \dfrac{y^2}{b^2} = 1, z = 0$, then the point $\left(\alpha - \dfrac{l\gamma}{n}, \beta - \dfrac{m\gamma}{n}\right)$ will satisfy the equation $\dfrac{x^2}{a^2} + \dfrac{y^2}{b^2} = 1$. Thus we have

$$\frac{1}{a^2}\left(\alpha - \frac{l\gamma}{n}\right)^2 + \frac{1}{b^2}\left(\beta - \frac{m\gamma}{n}\right)^2 = 1.$$

Using (1) in the above equation we get

$$\frac{1}{a^2}\left(\alpha - \frac{\gamma(x - \alpha)}{z - \gamma}\right)^2 + \frac{1}{b^2}\left(\beta - \frac{\gamma(y - \beta)}{z - \gamma}\right)^2 = 1$$
$$\text{or,} \quad b^2(\alpha z - \gamma x)^2 + a^2(\beta z - \gamma y)^2 = a^2b^2(z - \gamma)^2. \tag{2}$$

To get the section of the cone (2) by the plane $x = 0$, we put $x = 0$ in (2) and get

$$b^2\alpha^2z^2 + a^2(\beta z - \gamma y)^2 = a^2b^2(z - \gamma)^2.$$

If this represents a rectangular hyperbola in yz-plane, then we have

$$\text{Coeff. of } y^2 + \text{Coeff. of } z^2 = 0$$

i.e., $$b^2\alpha^2 + a^2\beta^2 - a^2b^2 + a^2\gamma^2 = 0$$

or, $$b^2\alpha^2 + a^2(\beta^2 + \gamma^2) = a^2b^2$$

or, $$\frac{\alpha^2}{a^2} + \frac{\beta^2 + \gamma^2}{b^2} = 1.$$

Generalizing, we get the required locus.]

18. Show that the bisectors of the angle between the lines in which the plane $ux + vy + wz = 0$ cuts the cone $ax^2 + by^2 + cz^2 = 0$ lie on the cone

$$\frac{u(b-c)}{x} + \frac{v(c-a)}{y} + \frac{w(a-b)}{z} = 0.$$

[*Hints*: Let the direction ratios of the lines of intersection be l, m, n and l', m', n' then we have

$$al^2 + bm^2 + cn^2 = 0 \quad \text{and} \quad al'^2 + bm'^2 + cn'^2 = 0 \tag{1}$$

and the direction ratios of the bisectors of the angle between them are $l \pm l'$, $m \pm m'$, $n \pm n'$. Since the normal to the given plane and these two bisectors from a set of three mutually perpendicular lines we have

$$\frac{l - l'}{(m+m')w - (n+n')v} = \frac{m - m'}{(n+n')u - (l+l')w} = \frac{n - n'}{(l+l')v - (m+m')u}.$$

Again from (1) we get

$$a(l+l')(l-l') + b(m+m')(m-m') + c(n+n')(n-n') = 0.$$

Therefore

$$a(l+l')[(m+m')w - (n+n')v] + b(m+m')[(n+n')u - (l+l')w]$$
$$+ c(n+n')[(l+l')v - (m+m')u] = 0$$

or, $$\frac{u(b-c)}{l+l'} + \frac{v(c-a)}{m+m'} + \frac{w(a-b)}{n+n'} = 0.$$

So the bisector whose equation is

$$\frac{x}{l+l'} + \frac{y}{m+m'} + \frac{z}{n+n'} = 0$$

generates the cone

$$\frac{u(b-c)}{x} + \frac{v(c-a)}{y} + \frac{w(a-b)}{z} = 0.]$$

Chapter 7

Cylinder

7.1 Definition

A cylinder is a surface generated by a moving line which is parallel to a fixed line and intersects a given curve or touches a fixed surface.

The fixed curve is called its guiding curve or base. The moving line which generates the cylinder is called its generator and the fixed line is called its axis.

7.2 Equation of Cylinder when the Axis and the Base is Given

Let $f(x, y) = 0$; $z = 0$ be the base of the cone. The equation of its axis be

$$\frac{x}{l} = \frac{y}{m} = \frac{z}{n}. \tag{1}$$

If (α, β, γ) be any point on the cylinder, then equations of its generators are

$$\frac{x-\alpha}{l} = \frac{y-\beta}{m} = \frac{z-\gamma}{n}. \tag{2}$$

Since the generator intersects the curve $f(x, y) = 0$; $z = 0$, then points $\left(\alpha - \frac{l\gamma}{n}, \beta - \frac{m\gamma}{n}\right)$ will satisfy the equation $f(x, y) = 0$. Thus we get

$$f\left(\alpha - \frac{l\gamma}{n}, \beta - \frac{m\gamma}{n}\right) = 0.$$

Hence locus of (α, β, γ) is

$$f\left(x - \frac{lz}{n}, y - \frac{mz}{n}\right) = 0. \tag{3}$$

This is the equation of the required cylinder.

If we take $f(x, y) = ax^2 + 2hxy + by^2 + 2gx + 2fy + c$, then the equation of the cylinder becomes

$$\begin{aligned} &a(nx - lz)^2 + 2h(nx - lz)(ny - mz) + b(ny - mz)^2 \\ &+ 2gn(nx - lz) + 2fn(ny - mz) + cn^2 = 0. \end{aligned}$$

Further if we take z-axis as the axis of the cylinder then from Eqn. (3) it is clear that the equation of the cylinder will be $f(x, y) = 0$.

7.3 Right Circular Cylinder

A right circular cylinder is a surface generated by a moving line which intersects a fixed circle and is perpendicular to the plane.

Evidently the length of the perpendicular from any point on the right circular cylinder to the axis of it is equal to the radius of the circle.

7.4 Equation of Right Circular Cylinder

Let the equation of the axis be

$$\frac{x-\alpha}{l} = \frac{y-\beta}{n} = \frac{z-\gamma}{n}$$

where $P(\alpha, \beta, \gamma)$ is any point on the axis.

Let $Q(x, y, z)$ be any point on the surface of the cylinder and R be the foot of the perpendicular drawn from Q on the axis of the cylinder, then

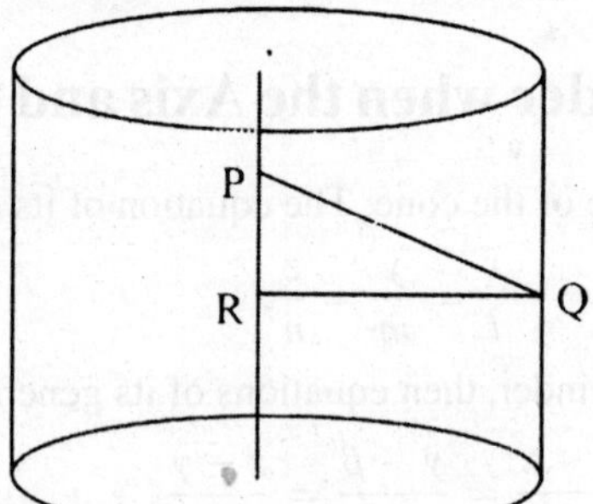

$$PQ^2 - PR^2 = QR^2$$

or, $$(x-\alpha)^2 + (y-\beta)^2 + (z-\gamma)^2 - \frac{[l(x-\alpha) + m(y-\beta) + n(z-\gamma)]^2}{l^2+m^2+n^2} = r^2,$$

where r is the radius of the circle.

This is the equation of the right circular cylinder.

Examples

▶▶ **1.** *Find the equation of the cylinder whose generators are parallel to the straight line* $\frac{x}{1} = \frac{y}{2} = \frac{z}{3}$ *and whose guiding curve is* $x^2 + y^2 = 4$, $z = 1$.

☐ If (α, β, γ) be any point on the cylinder, the equation of its generators can be written as

$$\frac{x-\alpha}{1} = \frac{y-\beta}{2} = \frac{z-\gamma}{3}.$$

If it meets curve is $x^2 + y^2 = 4, z = 1$, the point $\left(\alpha + \frac{1-\gamma}{3}, \beta + \frac{2(1-\gamma)}{3}\right)$ will satisfy the equation $x^2 + y^2 = 4$ i.e.

$$\left(\alpha + \frac{1-\gamma}{3}\right)^2 + \left(\beta + \frac{2(1-\gamma)}{3}\right)^2 = 4$$

$$\text{or,} \quad (3\alpha + 1 - \gamma)^2 + (3\beta + 2 - 2\gamma)^2 = 36.$$

Hence locus of the point (α, β, γ) is

$$(3x + 1 - z)^2 + (3y + 2 - 2z)^2 = 36.$$

▶▷▶ 2. *Find the equation of the right circular cylinder of radius 2 and whose axis is*

$$\frac{x-1}{4} = \frac{y-2}{5} = \frac{z-3}{3}.$$

☐ If (x, y, z) be any point on the cylinder, then length of the perpendicular from it to the axis is

$$(x-1)^2 + (y-2)^2 + (z-3)^2 - \frac{[4(x-1) + 5(y-2) + 3(z-3)]^2}{50}.$$

Hence equation of the cylinder is

$$(x-1)^2 + (y-2)^2 + (z-3)^2 - \frac{[4(x-1) + 5(y-2) + 3(z-3)]^2}{50} = 4$$

$$\text{or,} \quad 50[(x-1)^2 + (y-2)^2 + (z-3)^2] - [4(x-1) + 5(y-2) + 3(z-3)]^2 = 200.$$

▶▷▶ 3. *Obtain the equation of the right circular cylinder whose guiding curve is the circle passing through the points* $(1, 0, 0)$; $(0, 1, 0)$ *and* $(0, 0, 1)$.

☐ If (α, β, γ) be the centre of the circle passing through $A : (1, 0, 0)$; $B : (0, 1, 0)$ and $C : (0, 0, 1)$, then we have

$$(\alpha - 1)^2 + \beta^2 + \gamma^2 = \alpha^2 + (\beta - 1)^2 + \gamma^2 = \alpha^2 + \beta^2 + (\gamma - 1)^2$$

Solving we get $\alpha = \beta = \gamma = k$ (say). As it lies on the plane ABC, we have $k = 1/3$. So radius of the circle $= \sqrt{3k^2 - 2k + 1} = \sqrt{\frac{1}{3} - \frac{2}{3} + 1} = \sqrt{\frac{2}{3}}$. Again axis of the cylinder is perpendicular to the plane ABC and passes through the centre of the circle, so its equation is $x = y = z$. Hence equation of the right circular cylinder is

$$x^2 + y^2 + z^2 - \frac{[x + y + z]^2}{3} = \frac{2}{3}$$

$$\text{or,} \quad 3(x^2 + y^2 + z^2) - [x + y + z]^2 = 2$$

$$\text{or,} \quad x^2 + y^2 + z^2 - xy - yz - zx - 1 = 0.$$

▶▷▶ 4. *Find the equation of the right circular cylinder whose guiding curve is*

$$x^2 + y^2 + z^2 = 9; \quad x - y + z = 3.$$

□ Direction ratios of the axis of the cylinder are proportional to $1, -1, 1$. If (α, β, γ) be any point on the cylinder then equation of its generators are

$$\frac{x-\alpha}{1} = \frac{y-\beta}{-1} = \frac{z-\gamma}{1} = r.$$

Any point on it can be taken as $(\alpha + r, \beta - r, \gamma + r)$. If this point lies on the guiding curve then

$$(\alpha + r)^2 + (\beta - r)^2 + (\gamma + r)^2 = 9; \quad (\alpha + r) - (\beta - r) + (\gamma + r) = 3$$

$$\Rightarrow \quad \alpha^2 + \beta^2 + \gamma^2 + 2r(\alpha - \beta + \gamma) + 3r^2 = 9; \quad \alpha - \beta + \gamma + 3r = 3.$$

Eliminating r we get

$$\alpha^2 + \beta^2 + \gamma^2 - \frac{2(\alpha - \beta + \gamma - 3)(\alpha - \beta + \gamma)}{3} + \frac{(\alpha - \beta + \gamma - 3)^2}{3} = 9.$$

Hence locus of (α, β, γ) is

$$x^2 + y^2 + z^2 + yz - zx + xy - 9 = 0.$$

▶▷ 5. *Find the equation of the cylinder generated by the straight lines parallel to the x-axis and passing through the curve* $3y^2 - 5z^2 = 10x$; $2x - 3y + z = 2$.

□ Let (α, β, γ) be any point on the curve $3y^2 - 5z^2 = 10x$; $2x - 3y + z = 2$, then

$$3\beta^2 - 5\gamma^2 = 10\alpha; \quad 2\alpha - 3\beta + \gamma = 2. \tag{1}$$

If (x, y, z) be any point on the cylinder then equations of its generators are

$$\frac{x-\alpha}{1} = \frac{y-\beta}{0} = \frac{z-\gamma}{0}$$

which gives $y = \beta$ and $z = \gamma$. So from (1) we get

$$\frac{3y^2 - 5z^2}{10} = \alpha = \frac{2 + 3y - z}{2}$$

$$\text{or,} \quad 3y^2 - 5z^2 - 5(2 + 3y - z) = 0$$

which is the equation of the required cylinder.

EXERCISE

1. Obtain the equation of the cylinder whose generators intersect the curve $ax^2 + by^2 = 1$, $z = 0$ and are parallel to the straight line $\frac{x}{l} = \frac{y}{m} = \frac{z}{n}$.
2. Find the equation of the cylinder generated by the straight lines parallel to the z-axis and passing through the curve $5x^2 - 2y^2 + 7z^2 = 1$; $3x + 2y - z = 2$.
3. Find the equation of the cylinder generated by the straight lines parallel to the y-axis and passing through the curve $x^2 + y^2 + z^2 = 4$; $x + y + z = 4$.

4. Find the equation of the cylinder generated by the straight lines parallel to the z-axis and passing through the curve $ax^2 + by^2 + cz^2 = 1; lx + my + nz = p$.
[*Hints*: Let (α, β, γ) be any point on the curve $ax^2 + by^2 + cz^2 = 1; lx + my + nz = p$, then we have
$$a\alpha^2 + b\beta^2 + c\gamma^2 = 1; \quad l\alpha + m\beta + n\gamma = p. \tag{1}$$
If (x, y, z) be any point on the cylinder then equations of its generators are
$$\frac{x-\alpha}{0} = \frac{y-\beta}{0} = \frac{z-\gamma}{1}$$
which gives $y = \beta$ and $x = \alpha$. So from (1) we get
$$\frac{ax^2 + by^2 - 1}{-c} = \gamma^2 = \frac{(lx + my - p)^2}{n^2}$$
or, $$n^2(ax^2 + by^2 - 1) + c(lx + my - p)^2 = 0$$
which is the equation of the required cylinder.]
5. Find the equation of the right circular cylinder whose axis is $x = 2y = -z$ and radius is 4.
6. Find the equation of the right circular cylinder whose axis is x-axis and radius is 1.
7. Find the equation of the right circular cylinder of radius 3, whose axis is
$$\frac{x-1}{2} = \frac{y-2}{2} = \frac{z-3}{-1}.$$
8. Find the equation of the right circular cylinder whose guiding curve is
$$x^2 + y^2 + z^2 = 6; \quad x + y + z = 2.$$

Chapter 8

Surface of Revolution

8.1 Definition

The surface formed by the revolution of a plane curve or a straight line about a fixed line in its plane is called a surface of revolution.

For example, a sphere is generated by revolution of a circle about one of its diameters. A right circular cone is generated by a straight line by revolving it about a given line which intersects moving straight line.

The plane curve which generates the surface is called generatrix of the surface and the straight line about which the plane curve is revolved is called axis of the revolution.

8.2 Equation of the Surface of Revolution

(i) *Axis of revolution is x-axis and the generatrix is a curve in the x-y plane:*

Let the equation of the generatrix be

$$f(x, y) = 0; \quad z = 0. \tag{1}$$

If $P(h, k, 0)$ be any point on the curve given by (1) then

$$f(h, k) = 0. \tag{2}$$

As P revolves about the x-axis, its distance from the axis remains unchanged and its x-coordinate remains unchanged. Let at any subsequent stage coordinate becomes (α, β, γ), then we have

$$\alpha = h; \quad \beta^2 + \gamma^2 = k^2 \tag{3}$$

Using (3) in (2) we get

$$f(\alpha, \pm\sqrt{\beta^2 + \gamma^2}) = 0.$$

Hence the locus of (α, β, γ) i.e., equation of surface of revolution is

$$f(x, \pm\sqrt{y^2 + z^2}) = 0.$$

(ii) *Axis of revolution is parallel to x-axis on z-plane and the generatrix is a curve in the x-y plane*:

Let the equation of the generatrix be

$$f(x, y) = 0; \quad z = 0 \tag{1}$$

and the equation of the axis of revolution is

$$y = a; \quad z = 0. \tag{2}$$

If $P(h, k, 0)$ be any point on the curve given by (1) then

$$f(h, k) = 0. \tag{3}$$

As P revolves about the line $y = a$; $z = 0$, its distance from the line $y = a$; $z = 0$ remains unchanged and its x-coordinate remains unchanged. Let at any subsequent stage coordinate becomes (α, β, γ), then we have

$$\alpha = h; \quad (\beta - a)^2 + \gamma^2 = (k - a)^2. \tag{4}$$

Using (4) in (3) we get

$$f(\alpha, a \pm \sqrt{(\beta - a)^2 + \gamma^2}) = 0.$$

Hence the locus of (α, β, γ) i.e., equation of surface of revolution is

$$f(x, a \pm \sqrt{(y - a)^2 + z^2}) = 0.$$

Examples

▶▶ 1. *Find the equation of surface of revolution generated by $x^2 = 2y$, $z = 0$ about y-axis.*

☐ The generatrix is $x^2 - 2y = 0$, $z = 0$ and the axis of revolution is y-axis. So we replace x by $\pm\sqrt{x^2 + z^2}$ in $x^2 - 2y = 0$ and get the equation of surface of revolution:

$$x^2 + z^2 - 2y = 0.$$

▶▶ 2. *Find the equation of surface of revolution of the straight line $y = mx$, $z = 0$ about x-axis.*

☐ The generatrix is $y = mx$, $z = 0$ and the axis of revolution is x-axis. So we replace y by $\pm\sqrt{z^2 + y^2}$ in $y = mx$ and get the equation of surface of revolution:

$$z^2 + y^2 = m^2x^2.$$

▶▶ 3. *Find the equation of surface of revolution generated by $y^2 = 4z$, $x = 0$ about $z = 2$, $x = 0$.*

☐ Here the axis of revolution is $z = 2$, $x = 0$, so we replace z by $2 \pm \sqrt{(z-2)^2 + x^2}$ in $y^2 = 4z$ and get the equation of surface of revolution:

$$y^2 = 4[2 \pm \sqrt{(z-2)^2 + x^2}]$$

$$\text{or,} \quad (y^2 - 8)^2 = 16[(z-2)^2 + x^2].$$

▶▶ 4. *Show that the equation* $\frac{x^2}{4}+\frac{y^2}{4}-\frac{z^2}{9}=1$ *represents a surface of revolution.*

□ Here we see that the plane $z=k$ intersects the given surface in a circle given by

$$x^2+y^2=4\left(1+\frac{k^2}{9}\right)$$

whose centre lies on the z-axis. Hence the given equation is a surface of revolution whose axis of revolution is z-axis.

EXERCISE

1. Find the equation of surface of revolution generated by $z^2=2y$, $x=0$ about (i) y axis and (ii) z-axis.
2. Find the equation of surface of revolution generated by $x^2=4y$, $z=0$ about (i) y axis and (ii) x-axis.
3. Find the equation of surface of revolution generated by $y^2=4x$; $z=0$ about $x=2$, $z=0$.
4. Find the equation of surface of revolution generated by $x^2+2y^2=1$; $z=0$ about $x=3$, $z=0$.
5. Show that the equation $\frac{x^2}{4}-\frac{y^2}{6}+\frac{z^2}{4}=1$ represents a surface of revolution.
6. Show that the equation $\frac{x^2}{a^2}+\frac{y^2+z^2}{b^2}=1$ represents a surface of revolution.

Chapter 9

Conicoids

9.1 Central Conicoids

Conicoids having centres are called central conicoids. The general second degree equation in x, y and z is given by

$$ux^2 + by^2 + cz^2 + 2fyz + 2gxz + 2hxy + 2ux + 2vy + 2wz + d = 0$$

and represents a quadratic surface or a conicoid. In this chapter we shall discuss shapes and different results of conicoids. The standard form of a central conicoid is

$$ax^2 + by^2 + cz^2 = 1.$$

9.2 Shapes of Central Conicoids

The above equation of the central conicoid can also be written as

$$\pm\frac{x^2}{a^2} \pm \frac{y^2}{b^2} \pm \frac{z^2}{c^2} = 1.$$

We have four different cases depending on the $\pm$ signs in the above equation, which are discussed below:

(i) *Ellipsoid*: If all the signs are +ive, the surface is called ellipsoid, thus standard form of an ellipsoid is

$$\frac{x^2}{a^2} + \frac{y^2}{b^2} + \frac{z^2}{c^2} = 1.$$

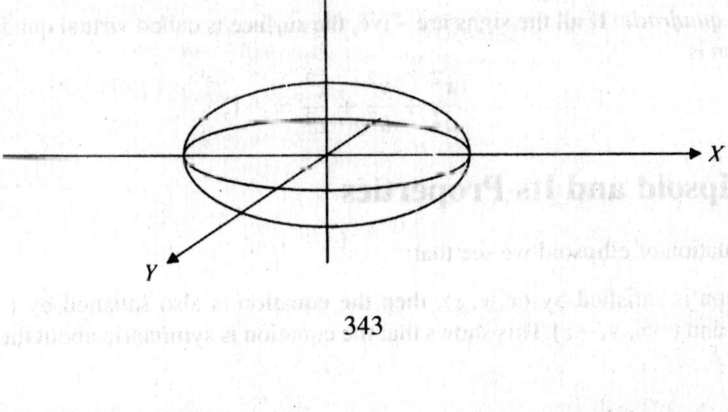

(ii) *Hyperboloid of one sheet*: If one of the signs is −ive, the surface is called hyperboloid of one sheet, thus standard form of a hyperboloid of one sheet is

$$\frac{x^2}{a^2}+\frac{y^2}{b^2}-\frac{z^2}{c^2}=1 \text{ or, } \frac{x^2}{a^2}-\frac{y^2}{b^2}+\frac{z^2}{c^2}=1 \text{ or, } -\frac{x^2}{a^2}+\frac{y^2}{b^2}+\frac{z^2}{c^2}=1.$$

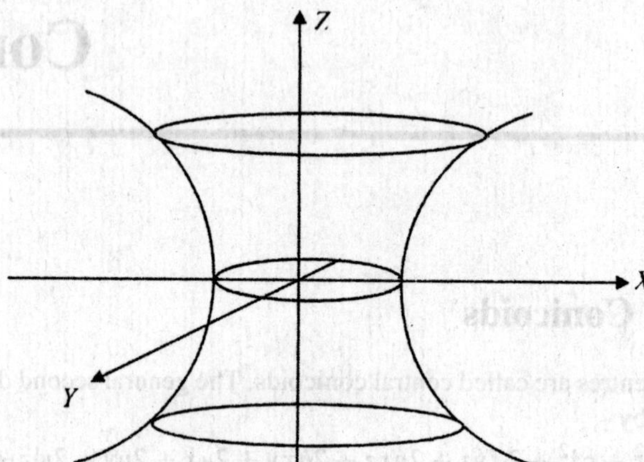

(iii) *Hyperboloid of two sheets*: If two of the signs are −ive, the surface is called hyperboloid of two sheets, thus standard form of a hyperboloid of two sheets is

$$\frac{x^2}{a^2}-\frac{y^2}{b^2}-\frac{z^2}{c^2}=1 \text{ or, } -\frac{x^2}{a^2}-\frac{y^2}{b^2}+\frac{z^2}{c^2}=1 \text{ or, } -\frac{x^2}{a^2}+\frac{y^2}{b^2}-\frac{z^2}{c^2}=1.$$

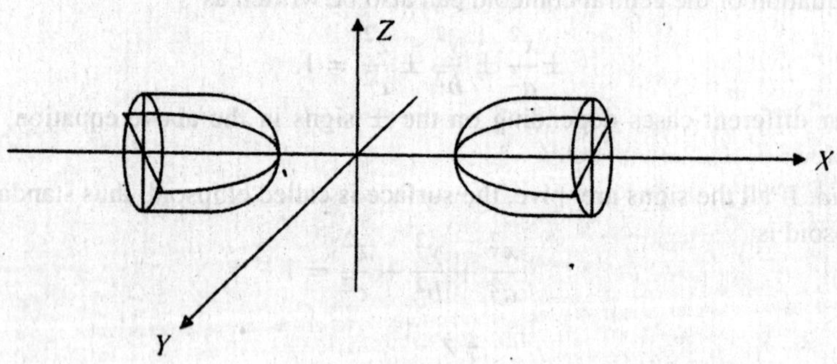

(iv) *Virtual quadratic*: If all the signs are −ive, the surface is called virtual quadratic and its equation is

$$\frac{x^2}{a^2}+\frac{y^2}{b^2}+\frac{z^2}{c^2}=-1.$$

9.3 Ellipsoid and Its Properties

From the equation of ellipsoid we see that:

If the equation is satisfied by (x, y, z), then the equation is also satisfied by $(-x, -y, z)$; $(x, -y, -z)$ and $(-x, y, -z)$. This shows that the equation is symmetric about the coordinate axes and these axes are called principal axes. Again if (x, y, z) lies on the ellipsoid then $(-x, -y, -z)$ also lies on that i.e., the equation of ellipsoid is symmetric about the origin. This shows that all the chords pass through the origin and are bisected by the origin implying that origin is the centre of the ellipsoid.

The coordinate axes meet the surface at $(\pm a, 0, 0)$; $(0, \pm b, 0)$ and $(0, 0, \pm c)$ respectively. These points are called the vertices of the ellipsoid. If $a > b > c$ then $2a$, $2b$, $2c$ are called the major, mean and minor axes. If any two of a, b, c are equal then the surface is called an ellipsoid of revolution and the axis corresponding to remaining length is the axis of revolution and if $a = b = c$ then the ellipsoid becomes a sphere.

Again, if the equation is satisfied by (x, y, z), then the equation is also satisfied by $(-x, y, z)$; $(x, -y, z)$ and $(x, y, -z)$. This shows that the equation is symmetric about the coordinate planes and these planes are called principal planes.

Also the equation of an ellipsoid is satisfied for all values $|x| \le a$; $|y| \le b$ and $|z| \le c$, this shows that the surface is bounded between planes $x = \pm a$; $y = \pm b$ and $z = \pm c$. So ellipsoid is a closed surface.

The section of the ellipsoid by the plane $z = k$ is given by

$$\frac{x^2}{a^2} + \frac{y^2}{b^2} = 1 - \frac{k^2}{c^2}; \quad z = k.$$

This represents an ellipse if $-c < k < c$. Similarly the sections of the ellipsoid by the planes $x = u$ and $y = v$ are also ellipses.

9.4 Hyperboloid of One Sheet and Its Properties

Like the equation of ellipsoid, the equation of hyperboloid of one sheet is symmetric about the coordinate axes, coordinate planes and also symmetric about the origin. So x-, y- and z-axes are the principal axes; $x = 0$, $y = 0$ and $z = 0$ are the principal planes and origin is the centre of the surface given by

$$\frac{x^2}{a^2} + \frac{y^2}{b^2} - \frac{z^2}{c^2} = 1.$$

The x- and y-axes meet the surface at $(\pm a, 0, 0)$, $(0, \pm b, 0)$ respectively, which are called vertices but the z-axis does not intersect the surface. Thus the surface is not closed. Two points are taken on z-axis as $(0, 0, \pm c)$ so that the lengths of the axes are $2a$, $2b$ and $2c$.

The section of the surface by the plane $z = k$ is given by

$$\frac{x^2}{a^2} + \frac{y^2}{b^2} = 1 + \frac{k^2}{c^2}; \quad z = k.$$

This represents an ellipse if $-\infty < k < \infty$ and the sections of the surface by the planes $x = u$ and $y = v$ are hyperbolas if $|u| \ne a$ and $|v| \ne b$. When $|u| = a$ and $|v| = b$, then the sections are pair of straight lines.

9.5 Hyperboloid of Two Sheets and Its Properties

The equation of hyperboloid of two sheets is also symmetric about the coordinate axes, coordinate planes and also symmetric about the origin. So x-, y- and z-axes are the principal axes; $x = 0$, $y = 0$ and $z = 0$ are the principal planes and origin is the centre of the surface given by

$$\frac{x^2}{a^2} - \frac{y^2}{b^2} - \frac{z^2}{c^2} = 1.$$

The x-axis meet the surface at $(\pm a, 0, 0)$, which are called vertices but the y- and z-axes do not intersect the surface. Thus the surface is not closed. Two points are taken on z-axis as $(0, 0, \pm c)$ and two points are taken on y-axis as $(0, \pm b, 0)$ so that the lengths of the axes are $2a$, $2b$ and $2c$.

The section of the surface by the plane $x = k$ is given by

$$\frac{z^2}{c^2} + \frac{y^2}{b^2} = \frac{k^2}{a^2} - 1; \quad x = k.$$

This represents an ellipse if $|k| > a$ and the sections of the surface by the planes $x = k$ are ellipses. The section by the plane $y = u$ or $z = v$ is hyperbola.

9.6 Shapes of Non-Central Conicoids

The general equation of non-central conicoids is of the form $ax^2 + by^2 = 2z$. These surfaces are called paraboloid.

9.7 Elliptic Paraboloid and Its Properties

The equation of an elliptic paraboloid is

$$\frac{x^2}{a^2} + \frac{y^2}{b^2} = \frac{2z}{c}.$$

From the equation it is clear that if (x, y, z) satisfies the equation then $(-x, -y, z)$ also satisfies the equation, i.e., the equation is symmetric about z-axis and the z-axis is called the principal axis of the above elliptic paraboloid. It is easy to see that the above equation is also symmetric about the planes $x = 0$ and $y = 0$ and these planes are called principal planes.

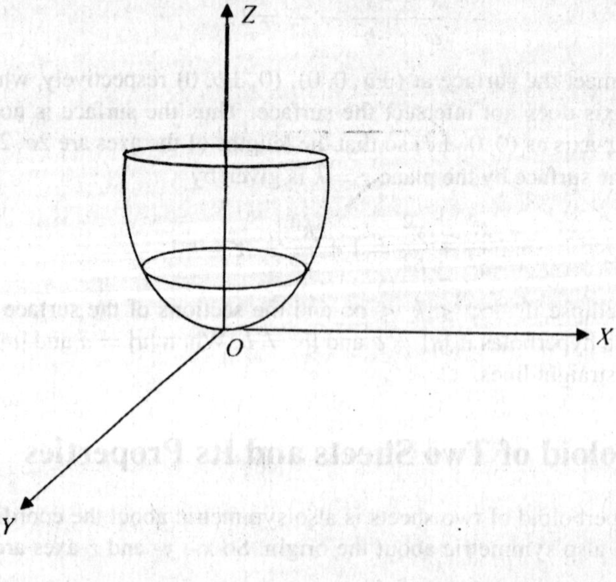

No negative value of z satisfies the equation, so the surface has no part below the x-y plane. The sections of the surface by the planes $x = k$ and $y = k$ are parabolas given by

$$y^2 = \frac{2b^2}{c}\left(z - \frac{ck^2}{2a^2}\right) \quad \text{and} \quad x^2 = \frac{2a^2}{c}\left(z - \frac{ck^2}{2b^2}\right)$$

and the section of the surface by the plane $z = k$ are ellipses given by

$$\frac{x^2}{a^2} + \frac{y^2}{b^2} = \frac{2k}{c}; \quad z = k.$$

If $k = 0$, we get two coincident points (0,0,0), thus the surface touches the x-y plane at the origin.

9.8 Hyperbolic Paraboloid and Its Properties

The equation of a hyperbolic paraboloid is

$$\frac{x^2}{a^2} - \frac{y^2}{b^2} = \frac{2z}{c}.$$

Like the equation of elliptic paraboloid, it is clear that if (x, y, z) satisfies the equation of hyperbolic paraboloid then $(-x, -y, z)$ also satisfies the equation i.e., the equation is symmetric about z-axis and the z-axis is called the principal axis of the above hyperbolic paraboloid. It is easy to see that the above equation is also symmetric about the planes $x = 0$ and $y = 0$ and these planes are called principal planes.

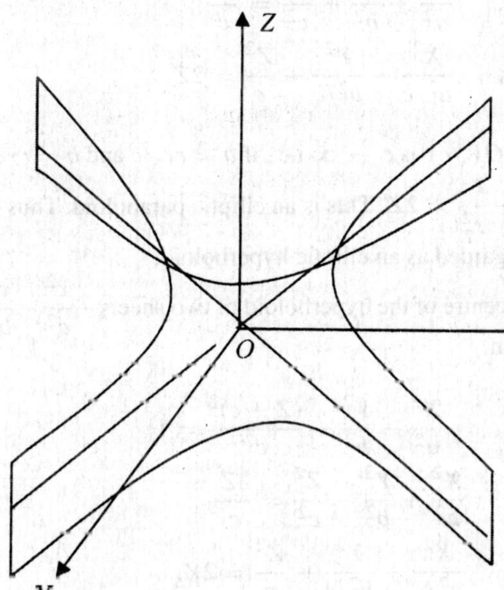

The sections of the surface by the planes $x = k$ and $y = k$ are parabolas given by

$$y^2 = -\frac{2b^2}{c}\left(z - \frac{ck^2}{2a^2}\right) \quad \text{and} \quad x^2 = \frac{2a^2}{c}\left(z + \frac{ck^2}{2b^2}\right)$$

and the section of the surface by the plane $z = k$ are hyperbolas given by

$$\frac{x^2}{a^2} - \frac{y^2}{b^2} = \frac{2k}{c}; \quad z = k.$$

If $k = 0$, we get a pair of straight lines passing through the origin i.e.,

$$\frac{x}{a} \pm \frac{y}{b} = 0; \quad z = 0.$$

9.9 Paraboloids are Limiting Case of Central Conicoids

Shifting the centre of a central conicoid at a distant point , it can be shown that paraboloids are limiting cases of central conicoids, i.e., elliptic paraboloid is the limiting case of ellipsoid or hyperboloid of two sheets and hyperbolic paraboloid is the limiting case of hyperboloid of one sheet.

Case I: Shifting the origin, i.e., centre of the ellipsoid $\frac{x^2}{a^2} + \frac{y^2}{b^2} + \frac{z^2}{c^2} = 1$ to the point $(0, 0, -c)$ we obtain

$$\frac{X^2}{a^2} + \frac{Y^2}{b^2} + \frac{(Z-c)^2}{c^2} = 1$$

$$\text{or,} \quad \frac{X^2}{a^2} + \frac{Y^2}{b^2} + \frac{Z^2}{c^2} = \frac{2Z}{c}$$

$$\text{or,} \quad \frac{X^2}{a^2/c} + \frac{Y^2}{b^2/c} + \frac{Z^2}{c} = 2Z.$$

Now if $a \sim O(\sqrt{c})$ and $b \sim O(\sqrt{c})$ as $c \to \infty$ i.e., if $a = r_1\sqrt{c}$ and $b = r_2\sqrt{c}$ for large values of c, then we get $\frac{X^2}{r_1^2} + \frac{Y^2}{r_2^2} = 2Z$. This is an elliptic paraboloid. Thus the limiting case of an ellipsoid can be regarded as an elliptic hyperboloid.

Case II: Shifting the origin, i.e., centre of the hyperboloid of two sheets $-\frac{x^2}{a^2} - \frac{y^2}{b^2} + \frac{z^2}{c^2} = 1$ to the point $(0, 0, c)$ we obtain

$$-\frac{X^2}{a^2} - \frac{Y^2}{b^2} + \frac{(Z+c)^2}{c^2} = 1$$

$$\text{or,} \quad \frac{X^2}{a^2} + \frac{Y^2}{b^2} - \frac{Z^2}{c^2} = \frac{2Z}{c}$$

$$\text{or,} \quad \frac{X^2}{a^2/c} + \frac{Y^2}{b^2/c} - \frac{Z^2}{c} = 2Z.$$

Now if $a \sim O(\sqrt{c})$ and $b \sim O(\sqrt{c})$ as $c \to \infty$ i.e., if $a = r_1\sqrt{c}$ and $b = r_2\sqrt{c}$ for large values of c, then we get $\frac{X^2}{r_1^2} + \frac{Y^2}{r_2^2} = 2Z$. This is an elliptic paraboloid. Thus the limiting case of hyperboloid of two sheets can be regarded as an elliptic hyperboloid.

Case III: Shifting the origin i.e., centre of the hyperboloid of one sheet $\frac{x^2}{a^2} - \frac{y^2}{b^2} + \frac{z^2}{c^2} = 1$ to the point $(0, 0, -c)$ we obtain

$$\frac{X^2}{a^2} - \frac{Y^2}{b^2} + \frac{(Z-c)^2}{c^2} = 1$$

$$\text{or,} \quad \frac{X^2}{a^2} - \frac{Y^2}{b^2} + \frac{Z^2}{c^2} = \frac{2Z}{c}$$

$$\text{or,} \quad \frac{X^2}{a^2/c} - \frac{Y^2}{b^2/c} + \frac{Z^2}{c} = 2Z.$$

Now if $a \sim O(\sqrt{c})$ and $b \sim O(\sqrt{c})$ as $c \to \infty$ i.e., if $a = r_1\sqrt{c}$ and $b = r_2\sqrt{c}$ for large values of c, then we get $\frac{X^2}{r_1^2} - \frac{Y^2}{r_2^2} = 2Z$. This is an hyperbolic paraboloid. Thus the limiting case of hyperboloid of one sheet can be regarded as an hyperbolic hyperboloid.

Examples

▶▷▶ 1. *Find the nature of the following surface given by the equation*

$$9x^2 + 36y^2 + 4z^2 - 36x + 216y + 32z + 388 = 0.$$

☐ The given equation can be written as

$$9(x-2)^2 + 36(y+3)^2 + 4(z+4)^2 = 36$$

$$\text{or,} \quad \frac{(x-2)^2}{2^2} + \frac{(y+3)^2}{1^2} + \frac{(z+4)^2}{3^2} = 1.$$

This is an ellipsoid with centre at $(2, -3, -4)$; semi-axes 2, 1, 3 and principal planes $x = 2$; $y = -3$ and $z = -4$.

▶▷▶ 2. *Find the nature of the following surface given by the equation*

$$4x^2 + 6y^2 - 16x + 12y - 24z + 22 = 0.$$

☐ The given equation can be written as

$$4(x-2)^2 + 6(y+1)^2 = 24z$$

$$\text{or,} \quad \frac{(x-2)^2}{6} + \frac{(y+1)^2}{4} = z.$$

This is an elliptic paraboloid with vertex $(2, -1, 0)$.

▶▷▶ 3. *Show that the plane $x = 2$ meets the ellipsoid $\frac{x^2}{16} + \frac{y^2}{12} + \frac{z^2}{4} = 1$ in an ellipse of area $3\sqrt{3}\pi$.*

☐ Putting $x = 2$ in the given equation we get

$$\frac{y^2}{12} + \frac{z^2}{4} = 1 - \frac{1}{4} = \frac{3}{4}$$

$$\text{or,} \quad \frac{y^2}{9} + \frac{z^2}{3} = 1,$$

this is an ellipse of area $3\sqrt{3}\pi$.

▶▷▶ 4. *Show that the plane* $3x - 3y + 4z + 2 = 0$ *meets the surface* $\frac{x^2}{2} - \frac{z^2}{3} = y$ *in a hyperbola.*

□ Eliminating y from the equations $3x - 3y + 4z + 2 = 0$ and $\frac{x^2}{2} - \frac{z^2}{3} = y$ we get

$$3x^2 - 2z^2 = 6y = 2(3x + 4z + 2)$$

$$\text{or,} \quad 3x^2 - 2z^2 - 6x - 8z - 4 = 0.$$

Here $h = 0$, $a = 3$ and $b = -2$, i.e. $h^2 > ab$, so it represents a hyperbola in x-z plane.

EXERCISE

1. Find the nature of the surface given by the equation

$$2x^2 + 5y^2 + 3z^2 - 4x + 20y - 6z = 5.$$

2. Find the nature of the surface given by the equation

$$2x^2 + 3y^2 - 8x + 6y - 12z + 11 = 0.$$

3. Show that the plane $z + 1 = 0$ meets the surface $\frac{x^2}{32} - \frac{y^2}{18} + \frac{z^2}{2} = 1$ in a hyperbola.
4. Show that the plane $x = k$ meets the surface $\frac{x^2}{a^2} + \frac{y^2}{b^2} + \frac{z^2}{c^2} = 1$ in an ellipse of area $\frac{\pi bc(a^2 - k^2)}{a^2}$.
5. Show that the plane $9x - 6y + 2z - 28 = 0$ meets the surface $\frac{x^2}{4} + \frac{y^2}{9} - \frac{z^2}{36} = 1$ in a hyperbola.
6. Find the values of k for which the plane $x + ky = 1$ intersects the conicoid $x^2 + y^2 - z^2 = 1$ in a hyperbola.
7. Show that the plane $3x - y + 6z - 14 = 0$ meets the surface $2x^2 + y^2 = 12z$ in an ellipse.
8. Find the values of k for which the plane $x + ky = 2$ intersects the paraboloid $\frac{x^2}{2} + \frac{z^2}{3} = y$ in (i) an ellipse; (ii) a parabola.

Chapter 10

Tangent, Normal, Pole and Polar

10.1 Equation of a Tangent to the Surface

Let the equation of the surface be

$$F(x, y, z) = ax^2 + by^2 + cz^2 + 2hxy + 2gxz + 2fyz + 2ux + 2vy + 2wz + d = 0.$$

The equation of a line passing through the point (x_1, y_1, z_1) and having direction cosines l, m, n is

$$\frac{x - x_1}{l} = \frac{y - y_1}{m} = \frac{z - z_1}{n} = r. \tag{1}$$

Any point on this line can be taken as $(x_1 + lr, y_1 + mr, z_1 + nr)$. If this point lies on the given surface then we have

$$\phi(l, m, n)r^2 + 2r[l(ax_1 + hy_1 + gz_1 + u) + m(hx_1 + by_1 + fz_1 + v) \\ + n(gx_1 + fy_1 + cz_1 + w)] + F(x_1, y_1, z_1) = 0.$$

where $\phi(x, y, z) = ax^2 + by^2 + cz^2 + 2hxy + 2gxz + 2fyz$.

The above quadratic equation in r gives two values of r indicating that the line (1) meets the given surface in two points. The line given by (1) will be a tangent to the surface if the equation in r has two equal roots and this will happen if

$$l(ax_1 + hy_1 + gz_1 + u) + m(hx_1 + by_1 + fz_1 + v) \\ + n(gx_1 + fy_1 + cz_1 + w) = 0 \tag{2}$$

and $$F(x_1, y_1, z_1) = 0. \tag{3}$$

If the line (1) touches the given surface at (x_1, y_1, z_1) then the condition (3) is identically satisfied. Now eliminating l, m, n from (2) with the help of (1) and (3) we get

$$(x - x_1)(ax_1 + hy_1 + gz_1 + u) + (y - y_1)(hx_1 + by_1 + fz_1 + v) \\ + (z - z_1)(gx_1 + fy_1 + cz_1 + w) = 0$$

or, $$axx_1 + byy_1 + czz_1 + f(yz_1 + y_1z) + g(xz_1 + x_1z) + h(xy_1 + x_1y) \\ + u(x + x_1) + v(y + y_1) + w(z + z_1) + d = 0.$$

10.2 Tangent Plane and Condition of Tangency

The locus of the tangent lines to a surface at a point is called tangent plane to the surface at that point. Next we will discuss under which condition the plane $lx + my + nz = p$ will touch

(i) the sphere: $x^2 + y^2 + z^2 + 2ux + 2vy + 2wz + d = 0$;
(ii) the central conicoid: $ax^2 + by^2 + cz^2 = 1$ and
(iii) the paraboloid: $ax^2 + by^2 = 2cz$.

(i) The plane will touch the sphere $x^2 + y^2 + z^2 + 2ux + 2vy + 2wz + d = 0$ if the length of the perpendicular from the centre to the plane is equal to the radius of the sphere i.e., if

$$\frac{-ul - vm - wn - p}{\sqrt{l^2 + m^2 + n^2}} = \sqrt{u^2 + v^2 + w^2 - d}$$

i.e., $$(ul + vm + wn + p)^2 = (l^2 + m^2 + n^2)(u^2 + v^2 + w^2 - d).$$

This is the required condition.

(ii) If the plane touches the conicoid at (α, β, γ) then it should be identical with

$$ax\alpha + by\beta + cz\gamma = 1.$$

Comparing this with $lx + my + nz = p$, we get

$$\frac{a\alpha}{l} = \frac{b\beta}{m} = \frac{c\gamma}{n} = \frac{1}{p}$$

This gives that the point of contact is $\left(\frac{l}{ap}, \frac{m}{bp}, \frac{n}{cp}\right)$ and we get

$$a\left(\frac{l}{ap}\right)^2 + b\left(\frac{m}{bp}\right)^2 + c\left(\frac{n}{cp}\right)^2 = 1$$

or, $$\frac{l^2}{a} + \frac{m^2}{b} + \frac{n^2}{c} = p^2.$$

This is the required condition.

(iii) If the plane touches the paraboloid at (α, β, γ) then it should be identical with

$$ax\alpha + by\beta = c(z + \gamma).$$

Comparing this with $lx + my + nz = p$, we get

$$\frac{a\alpha}{l} = \frac{b\beta}{m} = \frac{-c}{n} = \frac{c\gamma}{p}$$

This gives that the point of contact is $\left(\frac{-cl}{an}, \frac{-cm}{bn}, \frac{-p}{n}\right)$ and we get

$$a\left(\frac{cl}{an}\right)^2 + b\left(\frac{cm}{bn}\right)^2 = -\frac{2cp}{n}$$

or, $$\frac{l^2}{a} + \frac{m^2}{b} + \frac{2np}{c} = 0.$$

This is the required condition.

10.3 Condition of Tangency of the Plane $lx + my + nz = 0$ to the Cone $ax^2 + by^2 + cz^2 + 2hxy + 2gxz + 2fyz = 0$

Equation of the tangent to the cone at the point (α, β, γ) is

$$ax\alpha + by\beta + cz\gamma + h(x\beta + \alpha y) + g(x\gamma + \alpha z) + f(y\gamma + z\beta) = 0$$

or, $$(a\alpha + h\beta + g\gamma)x + (h\alpha + b\beta + f\gamma)y + (g\alpha + f\beta + c\gamma)z = 0.$$

Comparing this with $lx + my + nz = 0$ we get

$$\frac{a\alpha + h\beta + g\gamma}{l} = \frac{h\alpha + b\beta + f\gamma}{m} = \frac{g\alpha + f\beta + c\gamma}{n} = k, \quad \text{(say)}$$

i.e., $$a\alpha + h\beta + g\gamma - lk = 0; \quad h\alpha + b\beta + f\gamma - mk = 0$$

and $$g\alpha + f\beta + c\gamma - nk = 0.$$

We also have $l\alpha + m\beta + n\gamma = 0$.

Eliminating α, β, γ and k from the above equations we get

$$\begin{vmatrix} a & h & g & l \\ h & b & f & m \\ g & f & c & n \\ l & m & n & 0 \end{vmatrix} = 0.$$

This is the required condition. After expanding the above determinant we get

$$Al^2 + Bm^2 + Cn^2 + 2Fmn + 2Gnl + 2Hml = 0$$

where A, B, C, F, G, H are the cofactors of a, b, c, f, g, h in the determinant

$$\begin{vmatrix} a & h & g \\ h & b & f \\ g & f & c \end{vmatrix}.$$

10.4 Equation of the Plane of Contact of the Tangent Planes to a Conicoid

The plane containing the points of contact of the tangent planes through a point to a conicoid is called the plane of contact.

Equation of the tangent plane at the point (α, β, γ) to the conicoid $ax^2 + by^2 + cz^2 = 1$ is

$$ax\alpha + by\beta + cz\gamma = 1.$$

If it passes through the fixed point (f, g, h) then we have

$$af\alpha + bg\beta + ch\gamma = 1.$$

Hence the locus of the point (α, β, γ) is

$$afx + bgy + chz = 1.$$

This is the equation of the plane of contact of the tangent planes passing through the fixed point (f, g, h) to the conicoid $ax^2 + by^2 + cz^2 = 1$.

10.5 The Pole and Polar Plane

The locus of the points of intersection of the tangent planes drawn at the extremities of the chords passing through a fixed point to the conicoid is called polar plane of the fixed point with respect to the conicoid and the fixed point is called pole.

If we take the equation of conicoid as $ax^2 + by^2 + cz^2 = 1$ and (f, g, h) be the point of intersection of the tangent planes at the extremities of the chord passing through the fixed point (α, β, γ), then the equation of the plane of contact of the tangent planes passing through the point (f, g, h) is

$$axf + byg + czh = 1.$$

If it passes through the point (α, β, γ) then we have

$$a\alpha f + b\beta g + c\gamma h = 1.$$

Hence locus of the point (f, g, h) is

$$a\alpha x + b\beta y + c\gamma z = 1.$$

This is the polar plane of the point (α, β, γ) with respect to the conicoid and the point (α, β, γ) is called pole of the plane with respect to the conic.

Again, let $lx + my + nz = p$ be a polar plane of a pole (α, β, γ) with respect to the conicoid $ax^2 + by^2 + cz^2 = 1$, then comparing the equation of the plane with $a\alpha x + b\beta y + c\gamma z = 1$ we get

$$\frac{a\alpha}{l} = \frac{b\beta}{m} = \frac{c\gamma}{n} = \frac{1}{p}.$$

This gives pole of the plane $lx+my+nz = p$ with respect to the conicoid $ax^2+by^2+cz^2 = 1$ is

$$\left(\frac{l}{ap}, \frac{m}{bp}, \frac{n}{cp}\right).$$

If the polar plane of a point P passes through the point Q then the polar plane of Q passes through the point P and two such points are called conjugate points and the planes are called conjugate planes.

10.6 Director Sphere

The locus of the point of intersection of three mutually perpendicular tangent planes to a conicoid is called the director sphere.

(i) *Central conicoid*: From section 10.2, we see that for all values of l, m, n the tangent planes

$$lx + my + nz = \sqrt{\frac{l^2}{a} + \frac{m^2}{b} + \frac{n^2}{c}}$$

touch the conicoid $ax^2 + by^2 + cz^2 = 1$. If we take the three mutually perpendicular as

$$l_k x + m_k y + n_k z = \sqrt{\frac{l_k^2}{a} + \frac{m_k^2}{b} + \frac{n_k^2}{c}}, \quad k = 1, 2, 3. \tag{4}$$

Then we have $l_k^2 + m_k^2 + n_k^2 = 1$ and $l_i l_j + m_i m_j + n_i n_j = 0$ for $i, j, k = 1, 2, 3$. Now rotating the normals it can be shown that

$$\sum_{k=1}^{3} l_k^2 = 1 = \sum_{k=1}^{3} m_k^2 = \sum_{k=1}^{3} n_k^2 = 1 \quad \text{and} \quad \sum_{k=1}^{3} l_k m_k = 0 = \sum_{k=1}^{3} n_k m_k = \sum_{k=1}^{3} l_k n_k. \tag{5}$$

Now squaring and adding the equations given by (4) and using the results (5) we obtain

$$x^2 + y^2 + z^2 = \frac{1}{a} + \frac{1}{b} + \frac{1}{c}.$$

The above sphere is called director sphere of the conicoid $ax^2 + by^2 + cz^2 = 1$.

(ii) *Paraboloid*: From section 10.2, we also see that for all values of l, m, n the tangent planes

$$lx + my + nz + \frac{c}{2n}\left(\frac{l^2}{a} + \frac{m^2}{b}\right) = 0$$

touch the paraboloid $ax^2 + by^2 = 2cz$. If we take the three mutually perpendicular as

$$l_k x + m_k y + n_k z + \frac{c}{2n_k}\left(\frac{l_k^2}{a} + \frac{m_k^2}{b}\right) = 0, \quad k = 1, 2, 3. \tag{6}$$

Then we have $l_k^2 + m_k^2 + n_k^2 = 1$ and $l_i l_j + m_i m_j + n_i n_j = 0$ for $i, j, k = 1, 2, 3$. Proceeding as before we see that the locus of the point of intersection of three mutually perpendicular tangent planes to the paraboloid is

$$2z + c\left(\frac{1}{a} + \frac{1}{b}\right) = 0.$$

10.7 Reciprocal Cone and Its Equation

The locus of the lines through the vertex of a given cone and perpendicular to the tangent planes of the cone is called the reciprocal cone of the given cone.

Let the equation of the cone be $ax^2 + by^2 + cz^2 + 2hxy + 2gxz + 2fyz = 0$ and $lx + my + nz = 0$ be a tangent plane to it. Then we have

$$Al^2 + Bm^2 + Cn^2 + 2Fmn + 2Gnl + 2Hml = 0. \tag{7}$$

Equation of the line through the vertex of the cone i.e., $(0, 0, 0)$ and perpendicular to the tangent plane is

$$\frac{x}{l} = \frac{y}{m} = \frac{z}{n}. \tag{8}$$

Eliminating l, m, n from (7) and (8) we get the locus of the lines given by (8) is

$$Ax^2 + By^2 + Cz^2 + 2Fyz + 2Gxz + 2Hxy = 0 \tag{9}$$

where A, B, C, F, G, H are the cofactors of a, b, c, f, g, h in the determinant

$$\Delta = \begin{vmatrix} a & h & g \\ h & b & f \\ g & f & c \end{vmatrix}.$$

Hence equation of reciprocal cone of the given cone is given by Eqn. (9).

Further if $\Delta \neq 0$ then Δ^2 is also non-zero. Thus Eqn. (9) also represents a cone. Let the reciprocal cone of the cone given by (9) be

$$a'x^2 + b'y^2 + c'z^2 + 2f'yz + 2g'xz + 2h'xy = 0. \tag{10}$$

Then

$$\begin{aligned} a' = BC - F^2 &= (ac - g^2)(ab - h^2) - (af - hg)^2 \\ &= a(abc + 2fgh - af^2 - bg^2 - ch^2) = a\Delta. \end{aligned}$$

Similarly, $b' = b\Delta$; $c' = c\Delta$; $f' = f\Delta$; $g' = g\Delta$ and $h' = h\Delta$.

So the Eqn. (4) becomes

$$ax^2 + by^2 + cz^2 + 2hxy + 2gxz + 2fyz = 0 \quad \text{as} \quad \Delta \neq 0.$$

Hence the cones

$$\begin{aligned} &ax^2 + by^2 + cz^2 + 2hxy + 2gxz + 2fyz = 0 \\ \text{and} \quad &Ax^2 + By^2 + Cz^2 + 2Fyz + 2Gxz + 2Hxy = 0 \end{aligned}$$

are reciprocal to each other.

10.8 Enveloping Cone and Its Equation

The locus of the tangents to a surface drawn from a point is called enveloping cone to that surface with the fixed point as vertex. Being generated by the tangents to the surface it is also known as tangent cone.

Equation of a straight line passing through $P(\alpha, \beta, \gamma)$ and having direction cosines l, m, n is

$$\frac{x-\alpha}{l} = \frac{y-\beta}{m} = \frac{z-\gamma}{n} = r.$$

From section 10.1, we see that this line will touch the surface

$$F(x, y, z) = ax^2 + by^2 + cz^2 + 2hxy + 2gxz + 2fyz + 2ux + 2vy + 2wz + d = 0$$

if

$$\begin{aligned} &[l(a\alpha + h\beta + g\gamma + u) + m(h\alpha + b\beta + f\gamma + v) + n(g\alpha + f\beta + c\gamma + w)]^2 \\ &= \phi(l, m, n)F(\alpha, \beta, \gamma). \end{aligned}$$

Eliminating l, m, n we get

$$\begin{aligned} &[(x-\alpha)(a\alpha + h\beta + g\gamma + u) + (y-\beta)(h\alpha + b\beta + f\gamma + v) \\ &+ (z-\gamma)(g\alpha + f\beta + c\gamma + w)]^2 = \phi(x-\alpha, y-\beta, z-\gamma)F(\alpha, \beta, \gamma). \end{aligned}$$

If we write $S = F(x, y, z)$; $S_1 = F(\alpha, \beta, \gamma)$ and

$$\begin{aligned} T = &ax\alpha + by\beta + cz\gamma + f(y\gamma + \beta z) + g(x\gamma + \alpha z) + h(x\beta + \alpha y) \\ &+ u(x+\alpha) + v(y+\beta) + w(z+\gamma) + d \end{aligned}$$

then the above result can be simplified to

$$(T - S_1)^2 = S_1(S + S_1 - 2T) \quad \text{or} \quad SS_1 = T^2. \tag{11}$$

The Eqn. (11) is the equation of enveloping cone of the given surface with the point (α, β, γ) as vertex.

10.9 Enveloping Cylinder and Its Equation

The locus of the tangents to a surface drawn parallel to a given line is called enveloping cylinder to that surface.

If the given line be $\frac{x}{l} = \frac{y}{m} = \frac{z}{n}$, then equation of the tangent to surface

$$F(x, y, z) = ax^2 + by^2 + cz^2 + 2hxy + 2gxz + 2fyz + 2ux + 2vy + 2wz + d = 0.$$

at the point (α, β, γ) is

$$\frac{x-\alpha}{l} = \frac{y-\beta}{m} = \frac{z-\gamma}{n}.$$

From section 10.1, we see that this line will touch the surface $F(x, y, z) = 0$ if

$$[l(a\alpha + h\beta + g\gamma + u) + m(h\alpha + b\beta + f\gamma + v) + n(g\alpha + f\beta + c\gamma + w)]^2$$
$$= \phi(l, m, n)F(\alpha, \beta, \gamma).$$

Hence locus of the point (α, β, γ) i.e., equation of the enveloping cylinder of the surface $F(x, y, z) = 0$ is

$$[l(ax + hy + gz + u) + m(hx + by + fz + v) + n(gx + fy + cz + w)]^2$$
$$= \phi(l, m, n)F(x, y, z).$$

10.10 Equation of a Normal to a Conicoid

The normal to a surface at any point is a straight line through that point and perpendicular to the tangent plane at that point to the surface.

(i) *Central conicoid*: If we take the equation of central conicoid as

$$ax^2 + by^2 + cz^2 = 1$$

then equation of the tangent plane to it at the point (α, β, γ) is

$$ax\alpha + by\beta + cz\gamma = 1.$$

The direction ratios of the normal to the tangent plane are $a\alpha, b\beta, c\gamma$. Hence the equation of the normal to the conicoid at the point (α, β, γ) is

$$\frac{x-\alpha}{a\alpha} = \frac{y-\beta}{b\beta} = \frac{z-\gamma}{c\gamma}.$$

If p be the length of the perpendicular from origin to the plane then we have

$$p^2 = \frac{1}{(a\alpha)^2 + (b\beta)^2 + (c\gamma)^2}$$

$$\text{or,} \quad (ap\alpha)^2 + (bp\beta)^2 + (cp\gamma)^2 = 1.$$

This shows that the direction cosines of the normal to the tangent plane are $ap\alpha, bp\beta, cp\gamma$.

(ii) *Paraboloid*: In case of the paraboloid $ax^2 + by^2 = 2cz$ the equation of the tangent plane at the point (α, β, γ) is $ax\alpha + by\beta = c(z + \gamma)$, so the direction ratios of the normal to the tangent plane are $a\alpha, b\beta, -c$. Hence the equation of the normal to the paraboloid at the point (α, β, γ) is

$$\frac{x-\alpha}{a\alpha} = \frac{y-\beta}{b\beta} = \frac{z-\gamma}{-c}.$$

10.11 Normals from a Given Point

(i) *Central conicoid*: The equation of the normal to the conicoid $ax^2 + by^2 + cz^2 = 1$ at the point (α, β, γ) is

$$\frac{x-\alpha}{a\alpha} = \frac{y-\beta}{b\beta} = \frac{z-\gamma}{c\gamma}.$$

If it passes through the point (f, g, h) then we have

$$\frac{f-\alpha}{a\alpha} = \frac{g-\beta}{b\beta} = \frac{h-\gamma}{c\gamma} = k, \quad \text{(say)}.$$

Then

$$\alpha = \frac{f}{1+ak}; \quad \beta = \frac{g}{1+bk} \quad \text{and} \quad \gamma = \frac{h}{1+ck}.$$

Since (α, β, γ) lies on the conicoid we get

$$a\left(\frac{f}{1+ak}\right)^2 + b\left(\frac{g}{1+bk}\right)^2 + c\left(\frac{h}{1+ck}\right)^2 = 1.$$

Being a sixth degree equation in k, the above equation gives six values of k for a given values of (f, g, h). So from a given point six normals can be drawn to a conicoid.

(ii) *Paraboloid*: The equation of the normal to the paraboloid $ax^2 + by^2 = 2cz$ at the point (α, β, γ) is

$$\frac{x-\alpha}{a\alpha} = \frac{y-\beta}{b\beta} = \frac{z-\gamma}{-c}.$$

If it passes through the point (f, g, h) then we have

$$\frac{f-\alpha}{a\alpha} = \frac{g-\beta}{b\beta} = \frac{h-\gamma}{-c} = k, \quad \text{(say)}.$$

Then

$$\alpha = \frac{f}{1+ak}; \quad \beta = \frac{g}{1+bk} \quad \text{and} \quad \gamma = h + ck.$$

Since (α, β, γ) lies on the paraboloid we get

$$a\left(\frac{f}{1+ak}\right)^2 + b\left(\frac{g}{1+bk}\right)^2 = 2c(h+ck).$$

Being a fifth degree equation in k, the above equation gives five values of k for a given values of (f, g, h). So from a given point five normals can be drawn to a paraboloid.

10.12 Cubic Curve Through the Feet of the Normal

(i) *Central conicoid*: The equation of the normal to the conicoid $ax^2 + by^2 + cz^2 = 1$ at the point (α, β, γ) is

$$\frac{x-\alpha}{a\alpha} = \frac{y-\beta}{b\beta} = \frac{z-\gamma}{c\gamma}.$$

If it passes through the point (f, g, h) then we have

$$\frac{f-\alpha}{a\alpha} = \frac{g-\beta}{b\beta} = \frac{h-\gamma}{c\gamma}.$$

From this we see that the feet of the normals (α, β, γ) lie on the cones

$$by(f-x) = ax(g-y);$$
$$cz(f-x) = ax(h-z);$$
$$cz(g-y) = by(h-z).$$

The feet of the normals are common to the above cylinders and will lie on the common curve of intersection of the above cylinders whose parametric equations are

$$x = \frac{f}{1+ak}; \quad y = \frac{g}{1+bk} \quad \text{and} \quad z = \frac{h}{1+ck}.$$

Now the point of intersections of the curves with the plane $ax + by + cz = d$ are given by

$$\frac{af}{1+ak} + \frac{bg}{1+bk} + \frac{ch}{1+ck} = d.$$

The above cubic equation in k shows that the feet of the normals from a point to a conicoid are the intersection of the conicoid with a cubic curve.

(ii) *Paraboloid*: The equation of the normal to the paraboloid $ax^2 + by^2 = 2cz$ at the point (α, β, γ) is

$$\frac{x-\alpha}{a\alpha} = \frac{y-\beta}{b\beta} = \frac{z-\gamma}{-c}.$$

If it passes through the point (f, g, h) then we have

$$\frac{f-\alpha}{a\alpha} = \frac{g-\beta}{b\beta} = \frac{h-\gamma}{-c} = k, \quad \text{(say)}.$$

Then

$$\alpha = \frac{f}{1+ak}; \quad \beta = \frac{g}{1+bk} \quad \text{and} \quad \gamma = h + ck.$$

Thus the feet of the normals lie on the curve whose parametric equation is

$$x = \frac{f}{1+ak}; \quad y = \frac{g}{1+bk} \quad \text{and} \quad z = h + ck. \tag{12}$$

So the feet of the normal drawn through (f, g, h) to the paraboloid are the points of intersection of paraboloid with the curve given by (12). Now the points of intersection of the curve (12) with the plane $ax + by + cz = d$ are given by

$$\frac{af}{1+ak} + \frac{bg}{1+bk} + c(h+ck) = d.$$

The above cubic equation in k shows that the feet of the normals from a point to a paraboloid are the intersection of the paraboloid with a cubic curve.

10.13 Cone through the Six Normals to a Conicoid

Let the normals are drawn from the point (f, g, h) to the conicoid $ax^2 + by^2 + cz^2 = 1$, then the equation of the cubic curve through the feet of the normals is

$$x = \frac{f}{1+ak}; \quad y = \frac{g}{1+bk} \quad \text{and} \quad z = \frac{h}{1+ck}.$$

Equation of a line through the point (f, g, h) is

$$\frac{x-f}{l}=\frac{y-g}{m}=\frac{z-h}{n}. \tag{13}$$

If it intersects the cubic curve then we have

$$\frac{\dfrac{f}{1+ak}-f}{l}=\frac{\dfrac{g}{1+bk}-g}{m}=\frac{\dfrac{h}{1+ck}-h}{n}$$

or, $$\frac{af}{l(1+ak)}=\frac{bg}{m(1+bk)}=\frac{ch}{n(1+ck)}$$

or, $$\frac{\dfrac{af}{l}(b-c)}{(b-c)(1+ak)}=\frac{\dfrac{bg}{m}(c-a)}{(b-c)(1+bk)}=\frac{\dfrac{ch}{n}(a-b)}{(a-b)(1+ck)}$$

$$=\frac{\dfrac{af}{l}(b-c)+\dfrac{bg}{m}(c-a)\dfrac{ch}{n}(a-b)}{0}$$

i.e., $$\frac{af}{l}(b-c)+\frac{bg}{m}(c-a)\frac{ch}{n}(a-b)=0.$$

Eliminating l, m, n from the above equation with the help of (13) we get

$$\frac{af}{x-f}(b-c)+\frac{bg}{y-g}(c-a)+\frac{ch}{z-h}(a-b)=0$$

which is an equation of a cone generated by the lines given by (13) with guiding curve as the cubic curve.

10.14 Cone through the Five Normals to a Paraboloid

Let the normals are drawn from the point (f, g, h) to the paraboloid $ax^2+by^2=2cz$, then the equation of the cubic curve through the feet of the normals is

$$x=\frac{f}{1+ak};\quad y=\frac{g}{1+bk}\quad \text{and}\quad z=h+ck.$$

Equation of a line through the point (f, g, h) is

$$\frac{x-f}{l}=\frac{y-g}{m}=\frac{z-h}{n}. \tag{14}$$

If it intersects the cubic curve then we have

$$\frac{\dfrac{f}{1+ak}-f}{l}=\frac{\dfrac{g}{1+bk}-g}{m}=\frac{ck}{n}$$

or, $$\frac{af}{l(1+ak)}=\frac{bg}{m(1+bk)}=\frac{c}{n}$$

or, $$\frac{\dfrac{1}{a}+k}{\dfrac{f}{l}}=\frac{\dfrac{1}{b}+k}{\dfrac{g}{m}}=\frac{n}{c}=\frac{\dfrac{1}{a}-\dfrac{1}{b}}{\dfrac{f}{l}-\dfrac{g}{m}}$$

or, $$\frac{f}{l}-\frac{g}{m}+\frac{c}{n}\left(\frac{1}{b}-\frac{1}{a}\right)=0.$$

Eliminating l, m, n using (14) we get

$$\frac{f}{x-f} - \frac{g}{y-g} + \frac{c}{z-h}\left(\frac{1}{b} - \frac{1}{a}\right) = 0.$$

Which is an equation of a cone generated by the lines given by (14) with guiding curve as the cubic curve.

Examples

▶▷▶ 1. *Show that the straight line* $\frac{x-2}{2} = \frac{y-3}{-6} = \frac{z-1}{1}$ *meets the conicoid* $\frac{x^2}{16} + \frac{y^2}{9} - \frac{z^2}{4} = 1$ *and find the points of intersection.*

☐ Any point on the given line is $(2+2r, 3-6r, 1+r)$, where r is an arbitrary constant. If it lies on the conicoid then we have

$$\frac{(2+2r)^2}{16} + \frac{(3-6r)^2}{9} - \frac{(1+r)^2}{4} = 1$$

or, $$\frac{(3-6r)^2}{9} = 1$$

or, $$3 - 6r = \pm 3$$

or, $$r = 0,\ 1.$$

Two real values of r imply that the given straight line intersects the conicoid at two distinct points whose coordinates are (2, 3, 1) and (4, −3, 2).

▶▷▶ 2. *Show that the plane* $2x - 2y + 3z = 9$ *touches the ellipsoid* $x^2 + 2y^2 + 3z^2 = 9$.

☐ Let the point of contact be (f, g, h). Then the given equation must be identical with

$$fx + 2gy + 3zh = 9.$$

Comparing this with the given equation of the plane we get

$$\frac{f}{2} = \frac{2g}{-2} = \frac{3h}{3} = \frac{9}{9}$$

we get $f = 2$; $g = -1$ and $h = 1$. Now the point $(2, -1, 1)$ lies on the conicoid as

$$2^2 + 2(-1)^2 + 3(1)^2 = 9.$$

Hence the plane $2x - 2y + 3z = 9$ touches the ellipsoid $x^2 + 2y^2 + 3z^2 = 9$ at the point $(2, -1, 1)$.

▶▷▶ 3. *Find the equation of the tangent plane to the paraboloid* $2x^2 - y^2 = 4z$ *which passes through the straight line* $x + 1 = 0 = y + z - 1$.

☐ Equation of the plane through the given line can be written as

$$x + 1 + k(y + z - 1) = 0$$

or, $$x + ky + kz + 1 - k = 0. \tag{1}$$

Again equation of the tangent plane to the paraboloid at the point (f, g, h) is

$$2fx - yg - 2z - 2h = 0. \tag{2}$$

Comparing (1) and (2) we get

$$\frac{2f}{1} = \frac{-g}{k} = \frac{-2}{k} = \frac{-2h}{1-k}.$$

From the above ratios we get

$$f = -\frac{1}{k}; \quad g = 2; \quad h = \frac{1-k}{k}.$$

Since the point (f, g, h) lies on the conicoid we have

$$2f^2 - g^2 = 4h$$

or, $$\frac{2}{k^2} - 4 = 4\frac{1-k}{k}$$

or, $$\frac{1}{k}\left(1 - \frac{1}{2k}\right) = 0 \Rightarrow \frac{1}{k} = 0,\ 2.$$

Hence the equation of tangent planes are $y + z = 1$; $2x + y + z + 1 = 0$ and the points of contact are $(0, 2, -1)$ and $(-2, 2, 1)$.

▶▷▶ 4. *Find the equation of the tangent plane to the conicoid* $3x^2 + 2y^2 - 6z^2 = 6$ *which passes through the point* $(3, 4, -3)$ *and is parallel to the straight line* $x = y = -z$.

☐ Equation of the tangent plane to the conicoid $3x^2 + 2y^2 - 6z^2 = 6$ at the point (f, g, h) is

$$3xf + 2yg - 6zh = 6.$$

If it passes through the point $(3, 4, -3)$ then we have

$$9f + 8g + 18h = 6. \tag{1}$$

Again if the plane is parallel to the line $x = y = -z$ then we have

$$3f + 2g + 6h = 0. \tag{2}$$

As (f, g, h) lies on the conicoid we have

$$3f^2 + 2g^2 - 6h^2 = 6. \tag{3}$$

From (1) and (2) we get $g = 3$. Using this in (2) and (3) we get

$$f + 2h = -2 \quad \text{and} \quad f^2 - 2h^2 = -4.$$

Solving we get $f = 2$ and $h = -2$. Hence equation of the tangent plane is

$$6x + 6y + 12z = 6$$

i.e, $$x + y + 2z = 1.$$

▶▷▶ 5. *If* $2r$ *be the distance between two tangent planes to the ellipsoid* $\frac{x^2}{a^2} + \frac{y^2}{b^2} + \frac{z^2}{c^2} = 1$ *which are parallel to the plane* $lx + my + nz = 0$, *then show that*

$$(a^2 - r^2)l^2 + (b^2 - r^2)m^2 + (c^2 - r^2)n^2 = 0.$$

□ Equation of the plane parallel to the given plane is $lx + my + nz = p$. If it touches the ellipsoid then we have

$$a^2l^2 + b^2m^2 + c^2n^2 = p^2.$$

Thus the planes $lx + my + nz = \pm\sqrt{a^2l^2 + b^2m^2 + c^2n^2}$ touch the given ellipsoid for all values of l, m, n. If the distance between them be $2r$ then we have

$$\frac{2\sqrt{a^2l^2 + b^2m^2 + c^2n^2}}{\sqrt{l^2 + m^2 + n^2}} = 2r$$

or, $$a^2l^2 + b^2m^2 + c^2n^2 = r^2(l^2 + m^2 + n^2)$$

or, $$(a^2 - r^2)l^2 + (b^2 - r^2)m^2 + (c^2 - r^2)n^2 = 0.$$

▶▷▶ 6. *Show that the cones $ax^2 + by^2 + cz^2 = 0$ and $bcx^2 + acy^2 + abz^2 = 0$ are reciprocal to each other.*

□ In the equation of first cone $f = 0$; $g = 0$; $h = 0$. So if the equation of its reciprocal cone be

$$Ax^2 + By^2 + Cz^2 + 2Hxy + 2Gxz + 2Fyz = 0$$

then we have $A = bc - f^2 = bc$. Similarly, $B = ac$; $C = ab$. Also $F = hg - af = 0$ and $G = 0$; $H = 0$.

So the reciprocal cone $ax^2 + by^2 + cz^2 = 0$ is $bcx^2 + acy^2 + abz^2 = 0$ and conversely.

▶▷▶ 7. *The section of the enveloping cone of the ellipsoid $\frac{x^2}{a^2} + \frac{y^2}{b^2} + \frac{z^2}{c^2} = 1$ by the plane $z = 0$ is a parabola. Show that the locus of the vertex of the cone is a pair of planes $z = \pm c$.*

□ Enveloping cone of the ellipsoid with vertex (α, β, γ) is

$$\left(\frac{x^2}{a^2} + \frac{y^2}{b^2} + \frac{z^2}{c^2} - 1\right)\left(\frac{\alpha^2}{a^2} + \frac{\beta^2}{b^2} + \frac{\gamma^2}{c^2} - 1\right) = \left(\frac{x\alpha}{a^2} + \frac{y\beta}{b^2} + \frac{z\gamma}{c^2} - 1\right)^2.$$

To get its intersection with the plane $z = 0$ we put $z = 0$ in the above equation and get

$$\left(\frac{x^2}{a^2} + \frac{y^2}{b^2} - 1\right)\left(\frac{\alpha^2}{a^2} + \frac{\beta^2}{b^2} + \frac{\gamma^2}{c^2} - 1\right) = \left(\frac{x\alpha}{a^2} + \frac{y\beta}{b^2} - 1\right)^2.$$

If it be a parabola in x-y plane, then using the condition $h^2 = ab$ we get

$$\left(\frac{\alpha\beta}{a^2b^2}\right)^2 = \frac{1}{a^2b^2}\left(\frac{\beta^2}{b^2} + \frac{\gamma^2}{c^2} - 1\right)\left(\frac{\alpha^2}{a^2} + \frac{\gamma^2}{c^2} - 1\right)$$

or, $$\left(\frac{\gamma^2}{c^2} - 1\right)\left(\frac{\alpha^2}{a^2} + \frac{\beta^2}{b^2} + \frac{\gamma^2}{c^2} - 1\right) = 0$$

or, $$\gamma = \pm c \quad \text{as} \quad \frac{\alpha^2}{a^2} + \frac{\beta^2}{b^2} + \frac{\gamma^2}{c^2} \neq 1.$$

Hence, locus of the vertex (α, β, γ) is $z = \pm c$.

▶▷▶ 8. *The section of the enveloping cone of the ellipsoid $\frac{x^2}{a^2} + \frac{y^2}{b^2} + \frac{z^2}{c^2} = 1$ by the plane $z = 0$ is a circle. Find the locus of the vertex of the cone.*

□ Enveloping cone of the ellipsoid with vertex (α, β, γ) is

$$\left(\frac{x^2}{a^2}+\frac{y^2}{b^2}+\frac{z^2}{c^2}-1\right)\left(\frac{\alpha^2}{a^2}+\frac{\beta^2}{b^2}+\frac{\gamma^2}{c^2}-1\right)=\left(\frac{x\alpha}{a^2}+\frac{y\beta}{b^2}+\frac{z\gamma}{c^2}-1\right)^2.$$

To get its intersection with the plane $z = 0$ we put $z = 0$ in the above equation and get

$$\left(\frac{x^2}{a^2}+\frac{y^2}{b^2}-1\right)\left(\frac{\alpha^2}{a^2}+\frac{\beta^2}{b^2}+\frac{\gamma^2}{c^2}-1\right)=\left(\frac{x\alpha}{a^2}+\frac{y\beta}{b^2}-1\right)^2.$$

If it be a circle in x-y plane, then using the condition $a = b$ and $h = 0$ we get

$$\alpha\beta = 0 \quad \text{and} \quad \frac{1}{a^2}\left(\frac{\beta^2}{b^2}+\frac{\gamma^2}{c^2}-1\right)=\frac{1}{b^2}\left(\frac{\alpha^2}{a^2}+\frac{\gamma^2}{c^2}-1\right).$$

When $\alpha = 0$, we get

$$\frac{1}{a^2}\left(\frac{\beta^2}{b^2}+\frac{\gamma^2}{c^2}-1\right)=\frac{1}{b^2}\left(\frac{\gamma^2}{c^2}-1\right)$$

$$\text{or,}\quad \frac{\beta^2}{a^2b^2}+\left(\frac{\gamma^2}{c^2}-1\right)\left(\frac{1}{a^2}-\frac{1}{b^2}\right)=0$$

$$\text{or,}\quad \frac{\beta^2}{b^2-a^2}+\frac{\gamma^2}{c^2}=1, \quad \alpha = 0.$$

Hence locus of the vertex (α, β, γ) is

$$\frac{y^2}{b^2-a^2}+\frac{z^2}{c^2}=1, \quad x = 0.$$

When $\beta = 0$, we get

$$\frac{1}{b^2}\left(\frac{\alpha^2}{a^2}+\frac{\gamma^2}{c^2}-1\right)=\frac{1}{a^2}\left(\frac{\gamma^2}{c^2}-1\right)$$

$$\text{or,}\quad \frac{\alpha^2}{a^2b^2}+\left(\frac{\gamma^2}{c^2}-1\right)\left(\frac{1}{b^2}-\frac{1}{a^2}\right)=0$$

$$\text{or,}\quad \frac{\alpha^2}{a^2-b^2}+\frac{\gamma^2}{c^2}=1, \quad \beta = 0.$$

Hence locus of the of the vertex (α, β, γ) is

$$\frac{x^2}{a^2-b^2}+\frac{z^2}{c^2}=1, \quad y = 0.$$

▶▷▶ **9**. *Show that enveloping cylinder of the conicoid $ax^2 + by^2 + cz^2 = 1$ having generators perpendicular to the x-axis may meet the plane $x = 0$ in a pair of straight lines.*

□ Equation of any line perpendicular to x-axis is $\frac{x}{0} = \frac{y}{m} = \frac{z}{n}$. So the enveloping cylinder whose generator is perpendicular to x-axis is

$$(ax^2 + by^2 + cz^2 - 1)(bm^2 + cn^2) = (bmy + cnz)^2.$$

To get the curve of intersection of the cylinder by the plane $x = 0$ we put $x = 0$ in the above equation and get

$$(by^2 + cz^2 - 1)(bm^2 + cn^2) = (bmy + cnz)^2$$

or, $$bc(y^2n^2 + z^2m^2 - 2mnyz) = bm^2 + cn^2$$

or, $$(yn - zm)^2 = \frac{m^2}{c} + \frac{n^2}{b}$$

or, $$yn - zm = \pm\sqrt{\frac{m^2}{c} + \frac{n^2}{b}}$$

which are two straight lines in y-z plane if $\frac{m^2}{c} + \frac{n^2}{b} > 0$.

▶▷▶ **10**. *Prove that the enveloping cylinder of the ellipsoid* $\frac{x^2}{a^2} + \frac{y^2}{b^2} + \frac{z^2}{c^2} = 1$ *whose generators are parallel to the line* $\frac{x}{0} = \frac{y}{\pm\sqrt{a^2 - b^2}} = \frac{z}{c}$ *meet the plane* $z = 0$ *in a circle.*

☐ Equation of the enveloping cylinder is

$$\left(\frac{x^2}{a^2} + \frac{y^2}{b^2} + \frac{z^2}{c^2} - 1\right)\left(\frac{a^2 - b^2}{b^2} + \frac{c^2}{c^2}\right) = \left(\frac{\pm\sqrt{a^2 - b^2}}{b^2}y + \frac{z}{c}\right)^2.$$

To get the curve of intersection of the cylinder by the plane $z = 0$ we put $z = 0$ in the above equation and get

$$\left(\frac{x^2}{a^2} + \frac{y^2}{b^2} - 1\right)\frac{a^2}{b^2} = \left(\frac{\pm\sqrt{a^2 - b^2}}{b^2}y\right)^2; \quad z = 0$$

or, $$x^2 + y^2 = a^2; \quad z = 0.$$

This is a circle.

▶▷▶ **11**. *Show that the locus of the point from which three mutually perpendicular planes can be drawn to touch the ellipse* $\frac{x^2}{a^2} + \frac{y^2}{b^2} = 1$, $z = 0$ *is* $x^2 + y^2 + z^2 = a^2 + b^2$.

☐ Let the moving point be (α, β, γ). Then equation of the generator of the cone is

$$\frac{x - \alpha}{l} = \frac{y - \beta}{m} = \frac{z - \gamma}{n}.$$

As it intersects the given ellipse, we have

$$\frac{1}{a^2}\left(\alpha - \frac{l\gamma}{n}\right)^2 + \frac{1}{b^2}\left(\beta - \frac{m\gamma}{n}\right)^2 = 1.$$

Eliminating l, m, n we get

$$b^2(z\alpha - x\gamma)^2 + a^2(z\beta - y\gamma)^2 = a^2b^2(z - \gamma)^2$$

or, $$b^2\gamma^2x^2 + a^2\gamma^2y^2 + (b^2\alpha^2 + a^2\beta^2 - a^2b^2)z^2$$
$$- 2b^2\alpha\gamma xz - 2a^2\beta\gamma yz + 2a^2b^2\gamma z - a^2b^2\gamma^2 = 0.$$

The above cone will have three perpendicular tangent planes if its reciprocal cone has three mutually perpendicular generators i.e., if $A + B + C = 0$

$$\begin{aligned}
\text{or,} \quad & \text{if } bc - f^2 + ac - g^2 + ab - h^2 = 0 \\
\text{or,} \quad & b^2\gamma^2 a^2\gamma^2 + (b^2\alpha^2 + a^2\beta^2 - a^2b^2)a^2\gamma^2 + \\
& \quad (b^2\alpha^2 + a^2\beta^2 - a^2b^2)b^2\gamma^2 = (b^2\alpha\gamma)^2 + (a^2\beta\gamma)^2 \\
\text{or,} \quad & b^2\gamma^2 a^2 + (b^2\alpha^2 + a^2\beta^2 - a^2b^2)(a^2 + b^2) = b^4\alpha^2 + a^4\beta^2 \\
\text{or,} \quad & b^2\gamma^2 a^2 + b^2(a^2\beta^2 - a^2b^2) + a^2(b^2\alpha^2 - a^2b^2) = 0 \\
\text{or,} \quad & \alpha^2 + \beta^2 + \gamma^2 = a^2 + b^2 \quad \text{as} \quad a^2b^2 \neq 0.
\end{aligned}$$

Hence locus of the point (α, β, γ) is $x^2 + y^2 + z^2 = a^2 + b^2$.

▶▶ **12.** *Tangent planes are drawn to the ellipsoid* $\dfrac{x^2}{a^2} + \dfrac{y^2}{b^2} + \dfrac{z^2}{c^2} = 1$ *through the point* (α, β, γ). *Prove that the perpendicular to them through the origin generates the cone*

$$(\alpha x + \beta y + \gamma z)^2 = a^2x^2 + b^2y^2 + c^2z^2.$$

☐ Equation of the plane through the point (α, β, γ) is

$$lx + my + nz = l\alpha + m\beta + n\gamma.$$

If it touches the ellipsoid then we have

$$(l\alpha + m\beta + n\gamma)^2 = a^2l^2 + b^2m^2 + c^2n^2. \tag{1}$$

Equation of the line through the origin and perpendicular to it is

$$\frac{x}{l} = \frac{y}{m} = \frac{z}{n}. \tag{2}$$

Eliminating l, m, n from (1) with the help of (2) we get

$$(\alpha x + \beta y + \gamma z)^2 = a^2x^2 + b^2y^2 + c^2z^2.$$

▶▶ **13.** *A pair of perpendicular tangent planes to the ellipsoid* $\dfrac{x^2}{a^2} + \dfrac{y^2}{b^2} + \dfrac{z^2}{c^2} = 1$ *pass through the point* $(0, 0, k)$. *Show that their line of intersection lies on the cone*

$$x^2(b^2 + c^2 - k^2) + y^2(a^2 + c^2 - k^2) + (z - k)^2(a^2 + b^2) = 0.$$

☐ Let the line of intersection of the perpendicular tangent planes be

$$\frac{x}{l} = \frac{y}{m} = \frac{z - k}{n}. \tag{1}$$

and equation of one of the perpendicular tangent planes be $ux + vy + w(z - k) = 0$. This will contain the line (1) if

$$ul + vm + wn = 0 \tag{2}$$

and the plane will touch the ellipsoid if

$$u^2a^2 + v^2b^2 + w^2c^2 = w^2k^2. \tag{3}$$

Eliminating w from (2) with the help of (2) we get

$$u^2a^2 + v^2b^2 + \left(\frac{ul + vm}{n}\right)^2 (c^2 - k^2) = 0$$

or, $$[n^2a^2 + l^2(c^2 - k^2)]u^2 + 2lm(c^2 - k^2)uv + [n^2b^2 + m^2(c^2 - k^2)]v^2 = 0$$

or, $$[n^2a^2 + l^2(c^2 - k^2)]\left(\frac{u}{v}\right)^2 + 2lm(c^2 - k^2)\left(\frac{u}{v}\right) + [n^2b^2 + m^2(c^2 - k^2)] = 0.$$

The above second degree equation in u/v shows that two planes are possible to draw to the ellipsoid. If u_1/v_1 and u_2/v_2 be two roots of this equation then

$$\frac{u_1u_2}{v_1v_2} = \frac{n^2b^2 + m^2(c^2 - k^2)}{n^2a^2 + l^2(c^2 - k^2)}$$

or, $$\frac{u_1u_2}{n^2b^2 + m^2(c^2 - k^2)} = \frac{v_1v_2}{n^2a^2 + l^2(c^2 - k^2)}.$$

Similarly eliminating u from (2) and (3) we get

$$\frac{v_1v_2}{n^2a^2 + l^2(c^2 - k^2)} = \frac{w_1w_2}{m^2a^2 + l^2b^2}.$$

Hence the planes will be perpendicular if

$$n^2b^2 + m^2(c^2 - k^2) + n^2a^2 + l^2(c^2 - k^2) + a^2m^2 + l^2b^2 = 0$$

or, $$l^2(b^2 + c^2 - k^2) + m^2(a^2 + c^2 - k^2) + n^2(a^2 + b^2) = 0. \tag{4}$$

Eliminating l, m, n from (4) with the help of (1) we get the locus of the line (1) is

$$x^2(b^2 + c^2 - k^2) + y^2(a^2 + c^2 - k^2) + (z - k)^2(a^2 + b^2) = 0.$$

▶▶ **14.** *Show that the lines draw from origin parallel to the normals to $ax^2+by^2+cz^2 = 1$ at its points of intersection with the plane $lx + my + nz = p$, generates the cone*

$$p^2\left(\frac{x^2}{a} + \frac{y^2}{b} + \frac{z^2}{c}\right) = \left(\frac{lx}{a} + \frac{my}{b} + \frac{nz}{c}\right)^2.$$

□ Let (f, g, h) be any point on the curve of intersection of $ax^2 + by^2 + cz^2 = 1$ and $lx + my + nz = p$ then we have

$$af^2 + bg^2 + ch^2 = 1 \quad \text{and} \quad lf + mg + nh = p.$$

Making homogeneous the first equation with the help of (1) we get

$$af^2 + bg^2 + ch^2 = \left(\frac{lf + mg + nh}{p}\right)^2. \tag{1}$$

Equation of normal at the point (f, g, h) to the conicoid $ax^2 + by^2 + cz^2 = 1$ is

$$\frac{x - f}{af} = \frac{y - g}{bg} = \frac{z - h}{ch}.$$

So equation of the line passing through origin and parallel to the normal is

$$\frac{x}{af} = \frac{y}{bg} = \frac{z}{ch}. \tag{2}$$

Eliminating (f, g, h) from (1) with the help of (2) we get the surface generated by the line (2) is

$$p^2\left(\frac{x^2}{a} + \frac{y^2}{b} + \frac{z^2}{c}\right) = \left(\frac{lx}{a} + \frac{my}{b} + \frac{nz}{c}\right)^2.$$

▶▶ **15.** *If the normal at a point Q on the ellipsoid $\frac{x^2}{a^2} + \frac{y^2}{b^2} + \frac{z^2}{c^2} = 1$ meet the coordinate planes at F, G, H respectively such that $QF^2 + QG^2 + QH^2 = k^2$, prove that the locus of the point Q is*

$$\frac{x^2}{a^4} + \frac{y^2}{b^4} + \frac{z^2}{c^4} = \frac{k^2}{a^4 + b^4 + c^4}.$$

□ Equation of the normal at the point $Q(f, g, h)$ to the ellipsoid $\frac{x^2}{a^2} + \frac{y^2}{b^2} + \frac{z^2}{c^2} = 1$ is

$$\frac{x - f}{fp/a^2} = \frac{y - g}{gp/b^2} = \frac{z - h}{hp/c^2} = r$$

where $\frac{1}{p^2} = \frac{f^2}{a^4} + \frac{g^2}{b^4} + \frac{h^2}{c^4}$.

Any point on the normal is $\left(f + \frac{fpr}{a^2}, g + \frac{gpr}{b^2}, h + \frac{hpr}{c^2}\right)$. This point will be at the plane $x = 0$ if $r = -a^2/p = QF$. Similarly, $QG = -b^2/p$ and $QH = -c^2/p$.

Using the condition $QF^2 + QG^2 + QH^2 = k^2$, we get

$$\frac{a^4 + b^4 + c^4}{p^2} = k^2$$

$$\Rightarrow \quad \frac{1}{p^2} = \frac{k^2}{a^4 + b^4 + c^4} = \frac{f^2}{a^4} + \frac{g^2}{b^4} + \frac{h^2}{c^4}.$$

Hence locus of the point $Q(f, g, h)$ is

$$\frac{x^2}{a^4} + \frac{y^2}{b^4} + \frac{z^2}{c^4} = \frac{k^2}{a^4 + b^4 + c^4}.$$

▶▶ **16.** *Prove that the feet of the six normals from (f, g, h) to the ellipsoid $\frac{x^2}{a^2} + \frac{y^2}{b^2} + \frac{z^2}{c^2} = 1$ lie on the curve of intersection of the ellipsoid and the cone*

$$\frac{a^2(b^2 - c^2)f}{x} + \frac{b^2(c^2 - a^2)g}{y} + \frac{c^2(a^2 - b^2)h}{z} = 0.$$

□ Equation of the normal to the ellipsoid $\frac{x^2}{a^2} + \frac{y^2}{b^2} + \frac{z^2}{c^2} = 1$ at the point (α, β, γ) is

$$\frac{x - \alpha}{\alpha/a^2} = \frac{y - \beta}{\beta/b^2} = \frac{z - \gamma}{\gamma/c^2}.$$

If the normals are drawn from (f, g, h), then we have

$$\frac{f-\alpha}{\alpha/a^2} = \frac{g-\beta}{\beta/b^2} = \frac{h-\gamma}{\gamma/c^2} = r.$$

From it we get

$$r = \frac{fa^2}{\alpha} - a^2$$
$$r = \frac{gb^2}{\beta} - b^2$$
$$r = \frac{hc^2}{\gamma} - c^2.$$

Multiplying the above equations by $b^2 - c^2$, $c^2 - a^2$ and $a^2 - b^2$ and adding we get

$$\frac{a^2(b^2-c^2)f}{\alpha} + \frac{b^2(c^2-a^2)g}{\beta} + \frac{c^2(a^2-b^2)h}{\gamma} = 0.$$

hence locus of the point (α, β, γ) is

$$\frac{a^2(b^2-c^2)f}{x} + \frac{b^2(c^2-a^2)g}{y} + \frac{c^2(a^2-b^2)h}{z} = 0.$$

▶▷▶ 17. *Prove that the locus of the poles of the tangent planes of the conicoid $ax^2 + by^2 + cz^2 = 1$ with respect to the conicoid $\alpha x^2 + \beta y^2 + \gamma z^2 = 1$ is the conicoid*

$$\frac{\alpha^2 x^2}{a} + \frac{\beta^2 y^2}{b} + \frac{\gamma^2 z^2}{c} = 1.$$

☐ We know that for all values of l, m, n the plane

$$lx + my + nz = \sqrt{\frac{l^2}{a} + \frac{m^2}{b} + \frac{n^2}{c}} \qquad (1)$$

touches the conicoid $ax^2 + by^2 + cz^2 = 1$. If (f, g, h) be its pole with respect to the conicoid $\alpha x^2 + \beta y^2 + \gamma z^2 = 1$ then the plane (1) is identical with

$$\alpha f x + \beta g y + \gamma h z = 1. \qquad (2)$$

Comparing (1) and (2) we get

$$\frac{l}{f\alpha} = \frac{m}{g\beta} = \frac{n}{h\gamma} = \sqrt{\frac{l^2}{a} + \frac{m^2}{b} + \frac{n^2}{c}} = p, \quad \text{(say)}.$$

Eliminating l, m, n we get

$$\frac{f^2\alpha^2 p^2}{a} + \frac{g^2\beta^2 p^2}{b} + \frac{h^2\gamma^2 p^2}{c} = p^2$$

$$\text{or,} \quad \frac{f^2\alpha^2}{a} + \frac{g^2\beta^2}{b} + \frac{h^2\gamma^2}{c} = 1.$$

Hence locus of the pole (f, g, h) is

$$\frac{\alpha^2x^2}{a}+\frac{\beta^2y^2}{b}+\frac{\gamma^2z^2}{c}=1.$$

▶▷▶ **18.** *Prove that the planes which cut $ax^2+by^2+cz^2=0$ in perpendicular generators touch*

$$\frac{x^2}{b+c}+\frac{y^2}{a+c}+\frac{z^2}{a+b}=0.$$

☐ Let the plane $ux+vy+wz=0$ cuts the cone $ax^2+by^2+cz^2=0$ along the generators

$$\frac{x}{l}=\frac{y}{m}=\frac{z}{n}.$$

Then we have $ul+vm+wn=0$ and $al^2+bm^2+cn^2=0$. Eliminating n we get

$$al^2+bm^2+c\left(\frac{ul+vm}{w}\right)^2=0$$

or, $$(aw^2+cu^2)\left(\frac{l}{m}\right)^2+2uvc\left(\frac{l}{m}\right)+bw^2+cv^2=0.$$

If $\frac{l_1}{m_1}$ and $\frac{l_2}{m_2}$ be its two roots then we have

$$\frac{l_1l_2}{m_1m_2}=\frac{bw^2+cv^2}{aw^2+cu^2}$$

or, $$\frac{l_1l_2}{bw^2+cv^2}=\frac{m_1m_2}{aw^2+cu^2}=\frac{n_1n_2}{av^2+bu^2}.$$

So the lines will be perpendicular if

$$l_1l_2+m_1m_2+n_1n_2=0$$

or, $$(b+c)u^2+(a+c)v^2+(b+a)w^2=0.$$

Hence normal to the plane $ux+vy+wz=0$ generates the cone

$$(b+c)x^2+(a+c)y^2+(b+a)z^2=0.$$

So the plane will touch the reciprocal cone of the above cone whose equation is

$$\frac{x^2}{b+c}+\frac{y^2}{a+c}+\frac{z^2}{a+b}=0.$$

▶▷▶ **19.** *Prove that perpendicular tangent planes to the cone $ax^2+by^2+cz^2=0$ intersects along the generator of the cone $a(b+c)x^2+b(a+c)y^2+c(a+b)z^2=0$.*

☐ Let the plane $ux+vy+wz=0$ touches the cone $ax^2+by^2+cz^2=0$, then normal to the plane

$$\frac{x}{u}=\frac{y}{v}=\frac{z}{w}$$

generates the reciprocal cone

$$\frac{x^2}{a}+\frac{y^2}{b}+\frac{z^2}{c}=0$$

$$\text{or,}\quad bcx^2+acy^2+abz^2=0$$

and we also have

$$bcu^2+acv^2+abw^2=0. \tag{1}$$

This shows that there will be two tangent planes of the form $ux+vy+wz=0$ to the cone $ax^2+by^2+cz^2=0$. If their line of intersection be

$$\frac{x}{l}=\frac{y}{m}=\frac{z}{n}. \tag{2}$$

then we have

$$ul+vm+wn=0. \tag{3}$$

Eliminating w from (1) and (3) we get

$$bcu^2+acv^2+ab\left(\frac{ul+vm}{n}\right)^2=0$$

$$\text{or,}\quad b(cn^2+al^2)\left(\frac{u}{v}\right)^2+2ablm\left(\frac{u}{v}\right)+a(cn^2+bm^2)=0.$$

If $\dfrac{u_1}{v_1}$ and $\dfrac{u_2}{v_2}$ be its two roots then we have

$$\frac{u_1u_2}{v_1v_2}=\frac{a(bm^2+cn^2)}{b(al^2+cn^2)}$$

$$\text{or,}\quad \frac{u_1u_2}{a(bm^2+cn^2)}=\frac{v_1v_2}{b(al^2+cn^2)}=\frac{w_1w_2}{c(al^2+bm^2)}.$$

So the planes will be perpendicular if

$$u_1u_2+v_1v_2+w_1w_2=0$$

$$\text{or,}\quad a(b+c)l^2+b(a+c)m^2+c(b+a)n^2=0.$$

Hence the line (2) generates the cone

$$a(b+c)x^2+b(a+c)y^2+c(a+b)z^2=0.$$

▶▶ **20**. *Find the equation to the normal plane of the cone $ax^2+by^2+cz^2=0$ which passes through the generator $\frac{x}{l}=\frac{y}{m}=\frac{z}{n}$.*

☐ Any point on the generator is (lr, mr, nr) and equation of the tangent plane at this point is

$$alx+bmy+cnz=0. \tag{1}$$

Any plane through the given generator is

$$Ax+By+Cz=0 \tag{2}$$

$$\text{where}\quad Al+Bm+Cn=0. \tag{3}$$

If it be a normal plane then it will be perpendicular to the plane (1), so we have

$$Aal + Bbm + Ccn = 0. \tag{4}$$

Solving (3) and (4) we get

$$\frac{A}{mn(c-b)} = \frac{B}{nl(a-c)} = \frac{C}{ml(b-a)}. \tag{5}$$

Eliminating A, B, C from (2) with the help of (5) we get the equation of normal plane of the cone $ax^2 + by^2 + cz^2 = 0$ which passes through the generator $\frac{x}{l} = \frac{y}{m} = \frac{z}{n}$ is

$$\frac{b-c}{l}x + \frac{c-a}{m}y + \frac{a-b}{n}z = 0.$$

EXERCISE

1. Show that the plane $2x - 2y - z = 10$ touches the paraboloid $\frac{x^2}{9} + \frac{z^2}{4} = 2y$.
2. Show that the plane $x - 2y - 2z = 18$ touches the ellipsoid $\frac{x^2}{144} + \frac{y^2}{36} + \frac{z^2}{9} = 1$.
3. Find the equation of tangent planes to the surface $\frac{x^2}{42} - \frac{y^2}{18} + \frac{z^2}{3} = 1$ which passes through the line $\frac{x-8}{-7} = \frac{y+4}{5} = \frac{z-1}{1}$.
4. Find the equation of the tangent plane to the paraboloid $ax^2 + by^2 = 2z$ parallel to the plane $lx + my + nz = 0$.
5. Find the equation of the tangent planes to the surface $x^2 - 3y^2 + 2z^2 = 6$ perpendicular to the straight line $\frac{x}{1} = \frac{y}{3} = \frac{z}{4}$.
6. If the distance between two tangents planes to the ellipsoid $\frac{x^2}{a^2} + \frac{y^2}{b^2} + \frac{z^2}{c^2} = 1$ be $2r$, prove that the line through the origin perpendicular to the plane lie on the cone $(a^2 - r^2)x^2 + (b^2 - r^2)y^2 + (c^2 - r^2)z^2 = 0$.
   [*Hints*: See Example 5.]
7. Show that the cones $ayz + bxz + cxy = 0$ and $\sqrt{ax} + \sqrt{by} + \sqrt{cz} = 0$ are reciprocal to each other.
8. Find the equation of the cone which touches the coordinate planes.
9. Show that the locus of the luminous point which moves so that the sphere $x^2 + y^2 + z^2 = 2az$ castes a parabolic shadow on the plane $z = 0$ is $z = 2a$.

   [*Hints*: Let the luminous point be (f, g, h), then equation of the enveloping cone with vertex (f, g, h) is

   $$(x^2 + y^2 + z^2 - 2az)(f^2 + g^2 + h^2 - 2ah) = (fx + gy + hz - az - ah)^2.$$

   Putting $z = 0$ in the above equation we get

   $$(x^2 + y^2)(f^2 + g^2 + h^2 - 2ah) = [fx + gy - ah]^2.$$

If it be a parabola in x-y plane then we have

$$g^2 f^2 = (g^2 + h^2 - 2ah)(f^2 + h^2 - 2ah)$$

or, $$(f^2 + g^2 + h^2 - 2ah)(h^2 - 2ah) = 0.$$

or, $$h = 2a.$$

Hence locus of the point (f, g, h) is $z = 2a$.]

10. Find the locus of the point from which three mutually perpendicular tangents can be drawn to the surfaces (i) $ax^2 + by^2 + cz^2 = 1$, (ii) $ax^2 + by^2 = 2cz$.

[*Hints*: (i) Let from the point (f, g, h) tangents are drawn to the surface

$$ax^2 + by^2 + cz^2 = 1.$$

Equation of enveloping cone of the surface with vertex (f, g, h) is

$$(ax^2 + by^2 + cz^2 - 1)(af^2 + bg^2 + ch^2 - 1) = (axf + byg + czh - 1)^2.$$

If it has three mutually perpendicular generators then we have

$$\text{Coeff. of } x^2 + \text{Coeff. of } y^2 + \text{Coeff. of } z^2 = 0, \text{ etc.}$$

(ii) Proceed as (i).]

11. Prove that three mutually perpendicular tangent lines can be drawn to the sphere $x^2 + y^2 + z^2 = a^2$ from any point on the sphere $2(x^2 + y^2 + z^2) = 3a^2$.

[*Hints*: Proceed as hints of Ex. 10]

12. Prove that the plane $z = a$ meets the enveloping cone of the sphere $x^2 + y^2 + z^2 = a^2$ in a conic whose focus is at the point $(0, 0, a)$.

[*Hints*: Equation of enveloping cone with vertex (f, g, h) is

$$(x^2 + y^2 + z^2 - a^2)(f^2 + g^2 + h^2 - a^2) = (fx + gy + hz - a^2)^2.$$

Putting $z = a$ in the above equation we get

$$(x^2 + y^2)(f^2 + g^2 + h^2 - a^2) = (fx + gy + ah - a^2)^2$$

or, $$x^2(g^2 + h^2 - a^2) + y^2(f^2 + h^2 - a^2)$$
$$-2fgxy - 2fa(h - a)x - 2ga(h - a)y - a^2(h - a)^2 = 0$$

this is a conic in x-y plane, etc.]

13. The section of the enveloping cone of the ellipsoid $\frac{x^2}{a^2} + \frac{y^2}{b^2} + \frac{z^2}{c^2} = 1$ by the plane $z = 0$ is a rectangular hyperbola. Show that the locus of the vertex of the cone is

$$\frac{x^2 + y^2}{a^2 + b^2} + \frac{z^2}{c^2} = 1.$$

14. Prove that the locus of the foot of the perpendicular on the tangent planes to the ellipsoid $\frac{x^2}{a^2} + \frac{y^2}{b^2} + \frac{z^2}{c^2} = 1$ is the surface

$$(x^2 + y^2 + z^2)^2 = a^2x^2 + b^2y^2 + c^2z^2.$$

15. Find the locus of the perpendiculars from the origin to the tangent planes to the ellipsoid $\frac{x^2}{a^2}+\frac{y^2}{b^2}+\frac{z^2}{c^2}=1$ which cuts off from its axes intercepts the sum of whose reciprocals is equal to a constant $1/k$.

[*Hints*: Equation of tangent plane to ellipsoid is

$$lx+my+nz=\sqrt{a^2l^2+b^2m^2+c^2n^2}.$$

Its intercepts on coordinate axes are $\frac{1}{l}\sqrt{a^2l^2+b^2m^2+c^2n^2}$, $\frac{1}{m}\sqrt{a^2l^2+b^2m^2+c^2n^2}$ and $\frac{1}{n}\sqrt{a^2l^2+b^2m^2+c^2n^2}$ respectively. Given that

$$\frac{l+m+n}{\sqrt{a^2l^2+b^2m^2+c^2n^2}}=\frac{1}{k}$$

$$\text{or,}\quad k^2(l+m+n)^2=a^2l^2+b^2m^2+c^2n^2.$$

So the perpendiculars from origin to the tangent plane i.e., $\frac{x}{l}=\frac{y}{m}=\frac{z}{n}$ generates the surface $k^2(x+y+z)^2=a^2x^2+b^2y^2+c^2z^2$.]

16. If the normal at a point Q on the ellipsoid $\frac{x^2}{a^2}+\frac{y^2}{b^2}+\frac{z^2}{c^2}=1$ meet the coordinate planes at F, G, H respectively such that $QF:QG:QH=a^2:b^2:c^2$.

[*Hints*: See Example 15.]

17. If A, B, C, A', B', C' are the feet of the six normals from a given point on the ellipsoid $\frac{x^2}{a^2}+\frac{y^2}{b^2}+\frac{z^2}{c^2}=1$ and the plane ABC is given by $lx+my+nz=p$, then the plane $A'B'C'$ is given by

$$\frac{x}{a^2l}+\frac{y}{b^2m}+\frac{z}{c^2n}+\frac{1}{p}=0.$$

[*Hints*: Equation of the cone through the six normals of the ellipsoid $\frac{x^2}{a^2}+\frac{y^2}{b^2}+\frac{z^2}{c^2}=1$ drawn from the point is

$$\frac{f}{x-f}(b^2-c^2)+\frac{g}{y-g}(c^2-a^2)+\frac{h}{z-h}(a^2-b^2)=0$$

$$\text{or,}\quad f(b^2-c^2)yz+g(c^2-a^2)xz+h(a^2-b^2)yx$$
$$+x(\cdots)+y(\cdots)+z(\cdots)=0. \tag{1}$$

Now if equation of the plane $A'B'C'$ be $l'x+m'y+n'z=p'$ then

$$\frac{x^2}{a^2}+\frac{y^2}{b^2}+\frac{z^2}{c^2}-1+\lambda(lx+my+nz-p)(l'x+m'y+n'z-p')=0 \tag{2}$$

represents the cone containing the six normals for some values of λ. Comparing (1) and (2) we get

$$\lambda ll'+\frac{1}{a^2}=0;\quad \lambda mm'+\frac{1}{b^2}=0;\quad \lambda nn'+\frac{1}{c^2}=0\quad\text{and}\quad \lambda pp'=1$$

$$\text{or,}\quad l'=-\frac{1}{\lambda a^2l};\quad m'=-\frac{1}{\lambda b^2m};\quad n'=-\frac{1}{\lambda c^2n}\quad\text{and}\quad p'=\frac{1}{\lambda p}.$$

So equation of the plane $A'B'C'$ is

$$\frac{x}{a^2l} + \frac{y}{b^2m} + \frac{z}{c^2n} + \frac{1}{p} = 0.]$$

18. Show that enveloping cylinder of the ellipsoid $\frac{x^2}{a^2} + \frac{y^2}{b^2} + \frac{z^2}{c^2} = 1$ having generators perpendicular to the x-axis may meet the plane $x = 0$ in a pair of straight lines.

[*Hints*: Proceed as Example 9.]

19. Enveloping cylinders of the conicoid $ax^2 + by^2 + cz^2 = 1$ meet the plane $z = 0$ in rectangular hyperbolas. Prove that the central perpendiculars to their planes of contact generate the cone $b^2cx^2 + a^2cy^2 + ab(a+b)z^2 = 0$.

[*Hints*: Enveloping cylinders of the conicoid $ax^2 + by^2 + cz^2 = 1$ when the generators are parallel to the straight line with direction ratios l, m, n is

$$(ax^2 + by^2 + cz^2 - 1)(al^2 + bm^2 + cn^2) = (alx + bmy + cnz)^2.$$

It meets the plane $z = 0$ in the curve

$$(ax^2 + by^2 - 1)(al^2 + bm^2 + cn^2) = (alx + bmy)^2.$$

If it represents a rectangular hyperbola in x-y plane then we have

$$\text{Coeff. of } x^2 + \text{Coeff. of } y^2 = 0.$$

This gives

$$a(bm^2 + cn^2) + b(al^2 + cn^2) = 0$$

$$\text{or,} \quad abl^2 + abm^2 + c(a+b)n^2 = 0. \tag{1}$$

Now equation of the central perpendicular to the plane of contact $alx + bmy + cnz = 1$ is $\frac{x}{al} = \frac{y}{bm} = \frac{z}{cn}$ and the cone generated by it is obtained by eliminating l, m, n from the Eqn. (1) as $b^2cx^2 + a^2cy^2 + ab(a+b)z^2 = 0.$]

20. If the feet of the six normals from the point (α, β, γ) to the ellipsoid $\frac{x^2}{a^2} + \frac{y^2}{b^2} + \frac{z^2}{c^2} = 1$ are (x_r, y_r, z_r), $r = 1, 2, 3, 4, 5, 6$. Prove that

$$a^2\alpha \sum \frac{1}{x_r} + b^2\beta \sum \frac{1}{y_r} + c^2\gamma \sum \frac{1}{z_r} = 0.$$

[*Hints*: The feet of the normals drawn from (α, β, γ) are

$$x_k = \frac{a^2\alpha}{a^2 + \lambda_k}; \quad y_k = \frac{b^2\beta}{b^2 + \lambda_k} \quad \text{and} \quad z_k = \frac{c^2\gamma}{c^2 + \lambda_k}. \tag{1}$$

As (x_r, y_r, z_r) satisfy the equation of the ellipsoid we have

$$\frac{a^2\alpha^2}{(a^2+\lambda)^2} + \frac{b^2\beta^2}{(b^2+\lambda)^2} + \frac{c^2\gamma^2}{(c^2+\lambda)^2} = 1$$

$$\text{or,} \quad \lambda^6 + 2(a^2 + b^2 + c^2)\lambda^5 + \cdots = 0.$$

From which we get $\sum_{k=1}^{6} \lambda_k = -2(a^2 + b^2 + c^2)$, where λ_k's are the roots of the above equation in λ.
Now

$$
\begin{aligned}
& a^2\alpha \sum \frac{1}{x_r} + b^2\beta \sum \frac{1}{y_r} + c^2\gamma \sum \frac{1}{z_r} \\
&= a^2\alpha \sum \frac{a^2 + \lambda_r}{a^2\alpha} + b^2\beta \sum \frac{b^2 + \lambda_r}{b^2\beta} + c^2\gamma \sum \frac{c^2 + \lambda_r}{c^2\gamma} \\
&= 6(a^2 + b^2 + c^2) + 3 \sum \lambda_r \\
&= 0.]
\end{aligned}
$$

21. Show that feet of the normals from the point (α, β, γ) to the paraboloid $x^2 + y^2 = 2az$ lie on the sphere $x^2 + y^2 + z^2 - (a + \gamma)z - (\alpha^2 + \beta^2)\dfrac{y}{2\beta} = 0, \beta \neq 0$.

[*Hints*: If the normal at any point (f, g, h) to the paraboloid passes through (α, β, γ) then we have

$$\frac{\alpha - f}{f} = \frac{\beta - g}{g} = \frac{\gamma - h}{-a}$$

$$\text{or,} \quad \frac{\alpha}{f} = \frac{\beta}{g} = \frac{\gamma - h - a}{-a} = \frac{\alpha f + \beta g + 2\gamma h - 2h^2 - 2ah}{f^2 + g^2 - 2ah}.$$

$$\text{or,} \quad \alpha f + \beta g + 2\gamma h - 2h^2 - 2ah = 0, \quad \text{since } (f, g, h) \text{ lies on the paraboloid.}$$

$$\text{or,} \quad \frac{\alpha^2 g}{\beta} + \beta g + 2\gamma h - 2h^2 - 2ah = 0, \quad \text{if } \beta \neq 0,$$

$$\text{or,} \quad \frac{g}{2\beta}(\alpha^2 + \beta^2) + \gamma h - h^2 - ah = 0$$

$$\text{or,} \quad \frac{g}{2\beta}(\alpha^2 + \beta^2) + \gamma h - h^2 - f^2 - g^2 + 2ah - ah = 0.$$

Generalising we get the locus of the feet of the normals i.e., of (f, g, h) is

$$x^2 + y^2 + z^2 - (a + \gamma)z - (\alpha^2 + \beta^2)\frac{y}{2\beta} = 0, \quad \beta \neq 0.]$$

22. Show that the perpendiculars drawn from the origin to the normal planes of the cone $ax^2 + by^2 + cz^2 = 0$ generates the cone

$$\frac{a(b-c)^2}{x^2} + \frac{b(c-a)^2}{y^2} + \frac{c(a-b)^2}{z^2} = 0.$$

[*Hints*: Let $\dfrac{x}{l} = \dfrac{y}{m} = \dfrac{z}{n}$ be a generator of the cone $ax^2 + by^2 + cz^2 = 0$, then we have

$$al^2 + bm^2 + cn^2 = 0. \tag{1}$$

From Example 20, we see that the equation of the normal plane is

$$\frac{b-c}{l}x + \frac{c-a}{m}y + \frac{a-b}{n}z = 0. \tag{2}$$

Equation of a straight line passing through the origin and perpendicular to the normal plane (2) is

$$\frac{xl}{b-c} = \frac{ym}{c-a} = \frac{zn}{a-b}. \tag{3}$$

To get the locus of the line (3) we eliminate l, m, n from (1) with the help of (3) and get

$$\frac{a(b-c)^2}{x^2} + \frac{b(c-a)^2}{y^2} + \frac{c(a-b)^2}{z^2} = 0.]$$

23. Find the locus of the pole of the tangent plane $lx + my + nz = p$ with respect to the system of conicoids

$$\frac{x^2}{a^2+\lambda} + \frac{y^2}{b^2+\lambda} + \frac{z^2}{c^2+\lambda} = 1.$$

[*Hints*: Let (f, g, h) be the pole of the plane $lx + my + nz = p$ with respect to the conicoid $\frac{x^2}{a^2+\lambda} + \frac{y^2}{b^2+\lambda} + \frac{z^2}{c^2+\lambda} = 1$, then the plane is identical with the polar plane $\frac{fx}{a^2+\lambda} + \frac{gy}{b^2+\lambda} + \frac{hz}{c^2+\lambda} = 1$. Comparing we get

$$f = \frac{l(a^2+\lambda)}{p}; \quad g = \frac{m(b^2+\lambda)}{p} \quad \text{and} \quad h = \frac{n(c^2+\lambda)}{p}$$

or, $$\frac{f - \frac{la^2}{p}}{l} = \frac{g - \frac{mb^2}{p}}{m} = \frac{h - \frac{nc^2}{p}}{n}.$$

Hence locus of the point (f, g, h) is

$$\frac{x - \frac{la^2}{p}}{l} = \frac{y - \frac{mb^2}{p}}{m} = \frac{z - \frac{nc^2}{p}}{n}$$

which is a line perpendicular to the plane $lx + my + nz = p$.]

24. Prove that the locus of the line of intersection tangent planes to the cone $ax^2 + by^2 + cz^2 = 0$ which will touch along perpendicular generators is the cone

$$a^2(b+c)x^2 + b^2(a+c)y^2 + c^2(a+b)z^2 = 0.$$

25. Prove that for all values of λ, the normals to the conicoid

$$\frac{x^2}{a^2+\lambda} + \frac{y^2}{b^2+\lambda} + \frac{z^2}{c^2+\lambda} = 1$$

which pass through the point (f, g, h) meet the plane $z = 0$ in points on the conic

$$\frac{(c^2-a^2)f}{x} + \frac{(b^2-c^2)g}{y} + a^2 - b^2 = 0.$$

[*Hints*: From section 10.13 we have seen that the equation of the cone through the normals drawn to the conicoid $ax^2 + by^2 + cz^2 = 1$ from the point (f, g, h) is

$$\frac{af}{x-f}(b-c) + \frac{bg}{y-g}(c-a) + \frac{ch}{z-h}(a-b) = 0.$$

Changing a, b, c by a^{-2}, b^{-2}, c^{-2} and noting that $(a^2+\lambda)-(b^2+\lambda)=a^2-b^2$ we see that the equation of the cone through the normals drawn to the given conicoid from the point (f, g, h) is

$$\frac{f}{x-f}(b^2-c^2)+\frac{g}{y-g}(c^2-a^2)+\frac{h}{z-h}(a^2-b^2)=0.$$

It meets the plane $z=0$ in the following curve

$$\frac{f}{x-f}(b^2-c^2)+\frac{g}{y-g}(c^2-a^2)-a^2+b^2=0.$$

After simplifying we get the required equation.]

26. Show that the equation of two tangent planes to the paraboloid $ax^2+by^2=2cz$ which passes through the line $u\equiv lx+my+nz-p=0$; $u'\equiv l'x+m'y+n'z-p'=0$ is

$$u^2\left(\frac{l'^2}{a}+\frac{m'^2}{b}+2n'p'\right)-2uu'\left(\frac{ll'}{a}+\frac{mm'}{b}+np'+n'p\right)+u'^2\left(\frac{l^2}{a}+\frac{m^2}{b}+2np\right)=0.$$

27. If the feet of the three normals from P to the ellipsoid $\frac{x^2}{a^2}+\frac{y^2}{b^2}+\frac{z^2}{c^2}=1$ lie on the plane $\frac{x}{a}+\frac{y}{b}+\frac{z}{c}=1$, prove that feet of the other three lie on the plane $\frac{x}{a}+\frac{y}{b}+\frac{z}{c}+1=0$ and P lies on the line $a(b^2-c^2)x=b(c^2-a^2)y=c(a^2-b^2)z$.
[*Hints*: For the first part proceed as hints of exercise 17.
Second part: The feet of the normals lie on

$$\left(\frac{x}{a}+\frac{y}{b}+\frac{z}{c}-1\right)\left(\frac{x}{a}+\frac{y}{b}+\frac{z}{c}+1\right)=0$$

$$\text{or,}\quad \left(\frac{x}{a}+\frac{y}{b}+\frac{z}{c}\right)^2=1$$

$$\text{or,}\quad cxy+bxz+ayz=0, \tag{1}$$

since the feet also lie on the ellipsoid.
Again from example 16. we see that the feet of the six normals from (f, g, h) lie on the cone

$$\frac{a^2(b^2-c^2)f}{x}+\frac{b^2(c^2-a^2)g}{y}+\frac{c^2(a^2-b^2)h}{z}=0$$

$$\text{or,}\quad a^2(b^2-c^2)fyz+b^2(c^2-a^2)gxz+c^2(a^2-b^2)hxy=0. \tag{2}$$

Comparing (1) and (2) we get

$$\frac{a^2(b^2-c^2)f}{a}=\frac{b^2(c^2-a^2)g}{b}=\frac{c^2(a^2-b^2)h}{c}.$$

Hence locus of the point (f, g, h) is

$$a(b^2-c^2)x=b(c^2-a^2)y=c(a^2-b^2)z.]$$

28. Find the locus of the perpendicular from (α, β, γ) to its polar plane with respect to the paraboloid $\frac{x^2}{a^2} + \frac{y^2}{b^2} = 2z$.
[*Hints*: Equation of polar plane of the point (f, g, h) with respect to the paraboloid is

$$\frac{fx}{a^2} + \frac{gy}{b^2} = z + h.$$

Any line through the point (f, g, h) is

$$\frac{x-f}{l} = \frac{y-g}{m} = \frac{z-h}{n}. \tag{1}$$

If it is normal to the polar plane then we have

$$\frac{f/a^2}{l} = \frac{g/b^2}{m} = \frac{-1}{n} = k, \quad \text{(say)}.$$

From this we get

$$\frac{f}{l} - \frac{g}{m} = k(a^2 - b^2) = -\frac{1}{n}(a^2 - b^2). \tag{2}$$

To get the locus of the surface generated by the line (1) we eliminate l, m, n from (2) with the help of (1) and get

$$\frac{f}{x-f} - \frac{g}{y-g} + \frac{a^2 - b^2}{z-h} = 0.]$$

29. Prove that the normals from (α, β, γ) to the paraboloid $\frac{x^2}{a^2} + \frac{y^2}{b^2} = 2z$ lie on the cone

$$\frac{\alpha}{x-\alpha} - \frac{\beta}{y-\beta} + \frac{a^2 - b^2}{z-\gamma} = 0.$$

[*Hints*: Proceed as section 10.14.]

30. Two perpendicular tangent planes to the paraboloid $\frac{x^2}{a} + \frac{y^2}{b} = 2z$ intersect in a straight line lying in the plane $x = 0$. Show that the line touches the parabola

$$y^2 = (a + b)(2z + a); \quad x = 0.$$

31. Prove that the perpendiculars from the point (α, β, γ) to its polar plane with respect to a paraboloid lie on a cone.
32. Prove that the locus of the foot of the perpendicular on varying tangent planes to the ellipsoid $\frac{x^2}{a^2} + \frac{y^2}{b^2} + \frac{z^2}{c^2} = 1$ is the surface

$$(x^2 + y^2 + z^2)^2 = a^2x^2 + b^2y^2 + c^2z^2.$$

Chapter 11

Diameter

11.1 Diameter

A chord through the centre of a conicoid is called its diameter and a plane through the centre of a conicoid is called it's diametral plane.

(i) Section of a Central Conicoid with a Given Centre

Let us take the equation of a chord through the point (α, β, γ) as

$$\frac{x-\alpha}{l} = \frac{y-\beta}{m} = \frac{z-\gamma}{n} = r \tag{1}$$

Any point on it can be taken as $(\alpha + lr, \beta + mr, \gamma + nr)$. If this point lies on the conicoid $ax^2 + by^2 + cz^2 = 1$ then we have

$$a(\alpha + lr)^2 + b(\beta + mr)^2 + c(\gamma + nr)^2 = 1$$

$$\text{or,} \quad (al^2 + bm^2 + cn^2)r^2 + 2r(a\alpha l + b\beta m + c\gamma n) + a\alpha^2 + b\beta^2 + c\gamma^2 - 1 = 0.$$

The above equation will give two values of r and this implies that the chord intersects the conicoid in two points. If (α, β, γ) is the middle point of the chords, those points will be at equal distant from it and in the opposite sides of it. This implies that the above equation in r will two roots which are equal in magnitude but of opposite signs. Therefore

$$a\alpha l + b\beta m + c\gamma n = 0.$$

Hence locus of the middle points of a system of parallel chords i.e., of (α, β, γ) is

$$axl + bym + czn = 0.$$

This is called the diametral plane of the conicoid. If the chords are perpendicular to the diametral planes then the diametral planes are called principal planes and the line of intersection of two principal planes is called principal axis.

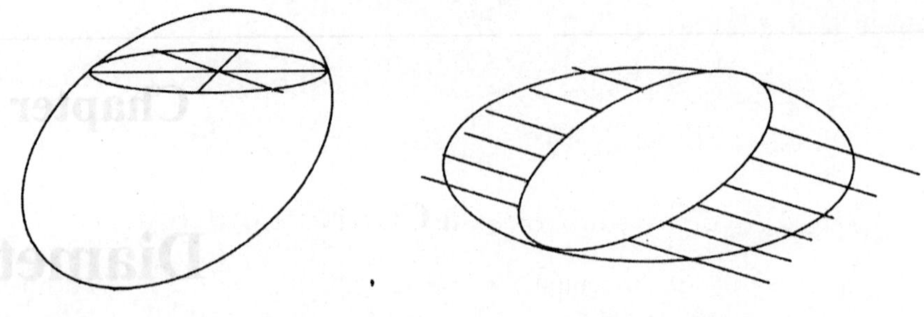

Further eliminating l, m, n, we get

$$a\alpha(x-\alpha)+b\beta(y-\beta)+c\gamma(z-\gamma)=0$$

$$\text{or,}\quad a\alpha x+b\beta y+c\gamma z=a\alpha^2+b\beta^2+c\gamma^2.$$

Hence the locus of the chords which are bisected at (α, β, γ) is a plane. Clearly this plane intersects the conicoid in a conic whose centre is at (α, β, γ).

The above equation can also be written as

$$S_1 = T.$$

Where $S_1 = a\alpha^2 + b\beta^2 + c\gamma^2 - 1$ and $T = a\alpha x + b\beta y + c\gamma z - 1$.

(ii) Section of a Paraboloid

Let us take the equation of a chord through the point (α, β, γ) as

$$\frac{x-\alpha}{l}=\frac{y-\beta}{m}=\frac{z-\gamma}{n}=r \tag{2}$$

Any point on it can be taken as $(\alpha + lr, \beta + mr, \gamma + nr)$. If this point lies on the paraboloid $ax^2 + by^2 = 2cz$ then we have

$$a(\alpha+lr)^2+b(\beta+mr)^2=2c(\gamma+nr)$$

$$\text{or,}\quad (al^2+bm^2)r^2+2r(a\alpha l+b\beta m-cn)+a\alpha^2+b\beta^2-2c\gamma=0.$$

Proceeding as before we see that the locus of the chords of the paraboloid which are bisected at (α, β, γ) is a plane whose equation is $a\alpha(x-\alpha)+b\beta(y-\beta)-c(z-\gamma)=0$ and this plane intersects the paraboloid in a conic whose centre is at (α, β, γ) and equation of the diametral plane is $axl + bym - cn = 0$ which is parallel to z-axis i.e., parallel to its axis.

(iii) Locus of the Centers of Parallel Plane Sections

Let (f, g, h) be the centre of a plane section of the conicoid $ax^2 + by^2 + cz^2 = 1$ and the plane section is parallel to the plane $lx + my + nz = p$. Now equation of the plane section with centre (f, g, h) is

$$afx+bgy+chz=af^2+bg^2+ch^2.$$

Since it is parallel to the plane $lx + my + nz = p$ we have

$$\frac{af}{l}=\frac{bg}{m}=\frac{ch}{n}.$$

Hence locus of (f, g, h) is

$$\frac{ax}{l} = \frac{by}{m} = \frac{cz}{n}$$

which is a line passing through the centre of the conicoid.

11.2 Conjugate Diameters of a Central Conicoid

Let $P(f, g, h)$ be a point on the central conicoid $ax^2 + by^2 + cz^2 = 1$, the equation of the diametral plane through the middle points of the chords parallel to OP is

$$axf + byg + czh = 0. \tag{3}$$

If the plane (3) passes through the point $Q(\alpha, \beta, \gamma)$ which lies on the conicoid, then the line OQ lies on the plane (3) and

$$a\alpha f + b\beta g + c\gamma h = 0. \tag{4}$$

Symmetry of the Eqn. (4) implies that the diametral plane of Q passes through the point P. Let the line of intersection of the diametral planes of OP and OQ meet the conicoid at the point $R(u, v, w)$. Since the point R is on the diametral planes of OP and OQ, the diametral plane $axu + byv + czw = 0$ passes through the points P and Q. From these we get

$$af^2 + bg^2 + ch^2 = 1; \quad a\alpha^2 + b\beta^2 + c\gamma^2 = 1; \quad au^2 + bv^2 + cw^2 = 1 \tag{5}$$

and $$a\alpha f + b\beta g + c\gamma h = 0; \quad a\alpha u + b\beta v + c\gamma w = 0; \quad auf + bvg + cwh = 0. \tag{6}$$

The lines OP, OQ, OR are called conjugate semi-diameters and the plane containing any two of those are the diametral plane of the rest and the diametral planes POQ, QOR, POR called conjugate planes.

Thus three diameters, such that the plane through any two of those bisect the chords parallel to the third, are called conjugate diameters.

From Eqns. (5) and (6) we further note that $\sqrt{a}f, \sqrt{b}g, \sqrt{c}h$; $\sqrt{a}\alpha, \sqrt{b}\beta, \sqrt{c}\gamma$; $\sqrt{a}u, \sqrt{b}v, \sqrt{c}w$ are the direction cosines of three mutually perpendicular straight lines. Then by rotation of axes it can be shown that direction cosines of three mutually perpendicular lines are

$$\sqrt{a}f, \sqrt{a}g, \sqrt{a}h; \quad \sqrt{b}\alpha, \sqrt{b}\beta, \sqrt{b}\gamma; \quad \sqrt{c}u, \sqrt{c}v, \sqrt{c}w.$$

So we get

$$f^2 + g^2 + h^2 = \frac{1}{a}; \quad \alpha^2 + \beta^2 + \gamma^2 = \frac{1}{b}; \quad u^2 + v^2 + w^2 = \frac{1}{c}$$

and $$\alpha f + \beta g + \gamma h = 0; \quad \alpha u + \beta v + \gamma w = 0; \quad uf + vg + wh = 0$$

From the above result we see that

$$OP^2 + OQ^2 + OR^2 = \frac{1}{a} + \frac{1}{b} + \frac{1}{c} = \text{constant}.$$

Examples

▶▶ 1. *Show that the chord through the point* $(2, 3, 4)$ *which is bisected by the plane* $10x - 24y = 21$ *of the paraboloid* $5x^2 - 6y^2 = 7z$ *is* $\frac{x-2}{1} = \frac{y-3}{2} = \frac{z-4}{3}$.

□ Let the chord through the point (2, 3, 4) be

$$\frac{x-2}{l}=\frac{y-3}{m}=\frac{z-4}{n}.$$

The equation of the diametral plane bisecting the chord parallel to the above line is

$$5lx-6my-\frac{7}{2}n=0.$$

Comparing this with the equation $10x-24y=21$ we get

$$\frac{5l}{10}=\frac{6m}{24}=\frac{7n}{42}\quad\text{i.e.,}\quad\frac{l}{1}=\frac{m}{2}=\frac{n}{3}.$$

So the required chord is

$$\frac{x-2}{1}=\frac{y-3}{2}=\frac{z-4}{3}.$$

▶▶ **2.** *Prove that the lines*

$$\frac{x}{1}=\frac{y}{4}=\frac{z}{3};\quad\frac{x}{4}=\frac{y}{1}=\frac{z}{-9}\quad\text{and}\quad\frac{x}{26}=\frac{y}{-28}=\frac{z}{45}$$

are three conjugate diameters of the ellipsoid

$$\frac{x^2}{2}+\frac{y^2}{4}+\frac{z^2}{9}=1.$$

□ Equation of the diametral plane conjugate to the line $\frac{x}{1}=\frac{y}{4}=\frac{z}{3}$ is

$$\frac{x}{2}+\frac{4y}{4}+\frac{3z}{9}=0$$

$$\text{or,}\quad 3x+6y+2z=0. \tag{1}$$

Now we note that the lines $\frac{x}{4}=\frac{y}{1}=\frac{z}{-9}$ and $\frac{x}{26}=\frac{y}{-28}=\frac{z}{45}$ lie on this plane as

$$3\times4+6\times1-2\times9=0\quad\text{and}\quad3\times26-6\times28+2\times45=0.$$

Again equation of the diametral plane conjugate to the line $\frac{x}{4}=\frac{y}{1}=\frac{z}{-9}$ is

$$2x+\frac{y}{4}-z=0$$

$$\text{or,}\quad 8x+y-4z=0. \tag{2}$$

Now we note that the lines $\frac{x}{1}=\frac{y}{4}=\frac{z}{3}$ and $\frac{x}{26}=\frac{y}{-28}=\frac{z}{45}$ lie on this plane as

$$1\times8+4\times1-3\times4=0\quad\text{and}\quad8\times26-1\times28-4\times45=0.$$

Further equation of the diametral plane conjugate to the line $\frac{x}{26}=\frac{y}{-28}=\frac{z}{45}$ is

$$13x-7y+5z=0. \tag{3}$$

Now we note that the lines $\frac{x}{4}=\frac{y}{1}=\frac{z}{-9}$ and $\frac{x}{1}=\frac{y}{4}=\frac{z}{3}$ lie on this plane as

$$13\times4-7\times1-5\times9=0\quad\text{and}\quad13\times1-7\times4+3\times5=0.$$

Thus the plane through any two of the given lines bisects the chords parallel to other and hence they are the three conjugate diameters of the given ellipsoid.

▶▷▶ 3. *Prove that the plane $lx + my + nz = p$ will pass through the extremities of the semi-conjugate diameters of the ellipsoid $\frac{x^2}{a^2} + \frac{y^2}{b^2} + \frac{z^2}{c^2} = 1$ if $a^2l^2 + b^2m^2 + c^2n^2 = 3p^2$.*

☐ If $P(x_1, y_1, z_1)$, $Q(x_2, y_2, z_2)$ and $R(x_3, y_3, z_3)$ be the extremities of semi-conjugate diameters of the given ellipsoid then we have

$$lx_1 + my_1 + nz_1 = p; \quad lx_2 + my_2 + nz_2 = p; \quad lx_3 + my_3 + nz_3 = p.$$

Multiplying the equations by x_1, x_2, x_3 respectively and adding we get

$$la^2 = p\sum_{i=1}^{3} x_i \tag{1}$$

and similarly we get

$$mb^2 = p\sum_{i=1}^{3} y_i \quad \text{and} \quad nc^2 = p\sum_{i=1}^{3} z_i \tag{2}$$

where we have used

$$\sum_{i=1}^{3} x_i^2 = a^2;\ \sum_{i=1}^{3} y_i^2 = b^2;\ \sum_{i=1}^{3} z_i^2 = c^2$$

$$\text{and} \quad \sum_{i=1}^{3} x_i y_i = 0;\ \sum_{i=1}^{3} z_i y_i = 0;\ \sum_{i=1}^{3} x_i z_i = 0. \tag{3}$$

From (1) and (2) we get

$$a^2l^2 + b^2m^2 + c^2n^2 = \frac{p^2}{a^2}\left(\sum_{i=1}^{3} x_i\right)^2 + \frac{p^2}{b^2}\left(\sum_{i=1}^{3} y_i\right)^2 + \frac{p^2}{c^2}\left(\sum_{i=1}^{3} z_i\right)^2 = 3p^2$$

as

$$\frac{x_1x_2}{a^2} + \frac{y_1y_2}{b^2} + \frac{z_1z_2}{c^2} = 0;\ \frac{x_3x_2}{a^2} + \frac{y_3y_2}{b^2} + \frac{z_3z_2}{c^2} = 0 \text{ and } \frac{x_1x_3}{a^2} + \frac{y_1y_3}{b^2} + \frac{z_1z_3}{c^2} = 0.$$

▶▷▶ 4. *Prove that $lx + my + p = 0$ and $l'x + m'y + p' = 0$ are conjugate diametral planes of the paraboloid $ax^2 + by^2 = 2cz$ if $\frac{ll'}{a} + \frac{mm'}{b} = 0$.*

☐ Equation of a diametral plane of the paraboloid $ax^2 + by^2 = 2cz$ which bisects the chords parallel to the line

$$\frac{x}{\lambda} = \frac{y}{\mu} = \frac{z}{\nu}$$

is

$$a\lambda x + b\mu y - c\nu = 0.$$

Comparing it with

$$lx + my + p = 0$$

we get

$$\frac{\lambda}{l/a} = \frac{\mu}{m/b} = \frac{\nu}{-p/c}.$$

Thus the diametral plane $lx + my + p = 0$ bisects the chord parallel to the line

$$\frac{x}{l/a} = \frac{y}{m/b} = \frac{z}{-p/c} \tag{1}$$

The plane $l'x + m'y + p' = 0$ will be a diametral plane if it is parallel to the line (1) i.e., if

$$\frac{ll'}{a} + \frac{mm'}{b} = 0.$$

▶▶ 5. *Prove that the locus of the equal conjugate semi-diameters of the ellipsoid* $\frac{x^2}{a^2} + \frac{y^2}{b^2} + \frac{z^2}{c^2} = 1$ *is* $\frac{x^2}{a^2} + \frac{y^2}{b^2} + \frac{z^2}{c^2} = \frac{3(x^2 + y^2 + z^2)}{a^2 + b^2 + c^2}$.

□ If OP, OQ, OR be three equi-conjugate semi-diameters of the ellipsoid then we have

$$OP^2 + OQ^2 + OR^2 = a^2 + b^2 + c^2$$

$$\text{or,} \quad OP^2 = OQ^2 = OR^2 = \frac{a^2 + b^2 + c^2}{3}. \tag{1}$$

If $P(f, g, h)$ be the end point of the semi-diameter then we have

$$OP^2 = f^2 + g^2 + h^2 \tag{2}$$

$$\frac{f^2}{a^2} + \frac{g^2}{b^2} + \frac{h^2}{c^2} = 1 \tag{3}$$

and equation of OP is

$$\frac{x}{f} = \frac{y}{g} = \frac{z}{h}. \tag{4}$$

From (1), (2) and (3) we get

$$\frac{f^2}{a^2} + \frac{g^2}{b^2} + \frac{h^2}{c^2} = \frac{3(f^2 + g^2 + h^2)}{a^2 + b^2 + c^2}.$$

So locus of the line (4) i.e., of OP is

$$\frac{x^2}{a^2} + \frac{y^2}{b^2} + \frac{z^2}{c^2} = \frac{3(x^2 + y^2 + z^2)}{a^2 + b^2 + c^2}.$$

EXERCISE

1. (a) Show that the locus of the middle points of the chords of the ellipsoid $\frac{x^2}{a^2} + \frac{y^2}{b^2} + \frac{z^2}{c^2} = 1$ passing through the point (α, β, γ) is

$$\frac{x^2}{a^2} + \frac{y^2}{b^2} + \frac{z^2}{c^2} = \frac{\alpha x}{a^2} + \frac{\beta y}{b^2} + \frac{\gamma z}{c^2}.$$

(b) Find the locus of the middle points of the chords of the conicoid $ax^2 + by^2 + cz^2 = 1$ passing through the point (α, β, γ).

2. Show that the chord through the point (2, 3, 4) which is bisected by the plane $3x+4y = 1$ of the paraboloid $5x^2 + 6y^2 = 2z$ is $\frac{x-2}{9} = \frac{y-3}{10} = \frac{z-4}{15}$.

   [*Hints*: Proceed as Example 1.]
3. Show that the planes $x + 3y = 3$ and $2x - y = 1$ are conjugate diametral planes of the paraboloid $2x^2 + 3y^2 = 4z$.

   [*Hints*: Proceed as Example 4.]
4. Find the equation of the plane through the end points of semi-conjugate diameters of the ellipsoid $\frac{x^2}{a^2} + \frac{y^2}{b^2} + \frac{z^2}{c^2} = 1$.

   [*Hints*: Let the end points of semi-conjugate diameters be

$$P(x_1, y_1, z_1), \quad Q(x_2, y_2, z_2) \text{ and } R(x_3, y_3, z_3).$$

   If the equation of the plane PQR be $lx + my + nz = p$, then we have

$$lx_1 + my_1 + nz_1 = p, \quad lx_2 + my_2 + nz_2 = p \quad \text{and} \quad lx_3 + my_3 + nz_3 = p.$$

   Multiplying the equations by x_1, x_2, x_3 respectively and adding we get

$$la^2 = p\sum_{i=1}^{3} x_i \tag{1}$$

   and similarly we get

$$mb^2 = p\sum_{i=1}^{3} y_i \quad \text{and} \quad nc^2 = p\sum_{i=1}^{3} z_i \tag{2}$$

   where we have used

$$\sum_{i=1}^{3} x_i^2 = a^2; \ \sum_{i=1}^{3} y_i^2 = b^2; \ \sum_{i=1}^{3} z_i^2 = c^2$$

$$\text{and} \quad \sum_{i=1}^{3} x_i y_i = 0; \ \sum_{i=1}^{3} z_i y_i = 0; \ \sum_{i=1}^{3} x_i z_i = 0. \tag{3}$$

   Using the results (1), (2) and (3) in the equation $lx + my + nz = p$ we get the results.]
5. If P, Q, R be the end points of semi-conjugate diameters of the ellipsoid $\frac{x^2}{a^2} + \frac{y^2}{b^2} + \frac{z^2}{c^2} = 1$, then show that the locus of the centre of the sections of the ellipsoid by the plane PQR is

$$\frac{x^2}{a^2} + \frac{y^2}{b^2} + \frac{z^2}{c^2} = \frac{1}{3}.$$

6. If P, Q, R be the end points of semi-conjugate diameters of the ellipsoid $\frac{x^2}{a^2} + \frac{y^2}{b^2} + \frac{z^2}{c^2} = 1$, then show that the plane PQR touches the ellipsoid

$$\frac{x^2}{a^2} + \frac{y^2}{b^2} + \frac{z^2}{c^2} = \frac{1}{3}.$$

7. If P, Q, R be the end points of semi-conjugate diameters of the ellipsoid $\frac{x^2}{a^2}+\frac{y^2}{b^2}+\frac{z^2}{c^2}$ $=1$ then find the locus of the foot of the perpendicular from the centre of the ellipsoid on the plane PQR.

[*Hints*: Let $L(\alpha, \beta, \gamma)$ be the foot of the perpendicular from O to the plane PQR, then Direction ratios of OL are $\frac{\alpha}{\sqrt{\sum\alpha^2}}, \frac{\beta}{\sqrt{\sum\alpha^2}}, \frac{\gamma}{\sqrt{\sum\alpha^2}}$ and $OL=\sqrt{\sum\alpha^2}$. So equation of the plane PQR is

$$\frac{\alpha}{\sqrt{\sum\alpha^2}}x+\frac{\beta}{\sqrt{\sum\alpha^2}}y+\frac{\gamma}{\sqrt{\sum\alpha^2}}z=\sqrt{\sum\alpha^2}$$

$$\text{or,}\quad \alpha x+\beta y+\gamma z=\sum\alpha^2.$$

Comparing this equation with

$$\frac{x}{a^2}\sum x_i+\frac{y}{b^2}\sum y_i+\frac{z}{c^2}\sum z_i=1$$

we get

$$\frac{a\alpha}{\sum\alpha^2}=\frac{\sum x_i}{a},\quad \frac{b\beta}{\sum\alpha^2}=\frac{\sum y_i}{b}\quad\text{and}\quad \frac{c\gamma}{\sum\alpha^2}=\frac{\sum z_i}{c}.$$

Squaring and adding the above results we get

$$a^2\alpha^2+b^2\beta^2+c^2\gamma^2=\left(\sum\alpha^2\right)^2\left[\frac{\left(\sum x_i\right)^2}{a^2}+\frac{\left(\sum y_i\right)^2}{b^2}+\frac{\left(\sum z_i\right)^2}{c^2}\right]=3\left(\sum\alpha^2\right)^2.$$

Generalising we get $a^2x^2+b^2y^2+c^2z^2=3(x^2+y^2+z^2)^2$.]

8. If P, Q, R be the end points of semi-conjugate diameters of the ellipsoid $\frac{x^2}{a^2}+\frac{y^2}{b^2}+\frac{z^2}{c^2}$ $=1$ then find the locus of the pole of the plane PQR.

[*Hints*: Let $T(\alpha, \beta, \gamma)$ be the pole of the plane PQR with respect to the ellipsoid, then equation of the plane PQR is

$$\frac{\alpha x}{a^2}+\frac{\beta y}{b^2}+\frac{\gamma z}{c^2}=1. \tag{1}$$

Again if the end points of semi-conjugate diameters be

$$P(x_1, y_1, z_1),\quad Q(x_2, y_2, z_2)\quad\text{and}\quad R(x_3, y_3, z_3).$$

Then equation of the plane PQR is

$$\frac{x}{a^2}\sum x_i+\frac{y}{b^2}\sum y_i+\frac{z}{c^2}\sum z_i=1. \tag{2}$$

Comparing (1) and (2) we get

$$\alpha=\sum x_i,\quad \beta=\sum y_i\quad\text{and}\quad \gamma=\sum z_i.$$

Squaring and adding we get

$$\frac{\alpha^2}{a^2} + \frac{\beta^2}{b^2} + \frac{\gamma^2}{c^2} = \left[\frac{\left(\sum x_i\right)^2}{a^2} + \frac{\left(\sum y_i\right)^2}{b^2} + \frac{\left(\sum z_i\right)^2}{c^2}\right] = 3.$$

Generalising we get of the locus of the pole of the plane PQR

$$\frac{x^2}{a^2} + \frac{y^2}{b^2} + \frac{z^2}{c^2} = 3.]$$

9. Prove that the locus of the intersection of three tangent planes to the ellipsoid $\frac{x^2}{a^2} + \frac{y^2}{b^2} + \frac{z^2}{c^2} = 1$, which are parallel to the conjugate diametral planes of the ellipsoid $\frac{x^2}{\alpha^2} + \frac{y^2}{\beta^2} + \frac{z^2}{\gamma^2} = 1$ is $\frac{x^2}{\alpha^2} + \frac{y^2}{\beta^2} + \frac{z^2}{\gamma^2} = \frac{a^2}{\alpha^2} + \frac{b^2}{\beta^2} + \frac{c^2}{\gamma^2}$.
10. Prove that the centres of the section of the conicoid $ax^2 + by^2 + cz^2 = 1$ by the planes which are at a constant distance p from the origin lie on the surface

$$(ax^2 + by^2 + cz^2)^2 = p^2(a^2x^2 + b^2y^2 + c^2z^2).$$

11. Prove that locus of the centres of the section of the ellipsoid $\frac{x^2}{a^2} + \frac{y^2}{b^2} + \frac{z^2}{c^2} = 1$ which touch the ellipsoid $\frac{x^2}{\alpha^2} + \frac{y^2}{\beta^2} + \frac{z^2}{\gamma^2} = 1$ is

$$\left(\frac{x^2}{a^2} + \frac{y^2}{b^2} + \frac{z^2}{c^2}\right)^2 = \frac{\alpha^2x^2}{a^4} + \frac{\beta^2y^2}{b^4} + \frac{\gamma^2z^2}{c^4}.$$

Chapter 12

. Generating Lines

12.1 Ruled Surface and Generating Lines

The line which, due to its movement, generates a surface is called generator or generating line of the surface and the locus of the line is called ruled surface. Thus if a line through a point on the surface lies wholly on the surface, the surface is called the ruled surface. For example, cones, cylinders, hyperboloid of one sheet and hyperbolic paraboloids are ruled surfaces.

A ruled surface is called developable surface if consecutive generating lines intersect. The cone is an example of developable surface. On the other hand if consecutive generating lines do not intersect, the ruled surface is called skew surface. Two skew surfaces namely, hyperboloid of one sheet and hyperbolic paraboloid will be discussed in this chapter.

12.2 Condition for a Straight Line to be a Generator of a Conicoid

Any point on the line

$$\frac{x-\alpha}{l}=\frac{y-\beta}{m}=\frac{z-\gamma}{n}=r \tag{1}$$

can be taken as $(\alpha+lr, \beta+mr, \gamma+nr)$. If it lies on the conicoid

$$ax^2+by^2+cz^2=1 \tag{2}$$

then we have

$$(al^2+bm^2+cn^2)r^2+2r(al\alpha+bm\beta+cn\gamma)+a\alpha^2+b\beta^2+c\gamma^2-1=0. \tag{3}$$

If the line (1) be a generator of the conicoid (2), i.e., if the line (1) lies wholly on the surface (2) then the Eqn. (3) will be satisfied for all values of r, which leads that

$$al^2+bm^2+cn^2=0 \tag{4}$$

$$al\alpha+bm\beta+cn\gamma=0 \tag{5}$$

$$a\alpha^2+b\beta^2+c\gamma^2-1=0. \tag{6}$$

From the Eqn. (6) we see that the point (α, β, γ) lies on the conicoid. Equation (5) shows that a generating line must lie on the tangent plane at (α, β, γ) to the conicoid. The Eqn. (4) shows that the lines through the centre of the conicoid and parallel to the generating lines generate the cone $ax^2 + by^2 + cz^2 = 0$ and this cone is called asymptotic cone.

From Eqns. (4) and (6) we get

$$(al^2 + bm^2)(a\alpha^2 + b\beta^2) + cn^2(1 - c\gamma^2) = 0$$

or, $$(al^2 + bm^2)(a\alpha^2 + b\beta^2) + cn^2 - (a\alpha l + b\beta m)^2 = 0$$

or, $$ab(\beta l - \alpha m)^2 = -cn^2$$

or, $$\beta l - \alpha m \pm n\sqrt{-\frac{c}{ab}} = 0. \tag{7}$$

The above results holds if any one of a, b, c is negative and other two are positive. This implies that of all central conicoids hyperboloid of one sheet is the only ruled surface. Solving (5) and (7) we get

$$\frac{l}{c\gamma\alpha \pm b\beta\sqrt{-\frac{c}{ab}}} = \frac{m}{c\gamma\beta \mp a\alpha\sqrt{-\frac{c}{ab}}} = \frac{n}{-(a\alpha^2 + b\beta^2)}.$$

The above result indicates that through a point on the hyperboloid of one sheet only two generators may be drawn.

12.3 Generating Lines of Hyperboloid of One Sheet and Their Properties

Rewriting the equation of hyperboloid of one sheet $\frac{x^2}{a^2} + \frac{y^2}{b^2} - \frac{z^2}{c^2} = 1$ as

$$\frac{x^2}{a^2} - \frac{z^2}{c^2} = 1 - \frac{y^2}{b^2}$$

or, $$\left(\frac{x}{a} + \frac{z}{c}\right)\left(\frac{x}{a} - \frac{z}{c}\right) = \left(1 + \frac{y}{b}\right)\left(1 - \frac{y}{b}\right)$$

we see that the above equation can be rewritten as

$$\frac{\left(1 + \frac{y}{b}\right)}{\left(\frac{x}{a} + \frac{z}{c}\right)} = \frac{\left(\frac{x}{a} - \frac{z}{c}\right)}{\left(1 - \frac{y}{b}\right)} = \lambda$$

and $$\frac{\left(1 - \frac{y}{b}\right)}{\left(\frac{x}{a} + \frac{z}{c}\right)} = \frac{\left(\frac{x}{a} - \frac{z}{c}\right)}{\left(1 + \frac{y}{b}\right)} = \mu$$

or,

$$\left(\frac{x}{a} + \frac{z}{c}\right) = \frac{1}{\lambda}\left(1 + \frac{y}{b}\right), \quad \left(\frac{x}{a} - \frac{z}{c}\right) = \lambda\left(1 - \frac{y}{b}\right) \tag{1}$$

and $$\left(\frac{x}{a} - \frac{z}{c}\right) = \mu\left(1 + \frac{y}{b}\right), \quad \left(\frac{x}{a} + \frac{z}{c}\right) = \frac{1}{\mu}\left(1 - \frac{y}{b}\right) \tag{2}$$

where λ, μ are two parameters.

The above planes represent two lines separately as no value of λ and μ can make Eqns. (1) and (2) identical, showing that there are two different sets of generators. We define the line represented by Eqn. (1) as first system of generators and the line represented by Eqn. (2) as second system of generators.

Properties

(i) *Every point on the lines of both systems lies on the hyperboloid*:
The result follows from the elimination of λ and μ from the system (1) and (2) respectively.

(ii) *Through every point of the hyperboloid there passes one generator of each system*:
If a generator of the first system passes through the point (α, β, γ) then we have

$$\lambda = \frac{\left(1+\frac{\beta}{b}\right)}{\left(\frac{\alpha}{a}+\frac{\gamma}{c}\right)} = \frac{\left(\frac{\alpha}{a}-\frac{\gamma}{c}\right)}{\left(1-\frac{\beta}{b}\right)}$$

i.e., $$\frac{\alpha^2}{a^2} - \frac{\gamma^2}{c^2} = 1 - \frac{\beta^2}{b^2}$$

or, $$\frac{\alpha^2}{a^2} + \frac{\beta^2}{b^2} - \frac{\gamma^2}{c^2} = 1.$$

This shows that the point (α, β, γ) lies on the hyperboloid

$$\frac{x^2}{a^2} + \frac{y^2}{b^2} - \frac{z^2}{c^2} = 1.$$

So through the point (α, β, γ) on the hyperboloid a generators of the first system passes. Similarly it can be shown by eliminating μ that through the point (α, β, γ) on the hyperboloid a generators of the second system passes.

(iii) *No two generators of the same system intersect*:
If possible that two generators of the first system

$$\left(\frac{x}{a}+\frac{z}{c}\right) = \frac{1}{\lambda_1}\left(1+\frac{y}{b}\right), \quad \left(\frac{x}{a}-\frac{z}{c}\right) = \lambda_1\left(1-\frac{y}{b}\right)$$

and $$\left(\frac{x}{a}+\frac{z}{c}\right) = \frac{1}{\lambda_2}\left(1+\frac{y}{b}\right), \quad \left(\frac{x}{a}-\frac{z}{c}\right) = \lambda_2\left(1-\frac{y}{b}\right) \quad \text{intersect.}$$

Then subtracting we get

$$\left(1+\frac{y}{b}\right)\left(\frac{1}{\lambda_1}-\frac{1}{\lambda_2}\right) = 0 \Rightarrow y = -b \text{ as } \lambda_1 \neq \lambda_2$$

and $$\left(1-\frac{y}{b}\right)(\lambda_1 - \lambda_2) = 0 \Rightarrow y = b \text{ as } \lambda_1 \neq \lambda_2.$$

Which is a contradiction, so no two generators of the first system intersect. Similarly, no two generators of the second system intersect.

(iv) *Any two generators of the two different systems intersect.*
Let us take two generators of two different systems as

$$\left(\frac{x}{a}-\frac{z}{c}\right) = \lambda\left(1-\frac{y}{b}\right), \quad \left(\frac{x}{a}+\frac{z}{c}\right) = \frac{1}{\lambda}\left(1+\frac{y}{b}\right) \tag{3}$$

and $$\left(\frac{x}{a}-\frac{z}{c}\right) = \mu\left(1+\frac{y}{b}\right), \quad \left(\frac{x}{a}+\frac{z}{c}\right) = \frac{1}{\mu}\left(1-\frac{y}{b}\right). \tag{4}$$

Solving the first equations of the above set of equations we get

$$\lambda\left(1-\frac{y}{b}\right)=\mu\left(1+\frac{y}{b}\right) \Rightarrow y=b\frac{\lambda-\mu}{\lambda+\mu}$$

Using this value of y in (3) we get

$$\frac{x}{a}-\frac{z}{c}=\lambda\left(1-\frac{\lambda-\mu}{\lambda+\mu}\right)=\frac{2\lambda\mu}{\lambda+\mu},\quad \frac{x}{a}+\frac{z}{c}=\frac{1}{\lambda}\left(1+\frac{\lambda-\mu}{\lambda+\mu}\right)=\frac{2}{\lambda+\mu}.$$

Solving we get

$$x=a\frac{\lambda\mu+1}{\lambda+\mu}\quad\text{and}\quad z=c\frac{1-\lambda\mu}{\lambda+\mu}.$$

It can be easily seen that these values of x, y, z satisfy the Eqns. (4). Thus the system of generators given by (3) and (4) intersect at the point

$$\left(a\frac{\lambda\mu+1}{\lambda+\mu}, b\frac{\lambda-\mu}{\lambda+\mu}, c\frac{1-\lambda\mu}{\lambda+\mu}\right). \tag{5}$$

(v) *Two intersecting generators belonging two different systems lie on the tangent plane to the hyperboloid at their point of intersection*:

Equation of the tangent plane at the point of intersection $\left(a\frac{\lambda\mu+1}{\lambda+\mu}, b\frac{\lambda-\mu}{\lambda+\mu}, c\frac{1-\lambda\mu}{\lambda+\mu}\right)$ of two different systems of generators given by Eqns. (1) and (2) is

$$\frac{x}{a}\frac{\lambda\mu+1}{\lambda+\mu}+\frac{y}{b}\frac{\lambda-\mu}{\lambda+\mu}-\frac{z}{c}\frac{1-\lambda\mu}{\lambda+\mu}=1$$

which can be rewritten as

$$\frac{x}{a}-\frac{z}{c}-\lambda\left(1-\frac{y}{b}\right)+\lambda\mu\left[\frac{x}{a}+\frac{z}{c}-\frac{1}{\lambda}\left(1+\frac{y}{b}\right)\right]=0 \tag{6}$$

as well as

$$\frac{x}{a}-\frac{z}{c}-\mu\left(1+\frac{y}{b}\right)+\lambda\mu\left[\frac{x}{a}+\frac{z}{c}-\frac{1}{\mu}\left(1-\frac{y}{b}\right)\right]=0. \tag{7}$$

These equations show that two different systems of generators lie on the tangent plane to the hyperboloid at their point of intersection.

12.4 Generating Lines of Hyperbolic Paraboloid and Their Properties

Rewriting the equation of hyperbolic paraboloid $\frac{x^2}{a^2}-\frac{y^2}{b^2}=2z$ as

$$\left(\frac{x}{a}+\frac{y}{b}\right)\left(\frac{x}{a}-\frac{y}{b}\right)=2z$$

we see that the above equation can be rewritten as

$$\frac{\frac{x}{a}-\frac{y}{b}}{z}=\frac{2}{\frac{x}{a}+\frac{y}{b}}=\lambda \quad \text{and} \quad \frac{\frac{x}{a}+\frac{y}{b}}{z}=\frac{2}{\frac{x}{a}-\frac{y}{b}}=\mu$$

$$\text{or,} \quad \frac{x}{a}-\frac{y}{b}=\lambda z,\ \frac{x}{a}+\frac{y}{b}=\frac{2}{\lambda} \quad \text{and} \quad \frac{x}{a}+\frac{y}{b}=\mu z,\ \frac{x}{a}-\frac{y}{b}=\frac{2}{\mu} \tag{1}$$

where λ, μ are two parameters.

The above planes represent two lines separately as no value of λ and μ can make the equations identical, showing that there are two different set of generators. Like earlier we define the line represented by first equation as first system of generators and the line represented by second equation as second system of generators. It is also clear from the Eqns. (1) that elimination of λ, μ from those yield the equation of the hyperbolic paraboloid showing that all the points on the generators i.e., the generators lie totally on the surface.

Properties

(i) *Every point on the lines of both systems lies on the hyperbolic paraboloid*:

(ii) *Through every point of the hyperboloid there passes one generator of each system*:

(iii) *No two generators of the same system intersect*:

(iv) *Any two generators of the two different systems intersect*:

Let us take two generators of two different systems as

$$\frac{x}{a}-\frac{y}{b}=\lambda z,\ \frac{x}{a}+\frac{y}{b}=\frac{2}{\lambda} \quad \text{and} \quad \frac{x}{a}+\frac{y}{b}=\mu z,\ \frac{x}{a}-\frac{y}{b}=\frac{2}{\mu}.$$

Solving the second equations of the above set of equations we get

$$x=a\frac{\lambda+\mu}{\lambda\mu} \quad \text{and} \quad y=b\frac{\mu-\lambda}{\lambda\mu}.$$

Using the values of x and y in first equation of the second system we get $z=\dfrac{2}{\lambda\mu}$.

It can be easily seen that these values of x, y, z satisfy the equation of the hyperbolic paraboloid. Thus the system of generators given by (1) intersect at the point

$$\left(a\frac{\lambda+\mu}{\lambda\mu},\ b\frac{\mu-\lambda}{\lambda\mu},\ \frac{2}{\lambda\mu}\right). \tag{2}$$

(v) *Two intersecting generators belonging two different systems lie on the tangent plane to the hyperbolic paraboloid at their point of intersection*:

Equation of the tangent plane at the point of intersection $\left(a\dfrac{\lambda+\mu}{\lambda\mu},\ b\dfrac{\mu-\lambda}{\lambda\mu},\ \dfrac{2}{\lambda\mu}\right)$ of two different systems of generators given by Eqns. (1) is

$$\frac{x}{a}\frac{\lambda+\mu}{\lambda\mu}-\frac{y}{b}\frac{\mu-\lambda}{\lambda\mu}=z+\frac{2}{\lambda\mu}$$

which can be rewritten as

$$\frac{x}{a}-\frac{y}{b}-\lambda z+\frac{\lambda}{\mu}\left[\frac{x}{a}+\frac{y}{b}-\frac{2}{\lambda}\right]=0 \tag{3}$$

as well as

$$\frac{x}{a} - \frac{y}{b} - \frac{2}{\mu} + \frac{\lambda}{\mu}\left[\frac{x}{a} + \frac{y}{b} - \mu z\right] = 0. \tag{4}$$

These equations show that two different systems of generators lie on the tangent plane to the hyperbolic paraboloid at their point of intersection.

12.5 Locus of the Point of Intersection of Perpendicular Generators

(a) *Hyperboloid of One Sheet:*

Let

$$\left(\frac{x}{a} + \frac{z}{c}\right) = \frac{1}{\lambda}\left(1 + \frac{y}{b}\right), \quad \left(\frac{x}{a} - \frac{z}{c}\right) = \lambda\left(1 - \frac{y}{b}\right) \tag{1}$$

and

$$\left(\frac{x}{a} - \frac{z}{c}\right) = \mu\left(1 + \frac{y}{b}\right), \quad \left(\frac{x}{a} + \frac{z}{c}\right) = \frac{1}{\mu}\left(1 - \frac{y}{b}\right) \tag{2}$$

be two systems of generator of the hyperboloid of one sheet

$$\frac{x^2}{a^2} + \frac{y^2}{b^2} - \frac{z^2}{c^2} = 1$$

which intersect at the point

$$\left(a\frac{\lambda\mu + 1}{\lambda + \mu},\ b\frac{\lambda - \mu}{\lambda + \mu},\ c\frac{1 - \lambda\mu}{\lambda + \mu}\right). \tag{3}$$

If l_1, m_1, n_1 be the direction cosines of the line of intersection of the plane given by (1) then we have

$$\frac{l_1}{a} - \frac{m_1}{\lambda b} + \frac{n_1}{c} = 0 \quad \text{and} \quad \frac{l_1}{a} + \frac{\lambda m_1}{b} - \frac{n_1}{c} = 0.$$

Solving we get

$$\frac{l_1}{\frac{1}{bc}\left(\frac{1}{\lambda} - \lambda\right)} = \frac{m_1}{\frac{2}{ac}} = \frac{n_1}{\frac{1}{ab}\left(\lambda + \frac{1}{\lambda}\right)}$$

$$\text{or,} \quad \frac{l_1}{\frac{1}{bc}\left(1 - \lambda^2\right)} = \frac{m_1}{\frac{2\lambda}{ac}} = \frac{n_1}{\frac{1}{ab}\left(\lambda^2 + 1\right)}.$$

Proceeding in the same way we see that the direction cosines l_2, m_2, n_2 of the plane given by (2) satisfy

$$\frac{l_2}{\frac{1}{bc}\left(1 - \mu^2\right)} = \frac{m_2}{-\frac{2\mu}{ac}} = \frac{n_2}{\frac{1}{ab}\left(\mu^2 + 1\right)}$$

So the generators given by (1) and (2) will be perpendicular if

$$l_1 l_2 + m_1 m_2 + n_1 n_2 = 0$$

$$\text{or,} \quad a^2(1 - \lambda^2)(1 - \mu^2) - 4b^2\lambda\mu + c^2(1 + \lambda^2)(1 + \mu^2) = 0$$

$$\text{or,} \quad a^2(1 + \lambda\mu)^2 + b^2(\lambda - \mu)^2 + c^2(1 - \lambda\mu)^2 = (\lambda + \mu)^2(a^2 + b^2 - c^2)$$

$$\text{or,} \quad a^2\left(\frac{1 + \lambda\mu}{\lambda + \mu}\right)^2 + b^2\left(\frac{\lambda - \mu}{\lambda + \mu}\right)^2 + c^2\left(\frac{1 - \lambda\mu}{\lambda + \mu}\right)^2 = a^2 + b^2 - c^2. \tag{4}$$

Equation (4) shows that the point of intersection of two systems of perpendicular generators given by (3) lie on the sphere $x^2+y^2+z^2=a^2+b^2-c^2$, which is the director sphere of the hyperboloid of one sheet. Hence the locus of the point of intersections of the perpendicular generators is the curve of intersection of the hyperboloid with its director sphere.

(b) *Hyperbolic Paraboloid:*

Let

$$\frac{x}{a}-\frac{y}{b}=\lambda z,\ \frac{x}{a}+\frac{y}{b}=\frac{2}{\lambda} \quad \text{and} \quad \frac{x}{a}+\frac{y}{b}=\mu z,\ \frac{x}{a}-\frac{y}{b}=\frac{2}{\mu} \tag{5}$$

be two systems of generators of the hyperbolic paraboloid

$$\frac{x^2}{a^2}-\frac{y^2}{b^2}=2z.$$

which intersect at the point $\left(a\dfrac{\lambda+\mu}{\lambda\mu},\ b\dfrac{\mu-\lambda}{\lambda\mu},\ \dfrac{2}{\lambda\mu}\right)$.

If l_1, m_1, n_1 be the direction cosines of the line of intersection of the planes given by first pair of equations given by (5) then we have

$$\frac{l_1}{a}-\frac{m_1}{b}-\lambda n_1=0 \quad \text{and} \quad \frac{l_1}{a}+\frac{m_1}{b}=0.$$

Solving we get

$$\frac{l_1}{a\lambda}=\frac{m_1}{-b\lambda}=\frac{n_1}{2}.$$

Similarly, the direction cosines l_2, m_2, n_2 of the line of intersection of planes given by second pair of equations given by (5) satisfy

$$\frac{l_1}{a\mu}=\frac{m_1}{b\mu}=\frac{n_1}{2}.$$

So the generators given by (5) will be perpendicular if

$$l_1l_2+m_1m_2+n_1n_2=0$$

$$\text{or,}\quad (a^2-b^2)\lambda\mu+4=0$$

$$\text{or,}\quad a^2-b^2+\frac{4}{\lambda\mu}=0.$$

This shows that the point of intersection of two systems of perpendicular generators given by (5) lie on the plane $a^2-b^2+2z=0$. Hence the locus of the point of intersection of the perpendicular generators is the curve of intersection of the hyperbolic paraboloid with the plane $a^2-b^2+2z=0$.

Examples

▶▷▶ 1. *Find the equation of the generators of the hyperboloid $x^2+4y^2-9z^2=25$ at the point* (3, 2, 0).

☐ Let the equation of the generator of the hyperboloid through the point (3, 2, 0) be

$$\frac{x-3}{l}=\frac{y-2}{m}=\frac{z}{n}=r.$$

Any point on the generator is $(3 + lr, 2 + mr, nr)$ and if it lies on the hyperboloid then we have

$$(3 + lr)^2 + 4(2 + mr)^2 - 9n^2r^2 = 25$$
$$\text{or,}\quad (l^2 + 4m^2 - 9n^2)r^2 + 2(3l + 8m)r = 0.$$

Now the line will be a generator of the hyperboloid if the above equation is satisfied for all values of r and this leads to

$$l^2 + 4m^2 - 9n^2 = 0 \quad \text{and} \quad 3l + 8m = 0. \tag{1}$$

From the second equation we get $\dfrac{l}{8} = \dfrac{m}{-3} = k$ and using this in the first equation we get $9n^2 = 100k^2$. Hence solutions of the Eqns. (1) are

$$\frac{l}{24} = \frac{m}{-9} = \frac{n}{10} \quad \text{and} \quad \frac{l}{-24} = \frac{m}{9} = \frac{n}{10}.$$

So the equations of generators are

$$\frac{x-3}{24} = \frac{y-2}{-9} = \frac{z}{10} \quad \text{and} \quad \frac{x-3}{-24} = \frac{y-2}{9} = \frac{z}{10}.$$

▶▶ **2.** *Find the equation of the generators of the paraboloid* $9x^2 - 25y^2 = 4z$ *at the point* $(1, 1, -4)$.

☐ Two systems of generators of the paraboloid $9x^2 - 25y^2 = 4z$ are

$$3x + 5y = 4\lambda;\ 3x - 5y = \frac{z}{\lambda} \quad \text{and} \quad 3x + 5y = \frac{z}{\mu};\ 3x - 5y = 4\mu.$$

If the generators pass through the point $(1, 1, -4)$ then we get from first pair of equations $\lambda = 2$ and from second pair of equations we get $\mu = -\dfrac{1}{2}$.

So the generators of the paraboloid at the point $(1, 1, -4)$ are

$$3x + 5y = 8;\ 6x - 10y = z \quad \text{and} \quad 3x + 5y + 2z = 0;\ 3x - 5y + 2 = 0.$$

▶▶ **3.** *Show that the line* $\dfrac{x+2}{2} = \dfrac{y}{3} = \dfrac{z-1}{-2}$ *is a generator of the paraboloid* $\dfrac{x^2}{4} - \dfrac{y^2}{9} = z$.

☐ Any point on the given line is $(-2 + 2r, 3r, 1 - 2r)$. Putting this values of x, y, z in the equation of paraboloid we see that

$$(r-1)^2 - r^2 = 1 - 2r$$

which is an identity in r implying that the given line totally lies on the paraboloid. Hence it is a generator of the paraboloid.

▶▶ **4.** *Find the equations of the generating lines of the hyperboloid* $yz + 2zx + 3xy + 6 = 0$ *at the point* $(-1, 0, 3)$.

☐ Let the equation of the generator of the hyperboloid through the point $(-1, 0, 3)$ be

$$\frac{x+1}{l} = \frac{y}{m} = \frac{z-3}{n} = r.$$

Any point on the generator is $(-1+lr, mr, 3+nr)$ and if it lies on the hyperboloid then we have

$$mr(3+nr)+2(-1+lr)(3+nr)+3mr(-1+lr)+6=0$$
$$\text{or,}\quad (nm+2nl+3lm)r^2+(6l-2n)r=0.$$

Now the line will be a generator of the paraboloid if the above equation is satisfied for all values of r and this leads to

$$2nl+3lm+nm=0 \quad\text{and}\quad 6l-2n=0. \tag{1}$$

From the second equation we get $n=3l$. Using this in the first equation we get

$$6l^2+6lm=0 \;\Rightarrow\; l=0,-m.$$

Hence the solutions of the Eqns. (1) are

$$\frac{l}{0}=\frac{m}{k}=\frac{n}{0} \quad\text{and}\quad \frac{l}{1}=\frac{m}{-1}=\frac{n}{3}.$$

So the equations of generators are

$$x+1=0=z-3 \quad\text{and}\quad \frac{x+1}{1}=\frac{y}{-1}=\frac{z-3}{3}.$$

▶▷▶ 5. *Show that the perpendiculars from origin on the generators of the paraboloid* $\dfrac{x^2}{a^2}-\dfrac{y^2}{b^2}=\dfrac{2z}{c}$ *lie on the cone* $\left(\dfrac{x}{a}-\dfrac{y}{b}\right)(ax-by)+2z^2=0.$

☐ Equation of the generator in the first system of the paraboloid is

$$\frac{x}{a}-\frac{y}{b}=\frac{\lambda z}{c},\quad \frac{x}{a}+\frac{y}{b}=\frac{2}{\lambda}.$$

The symmetrical form of the above equation is

$$\frac{x-\dfrac{a}{\lambda}}{a\lambda}=\frac{y-\dfrac{b}{\lambda}}{-b\lambda}=\frac{z}{2c}=r. \tag{1}$$

Let equation of the straight line through origin and perpendicular to the line (1) be

$$\frac{x}{l}=\frac{y}{m}=\frac{z}{n}=R \tag{2}$$

then we have

$$a\lambda l-b\lambda m+2cn=0. \tag{3}$$

At the point of intersection of the lines (1) and (2) we have

$$\frac{a}{\lambda}+a\lambda r=lR;\quad \frac{b}{\lambda}-b\lambda r=mR \quad\text{and}\quad 2cr=nR.$$

From the first two equations we get $2ab\lambda r=(lb-am)R$. Using the second equation in it we get

$$\frac{ab}{c}\lambda=\frac{lb-am}{n} \;\Rightarrow\; \lambda=\frac{c(lb-am)}{nab}.$$

Putting this value of λ in the Eqn. (3) we get

$$\frac{c(lb - am)}{nab}(al - bm) + 2cn = 0$$

$$\text{or,} \quad ab(l^2 + m^2 + 2n^2) = (a^2 + b^2)lm.$$

Therefore the locus of the line (2) is

$$ab(x^2 + y^2 + 2z^2) = (a^2 + b^2)xy$$

$$\text{or,} \quad \left(\frac{x}{a} - \frac{y}{b}\right)(ax - by) + 2z^2 = 0.$$

▶▷▶ 6. *Prove that any point on the lines $x + 1 = \mu y = -(\mu + 1)z$ lies on the surface*

$$yz + zx + xy + y + z = 0$$

and find equations to determine the other system of lines which lies on the surface.

☐ Eliminating μ from the equations $x + 1 = \mu y$ and $x + z + 1 = -\mu z$ we get the equation of the surface is

$$z(x + 1) + y(x + z + 1) = 0$$

$$\text{or,} \quad yz + zx + xy + y + z = 0.$$

The above equation can also be written as

$$z(x + 1) = -y(x + z + 1).$$

So, if one of the systems of generators be $x + 1 = \mu y$; $\mu z = -(x + z + 1)$, then other system of generators will be $x + 1 = \lambda(x + z + 1)$; $y = -\lambda z$.

▶▷▶ 7. *Show that the polar lines with respect to the sphere $x^2 + y^2 + z^2 = a^2$ of the generators of the surface $x^2 - y^2 = 2az$ lie on the surface $x^2 - y^2 = -2az$.*

☐ Two systems of generators of the surface $x^2 - y^2 = 2az$ can be written as

$$x - y = 2a\lambda; \ x + y = \frac{z}{\lambda} \quad \text{and} \quad x + y = 2\mu z; \ x - y = \frac{a}{\mu}$$

In symmetrical form the first line can be written as

$$\frac{x - \lambda a}{1} = \frac{y + \lambda a}{1} = \frac{z}{2\lambda} = r.$$

Any point on the line can be taken as $(\lambda a + r, -\lambda a + r, 2\lambda r)$. Equation of the polar plane of this point with respect to the sphere is

$$(\lambda a + r)x + (-\lambda a + r)y + 2\lambda rz = a^2$$

$$\text{or,} \quad r(x + y + 2\lambda z) + a(\lambda x - \lambda y - a) = 0.$$

So the polar line is the line of intersection of the planes

$$x + y + 2\lambda z = 0 \quad \text{and} \quad \lambda x - \lambda y - a = 0.$$

To get the locus of the polar line we eliminate the parameter λ from the above equations and get

$$x^2 - y^2 = -2az.$$

Proceeding in the same way we also get the same locus for the second system of generators.

▶▷▶ **8.** *Find the equation of generators of the hyperboloid* $\frac{x^2}{a^2} + \frac{y^2}{b^2} - \frac{z^2}{c^2} = 1$ *through the point* $(a\cos\theta, b\sin\theta, 0)$. *Hence show that the projections of generators of a hyperboloid on the principal planes are tangents to the section of the hyperboloid with those planes.*

☐ Any point on the generator at the point $(a\cos\theta, b\sin\theta, 0)$

$$\frac{x - a\cos\theta}{l} = \frac{y - b\sin\theta}{m} = \frac{z}{n} = r \tag{1}$$

of the hyperboloid can be taken as $(a\cos\theta + lr, b\sin\theta + mr, nr)$. So we have

$$\frac{(a\cos\theta + lr)^2}{a^2} + \frac{(b\sin\theta + mr)^2}{b^2} - \frac{n^2r^2}{c^2} = 1$$

or, $$\left(\frac{l^2}{a^2} + \frac{m^2}{b^2} - \frac{n^2}{c^2}\right)r^2 + 2\left(\frac{l\cos\theta}{a} + \frac{m\sin\theta}{b}\right)r = 0.$$

Now the line (1) will be a generator of the hyperboloid if the above equation is satisfied for all values of r and this leads to

$$\frac{l^2}{a^2} + \frac{m^2}{b^2} - \frac{n^2}{c^2} = 0 \quad \text{and} \quad \frac{l\cos\theta}{a} + \frac{m\sin\theta}{b} = 0.$$

From these equations we get

$$\frac{l/a}{\sin\theta} = \frac{m/b}{-\cos\theta} = \frac{\sqrt{\frac{l^2}{a^2} + \frac{m^2}{b^2}}}{1} = \pm\frac{n}{c}.$$

Hence the generators of the hyperboloid through the point $(a\cos\theta, b\sin\theta, 0)$ are

$$\frac{x - a\cos\theta}{a\sin\theta} = \frac{y - b\sin\theta}{-b\cos\theta} = \frac{z}{\pm c}. \tag{2}$$

Now the plane $z = 0$ intersects the hyperboloid in an ellipse given by

$$\frac{x^2}{a^2} + \frac{y^2}{b^2} = 1; \quad z = 0 \tag{3}$$

From Eqn. (2) we see that the projection of the generators in the plane $z = 0$ is

$$\frac{x - a\cos\theta}{a\sin\theta} = \frac{y - b\sin\theta}{-b\cos\theta}; \quad z = 0$$

or, $$\frac{x}{a\sin\theta} + \frac{y}{b\cos\theta} = \frac{\cos\theta}{\sin\theta} + \frac{\sin\theta}{\cos\theta}; \quad z = 0$$

or, $$\frac{x}{a}\cos\theta + \frac{y}{b}\sin\theta = 1; \quad z = 0$$

which is a tangent to ellipse at the point $(a\cos\theta, b\sin\theta, 0)$. Similarly, it can be shown that the projection of the generators in the planes $x = 0$ and $y = 0$ are the tangents to the corresponding sections of the hyperboloid made by the planes respectively.

▶▶ **9.** *Show that if two generators of the hyperboloid* $\frac{x^2}{a^2}+\frac{y^2}{b^2}-\frac{z^2}{c^2}=1$ *through the points* $P(a\cos\theta, b\sin\theta, 0)$ *and* $Q(a\cos\phi, b\sin\phi, 0)$ *are perpendicular then their projections in the plane* $z = 0$ *intersect at an angle* ω *given by*

$$\tan\omega = \frac{ab}{c^2}\sin(\theta-\phi).$$

□ Equation of the generator of the one system of the hyperboloid through the point $P(a\cos\theta, b\sin\theta, 0)$ is

$$\frac{x-a\cos\theta}{a\sin\theta}=\frac{y-b\sin\theta}{-b\cos\theta}=\frac{z}{c} \tag{1}$$

Equation of the generator of the other system of the hyperboloid through the point $Q(a\cos\phi, b\sin\phi, 0)$ is

$$\frac{x-a\cos\phi}{-a\sin\phi}=\frac{y-b\sin\phi}{b\cos\phi}=\frac{z}{c}. \tag{2}$$

They will be perpendicular if

$$a^2\sin\phi\sin\theta+b^2\cos\phi\cos\theta=c^2. \tag{3}$$

Now the projections of the generators in the plane $z = 0$ are

$$\frac{x-a\cos\theta}{a\sin\theta}=\frac{y-b\sin\theta}{-b\cos\theta} \quad \text{and} \quad \frac{x-a\cos\phi}{-a\sin\phi}=\frac{y-b\sin\phi}{b\cos\phi}$$

$$\text{i.e.,} \quad \frac{x}{a}\cos\theta+\frac{y}{b}\sin\theta=1;\ z=0 \quad \text{and} \quad \frac{x}{a}\cos\phi+\frac{y}{b}\sin\phi=1;\ z=0$$

respectively.

If the angle between them be ω then

$$\begin{aligned}\tan\omega &= \frac{\frac{b}{a}\cot\phi-\frac{b}{a}\cot\theta}{1+\frac{b^2}{a^2}\cot\theta\cot\phi}\\ &= \frac{ab\sin(\theta-\phi)}{a^2\sin\theta\sin\phi+b^2\cos\theta\cos\phi}\\ &= \frac{ab}{c^2}\sin(\theta-\phi).\end{aligned}$$

▶▶ **10.** *Show that the perpendiculars from the origin on the generator of the hyperboloid* $\frac{x^2}{a^2}+\frac{y^2}{b^2}-\frac{z^2}{c^2}=1$ *lie on the cone*

$$\frac{a^2(b^2+c^2)^2}{x^2}+\frac{b^2(a^2+c^2)^2}{y^2}=\frac{c^2(a^2-b^2)^2}{z^2}.$$

□ Equation of the generator of the one system of the hyperboloid through the point $P(a\cos\theta, b\sin\theta, 0)$ is

$$\frac{x-a\cos\theta}{a\sin\theta}=\frac{y-b\sin\theta}{-b\cos\theta}=\frac{z}{c}. \tag{1}$$

We take the straight line through origin as

$$\frac{x}{l}=\frac{y}{m}=\frac{z}{n}. \tag{2}$$

The lines given by (1) and (2) will be perpendicular if

$$al\sin\theta-bm\cos\theta+cn=0. \tag{3}$$

Being intersecting lines, the lines given by (1) and (2) are also coplanar. So we have

$$\begin{vmatrix} a\cos\theta & b\sin\theta & 0\\ a\sin\theta & -b\cos\theta & c\\ l & m & n\end{vmatrix}=0$$

$$\text{or,}\quad lbc\sin\theta-amc\cos\theta-anb=0. \tag{4}$$

From (3) and (4) we get

$$\frac{\sin\theta}{amn(b^2+c^2)}=\frac{\cos\theta}{bnl(a^2+c^2)}=\frac{1}{mlc(b^2-a^2)}$$

$$\text{or,}\quad \frac{\sin\theta}{\frac{a}{l}(b^2+c^2)}=\frac{\cos\theta}{\frac{b}{m}(a^2+c^2)}=\frac{1}{\frac{c}{n}(b^2-a^2)}.$$

Eliminating θ we get

$$\frac{a^2(b^2+c^2)^2}{l^2}+\frac{b^2(a^2+c^2)^2}{m^2}=\frac{c^2(a^2-b^2)^2}{n^2} \tag{5}$$

To get the locus of the line (2) we eliminate l, m, n from (5) with the help of (2) and get

$$\frac{a^2(b^2+c^2)^2}{x^2}+\frac{b^2(a^2+c^2)^2}{y^2}=\frac{c^2(a^2-b^2)^2}{z^2}.$$

EXERCISE

1. Find the equations of the generators of the following surfaces at the given points:
   (i) $x^2+y^2-4z^2=9$ at the point $(-3, 4, 2)$.
   (ii) $9x^2-y^2+z^2=9$ at the point $(1, 1, 1)$.
   (iii) $36x^2+16y^2-9z^2=144$ at the points $(2, 3, 4)$ and $(2, -1, 4/3)$.
   (iv) $4x^2-y^2=7z$ at the point $(2, 3, 1)$.
   (v) $9x^2-16y^2=4z$ at the point $(4, 3, 0)$.
   (vi) $16x^2-9y^2=4z$ at the point $(1, 0, 4)$.
   (vii) $yz+2zx+3xy+6=0$ at the point $(0, -3, 2)$.
2. Find the equations of the generating lines of the paraboloid

$$(x+y+z)(2x+y-z)=6z,$$

which passes through the point $(1, 1, 1)$.

3. Find the equations of the generating lines of the hyperboloid

$$x^2 + y^2 + z^2 - 2xy - 2yz - 2xz = 1.$$

4. Show that the angle between the generators of the paraboloid $\dfrac{x^2}{a} - \dfrac{y^2}{b} = 4z$ at the point $(\alpha, 0, \gamma)$ is $\cos^{-1}\left(\dfrac{a-b+\gamma}{a+b+\gamma}\right)$.

5. Find the equations of the generators of the hyperboloid $\dfrac{x^2}{4} + \dfrac{y^2}{9} - \dfrac{z^2}{16} = 1$ in which the plane $6x + 4y + 3z = 12$ intersects it.

6. Prove that the points of intersection of generators of $xy = az$ which are inclined at an angle α lie on the curve of intersection of the paraboloid and the hyperboloid

$$x^2 + y^2 - z^2 \tan^2 \alpha + a^2 = 0.$$

[*Hints*: Generators of two systems of $xy = az$ are

$$x = \lambda z;\ y = \frac{a}{\lambda} \quad \text{and} \quad x = \frac{a}{\mu};\ y = \mu z. \tag{1}$$

Point of intersection of the above generators is $\left(\dfrac{a}{\mu}, \dfrac{a}{\lambda}, \dfrac{a}{\lambda\mu}\right)$. Direction ratios of the line of intersections of the planes given by (1) are $(\lambda, 0, 1)$ and $(0, \mu, 1)$ respectively. If the angle between the generators be α, then we have

$$\cos\alpha = \frac{1}{\sqrt{(1+\lambda^2)(1+\mu^2)}}$$

$$\text{or,} \quad \tan^2\alpha = (1+\lambda^2)(1+\mu^2) - 1 = \lambda^2 + \mu^2 + \lambda^2\mu^2.$$

Eliminating λ, μ we get from the above equation

$$\tan^2\alpha = a^2\left(\frac{1}{x^2} + \frac{1}{y^2} + \frac{1}{z^2}\right)$$

$$\text{or,} \quad x^2y^2z^2\tan^2\alpha = a^2(y^2z^2 + x^2z^2 + y^2x^2)$$

$$\text{or,} \quad x^2y^2(z^2\tan^2\alpha - a^2) = a^2(y^2z^2 + x^2z^2)$$

$$\text{or,} \quad z^2\tan^2\alpha - a^2 = y^2 + x^2 \quad \text{as} \quad xy = az$$

$$\text{or,} \quad x^2 + y^2 - z^2\tan^2\alpha + a^2 = 0.]$$

7. Prove that the generators of the surface $yz + xz + xy + a^2 = 0$ through the point $\left(0, am, -\dfrac{a}{m}\right)$ are $x(1 \pm m) = am - y = \mp(mz + a)$.

8. Find the equations of the generators of the hyperboloid $\dfrac{x^2}{25} + \dfrac{y^2}{16} - \dfrac{z^2}{4} = 1$ which are parallel to the plane $8x + 10y + 20z = 11$.

9. Show that the most general quadric surface which has the lines

$$x = 0,\ y = 0;\quad x = 0,\ z = c;\quad y = 0,\ z = -c.$$

as generators is

$$fy(z - c) + gx(z + c) + hxy = 0.$$

[*Hints*: The general equation of the surface is

$$Ax^2 + By^2 + Cz^2 + 2Fyz + 2Gxz + 2Hxy + 2Ux + 2Vy + 2Wz + D = 0.$$

As the line $x = 0$; $y = 0$ lies on it we have $Cz^2 + 2Wz + D = 0$, This must be an identity in z, which leads that $C = 0$, $W = 0$, $D = 0$.
Again the lines $x = 0, z = c$; $y = 0, z = -c$ also lie on the surface, so we have

$$By^2 + 2Fcy + 2Vy = 0 \quad \text{and} \quad Ax^2 - 2Gcx + 2Ux = 0.$$

These are also identities in y and x respectively, so we get

$$B = 0;\ V = -Fc \quad \text{and} \quad A = 0;\ U = Gc.$$

So the above general equation of the surface becomes

$$2Fyz + 2Gxz + 2Hxy + 2Gcx - 2Fcy = 0$$

or, $$Fy(z - c) + Gx(z + c) + Hxy = 0.]$$

10. Show that the generators of the hyperboloid $x^2 + y^2 - z^2 = 1$ intersects on the plane $z = 0$ are at right angles.
11. Prove that the equations $2x = ae^{2\phi}$, $y = be^{\phi}\cosh\theta$, $z = ce^{\phi}\sinh\theta$ determines a hyperbolic paraboloid and that $\theta + \phi$ is constant for points of a given generator of one system and $\theta - \phi$ is constant for points of a given generator of other system.

[*Hints*: Eliminating θ and ϕ we get $\dfrac{y^2}{b^2} - \dfrac{z^2}{c^2} = \dfrac{2x}{a}$. The generators of it are

$$\frac{y}{b} - \frac{z}{c} = 2\lambda;\ \frac{y}{b} + \frac{z}{c} = \frac{x}{a\lambda} \quad \text{and} \quad \frac{y}{b} + \frac{z}{c} = 2\mu;\ \frac{y}{b} - \frac{z}{c} = \frac{x}{a\mu}.$$

As the generators pass through a point in the paraboloid we use $2x = ae^{2\phi}$, $y = be^{\phi}\cosh\theta$, $z = ce^{\phi}\sinh\theta$ in the above equation of generators and get for the first system $e^{\phi-\theta} = 2\lambda$. For a given generator of the first system λ is constant so $\theta - \phi$ is constant. Similarly using $2x = ae^{2\phi}$, $y = be^{\phi}\cosh\theta$, $z = ce^{\phi}\sinh\theta$ in the second system of generators we get $\theta + \phi$ is constant.]

12. Planes are drawn through the origin O and the generators through any point P of the paraboloid $x^2 - y^2 = az$. Prove that the angle between them is $\tan^{-1}\left(\frac{2r}{a}\right)$, where $OP = r$.

[*Hints*: Equation of the planes through the origin and the following two systems of generators of the given paraboloid

$$x - y = \lambda z;\quad x + y = \frac{a}{\lambda}$$

and $$x - y = \frac{a}{\mu};\quad x + y = \mu z$$

are $x - y - \lambda z = 0$ and $x + y - \mu z = 0$. If α be the angle between them, then we have

$$\cos\alpha = \frac{1 - 1 + \lambda\mu}{\sqrt{(1 + 1 + \lambda^2)(1 + 1 + \mu^2)}}$$

or, $$\tan^2\alpha = \frac{(2 + \lambda^2)(2 + \mu^2) - \lambda^2\mu^2}{\lambda^2\mu^2} = \frac{4 + 2\lambda^2 + 2\mu^2}{\lambda^2\mu^2}.$$

Now the point of intersection of the generators is $\left(\frac{a}{2}\frac{\lambda+\mu}{\lambda\mu}, \frac{a}{2}\frac{\mu-\lambda}{\lambda\mu}, \frac{a}{\lambda\mu}\right)$, so

$$OP^2 = r^2 = \frac{a^2}{4\lambda^2\mu^2}[(\lambda+\mu)^2+(\lambda-\mu)^2+4] = \frac{a^2}{4\lambda^2\mu^2}[2(\lambda^2+\mu^2)+4] = \frac{a^2\tan^2\alpha}{4}$$

gives $\alpha = \tan^{-1}\left(\frac{2r}{a}\right)$.]

13. If the generators through a point P of the hyperboloid $\frac{x^2}{a^2}+\frac{y^2}{b^2}-\frac{z^2}{c^2}=1$ meet the principal elliptic section in two points such that eccentric angle of one is three times that of other, prove that P lies on the curve of intersection of the hyperboloid with the cylinder $y^2(z^2+c^2)=4b^2z^2$.

[*Hints*: If $P(\alpha, \beta, \gamma)$ be the point such that the generators meet the elliptic section at the points $(a\cos\theta, b\sin\theta, 0)$ and $(a\cos\phi, b\sin\phi, 0)$ then we have

$$\alpha = a\frac{\cos\frac{\theta+\phi}{2}}{\cos\frac{\theta-\phi}{2}},\ \beta = b\frac{\sin\frac{\theta+\phi}{2}}{\cos\frac{\theta-\phi}{2}},\ \gamma = \pm c\frac{\sin\frac{\theta-\phi}{2}}{\cos\frac{\theta-\phi}{2}}$$

Given that $\theta = 3\phi$, so

$$\alpha = a\frac{\cos 2\phi}{\cos\phi},\ \beta = b\frac{\sin 2\phi}{\cos\phi},\ \gamma = \pm c\frac{\sin\phi}{\cos\phi}$$

Eliminating ϕ we get the required result.]

14. If the generators through a point P of the hyperboloid $\frac{x^2}{a^2}+\frac{y^2}{b^2}-\frac{z^2}{c^2}=1$ meet the principal elliptic section in two points such that the difference of their eccentric angles is 2α, prove that P lies on the curve of intersection of the hyperboloid with the cone

$$\frac{x^2}{a^2}+\frac{y^2}{b^2}-\frac{z^2}{c^2}\operatorname{cosec}^2\alpha = 0.$$

[*Hints*: Proceed as hints of exercise 13 and take $\theta-\phi = 2\alpha$.]

15. Prove that the projection of the generators of the hyperboloid $\frac{x^2}{a^2}+\frac{y^2}{b^2}-\frac{z^2}{c^2}=1$ on the plane $y=0$ is a tangent to the hyperbola $\frac{x^2}{a^2}-\frac{z^2}{c^2}=1;\ y=0$.

[*Hints*: Proceed as Example 8.]

Chapter 13

The General Second Degree Equation

13.1 The General Equation

The general equation of a surface is

$$F(x, y, z) = ax^2 + by^2 + cz^2 + 2hxy + 2gxz + 2fyz + 2ux + 2vy + 2wz + d = 0$$
$$= \phi(x, y, z) + 2ux + 2vy + 2wz + d = 0 \qquad (1)$$

where $\phi(x, y, z) = ax^2 + by^2 + cz^2 + 2hxy + 2gxz + 2fyz$ is the homogeneous part of $F(x, y, z)$.

Dividing the Eqn. (1) by d we see that the number constants in it is nine implying that nine equations involving the constants are required to solve for those nine constants and hence nine conditions are needed to form those nine equations. The conditions may be:

(i) the surface passes through nine points, no four of which are coplanar;

(ii) the surface passes through three non-intersecting straight lines, etc.

The equation of a line passing through the point (x_1, y_1, z_1) and having direction cosines l, m, n is

$$\frac{x - x_1}{l} = \frac{y - y_1}{m} = \frac{z - z_1}{n} = r. \qquad (2)$$

Any point on this line can be taken as $(x_1 + lr, y_1 + mr, z_1 + nr)$. If this point lies on the given surface then we have

$$\phi(l, m, n)r^2 + 2r[l(ax_1 + hy_1 + gz_1 + u) + m(hx_1 + by_1 + fz_1 + v)$$
$$+ n(gx_1 + fy_1 + cz_1 + w)] + F(x_1, y_1, z_1) = 0.$$

The above quadratic equation in r gives two values of r indicating that the line (2) lying on a plane meets the given surface (1) in two points.

13.2 Principal Planes and Principal Directions

A diametral plane conjugate to the direction l, m, n is said to be a principal plane if the chords, parallel to the direction l, m, n, bisected by it are perpendicular to it and the direction is called principal direction. For example in case of the ellipsoid $\frac{x^2}{a^2} + \frac{y^2}{b^2} + \frac{z^2}{c^2} = 1$ the coordinate planes are the only principal planes and the coordinate axes are the only principal directions but in the case of sphere $x^2 + y^2 + z^2 = a^2$ any triad of mutually perpendicular planes are principal planes and the directions of the lines perpendicular to them are the principal directions.

The diametral plane bisecting the chords parallel to the direction l, m, n is (from section 10.1)

$$l(ax + hy + gz + u) + m(hx + by + fz + v) + n(gx + fy + cz + w) = 0$$

$$\text{or,} \quad (al + hm + gn)x + (hl + bm + fn)y + (gl + fm + cn)z + lu + mv + nw = 0$$

where

$$\frac{al + hm + gn}{l} = \frac{hl + bm + fn}{m} = \frac{gl + fm + cn}{n} = k, \quad \text{(say)} \tag{3}$$

which gives

$$\begin{aligned} (a - k)l + hm + gn &= 0 \\ hl + (b - k)m + fn &= 0 \\ gl + fm + (c - k)n &= 0. \end{aligned}$$

The above system of equations in l, m, n will have a non-trivial solution if the coefficient determinant vanishes i.e., if

$$\begin{vmatrix} a-k & h & g \\ h & b-k & f \\ g & f & c-k \end{vmatrix} = 0.$$

Expanding the determinant we get

$$k^3 - (a + b + c)k^2 + (A + B + C)k - \Delta = 0 \tag{4}$$

where A, B, C are the cofactors of a, b, c in Δ and

$$\Delta = \begin{vmatrix} a & h & g \\ h & b & f \\ g & f & c \end{vmatrix} = abc + 2fgh - af^2 - bg^2 - ch^2. \tag{5}$$

The above cubic equation in k is called discriminating cubic and the roots of that equation are called characteristic root. As the imaginary roots of an equation with real coefficients always occur in pair, the discriminating cubic has at least one real root implying that there exists at least one principal plane to a conicoid.

13.3 Principal Directions and Their Properties

If k be a root of the discriminating cubic, then

$$k = \phi(l, m, n)$$

where l, m, n are the direction cosines of the principal direction.

The direction cosines of a principal direction l, m, n for a non-zero root of the discriminating cubic are given by Eqn. (3) i.e., by

$$\frac{\frac{\partial\phi}{\partial l}}{2l} = \frac{\frac{\partial\phi}{\partial m}}{2m} = \frac{\frac{\partial\phi}{\partial n}}{2n} = k \tag{6}$$

where the function $\phi(l, m, n)$ is defined in section 13.1.

From (6) we get

$$k = \frac{l\frac{\partial\phi}{\partial l} + m\frac{\partial\phi}{\partial m} + n\frac{\partial\phi}{\partial n}}{2(l^2 + m^2 + n^2)} = \frac{2\phi(l, m, n)}{2} = \phi(l, m, n).$$

Properties

(i) *Two principal directions corresponding to two distinct root of the discriminating cubic are perpendicular.*

If corresponding to the roots k_1 and k_2 of the discriminating cubic, the direction cosines of principal directions are l_1, m_1, n_1 and l_2, m_2, n_2 then

$$\begin{aligned} l_1\frac{\partial\phi}{\partial l_2} + m_1\frac{\partial\phi}{\partial m_2} + n_1\frac{\partial\phi}{\partial n_2} &= 2[al_1l_2 + bm_1m_2 + cn_1n_2 + h(l_1m_2 + l_2m_1) \\ &\quad + g(l_1n_2 + l_2n_1) + f(n_1m_2 + n_2m_1)] \qquad (7) \\ &= l_2\frac{\partial\phi}{\partial l_1} + m_2\frac{\partial\phi}{\partial m_1} + n_2\frac{\partial\phi}{\partial n_1}. \end{aligned}$$

Again from (6)

$$\frac{\partial\phi}{\partial l_1} = 2k_1l_1; \quad \frac{\partial\phi}{\partial m_1} = 2k_1m_1; \quad \frac{\partial\phi}{\partial n_1} = 2k_1n_1$$

and

$$\frac{\partial\phi}{\partial l_2} = 2k_2l_2; \quad \frac{\partial\phi}{\partial m_2} = 2k_2m_2; \quad \frac{\partial\phi}{\partial n_2} = 2k_2n_2.$$

Using the above results in (7) we get

$$2(l_1l_2 + m_1m_2 + n_1n_2)(k_1 - k_2) = 0$$

which gives

$$l_1l_2 + bm_1m_2 + cn_1n_2 = 0 \quad \text{as} \quad k_1 \neq k_2.$$

Hence the proposition is proved.

(ii) *If all the roots of the discriminating cubic be distinct, then the principal directions are unique.*

(iii) *If the discriminating cubic has two equal roots then the principal directions are not unique.*

(iv) *If the discriminating cubic has three equal roots then any set of three mutually perpendicular directions are principal directions.*

13.4 Centre and Reduction of Central Conicoid

Let (α, β, γ) be the centre of the conicoid given by Eqn. (1). Shifting origin to the centre of the conicoid i.e., putting

$$x = X + \alpha; \quad y = Y + \beta \text{ and } z = Z + \gamma$$

in Eqn. (1) we get

$$\begin{aligned}\phi(X, Y, Z) &+ 2X(a\alpha + h\beta + g\gamma + u) + 2Y(h\alpha + b\beta + f\gamma + v) \\ &+ 2Z(g\alpha + f\beta + c\gamma + w) + F(\alpha, \beta, \gamma) = 0.\end{aligned}$$

As centre of the transformed equation is the origin of the new coordinate system there would no term in X, Y, Z only, which leads that

$$\begin{aligned} a\alpha + h\beta + g\gamma + u &= 0 \\ h\alpha + b\beta + f\gamma + v &= 0 \\ g\alpha + f\beta + c\gamma + w &= 0. \end{aligned} \tag{8}$$

If Δ, defined by Eqn. (5), does not vanish then multiplying the equations of (8) respectively by A, H, G and adding we get

$$\alpha = -\frac{Au + Hv + Gw}{\Delta}.$$

Similarly, multiplying the equations of (8) respectively by H, B, F and G, F, C and adding we get

$$\beta = -\frac{Hu + Bv + Fw}{\Delta}$$

$$\text{and} \quad \gamma = -\frac{Gu + Fv + Cw}{\Delta}$$

where A, B, C, F, G, H are the cofactors of a, b, c, f, g, h in Δ.

Thus we see that the centre lies on the planes

$$\begin{aligned} ax + hy + gx + u &= 0 \\ hx + by + fz + v &= 0 \\ gx + fy + cz + w &= 0 \end{aligned}$$

which are called central planes of the conicoid given by Eqn. (1) and the equation of the conicoid reduces to

$$\phi(X, Y, Z) + F(\alpha, \beta, \gamma) = 0.$$

We further note that

$$\begin{aligned} F(\alpha, \beta, \gamma) &= a\alpha^2 + b\beta^2 + c\gamma^2 + 2h\alpha\beta + 2g\alpha\gamma + 2f\beta\gamma + 2u\alpha + 2v\beta + 2w\gamma + d \\ &= \alpha(a\alpha + h\beta + g\gamma + u) + \beta(h\alpha + b\beta + f\gamma + v) \\ &\quad + \gamma(g\alpha + f\beta + c\gamma + w) + u\alpha + v\beta + w\gamma + d \\ &= u\alpha + v\beta + w\gamma + d. \end{aligned}$$

Thus the general equation of the conicoid given by (1) with referred to the centre (α, β, γ) is reduced to

$$\phi(X, Y, Z) + u\alpha + v\beta + c\gamma + d = 0. \tag{9}$$

Next let $l_i, m_i, n_i;\ i = 1, 2, 3$ be the direction cosines of three principal directions corresponding to the non-zero roots k_1, k_2, k_3 of the discriminating cubic. If we denote the principal directions as ξ, η, ζ-axes then using the transformation laws

	X	Y	Z
ξ	l_1	m_1	n_1
η	l_2	m_2	n_2
ζ	l_3	m_3	n_3

in Eqn. (8) we get

$$\begin{aligned} &a(l_1\xi + l_2\eta + l_3\zeta)^2 + b(m_1\xi + m_2\eta + m_3\zeta)^2 \\ &+ c(n_1\xi + n_2\eta + n_3\zeta)^2 + 2f(m_1\xi + m_2\eta + m_3\zeta)(n_1\xi + n_2\eta + n_3\zeta) \\ &+ 2g(l_1\xi + l_2\eta + l_3\zeta)(n_1\xi + n_2\eta + n_3\zeta) + 2h(l_1\xi + l_2\eta + l_3\zeta)(m_1\xi + m_2\eta + m_3\zeta) \\ &+ u\alpha + v\beta + w\gamma + d - 0. \end{aligned} \tag{10}$$

In which the coefficient of ξ^2 is

$$\begin{aligned} &al_1^2 + bm_1^2 + cn_1^2 + 2fm_1n_1 + 2gl_1n_1 + 2hl_1m_1 \\ &= l_1(al_1 + hm_1 + gn_1) + m_1(hl_1 + bm_1 + fn_1) + n_1(gl_1 + fm_1 + cn_1) \\ &= k_1(l_1^2 + m_1^2 + n_1^2) = k_1 \end{aligned}$$

Similarly, the coefficients of η^2 and ζ^2 in the above expression are k_2 and k_3. Now the coefficient of $2\xi\eta$ in (9) is

$$\begin{aligned} &al_1l_2 + bm_1m_2 + cn_1n_2 + f(m_1n_2 + m_2n_1) + g(l_1n_2 + l_2n_1) + h(l_1m_2 + l_2m_1) \\ &= l_1(al_2 + hm_2 + gn_2) + m_1(hl_2 + bm_2 + fn_2) + n_1(gl_2 + fm_2 + cn_2) \\ &= k_2(l_1l_2 + m_1m_2 + n_1n_2) = 0. \end{aligned}$$

Similarly the coefficients of $2\zeta\eta$ and $2\xi\zeta$ in (10) are also zero. Thus the Eqn. (10) reduces to

$$k_1\xi^2 + k_2\eta^2 + k_3\zeta^2 + d' = 0 \tag{11}$$

where $d' = u\alpha + v\beta + w\gamma + d$.

Hence by shifting the origin to the centre of the conicoid and rotating the coordinate axes properly the general equation of a central conicoid given by (1) can be reduced to (11) which is called the canonical form of the central conicoid.

13.5 Classification of Quadrics

The quadrics given by the Eqn. (1) depends on nature of the solutions of system of equations given by (8). Here we discuss the following cases:

(i) If $\Delta \neq 0$, the surface has single centre at a finite distance and hence the surface may be an ellipsoid, hyperboloid or cone.

(ii) If $\Delta = 0$ and $Au + Hv + Gw \neq 0$, the surface has a centre at infinite distance and hence it represents a paraboloid.

(iii) If $\Delta = 0$, $Au + Hv + Gw = 0$ and $A \neq 0$, the surface has a line of centres at a finite distance and it represents an elliptic or hyperbolic cylinder, pair of intersecting planes.

(iv) If A, B, C, F, G, H are all zero and $fu \neq gv$, the surface has a line of centres at an infinite distance and it represents a parabolic cylinder.

(v) If A, B, C, F, G, H are all zero and $fu = gv = hw$, central planes are coincident and hence the surface represents a pair of parallel planes.

13.6 Canonical Form of the Equation of Quadrics

Here we give the canonical forms of different quadric surfaces according to the roots of the discriminating cubic and position of their centres.

Centre at Finite Distance

Case I: None of the roots of the discriminating cubic vanishes:

(i) case of three unequal roots: (single centre)

Equation	Name of the surface
1. $Ax^2 + By^2 + Cz^2 = 1$	Ellipsoid
2. $Ax^2 + By^2 - Cz^2 = 1$	Hyperboloid of one sheet
3. $Ax^2 - By^2 - Cz^2 = 1$	Hyperboloid of two sheets
4. $Ax^2 + By^2 + Cz^2 = 0$	Cone

(ii) case of two equal roots: (line of centres)

1. $A(x^2 + y^2) + Cz^2 = 1$	Ellipsoid of revolution
2. $A(x^2 + y^2) - Cz^2 = 1$	Hyperboloid of revolution
3. $A(x^2 + y^2) + Cz^2 = 0$	Cone of revolution

Case II: One of the roots of the discriminating cubic vanishes

(i) Remaining two roots are unequal: (line of centres)

Equation	Name of the surface
1. $Ax^2 + By^2 = C$	Elliptic cylinder
2. $Ax^2 - By^2 = C$	Hyperbolic cylinder
3. $Ax^2 - By^2 = 0$	Pair of intersecting planes

(ii) Remaining two roots are equal: (line of centres)

1. $A(x^2 + y^2) = C$	Right circular cylinder

Case III: Two of the roots of the discriminating cubic vanishes

Equation	Name of the surface
1. $Ax^2 = C$	Pair of parallel planes

Centre at Infinite Distance

Case I: One of the roots of the discriminating cubic vanishes

(i) Remaining two roots are unequal: (Single centre)

Equation	Name of the surface
1. $Ax^2 + By^2 + Cz = 0$	Elliptic paraboloid
2. $Ax^2 - By^2 + Cz = 0$	Hyperbolic paraboloid

(ii) Remaining two roots are equal: (Single centre)

1. $A(x^2 + y^2) + Cz = 0$	Paraboloid of revolution

Case II: Two of the roots of the discriminating cubic vanishes

Equation	Name of the surface
1. $Ax^2 + Cz = 0$	Parabolic cylinder

13.7 Working Rule

Let the general equation of the surface be $F(x, y, z) = 0$, defined by Eqn. (1). For reduction of the general equation to one of the above cases we proceed as follows:

When $\varphi(x, y, z)$ is not a Perfect Square

Case I: None of the roots of the discriminating cubic vanishes

(a) Solve the Eqns. (8) for centre,
(b) Reduce the equation to $\phi(x, y, z) + u\alpha + \upsilon\beta + c\gamma + d = 0$
(c) Rotating the axes properly reduce the above equation to $k_1x^2 + k_2y^2 + k_3z^2 + d' = 0$, where $d' = u\alpha + \upsilon\beta + w\gamma + d$.
(d) For direction ratios of the principal directions solve any two of the following equations:

$$\begin{aligned}(a - k_i)l + hm + gn &= 0\\ hl + (b - k_i)m + fn &= 0, \quad i = 1, 2, 3\\ gl + fm + (c - k_i)n &= 0.\end{aligned}$$

(e) Principal planes are given by the equations

$$k_i(lx + my + nz) + ul + \upsilon m + wn = 0.$$

Case II: One of the roots of the discriminating cubic vanishes

(a) If $k_1 = 0$, the direction ratios of the direction corresponding to this value of k are obtained by solving any two of the following equations:

$$\begin{aligned}al_1 + hm_1 + gn_1 &= 0\\ hl_1 + bm_1 + fn_1 &= 0\\ gl_1 + fm_1 + cn_1 &= 0.\end{aligned}$$

(b) Next calculate $\delta = ul_1 + vm_1 + wn_1$. If $\delta \neq 0$, the reduced equation is

$$k_2y^2 + k_3z^2 + 2\delta x = 0.$$

The vertex of the above surface is obtained by solving any two of the following equations

$$\begin{aligned} ax + hy + gx + u - \delta l_1 &= 0 \\ hx + by + fz + v - \delta m_1 &= 0 \\ gx + fy + cz + w - \delta n_1 &= 0 \end{aligned}$$

with the equation

$$\delta(l_1x + m_1y + n_1z) + ux + vy + wz + d = 0.$$

(c) If $\delta = 0$, then solve any two of the following equations

$$\begin{aligned} ax + hy + gx + u &= 0 \\ hx + by + fz + v &= 0 \\ gx + fy + cz + w &= 0. \end{aligned}$$

i.e., take any point on the line of intersection of any two of the above planes. Let it be (α, β, γ). Shifting origin to this point and rotating the axes the general equation of the surface we get reduced equation of the surface as

$$k_2y^2 + k_3z^2 + d' = 0.$$

Case III: Two of the roots of the discriminating cubic vanish:

Proceeding as Case I, we get the reduced equation as $k_3z^2 + d' = 0$.

When $\varphi(x, y, z)$ is a Perfect Square

(a) The equation $F(x, y, z) = 0$ can be expressed as

$$(\sqrt{a}x + \sqrt{b}y + \sqrt{c}z)^2 + k(\sqrt{a}x + \sqrt{b}y + \sqrt{c}z) + d = 0.$$

Or,

(b) The equation $F(x, y, z) = 0$ can be expressed as.

$$(\sqrt{a}x + \sqrt{b}y + \sqrt{c}z + \mu)^2 = k(Ax + By + Cz + D).$$

such that the planes $\sqrt{a}x + \sqrt{b}y + \sqrt{c}z + \mu = 0$ and $Ax + By + Cz + D = 0$ are perpendicular.

Examples .

▶▷▶ 1. *Reduce the equation* $11x^2+10y^2+6z^2-8yz+4xz-12xy+72x-72y+36z+150 = 0$ *to its canonical form. Find the equation of its axis.*

☐ Here we see that $abc+2fgh-af^2-bg^2-ch^2 = 660+96-176-40-216 = 324 \neq 0$, so the given equation is of central quadric. The discriminating cubic is

$$\begin{vmatrix} 11-k & -6 & 2 \\ -6 & 10-k & -4 \\ 2 & -4 & 6-k \end{vmatrix} = 0$$

or, $\quad k^3 - 27k^2 + 180k - 324 = 0.$

Solving we get $k = 3, 6, 18$. Since all the roots of discriminating cubic are unequal, the given quadric is either ellipsoid or, hyperboloid or cone. Centre (α, β, γ) of the quadric is given by

$$11\alpha - 6\beta + 2\gamma + 36 = 0$$
$$-6\alpha + 10\beta - 4\gamma - 36 = 0$$
$$2\alpha - 4\beta + 6\gamma + 18 = 0.$$

Solving we get $\alpha = -2$, $\beta = 2$ and $\gamma = -1$. So shifting the origin to the point $(-2, 2, -1)$ and rotating the axes properly the given equation can be reduced to

$$3x^2 + 6y^2 + 18z^2 + 36(-2) - 36(2) + 18(-1) + 150 = 0$$

or, $\dfrac{x^2}{4} + \dfrac{y^2}{2} + \dfrac{z^2}{2/3} = 1,$ which is an ellipsoid.

The direction ratios of the principal axis corresponding to $k = 3$ are given by

$$8l - 6m + 2n = 0$$
$$-6l + 7m - 4n = 0$$
$$2l - 4m + 3n = 0.$$

Solving any two of the following equations we get

$$\frac{l}{1} = \frac{m}{2} = \frac{n}{2} = \frac{1}{3}.$$

Similarly the direction ratios of the principal axis corresponding to $k = 6$ and 18 are given by

$$5l - 6m + 2n = 0$$
$$-6l + 4m - 4n = 0$$
$$2l - 4m = 0.$$

and

$$-7l - 6m + 2n = 0$$
$$-6l - 8m - 4n = 0$$
$$2l - 4m - 12n = 0.$$

Solving the above equations we get the direction cosines of the other two principal axes are $\frac{2}{3}, \frac{1}{3}, -\frac{2}{3}$ and $-\frac{2}{3}, \frac{2}{3}, -\frac{1}{3}$ respectively.

Hence the equations of principal axes are

$$\frac{x+2}{1} = \frac{y-2}{2} = \frac{z+1}{2}, \quad \frac{x+2}{2} = \frac{y-2}{1} = \frac{z+1}{-2} \quad \text{and} \quad \frac{x+2}{2} = \frac{y-2}{-2} = \frac{z+1}{1}.$$

▶▷▶ 2. *Reduce the equation $x^2 + y^2 + z^2 + yz + xz + xy + 3x + y + 4z + 4 = 0$ to its canonical form.*

☐ Here we see that $abc + 2fgh - af^2 - bg^2 - ch^2 = \frac{1}{2} \neq 0$, so the given equation is of central quadric. The discriminating cubic is

$$\begin{vmatrix} 1-k & 1/2 & 1/2 \\ 1/2 & 1-k & 1/2 \\ 1/2 & 1/2 & 1-k \end{vmatrix} = 0$$

or, $\quad 4k^3 - 12k^2 + 9k - 2 = 0.$

Solving we get $k = 2, \frac{1}{2}, \frac{1}{2}$. Since two of the roots of discriminating cubic are equal, the given quadric is a surface of revolution. Coordinates of the centre are given by

$$2\alpha + \beta + \gamma + 3 = 0$$
$$\alpha + 2\beta + \gamma + 1 = 0$$
$$\alpha + \beta + 2\gamma + 4 = 0.$$

Solving we get $\alpha = -1, \beta = 1, \gamma = -2$. So $d' = \frac{3}{2}(-1) + \frac{1}{2}(1) + 2(-2) + 4 = -1$. Hence the reduced equation is $x^2 + y^2 + 4z^2 = 2$, which is an ellipsoid of revolution .The direction ratios of the axis of revolution are the direction ratios of the axis corresponding to the unequal value of k i.e., for $k = 2$. So the direction ratios of the axis of revolution satisfy the following equations

$$-2l + m + n = 0$$
$$l - 2m + n = 0$$
$$l + m - 2n = 0.$$

Solving which we get

$$\frac{l}{1} = \frac{m}{1} = \frac{n}{1}.$$

Hence the equation of the axis of revolution is $x + 1 = y - 1 = z + 2$.

▶▶ **3.** *Reduce the equation* $6y^2 - 18yz - 6xz + 2xy - 9x + 5y - 5z + 2 = 0$ *to its canonical form.*

☐ Here we see that $abc + 2fgh - af^2 - bg^2 - ch^2 = 54 - 54 = 0$ and $Au + Bv + Cw \neq 0$. So the given quadric is a paraboloid. The discriminating cubic is

$$\begin{vmatrix} -k & 1 & -3 \\ 1 & 6-k & -9 \\ -3 & -9 & -k \end{vmatrix} = 0$$

$$\text{or,} \quad k^3 - 6k^2 - 91k = 0.$$

Solving we get

$$k = -7, 0, 13.$$

The direction ratios corresponding to $k = 0$ are given by

$$0 \cdot l + m - 3n = 0$$
$$l + 6m - 9n = 0$$
$$-3l - 9m + 0 \cdot n = 0.$$

Solving any two of these equations we get

$$\frac{l}{-9} = \frac{m}{3} = \frac{n}{1} = \frac{1}{\sqrt{91}}$$

Now

$$\delta = ul + vm + wn = -\frac{9}{2\sqrt{91}}(-9) + \frac{5}{2\sqrt{91}}(3) - \frac{5}{2\sqrt{91}}(1) = \frac{\sqrt{91}}{2} \neq 0$$

Hence the reduced equation is

$$-7x^2 + 13y^2 + \sqrt{91}z = 0.$$

▶▷▶ 4. *Reduce the equation* $13x^2 + 45y^2 + 40z^2 + 12yz + 36xz - 24xy - 49 = 0$ *to its canonical form.*

□ Here we see that $abc + 2fgh - af^2 - bg^2 - ch^2 = 0$ and $Au + Bv + Cw = 0$ with $A \neq 0$. The discriminating cubic is

$$\begin{vmatrix} 13-k & -12 & 18 \\ -12 & 45-k & 6 \\ 18 & 6 & 40-k \end{vmatrix} = 0$$

$$\text{or,} \quad k^3 - 98k^2 + 2401k = 0.$$

Solving we get $k = 0, 49, 49$. Since two roots are equal, the given quadric is a surface of revolution.

The direction ratios corresponding to $k = 0$ are given by

$$13l - 12m + 18n = 0$$
$$-12l + 45m + 6n = 0$$
$$18l + 6m + 40n = 0.$$

Solving any two of these equations we get

$$\frac{l}{6} = \frac{m}{2} = \frac{n}{-3} = \frac{1}{7}.$$

Now $\delta = ul + vm + wn = 0$. The line of centres of the quadric is the line of intersection of the following planes

$$13x - 12y + 18z = 0,$$
$$12x - 45y - 6z = 0,$$
$$18x + 6y + 40z = 0.$$

We choose any point on the line of centres as $(0, 0, 0)$ which satisfies all the three equations. So $d' = d = -49$.

Hence the reduced equation is $49x^2 + 49y^2 - 49 = 0$ or, $x^2 + y^2 = 1$. Which is a circular cylinder whose axis is $\dfrac{x}{6} = \dfrac{y}{2} = \dfrac{z}{-3}$.

▶▷▶ 5. *Reduce the equation* $26x^2 + 20y^2 + 10z^2 - 4yz - 16xz - 36xy + 52x - 36y - 16z + 25 = 0$ *to its canonical form. Find the equation of its axis.*

□ Here we see that $abc+2fgh-af^2-bg^2-ch^2 = 5200-576-104-1280-3240 = 0$ and $Au + Bv + Cw = 196(26) + 196(-18) + 196(-8) = 0$ with $A = 196 \neq 0$. The discriminating cubic is

$$\begin{vmatrix} 26-k & -18 & -8 \\ -18 & 20-k & -2 \\ -8 & -2 & 10-k \end{vmatrix} = 0$$

$$\text{or,} \quad k^3 - 56k^2 + 588k = 0.$$

Solving we get $k = 0, 14, 42$. Since the roots of the discriminating cubic are unequal, the given quadric is either a pair of intersecting planes or elliptic or hyperbolic cylinder. The direction ratios corresponding to $k = 0$ are given by

$$26l - 18m - 8n = 0$$
$$-18l + 20m - 2n = 0$$
$$-8l - 2m + 10n = 0.$$

Solving any two of these equations we get

$$\frac{l}{1} = \frac{m}{1} = \frac{n}{1} = \frac{1}{\sqrt{3}}.$$

Now $\delta = ul + vm + wn = \frac{1}{\sqrt{3}}(26 - 18 - 8) = 0$. The line of centres of the quadric is the line of intersection of the following planes

$$26x - 18y - 8z + 26 = 0,$$
$$18x - 20y + 2z + 18 = 0,$$
$$8x + 2y - 10z + 8 = 0.$$

We choose any point on the line of centres as $(-1, 0, 0)$ which satisfies all the three equations.

Now $d' = 26(-1) - 18(0) - 8(0) + 25 = -1$. Hence the reduced equation is

$$14x^2 + 42y^2 = 1$$

which is an elliptic cylinder whose axis is

$$\frac{x+1}{1} = \frac{y}{1} = \frac{z}{1}.$$

▶▶ 6. *Reduce the equation* $x^2 + y^2 + z^2 - 2yz + 2xz - 2xy + 2x - 2y + 2z + 3 = 0$ *to its canonical form.*

□ Here we see that the second degree term is a perfect square term. We rewrite the given equation as

$$(x - y + z)^2 + 2(x - y + z) + 3 = 0. \tag{1}$$

If we choose the plane $x - y + z = 0$ as Y-Z plane then the perpendicular distance from any point (x, y, z) to the plane $x - y + z = 0$ is given by

$$X = \frac{x - y + z}{\sqrt{3}}.$$

So the Eqn. (1) becomes

$$3X^2 + 2\sqrt{3}X + 3 = 0,$$

which represents a pair of parallel planes.

▶▷▶ 7. *Reduce the equation* $9x^2+4y^2+4z^2+8yz+12xz+12xy+4x+y+10z+1=0$ *to its canonical form.*

□ Here we see that the second degree term is a perfect square term. We rewrite the given equation as

$$(3x+2y+2z)^2 + 4x + y + 10z + 1 = 0$$

$$\text{or,} \quad (3x+2y+2z+k)^2 = 2x(3k-2) + y(4k-1) + 2z(2k-5) + k^2 - 1. \quad (1)$$

Now we choose the constant k in such a way that the planes

$$3x+2y+2z+k=0 \quad \text{and} \quad 2x(3k-2)+y(4k-1)+2z(2k-5)+k^2-1=0$$

become perpendicular. Which leads that

$$6(3k-2)+2(4k-1)+4(2k-5)=0 \quad \text{i.e.,} \quad k=1.$$

Putting this value of k in the Eqn. (1) we get

$$(3x+2y+2z+1)^2 = 2x+3y-6z. \quad (2)$$

Now if we take the planes $3x+2y+2z+1=0$ and $2x+3y-6z=0$ as the plane $Y=0$ and $X=0$ respectively then the perpendicular distance from any point (x, y, z) to the planes $3x+2y+2z+1=0$ and $2x+3y-6z=0$ are given by

$$Y = \frac{3x+2y+2z+1}{\sqrt{3^2+2^2+2^2}} \quad \text{and} \quad X = \frac{2x+3y-6z}{\sqrt{2^2+3^2+6^2}}.$$

Using the results in Eqn. (2) we obtain

$$17Y^2 = 7X$$

$$\text{or,} \quad Y^2 = \frac{7}{17}X$$

which is a parabolic cylinder. The vertices lie on the line of intersection of the planes $3x+2y+2z+1=0$ and $2x+3y-6z=0$.

▶▷▶ 8. *Determine the nature of the quadric* $ax^2+by^2+cz^2+2fyz+2gxz+2hxy=1$ *if*

$$a-\frac{gh}{f} = b-\frac{hf}{g} = c-\frac{fg}{h} = 0.$$

□ We first note that $u=v=w=0$, so

$$fu = gv = hw = 0. \quad (1)$$

Again from $a-\dfrac{gh}{f} = b-\dfrac{hf}{g}$ we get

$$g(af-gh) = f(bg-hf) \Rightarrow fg(a-b) = h(g^2-f^2) \quad (2)$$

From $c - \dfrac{fg}{h} = 0$ we get $fg = hc$. So from (2) we get

$$hc(a-b) = h(g^2 - f^2)$$
$$\text{or,} \quad ac - g^2 = bc - f^2 \quad \Rightarrow \quad A = B.$$

Similarly, it can be shown that $B = C$ and $C = A$.

Again from $a - \dfrac{gh}{f} = 0$ we get

$$af = gh \;\Rightarrow\; abf = bgh \;\Rightarrow\; abf = fh^2 \quad \text{as} \quad bg = hf,$$

i.e., $ab = h^2 = 0$ or, $A = 0$. Hence we get

$$A = B = C = 0. \tag{3}$$

We also get from the given condition that

$$F = G = H = 0. \tag{4}$$

From the results (1), (3) and (4) we see that the given equation is of a pair of parallel planes.

▶▷ 9. *Show that* $\sqrt{x} + \sqrt{y} + \sqrt{z} = 0$ *represents a cone of revolution whose axis is* $x = y = z$.

□ $\sqrt{x} + \sqrt{y} + \sqrt{z} = 0 \;\Rightarrow\; \sqrt{x} + \sqrt{y} = -\sqrt{z}$, Squaring both sides we get

$$x + y - z = -2\sqrt{xy}.$$

Again squaring both sides we get

$$x^2 + y^2 + z^2 - 2yz - 2xz - 2xy = 0.$$

Here we see that $abc + 2fgh - af^2 - bg^2 - ch^2 \neq 0$, so the given equation is of central quadric. The discriminating cubic is

$$\begin{vmatrix} 1-k & -1 & -1 \\ -1 & 1-k & -1 \\ -1 & -1 & 1-k \end{vmatrix} = 0$$
$$\text{or,} \quad k^3 - 3k^2 + 4 = 0.$$

Solving we get $k = 2, 2, -1$. Since two of the roots of discriminating cubic are equal the given quadric is a surface of revolution. Coordinates of the centre are given by

$$\alpha - \beta - \gamma = 0$$
$$\alpha - \beta + \gamma = 0$$
$$\alpha + \beta - \gamma = 0.$$

Solving we get $\alpha = 0, \beta = 0, \gamma = 0$. So $d' = 0$. Hence the reduced equation is $2(x^2 + y^2) - z^2 = 0$, which is a cone of revolution. The direction ratios of the axis of the cone are the direction ratios of the axis corresponding to the unequal value of k i.e., for $k = -1$. So the direction ratios of the axis of the cone satisfy the following equations

$$-2l + m + n = 0$$
$$l - 2m + n = 0$$
$$l + m - 2n = 0.$$

Solving which we get
$$\frac{l}{1} = \frac{m}{1} = \frac{n}{1}.$$

Hence the axis of the cone is $x = y = z$.

EXERCISE

1. Find the principal directions and principle planes of the following quadrics:
   (i) $3x^2 + 5y^2 + 3z^2 + 2yz + 2xz + 2xy - 4x - 8z + 5 = 0$
   (ii) $x^2 + 3y^2 + 3z^2 - 2yz - 2x - 2y + 6z + 3 = 0$
   (iii) $3x^2 - y^2 - z^2 + 6yz - 6x + 6y - 2z - 2 = 0$
2. Find the centres of the following quadrics:
   (i) $3x^2 + 5y^2 + 3z^2 - 2yz + 2xz - 2xy + 3x + 5y + 8z + 18 = 0$
   (ii) $x^2 + y^2 + z^2 - 2yz + 2xz - 2xy - 3x + 3y - 3z + 5 = 0$
   (iii) $z^2 - yz + xz + xy - 2y + 2z + 3 = 0$
   (iv) $2x^2 - 5y^2 + 2z^2 - 2yz + 4xz - 2xy + 14x - 16y + 14z + 26 = 0$
3. Reduce the following equations to their canonical forms:
   (i) $x^2 + 3y^2 + z^2 + 4yz + 2x + 12y - 2z + 9 = 0$
   (ii) $2x^2 + 2y^2 + z^2 + 2yz - 2xz - 4xy + x + y = 0$
   (iii) $x^2 + 6y^2 - z^2 - yz + 5xy + 2x + 5y = 0$
   (iv) $2y^2 + 4xz + 2x - 4y + 6z + 5 = 0$
   (v) $x^2 + 4y^2 + 9z^2 - 12yz + 6xz - 4xy + 4x - 8y + 12z + 4 = 0$
   (vi) $2x^2 + 2y^2 - 4z^2 - 2yz - 2xz - 5xy - 2x - 2y + z = 0$
   (vii) $x^2 + y^2 + z^2 - yz - xz - xy - 3x - 6y - 9z + 21 = 0$
   (viii) $x^2 + 3y^2 + 3z^2 - 2yz - 2x - 2y + 6z + 3 = 0$
   (ix) $3x^2 - y^2 - z^2 + 6yz - 6x + 6y - 2z - 2 = 0$
   (x) $4y^2 + 4z^2 + 4yz - 2x - 14y - 22z + 33 = 0$
   (xi) $14x^2 + 14y^2 + 8z^2 - 4yz - 4xz - 8xy + 18x - 18y + 5 = 0$
   (xii) $5x^2 + 26y^2 + 10z^2 + 4yz + 14xz + 6xy - 8x - 18y - 10z + 4 = 0$
   (xiii) $16x^2 + 9y^2 + 4z^2 - 12yz - 16xz + 24xy + 7x + 2y - 12z = 0$
   (xiv) $11y^2 + 14yz + 8xz + 14xy - 6x - 16y + 2z - 2 = 0$
   (xv) $4x^2 + 9y^2 + 36z^2 - 36yz + 24xz - 12xy - 10x + 15y - 30z + 6 = 0$
   (xvi) $3x^2 + 5y^2 + 3z^2 - 2yz + 2xz - 2xy + 2x + 12y + 10z + 20 = 0$
   (xvii) $2x^2 + 5y^2 + 10z^2 + 12yz + 6xz + 4xy = 1$
   (xviii) $2x^2 + 5y^2 + z^2 - 4xy - 8x + 14y + 3 = 0$
   (xix) $4y^2 + 4z^2 + 4yz - 2x - 11y - 22z + 33 = 0$
   (xx) $x^2 - 3y^2 + z^2 - 10y - 6x + 15 = 0.$

Chapter 14

Other Coordinate Systems

14.1 Introduction

In this part of the book, so far we have discussed the rectangular coordinate system. Apart from these coordinate systems we have other coordinate systems which are required to define the position of a point in the space. Let a point $P(x, y, z)$ be associated uniquely to a set of values (u, v, w), then under some specified conditions (u, v, w) can also be expressed in terms of (x, y, z) and the coordinates (u, v, w) are called the curvilinear coordinates of the point P. The equations $u = c_1$; $v = c_2$ and $w = c_3$, where c_1, c_2, c_3 are constants, are called coordinate surfaces and pair-wisely those intersect along a curve or straight line which are called coordinate curves. If the angle of intersections of the coordinate surfaces be right angle then the system of coordinate is called orthogonal curvilinear coordinate system.

14.2 Special Curvilinear Coordinate Systems

Followings are the special curvilinear coordinates systems:

I. Cylindrical coordinate system: (r, θ, z)
II. Spherical coordinate system: (r, θ, ϕ)
III. Parabolic cylindrical coordinate system: (u, v, z)
IV. Paraboloidal coordinate system: (u, v, ϕ)
V. Elliptic cylindrical coordinate system: (u, v, z)
VI. Prolate spheroidal coordinate system: (ξ, η, ϕ)
VII. Oblate spheroidal coordinate system: (ξ, η, ϕ)
VIII. Ellipsoidal Coordinate System: (λ, μ, ν).

Here we shall discuss about cylindrical and spherical coordinate systems.

14.3 Cylindrical Coordinate System

Let $P(x, y, z)$ be a point in the rectangular coordinate system. We draw PN, a perpendicular from P on xy-plane and the line MN, parallel to Oy. If we take $ON = r$; $\angle MON = \theta$ and $NP = z$ then the coordinates (r, θ, z) called the cylindrical coordinates of the point P with respect to O (from which r is measured), the fixed line Ox (with which the line ON makes the angle θ) and the fixed plane xOy (from which the distance of P is measured). The numbers $r(\geq 0), \theta(-\pi < \theta \leq \pi)$, z are called the radius vector, the vectorial angle and the applicate of the point P respectively. Thus for a point P in space there exists a ordered triplet of the numbers r, θ, z and conversely.

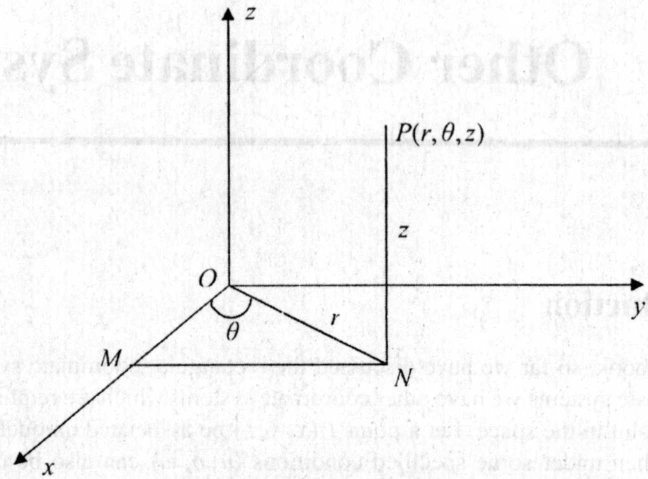

Cylindrical coordinate system

If (x, y, z) and (r, θ, z) be the cartesian and cylindrical coordinates of a point P as shown in the figure then we have

$$x = r\cos\theta;\ y = r\sin\theta;\ z = z \quad \text{and} \quad r = \sqrt{x^2 + y^2};\ \theta = \tan^{-1}\left(\frac{y}{x}\right);\ z = z.$$

14.4 Coordinate Surfaces and Curves for Cylindrical Coordinate System

(a) Coordinate Surfaces

(i) From $r = a$ we get $x^2 + y^2 = a^2$ which is a right circular cylinder (when $a \neq 0$) whose axis is z-axis and if $a = 0$ then we get $x = 0 = y$ which is the equation of z-axis.

(ii) From $\theta = \alpha$ (constant) we get $y = x\tan\alpha$ which is a equation of plane through z-axis making an angle α with the plane xOz.

(iii) The equation $z = k$ represents a plane perpendicular to z-axis and at a distance k from the xOy plane.

(b) Coordinate Curves

(i) The curve of intersection of the surfaces $r = a$ and $\theta = \alpha$ is called z-curve which is a straight line in this case.

(ii) The curve of intersection of the surfaces $r = a$ and $z = k$ is called θ-curve which is a circle or a point according as a is non-zero or zero.

(iii) The curve of intersection of the surfaces $\theta = \alpha$ and $z = k$, which are two perpendicular planes, is called r-curve which is a straight line.

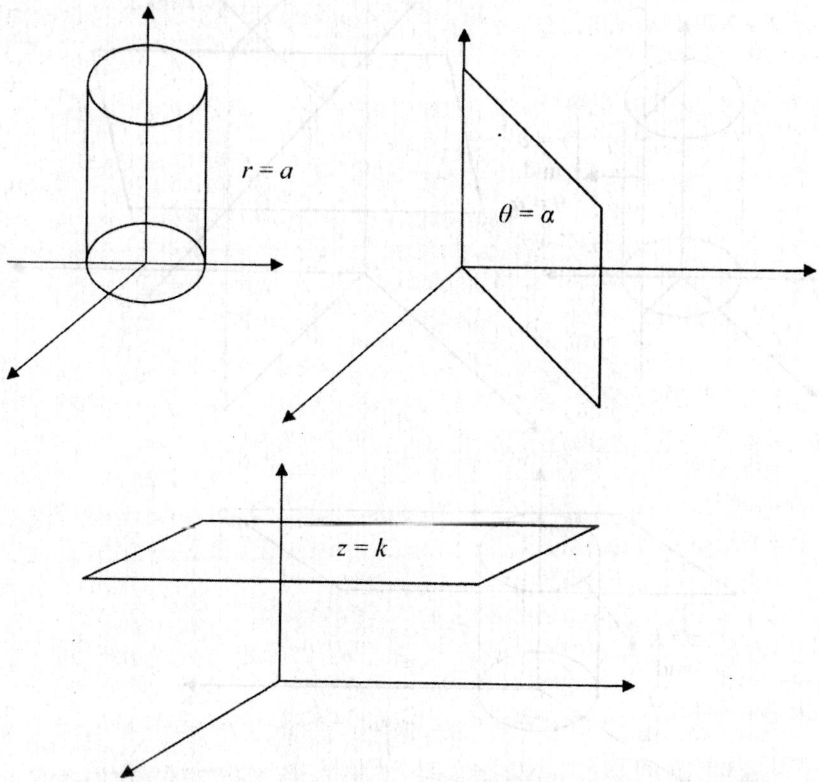

Coordinate surfaces of cylindrical coordinate system

14.5 Spherical Coordinate System

Let $P(x, y, z)$ be a point in the rectangular coordinate system. We draw PN, a perpendicular from P on x-y plane and the line MN, parallel to Oy. If we take $ON = r$; $\angle zOP = \theta$ and $\angle MON = \phi$ then the coordinates (r, θ, ϕ) called the spherical polar coordinates of the point P with respect to O (from which r is measured), the fixed line Oz (with which the line OP makes the angle θ) and the fixed line Ox (with which the line ON makes the angle ϕ). The numbers r (≥ 0), θ $(0 \leq \theta \leq \pi)$, ϕ $(-\pi < \phi \leq \pi)$ are called the radius vector, the colatitude and the latitude of the point P respectively. Thus for a point P in space there exists a ordered triplet of the numbers r, θ, ϕ and conversely.

If (x, y, z) and (r, θ, z) be the cartesian and spherical coordinates of a point P as shown in the figure then we have

$$x = r \sin\theta \cos\phi; \; y = r \sin\theta \sin\phi; \; z = r \cos\theta$$

and $$r = \sqrt{x^2 + y^2 + z^2}; \; \theta = \cos^{-1}\left(\frac{z}{\sqrt{x^2 + y^2 + z^2}}\right); \; \phi = \tan^{-1}\left(\frac{y}{x}\right).$$

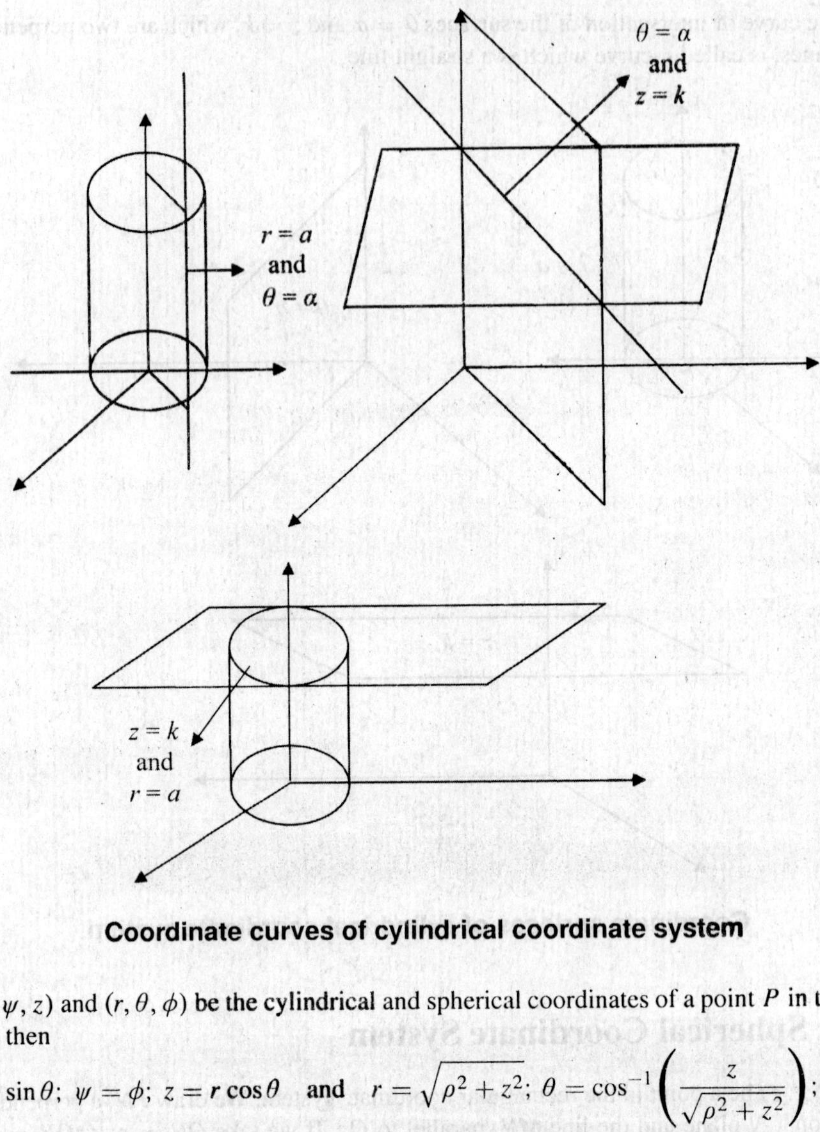

Coordinate curves of cylindrical coordinate system

If (ρ, ψ, z) and (r, θ, ϕ) be the cylindrical and spherical coordinates of a point P in the same frame then

$$\rho = r\sin\theta;\ \psi = \phi;\ z = r\cos\theta \quad \text{and} \quad r = \sqrt{\rho^2 + z^2};\ \theta = \cos^{-1}\left(\frac{z}{\sqrt{\rho^2 + z^2}}\right);\ \phi = \psi.$$

14.6 Coordinate Surfaces and Curves for Spherical Coordinate System

(a) Coordinate Surfaces

(i) From $r = a$ we get $x^2 + y^2 + z^2 = a^2$ which is a sphere (when $a \neq 0$) whose centre is at the origin and if $a = 0$ then we get the origin.

(ii) From $\theta = \alpha\ (\neq 0, \pi/2, \pi)$ we get $z = \sqrt{x^2 + y^2 + z^2}\cos\alpha \ \Rightarrow\ x^2 + y^2 - z^2\tan^2\alpha = 0$ which is a cone whose vertex is at the origin. For $\alpha = 0$ or π we get $x^2 + y^2 = 0 \ \Rightarrow\ x = 0 = y$ which is the z-axis and for $\alpha = \pi/2$, we get $z = 0$ i.e., the surface is x-y plane.

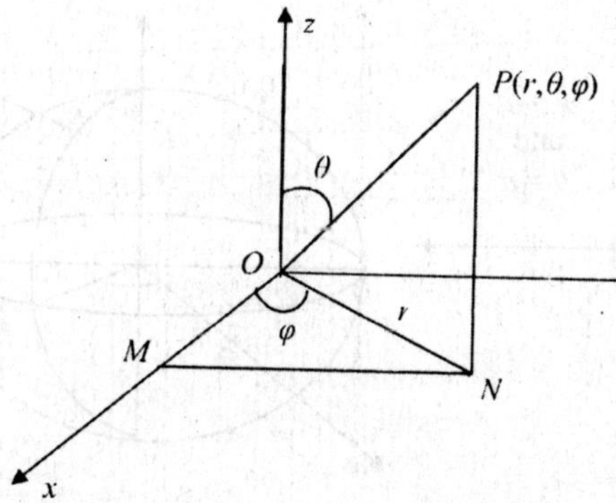

Spherical coordinate system

(iii) The equation $\phi = k$ gives $y = x \tan k$ which is a equation of plane through z-axis making an angle k with the plane xOz.

(a) Coordinate Curves

(i) The curve of intersection of the surfaces $r = a$ and $\theta = \alpha$ is called ϕ-curve which is a circle in this case.

(ii) The curve of intersection of the surfaces $r = a$ and $\phi = k$ is called θ-curve which is a semi-circle if a is non-zero.

(iii) The curve of intersection of the surfaces $\theta = \alpha$ and $\phi = k$, is called r-curve which is a straight line.

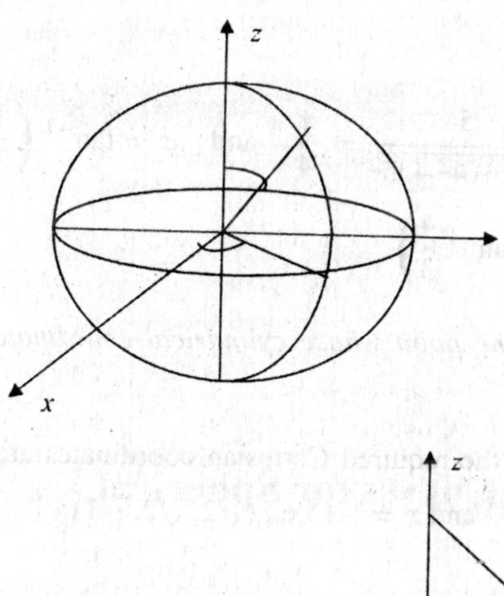

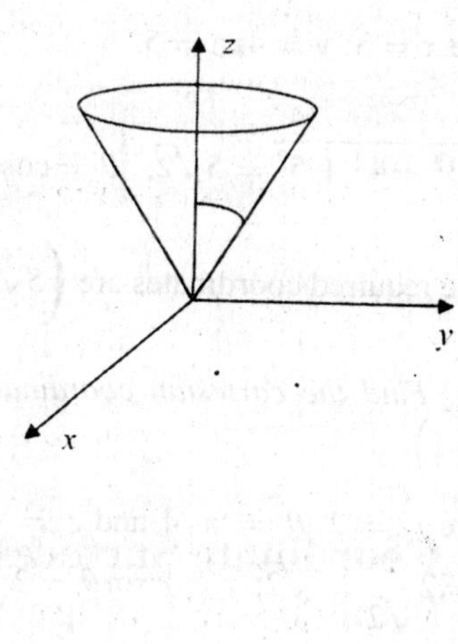

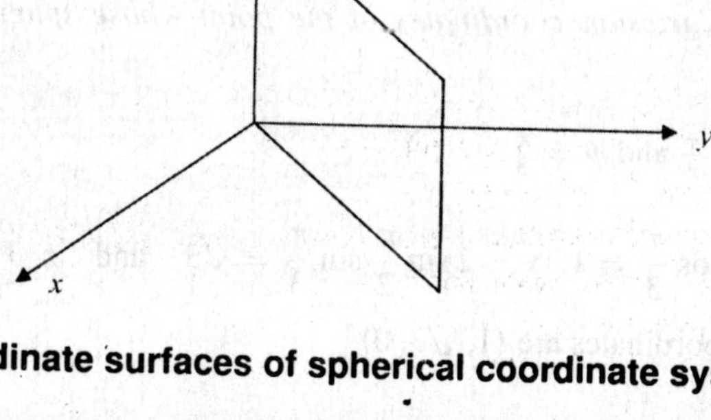

Coordinate surfaces of spherical coordinate system

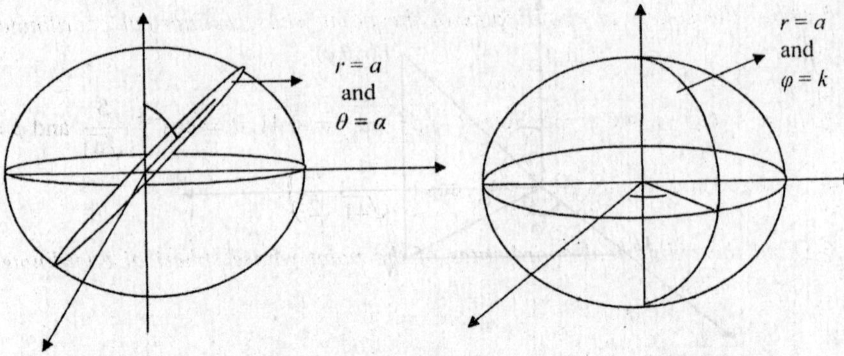

Coordinate curves in spherical coordinate system

Examples

▶▷ 1. *Find the cylindrical coordinates of the point whose cartesian coordinates are* (3, 4, 5).

☐ Here $r\cos\theta = 3$; $r\sin\theta = 4$ and $z = 5$. So $r = \sqrt{3^2+4^2} = 5$ and $\theta = \tan^{-1}\left(\frac{4}{3}\right)$,

Hence the required coordinates are $\left(5, \tan^{-1}\frac{4}{3}, 5\right)$.

▶▷ 2. *Find the spherical coordinates of the point whose cartesian coordinates are* (3, 4, 5).

☐ Here $x = 3$; $y = 4$; $z = 5$.
So

$$r = \sqrt{3^2+4^2+5^2} = 5\sqrt{2};\quad \theta = \cos^{-1}\frac{5}{\sqrt{3^2+4^2+5^2}} = \frac{\pi}{4} \quad\text{and}\quad \phi = \tan^{-1}\left(\frac{4}{3}\right).$$

Hence the required coordinates are $\left(5\sqrt{2}, \frac{\pi}{4}, \tan^{-1}\frac{4}{3}\right)$.

▶▷ 3. *Find the cartesian coordinates of the point whose cylindrical coordinates are* $\left(2, \frac{\pi}{4}, -1\right)$.

☐ Here $r = 2$; $\theta = \pi/4$ and $z = -1$, so the required Cartesian coordinates are $x = r\cos\theta = 2\frac{1}{\sqrt{2}} = \sqrt{2}$; $y = r\sin\theta = 2\frac{1}{\sqrt{2}} = \sqrt{2}$ and $z = -1$ i.e., $(\sqrt{2}, \sqrt{2}, -1)$.

▶▷ 4. *Find the Cartesian coordinates of the point whose spherical coordinates are* $\left(2, \frac{\pi}{2}, \frac{\pi}{3}\right)$.

☐ Here $r = 2, \theta = \frac{\pi}{2}$ and $\phi = \frac{\pi}{3}$.
So

$$x = 2\sin\frac{\pi}{2}\cos\frac{\pi}{3} = 1;\quad y = 2\sin\frac{\pi}{2}\sin\frac{\pi}{3} = \sqrt{3} \quad\text{and}\quad z = 2\cos\frac{\pi}{2} = 0.$$

Hence the required coordinates are $(1, \sqrt{3}, 0)$.

▶▶ 5. *Find the spherical coordinates of the point whose cylindrical coordinates are* $\left(4, \frac{\pi}{2}, 5\right)$.

□ Here $\rho = 4, \psi = \frac{\pi}{2}$ and $z = 5$. So $r = \sqrt{\rho^2 + z^2} = \sqrt{41}; \theta = \cos^{-1}\frac{5}{\sqrt{41}}$ and $\phi = \frac{\pi}{2}$.

Hence the required coordinates are $\left(\sqrt{41}, \cos^{-1}\frac{5}{\sqrt{41}}, \frac{\pi}{2}\right)$.

▶▶ 6. *Find the cylindrical coordinates of the point whose spherical coordinates are* $\left(8, \frac{2\pi}{3}, \frac{\pi}{3}\right)$.

□ Here $r = 8, \theta = \frac{2\pi}{3}$ and $\phi = \frac{\pi}{3}$.

So $\rho = r\sin\theta = 8\frac{\sqrt{3}}{2} = 4\sqrt{3}; \psi = \frac{\pi}{3}$ and $z = r\cos\theta = -8\frac{1}{2} = -4$. Hence the required coordinates are $\left(4\sqrt{3}, \frac{\pi}{3}, -4\right)$.

▶▶ 7. *Find the square of an element of arc length in cylindrical coordinates.*

□ Transformation formulae from cylindrical to cartesian coordinates are

$$x = r\cos\theta;\ y = r\sin\theta;\ z = z.$$

So

$$dx = dr\cos\theta - r\sin\theta d\theta$$

$$dy = dr\sin\theta + r\cos\theta d\theta$$

and

$$dz = dz.$$

Now

$$\begin{aligned}(ds)^2 &= (dx)^2 + (dy)^2 + (dz)^2 \\ &= (\cos\theta dr - r\sin\theta d\theta)^2 + (\sin\theta dr + r\cos\theta d\theta)^2 + (dz)^2 \\ &= (dr)^2 + r^2(d\theta)^2 + (dz)^2.\end{aligned}$$

▶▶ 8. *Find the square of an element of arc length in spherical coordinates.*

□ Transformation formulae from spherical to cartesian coordinates are

$$x = r\sin\theta\cos\phi;\ y = r\sin\theta\sin\phi;\ z = r\cos\theta.$$

So

$$\begin{aligned}dx &= dr\sin\theta\cos\phi + r\cos\phi\cos\theta d\theta - r\sin\theta\sin\phi d\phi \\ dy &= dr\sin\theta\sin\phi + r\sin\phi\cos\theta d\theta + r\sin\theta\cos\phi d\phi \\ dz &= \cos\theta dr - r\sin\theta d\theta.\end{aligned}$$

Now

$$\begin{aligned}(ds)^2 &= (dx)^2 + (dy)^2 + (dz)^2 \\ &= (dr\sin\theta\cos\phi + r\cos\phi\cos\theta d\theta - r\sin\theta\sin\phi d\phi)^2 \\ &\quad + (dr\sin\theta\sin\phi + r\sin\phi\cos\theta d\theta + r\sin\theta\cos\phi d\phi)^2 \\ &\quad + (\cos\theta dr - r\sin\theta d\theta)^2 \\ &= (dr)^2 + r^2(d\theta)^2 + r^2\sin^2\theta(d\phi)^2.\end{aligned}$$

EXERCISE

1. Find the Cartesian coordinates of the points whose cylindrical coordinates are
   (i) $\left(3, \frac{\pi}{2}, 3\right)$; (ii) $\left(2, \frac{\pi}{6}, 4\right)$; (iii) $\left(4, \frac{\pi}{3}, 0\right)$
2. Find the Cartesian coordinates of the points whose spherical coordinates are
   (i) $\left(3, \frac{\pi}{3}, \frac{\pi}{2}\right)$; (ii) $\left(2, \frac{\pi}{3}, \frac{\pi}{6}\right)$; (iii) $\left(4, \cos^{-1}\frac{1}{3}, \frac{\pi}{4}\right)$
3. Find the spherical coordinates of the points whose cylindrical coordinates are
   (i) $\left(3, \frac{\pi}{2}, 3\right)$; (ii) $\left(2, \frac{\pi}{6}, 4\right)$; (iii) $\left(4, \frac{\pi}{3}, 0\right)$
4. Find the cylindrical coordinates of the points whose spherical coordinates are
   (i) $\left(3, \frac{\pi}{3}, \frac{\pi}{2}\right)$; (ii) $\left(2, \frac{\pi}{3}, \frac{\pi}{6}\right)$; (iii) $\left(4, \cos^{-1}\frac{1}{3}, \frac{\pi}{4}\right)$
5. Find the cylindrical coordinates of the points whose cartesian coordinates are
   (i) (2, 2, 1); (ii) $\left(\frac{3}{2}, \frac{\sqrt{3}}{2}, 1\right)$; (iii) $(2, 2\sqrt{3}, 0)$
6. Find the spherical coordinates of the points whose cartesian coordinates are
   (i) (1, 2, 1); (ii) (0, 1, 1); (iii) (3, 3, 4)
7. Find the coordinate surfaces of cylindrical coordinate.
8. Find the coordinate surfaces of spherical coordinate.
9. Find the coordinate curves of cylindrical coordinate.
10. Find the coordinate curves of spherical coordinate.

FEW OBJECTIVE QUESTIONS

ON

COORDINATE GEOMETRY

FEW OBJECTIVE QUESTIONS ON COORDINATE GEOMETRY

1. Find the equation of the pair of straight lines joining origin to the point of intersection of the line $y = 2x + 1$ with the parabola $y^2 = 4(x + 2)$.

☐ Making homogeneous $y^2 = 4(x + 2)$ with the help of $y = 2x + 1$ we get

$$y^2 = 4x(y - 2x) + 8(y - 2x)^2$$

$$\text{or,} \quad 7y^2 - 28xy + 24x^2 = 0.$$

2. Prove that the angle between the lines $(x^2 + y^2)\sin^2\theta = (x\cos\theta - y\sin\theta)^2$ is 2θ.

☐ Rewrite the equation as $ax^2 + 2hxy + by^2 = 0$ and use the formula $\tan\varphi = \dfrac{2\sqrt{h^2 - ab}}{a + b}$.

3. Show that the difference of the tangents of the angles which the lines

$$x^2(\sec^2\theta - \sin^2\theta) - 2xy\tan\theta + y^2\sin^2\theta = 0$$

makes with x-axis is 2.

☐ Use the results $m_1 + m_2 = -\dfrac{2h}{b} = \dfrac{2\tan\theta}{\sin^2\theta}$; $m_1m_2 = \dfrac{a}{b} = \dfrac{\sec^2\theta - \sin^2\theta}{\sin^2\theta}$ in

$$\tan\theta_1 - \tan\theta_2 = m_1 - m_2 = \sqrt{(m_1 + m_2)^2 - 4m_1m_2}.$$

4. Find the condition for which one of the lines $ax^2 + 2hxy + by^2 = 0$ coincides with one of the lines of $a'x^2 + 2h'xy + b'y^2 = 0$.

☐ If $y = mx$ be the common line then we have

$$a + 2hm + bm^2 = 0;$$
$$a' + 2h'm + b'm^2 = 0.$$

Solving we get

$$\frac{m^2}{2(ah' - ha')} = \frac{m}{a'b - ab'} = \frac{1}{2(hb' - bh')}.$$

Eliminating m we get the required result.

5. Find the condition for which one of the lines $ax^2 + 2hxy + by^2 = 0$ is perpendicular to one of the lines of $a'x^2 + 2h'xy + b'y^2 = 0$.

☐ Put $y = mx$ in $ax^2 + 2hxy + by^2 = 0$ and $y = -\dfrac{1}{m}x$ in $a'x^2 + 2h'xy + b'y^2 = 0$ and solve the equations in m by cross multiplication method. Then eliminate m.

6. Find the equation of the pair of straight lines which make an angle θ with the line $y = x$.

□ Let $y = mx$ be one of the pair of st. lines, then by given condition

$$\tan(\pm\theta) = \frac{1-m}{1+m} \Rightarrow \tan^2\theta = \frac{(1-y/x)^2}{(1+y/x)^2} \Rightarrow (x-y)^2 = (x+y)^2\tan^2\theta,$$

this is the equation of the required pair of straight lines.

7. Show that the straight lines joining origin to the points of intersection of $ax^2 + 2hxy + by^2 + 2gx = 0$ and $a'x^2 + 2h'xy + b'y^2 + 2g'x = 0$ will be at right angle if

$$g(a' + b') = g'(a + b).$$

□ Multiplying the first equation by g' and second by g and subtracting one from other we get a second degree homogeneous equation in x, y which represents the straight lines joining origin to the points of intersection of $ax^2 + 2hxy + by^2 + 2gx = 0$ and $a'x^2 + 2h'xy + b'y^2 + 2g'x = 0$. Now the st. lines will be at right angle if coeff. of x^2 + coeff. of $y^2 = 0$.

8. Show that the straight lines joining origin to the points of intersection of $x^2 + hxy - y^2 + gx + fy = 0$ and $fx - gy = k$ are at right angle for all values of k.

□ Making homogeneous $x^2 + hxy - y^2 + gx + fy = 0$ with the aid of $fx - gy = k$ as follows

$$x^2 + hxy - y^2 + \frac{1}{k}(gx + fy)(fx - gy) = 0$$

$$\text{or,} \quad x^2(k + fg) + (hk + f^2 - g^2)xy - y^2(k + fg) = 0.$$

In which we see that for all values of k, coeff. of x^2 + coeff. of $y^2 = 0$. So the lines are at right angle for all values of k.

9. Find the condition that line $lx + my + n = 0$ may be a normal to the circle $x^2 + y^2 + 2gx + 2fy + c = 0$.

□ Centre of the given circle is $(-g, -f)$. The line $lx + my + n = 0$ will be a normal to the circle if it passes through the centre of the circle i.e., if $gl + fm = n$.

10. Find the pole of the line $\frac{x}{c} + \frac{y}{b} = 1$ with respect to the circle $x^2 + y^2 = a^2$.

□ If $P(h, k)$ be the pole then its polar with respect to the circle $x^2 + y^2 = a^2$ is

$$xh + yk = a^2.$$

Comparing this with the given line we get

$$ch = bk = a^2.$$

Hence the pole is $\left(\frac{a^2}{c}, \frac{a^2}{b}\right)$.

11. Two tangents form a point P to the parabola $y^2 = 4ax$ makes angles α, β with x-axis. Find the locus of the point when

$$\text{(i)} \quad \tan\alpha + \tan\beta = c \quad \text{and} \quad \text{(ii)} \quad \tan^2\alpha + \tan^2\beta = c.$$

☐ We know that for all values of m the line $y = mx + \frac{a}{m}$ touches the parabola $y^2 = 4ax$. If the tangent passes through the point $P(h, k)$ then we have

$$k = mh + \frac{a}{m}$$

$$\text{or,} \quad hm^2 - km + a = 0.$$

If m_1, m_2 be its two roots then we get

$$m_1 + m_2 = \frac{k}{h} \quad \text{and} \quad m_1 m_2 = \frac{a}{h}.$$

So in the first case locus of the point is $y = cx$ and in the second case we get

$$c = (m_1 + m_2)^2 - 2m_1 m_2 = \frac{k^2}{h^2} - 2\frac{a}{h}.$$

So in the second case locus is $y^2 - cx^2 - 2ax = 0$.

12. If the parabola $y = x^2 + bx + c$ touches the line $y = x$ at the point $(1, 1)$, find the values of b, c.

☐ The parabola passes through the point $(1, 1)$, so we have $b + c = 0$. Solving the given equations we get $x^2 + (b - 1)x + c = 0$ which must have repeated roots. This leads that

$$(b-1)^2 = 4c$$

$$\text{or,} \quad (b-1)^2 + 4b = 0$$

$$\text{or,} \quad b = -1 \quad \text{and hence} \quad c = 1.$$

13. Prove that the locus of the middle points of the chords of the ellipse $\frac{x^2}{a^2} + \frac{y^2}{b^2} = 1$ passing through a fixed point (h, k) is

$$\frac{x(x-h)}{a^2} + \frac{y(y-k)}{b^2} = 0.$$

☐ If (α, β) be the middle point of a chord then equation of that chord is

$$\frac{x\alpha}{a^2} + \frac{y\beta}{b^2} = \frac{\alpha^2}{a^2} + \frac{\beta^2}{b^2}.$$

If it passes through the point (h, k) then we have

$$\frac{h\alpha}{a^2} + \frac{k\beta}{b^2} = \frac{\alpha^2}{a^2} + \frac{\beta^2}{b^2}.$$

Generalizing we get

$$\frac{x(x-h)}{a^2} + \frac{y(y-k)}{b^2} = 0.$$

14. Find the director circles of (a) $x^2 + y^2 = a^2$ and (b) $\dfrac{x^2}{a^2} + \dfrac{y^2}{b^2} = 1$.

□ (b) We know that for all values of m, the straight line $y = mx + \sqrt{a^2m^2 + b^2}$ touches the ellipse $\dfrac{x^2}{a^2} + \dfrac{y^2}{b^2} = 1$. If the tangents meet at (h, k) then we have $k = mh + \sqrt{a^2m^2 + b^2}$ i.e.,

$$(k - mh)^2 = a^2m^2 + b^2 \quad \text{i.e.,} \quad m^2(h^2 - a^2) - 2mkh + k^2 - b^2 = 0.$$

For two perpendicular tangents we have

$$m_1m_2 = -1 = \frac{k^2 - b^2}{h^2 - a^2} \quad \text{i.e.,} \quad k^2 + h^2 = a^2 + b^2.$$

Hence the locus of two perpendicular tangents to the ellipse i.e., equation of the director circle to the ellipse is $x^2 + y^2 = a^2 + b^2$.

Putting $a = b$ in this equation we get the director circle of the circle given by (a) is $x^2 + y^2 = 2a^2$.

15. Find the parametric equation of the parabola $(y - 1)^2 = 4(x + 2)$.

□ Here $a = 1$. Parametric equation of the parabola $Y^2 = 4aX$ is $X = at^2$; $Y = 2at$, t being the parameter. So parametric equation of the parabola $(y - 1)^2 = 4(x + 2)$ is $x = -2 + t^2$; $y = 1 + 2t$.

16. Prove that polar of a point w.r.t. $x^2 + y^2 - 2cx = a^2$ passes through a fixed point for all values of c.

□ Polar of a point (h, k) w.r.t. $x^2 + y^2 - 2cx = a^2$ is $xh + yk - c(x + h) - a^2 = 0 \Rightarrow xh + yk - a^2 - c(x + h) = 0$, which shows that the polar passes through the point of intersection of $xh + yk - a^2 = 0$ and $x + h = 0$ which is fixed.

17. Find the locus of the middle points of the chords of $x^2 + y^2 = a^2$ which passes through (p, q).

□ Equation of the chord whose middle point is (h, k) is $xh + yk = h^2 + k^2$, if it passes through (p, q) then we get $ph + qk = h^2 + k^2$. So locus of (h, k) is $px + qy = x^2 + y^2$.

18. Find the equation of the circle describing the chord $x\cos\theta + y\sin\theta = k$ of $x^2 + y^2 = a^2$ as diameter.

□ Equation of the new circle can be written as $x^2 + y^2 - a^2 + \lambda(x\cos\theta + y\sin\theta - k) = 0$, whose centre is $(-\lambda\cos\theta/2, -\lambda\sin\theta/2)$ and it will lie on the line $x\cos\theta + y\sin\theta = k$, which gives $\lambda = -2k$. So equation of the new circle is

$$x^2 + y^2 - a^2 - 2k(x\cos\theta + y\sin\theta - k) = 0.$$

19. Find the condition that the conics $ax^2 + 2hxy + by^2 + 2gx + 2fy + c = 0$ and $a'x^2 + 2h'xy + b'y^2 + 2g'x + 2f'y + c' = 0$ may intersect at four concyclic points.

□ Equation of the curve which passes through the given conics can be written as

$$\begin{aligned} &ax^2 + 2hxy + by^2 + 2gx + 2fy + c \\ &\quad + k(a'x^2 + 2h'xy + b'y^2 + 2g'x + 2f'y + c') = 0 \\ \text{or,} \quad &(a + ka')x^2 + 2(h + kh')xy + (b + kb')y^2 \\ &\quad + 2(g + kg')x + 2(f + kf')y + (c + kc') = 0. \end{aligned}$$

Now if this is a circle then we have $(a + ka') = (b + kb')$ and $(h + kh') = 0$. These two equations give $\dfrac{a-b}{h} = \dfrac{a'-b'}{h'}$.

20. If the normal at the end of a latus rectum of an ellipse passes through one extremity of the minor axis, show that $e^4 + e^2 = 1$, e being the eccentricity of the ellipse.

☐ Write the equation of normal at $\left(ae, \frac{b^2}{a}\right)$. In that equation put $x = 0$, $y = -b$ and use $e^2 = 1 - b^2/a^2$.

21. Show that the tangents at the end points of conjugate diameters of the ellipse $\dfrac{x^2}{a^2} + \dfrac{y^2}{b^2} = 1$, intersect on the ellipse $\dfrac{x^2}{a^2} + \dfrac{y^2}{b^2} = 2$.

☐ Equation of the tangent at $(a\cos\theta, b\sin\theta)$ is $\dfrac{x}{a}\cos\theta + \dfrac{y}{b}\sin\theta = 1$, changing θ by $90° + \theta$ in this equation $-\dfrac{x}{a}\sin\theta + \dfrac{y}{b}\cos\theta = 1$. To get the locus of the point of intersection of the tangents at the end points of conjugate diameters we eliminate θ by squaring and adding and we get $\dfrac{x^2}{a^2} + \dfrac{y^2}{b^2} = 2$.

22. Prove that the distances of two points $P(x_1, y_1)$ and $Q(x_2, y_2)$ from the centre of a circle are proportional to the distance of each from the polar of the other.

☐ Let the circle be $x^2 + y^2 = a^2$, then $OP = \sqrt{x_1^2 + y_1^2}$ and $OQ = \sqrt{x_2^2 + y_2^2}$. Polars of P and Q are $xx_1 + yy_1 = a^2$ and $xx_2 + yy_2 = a^2$. If d_1 and d_2 be the distances of P and Q from the polars of Q and P, then $\dfrac{d_1}{d_2} = \dfrac{OP}{OQ}$.

23. If 'O' be the centre of a circle and P be any point, then OP is perpendicular to the polar of P.

☐ Let $P(x_1, y_1)$ be any point, then it's polar w.r.t. the circle $x^2+y^2 = a^2$ is $xx_1+yy_1 = a^2$. Now slope of OP × slope of the polar $= \left(\dfrac{y_1}{x_1}\right) \times \left(-\dfrac{x_1}{y_1}\right) = -1$.

24. Find the equation of the circle through the intersection of the straight line $\dfrac{x}{a} + \dfrac{y}{b} = 1$ with the coordinate axes and having small radius.

☐ Obviously, the circle describing the line segment joining $(a, 0)$ and $(0, b)$ as diameter will have the least radius and the equation of this circle is $x(x - a) + y(y - b) = 0$.

25. Adjust the parameter λ so that the equation may represent a circle

$$\frac{x^2}{25} + \frac{y^2}{16} - 1 - \lambda(x-1)^2 = 0.$$

□ The given equation will represent a circle if coeff. of x^2 = coeff. of y^2 (as coeff. of xy = 0 is identically satisfied). So $\lambda - 1/25 = -1/16 \Rightarrow \lambda = -9/400$.

26. Find the equation of the hyperbola whose asymptotes are $2x - y = 3$ and $3x + y = 7$ and which passes through $(1, 1)$.

□ Equation of the hyperbola is $(2x - y - 3)(3x + y - 7) = k$. As it passes through $(1, 1)$, $k = 6$. So the equation of the hyperbola is $(2x - y - 3)(3x + y - 7) = 6$.

27. Show that $16x^2 + \mu xy - 25y^2 = 0$ represents a pair of conjugate diameters of $16x^2 + 25y^2 = 1$ for any value of μ.

□ If m_1 and m_2 are the slopes of the individual lines of $16x^2 + \mu xy - 25y^2 = 0$, then $m_1 m_2 = -16/25$ and if $y = m_1 x$ and $y = m_2 x$ are the pair of conjugate diameters of $16x^2 + 25y^2 = 1$ then $m_1 m_2 = -16/25$. Hence the given pair of st. lines will be the conjugate diameters of the given ellipse.

28. Prove that the length of the semi-axes of the conic $ax^2 + 2hxy + ay^2 = d$ are $\sqrt{\dfrac{d}{a+h}}$ and $\sqrt{\dfrac{d}{a-h}}$.

□ Let the equation $ax^2 + 2hxy + ay^2 = d$ be reduced to $Ax'^2 + By'^2 = d$ after rotating axes, then by properties of invariant $A + B = a + a = 2a$ and $AB = a^2 - h^2 \Rightarrow A - B = 2h$. Solving we get $A = a + h$ and $B = a - h$. So, the length of the axes are $\sqrt{\dfrac{d}{A}}$ and $\sqrt{\dfrac{d}{B}}$ i.e., $\sqrt{\dfrac{d}{a+h}}$ and $\sqrt{\dfrac{d}{a-h}}$.

29. Find the angle through the axes are to be rotated, so that the equation $17x^2 + 18xy - 7y^2 = 1$ may be reduced to the form $Ax^2 + By^2 = 1$, $A > 0$.

□ The required angle is given by $\dfrac{1}{2}\tan^{-1}\left(\dfrac{2h}{a-b}\right) = \dfrac{1}{2}\tan^{-1}\left(\dfrac{18}{17+7}\right) = \dfrac{1}{2}\tan^{-1}\dfrac{3}{4}$.

30. Find the area of the ellipse represented by $(a^2 + 1)x^2 + 2(a + b)xy + (b^2 + 1)y^2 = c$.

□ To find the area of the ellipse we reduce the given equation to standard form, say

$$Ax^2 + By^2 = c.$$

Then by the properties of invariant we have $AB = (a^2 + 1)(b^2 + 1) - (a + b)^2 = (ab - 1)^2$. Now the area of the ellipse $Ax^2 + By^2 = c$ is $\dfrac{\pi c}{\sqrt{AB}} = \dfrac{\pi c}{ab - 1}$.

31. Find the locus of the feet of the perpendiculars from the origin to the lines passing through a fixed point (h, k).

☐ Equation of a line passing through the point (h, k) is $y - k = m(x - h)$, m being the slope of the line. Now equation of the line passing through origin and perpendicular to the above line is $y = -x/m$. Eliminating m from the above lines we get $y - k = -\dfrac{x(x-h)}{y}$ or $x^2 + y^2 - xh - yk = 0$, which is the required locus.

32. Show that the equation $y^2 + 2y - 8x + 25 = 0$ can be reduced to $y^2 = 4ax$, find a.

☐ $y^2 + 2y - 8x + 25 = 0 \Rightarrow (y+1)^2 = 8(x-3)$. So shifting origin to the point $(3, -1)$ we get $y'^2 = 8x'$. So $a = 2$.

33. **Show that** the equation $x^2 + 6xy + 9y^2 + 10x + 30y + 25 = 0$ represents a pair of coincident straight lines.

☐ Here we see that $\Delta = 0$ and $D = 0$ i.e., the given equation represents a pair of parallel straight lines. Now we see that the given equation is a perfect square, i.e. $(x + 3y + 5)^2 = 0$. Hence the given equation represents a pair of coincident straight lines.

34. Prove that the latus rectum is the shortest focal chord of a conic.

☐ Length of a focal chord of the conic $l/r = 1 + e\cos\theta$ inclined at an angle α to the axis $= \dfrac{2l}{1 - e^2\cos^2\alpha}$. This will be minimum when $1 - e^2\cos^2\alpha$ is maximum i.e., when $\cos^2\alpha$ is minimum and minimum value of $\cos^2\alpha$ is 0. Hence the latus rectum is the shortest focal chord of a conic.

35. How many other conditions can a conic section satisfy when you are given it's focus?

☐ Let S be the focus of the conic . Then there must be a straight line which is perpendicular to the axis of the conic such that $SP = ePM$, where $P(x, y)$ be a point on the conic and it's perpendicular distance from the fixed straight line is PM and e can have any finite value.

36. Find the position of the straight line $x = 7$ w.r.t. the circle $x^2 + y^2 - 4x - 6y - 12 = 0$.

☐ Putting $x = 7$ in the given equation of circle we get $49 + y^2 - 28 - 6y - 12 = 0$, i.e., $(y-3)^2 = 0$. So the straight line $x = 7$ touches the circle at the point $(7, 3)$.

37. If the axes are turned through an angle $\tan^{-1} 2$, what does the equation $4xy = 1$ become?

☐ Given $\tan\alpha = 2 \Rightarrow \sin\alpha = \dfrac{2}{\sqrt{5}}$ and $\cos\alpha = \dfrac{1}{\sqrt{5}}$. Now $x = x'\cos\alpha - y'\sin\alpha = \dfrac{x' - 2y'}{\sqrt{5}}$ and $y = x'\sin\alpha + y'\cos\alpha = \dfrac{2x' + y'}{\sqrt{5}}$. Hence the equation $4xy = 1$ becomes $4(x' - 2y')(2x' + y') = 5$.

38. If (x_k, y_k): $k = 1, 2, 3$ be three non-collinear points in a plane, then show that

$$x_1(y_2 - y_3) + x_2(y_3 - y_1) + x_3(y_1 - y_2)$$

remains an invariant under orthogonal transformation.

□ The given expression is twice the area of the triangle whose vertices are (x_k, y_k): $k = 1, 2, 3$ and area of a triangle is an invariant under orthogonal transformation, so

$$x_1(y_2 - y_3) + x_2(y_3 - y_1) + x_3(y_1 - y_2)$$

remains an invariant under orthogonal transformation.

39. Find the pole of the line $x + 2y + 3 = 0$ w.r.t. the circle $x^2 + y^2 - 2x + 5 = 0$.

□ Let (α, β) be the pole, then it's polar w.r.t. $x^2+y^2-2x+5 = 0$ is $x(\alpha-1)+y\beta+5-\alpha = 0$ comparing this with $x + 2y + 3 = 0$ we get $\alpha = 2; \beta = 2$.

40. Prove that the difference of focal distances of any point on the hyperbola is equal to it's transverse axis.

□ Focal distance of a point $P(x, y)$ is $SP = ePM = e\left(x - \frac{a}{e}\right) = ex - a$ and $S'P = ePM' = e\left(x + \frac{a}{e}\right) = ex + a$. So, $S'P - SP = 2a =$ length of transverse axis.

41. Determine whether the following curve has one centre, no centre or infinitely many centres $4x^2 + 4xy + y^2 + 4x + 2y + 20 = 0$.

□ Here $\Delta = 0$ (Since two columns are identical), $D = 4.1 - 22 = 0$. So, the given equation represents a pair of parallel straight lines. Now the given equation can be written as $(2x + y + 1)^2 + 19 = 0 \Rightarrow$ the given equation represents two imaginary lines, so the curve has no centre.

42. Show that the diameter of the parabola is parallel to it's axis.

□ The equation of the chord of the parabola $y^2 = 4ax$ whose mid-point is (h, k) is

$$yk - 2ax = k^2 - 2ah.$$

If it is parallel to $y = mx$ then $k = 2a/m$, so the equation of the diameter corresponding to the system of chords is $y = 2a/m$, it is a straight line parallel to the axis of the parabola.

43. Reduce the equation of the rectangular hyperbola $(3x + 4y - 1)(4x - 3y + 1) = 1$ to standard form taking $3x + 4y - 1 = 0$ and $4x - 3y + 1 = 0$ as axes.

□ Taking $3x+4y-1 = 0$ as X-axis and $4x-3y+1 = 0$ as Y-axis which are perpendicular, we see that the equation $(3x + 4y - 1)(4x - 3y + 1) = 1$ reduces to

$$\frac{5(3x + 4y - 1)}{\sqrt{3^2 + 4^2}} \times \frac{5(4x - 3y + 1)}{\sqrt{3^2 + 4^2}} = 1$$

$$\Rightarrow \quad 25XY = 1. \Rightarrow XY = (1/5)^2.$$

44. Transfer the origin of coordinate suitably and then rotate the axes so that referred to new axes the equation $y = 2x + 5$ may reduce to $Y - X = 0$.

□ On shifting the origin to the point (0, 5), the equation $y = 2x + 5$ reduced to $y' = 2x'$. Now putting $X = x'\cos\alpha + y'\sin\alpha$ and $Y = y'\cos\alpha - x'\sin\alpha$ in $Y = X$, we get $y'(\cos\alpha - \sin\alpha) = x'(\cos\alpha + \sin\alpha)$. Comparing this with $y' = 2x'$, we get $\tan\alpha = 1/3$. So, by shifting origin to (0, 5) and then rotating the axes through an angle $\tan^{-1}(1/3)$, the equation $y = 2x + 5$ becomes $Y = X$

45. Find the condition for the lines joining the origin to the point of intersection of the line $lx + my = n$ and the curve $y^2 = 4a(x + a)$ to be at right angle.

□ Making homogeneous $y^2 = 4a(x + a)$ with the help of $lx + my = n$ and using coeff. of x^2 + coeff. of $y^2 = 0$ we get $4a(al^2 + am^2 + ln) = n^2$.

46. The origin is shifted to the point (3, −1) and the axes are rotated through an angle $\tan^{-1}(3/4)$. Find the coordinates of the point (4, 1) in the new system.

□ Given $\tan\alpha = 3/4 \Rightarrow \sin\alpha = 3/5, \cos\alpha = 4/5$.
Now from $4 = 3 + x\cos\alpha - y\sin\alpha$ and $1 = -1 + x\sin\alpha + y\cos\alpha$ we get $x = 2$, $y = 1$. So, the point (4, 1) becomes (2, 1).

47. The coordinate axes are rotated through an angle 60° without change of origin. If the coordinate of a point in new system be $(2\sqrt{3}, -6)$, find it's coordinates in the old system.

□ $x = 2\sqrt{3}\cos 60° + 6\sin 60° = 4\sqrt{3}$ and $y = 2\sqrt{3}\sin 60° - 6\cos 60° = 0$. Coordinate was $(4\sqrt{3}, 0)$.

48. The coordinate of a point w.r.t. to rectangular Cartesian coordinate system are (2, 3). What will be it's coordinate if axes are rotated through 30°?

□ $2 = x'\cos 30° - y'\sin 30° = \dfrac{\sqrt{3}x' - y'}{2}$ and $3 = x'\sin 30° + y'\cos 30° = \dfrac{\sqrt{3}y' + x'}{2}$.
Solving we get $x' = \dfrac{3 + 2\sqrt{3}}{2}$ and $y' = \dfrac{3\sqrt{3} - 2}{2}$.

49. What does the equation $(a - b)(x^2 + y^2) = 2abx$ become if the origin is transferred to the point $\left(\frac{ab}{a-b}, 0\right)$?

□ Putting $x = X + \dfrac{ab}{a - b}$, $y = Y$ in the given equation we get

$$(a - b)\left[\left(X + \frac{ab}{a - b}\right)^2 + Y^2\right] - 2ab\left(X + \frac{ab}{a - b}\right) \Rightarrow X^2 + Y^2 = \left(\frac{ab}{a - b}\right)^2.$$

50. If $ax + by$ and $cx + dy$ are changed to $AX + BY$ and $CX + DY$ respectively for rotation of axes, show that $ad - bc = AD - BC$.

□ If the x, y axes are rotated through an angle θ then $X = x\cos\theta + y\sin\theta$ and $Y = -x\sin\theta + y\cos\theta$. Now

$$\begin{aligned} AX + BX &= A(x\cos\theta + y\sin\theta) + B(-x\sin\theta + y\cos\theta) \\ &= x(A\cos\theta - B\sin\theta) + y(A\sin\theta + B\cos\theta) = ax + by \\ \Rightarrow \quad a &= A\cos\theta - B\sin\theta \quad \text{and} \quad b = A\sin\theta + B\cos\theta. \end{aligned}$$

Similarly, $c = C\cos\theta - D\sin\theta$ and $d = C\sin\theta + D\cos\theta$. Now,

$$ad - bc = (A\cos\theta - B\sin\theta)(C\sin\theta + D\cos\theta) - (A\sin\theta + B\cos\theta)(C\cos\theta - D\sin\theta)$$
$$= AD - BC.$$

51. What type of curve represented by $2x^2 + 4xy + y^2 = 0$?

☐ **Being a secord degree homogeneous equation in x, y it represents a pair of straight lines.**

52. Find the polar coordinates of the centre of the circle $r = 12\cos\theta + 5\sin\theta$.

☐ $r = 13\left(\frac{12}{13}\cos\theta + \frac{5}{13}\sin\theta\right) = 13\cos(\theta - \alpha)$ where $\frac{12}{13} = \cos\alpha$ and $\frac{5}{13} = \sin\alpha$.

Hence the coordinate of the centre is $\left(\frac{13}{2}, \alpha\right)$ where α is given by $\tan\alpha = 5/12$.

53. Write down the equation of parabola $y^2 = 8x$ in polar form.

☐ Putting $y = r\sin\theta$ and $x = 2 + r\cos\theta$ in $y^2 = 8x$ we get

$$r^2\sin^2\theta = 8(2 + r\cos\theta) \Rightarrow r^2 = (4 + r\cos\theta)^2 \Rightarrow \frac{4}{r} = 1 - \cos\theta.$$

54. Determine the nature of the conic $\frac{l}{r} = 4 - 5\cos\theta$. Find its vertex and length of the latus rectum.

☐ $\frac{l}{r} = 4 - 5\cos\theta \Rightarrow \frac{l}{4r} = 1 - \frac{5}{4}\cos\theta \Rightarrow e > 1 \Rightarrow$ **the conic is hyperbola. Whose** vertex is $\left(\frac{l}{9}, \pi\right)$ and L.R. $= 2\frac{l}{4} = \frac{l}{2}$.

55. Find a point on the conic $\frac{l}{r} = 1 - e\cos\theta$ having the least radius vector.

☐ $\frac{l}{r} = 1 - e\cos\theta \Rightarrow r = \frac{l}{1 - e\cos\theta}$ **will be least when $1 - e\cos\theta$ is maximum i.e.,** when $e\cos\theta$ is least i.e., **when $\theta = \pi$ and then** $r = \frac{l}{1+e}$. **So the required point is** $\left(\frac{l}{1+e}, \pi\right)$.

56. Find the point of intersection of the lines $r\cos(\theta - \alpha) = p$ and $r\cos(\theta - \beta) = p$.

☐ At the point of intersection $\cos(\theta - \alpha) = \cos(\theta - \beta) \Rightarrow (\theta - \alpha) = \pm(\theta - \beta) \Rightarrow \theta = \frac{\alpha + \beta}{2}$. Putting this value in $r\cos(\theta - \alpha) = p$, we get $r = \frac{p}{\cos(\theta - \alpha)} = p\sec\left(\frac{\alpha - \beta}{2}\right)$.

So the point of intersection is

$$\left(p\sec\left(\frac{\alpha - \beta}{2}\right), \frac{\alpha + \beta}{2}\right)$$

57. If a circle passes through the pole and has centre at (a, α), show that the equation of the circle is $r = 2a\cos(\theta - \alpha)$.

☐ In polar coordinates, the general equation of the circle is $\cos(\theta - \alpha) = \dfrac{r^2 + a^2 - CP^2}{2ar}$ where C: (a, α) and P: (r, θ). If the circle passes through the pole then $CP = a$. Hence equation of the required circle is $r = 2a\cos(\theta - \alpha)$.

58. Show that the equation $r^2 \cos\theta - ar\cos 2\theta - 2a^2\cos\theta = 0$ represents a circle and a straight line.

☐
$$\begin{aligned} & r^2\cos\theta - ar\cos 2\theta - 2a^2\cos\theta = 0 \\ \Rightarrow\quad & r^2\cos\theta - ar(2\cos^2\theta - 1) - 2a^2\cos\theta = 0 \\ \Rightarrow\quad & r^2\cos\theta - 2ar\cos^2\theta + ar - 2a^2\cos\theta = 0 \\ \Rightarrow\quad & (r - 2a\cos\theta)(r\cos\theta + a) = 0. \end{aligned}$$

Hence, the given equation represents a circle: $r = 2a\cos\theta$ and a straight line: $r\cos\theta + a = 0$.

59. What does the equation $x^2 + y^2 = 4$ represent in 3-dimensional geometry?

☐ $x^2 + y^2 = 4$ and z can have any value i.e., the given equation represents a right circular cylinder whose axis is z-axis.

60. A variable point moves in such a way that the sum of the reciprocals of the intercepts made by the plane on the axes is constant. Prove that the plane passes through a fixed point.

☐ Let the variable plane is $\frac{x}{a} + \frac{y}{b} + \frac{z}{c} = 1$ where it is given that $\frac{1}{a} + \frac{1}{b} + \frac{1}{c} = \frac{1}{k}$ (constant) or $\frac{k}{a} + \frac{k}{b} + \frac{k}{c} = 1$. Comparing this result with the equation of the plane we see that the plane passes through the fixed point (k, k, k).

61. What type of cone represented by the equation $ayz + bzx + cxy = 0$?

☐ The equation $ayz + bzx + cxy = 0$ is a second degree homogeneous equation in x, y, z in which coeff. of x^2 + coeff. of y^2 + coeff. of $z^2 = 0$, so it represents a cone whose vertex is $(0, 0, 0)$ and which touches the coordinate axes.

62. What does the equation $9x^2 + 24xy + 16y^2 = 16$ represent in 3D geometry?

☐ $9x^2 + 24xy + 16y^2 = 16 \Rightarrow af^2 + bg^2 + ch^2 - abc - 2fgh = 0$ and $h^2 = ab$, so the given equation represents a pair of parallel planes.

63. What does the equation $x^2 = 4y$ represent in 3D geometry?

☐ $x^2 = 4y$ represents a parabola in 2D. As z can have any value, so this represents a parabolic cylinder in 3D.

64. Show that through the line $3x - 4y - 8 = 0 = y - 3z + 2$ only one tangent plane can be drawn to the sphere $x^2 + y^2 + z^2 - 2x + 6y + 2z + 8 = 0$.

☐ Equation of the plane through the line $3x - 4y - 8 + k(y - 3z + 2) = 0$ or,

$$3x + (k - 4)y - 3kz + 2k - 8 = 0.$$

This plane will touch the sphere if the length of the perpendicular from the centre of it to the plane is equal to the radius of the sphere, i.e., if

$$\frac{3(1) + (k - 4)(-3) - 3k(-1) + 2k - 8}{\sqrt{9 + (k - 4)^2 + 9k^2}} = \sqrt{1 + 9 + 1 - 8}$$

or, $\quad (2k + 7)^2 = 3(10k^2 - 8k + 25)$

or, $\quad k^2 - 2k + 1 = 0 \Rightarrow k = 1.$

This shows that only one tangent plane can be drawn to the sphere.

65. Find the value of k for which the plane $x + kz = 2$ intersects $x^2 + y^2 - z^2 + 1 = 0$ in a hyperbola.

☐ Putting $x = 2 - kz$ in $x^2 + y^2 - z^2 + 1 = 0$ we get

$$(2 - kz)^2 + y^2 - z^2 + 1 = 0$$

or, $\quad y^2 - (1 - k^2)z^2 - 4kz + 5 = 0.$

This will be a hyperbola in y-z plane if $-(1 - k^2) < 0$ i.e., if $-1 < k < 1$.

66. Find the equation of a circle through the points $(a, 0, 0)$, $(0, b, 0)$ and $(0, 0, c)$.

☐ Equation of the sphere passing through $(0, 0, 0)$, $(a, 0, 0)$, $(0, b, 0)$ and $(0, 0, c)$ is

$$x^2 + y^2 + z^2 - ax - by - cz = 0$$

and equation of the plane through $(a, 0, 0)$, $(0, b, 0)$ and $(0, 0, c)$ is $\frac{x}{a} + \frac{y}{b} + \frac{z}{c} = 1$. Obviously the plane will cut the sphere in a circle which will pass through $(a, 0, 0)$, $(0, b, 0)$ and $(0, 0, c)$. Hence the equation of circle through $(a, 0, 0)$, $(0, b, 0)$ and $(0, 0, c)$ is

$$x^2 + y^2 + z^2 - ax - by - cz = 0; \quad \frac{x}{a} + \frac{y}{b} + \frac{z}{c} = 1.$$

67. Find the distance of the point $(3, -4, 5)$ from the plane $2x + 5y - 6z = 16$ measured parallel to a line with direction ratios $(2, 1, -2)$.

☐ Equation of the line through $(3, -4, 5)$ and having d.r.^{s} $(2, 1, -2)$ is $\frac{x - 3}{2} = \frac{y + 4}{1} = \frac{z - 5}{-2}$. If it intersects the plane $2x + 5y - 6z = 16$ then for some value of r, $(3 + 2r, -4 + r, 5 - 2r)$ will lie on the plane. Which gives that $r = 20/7$. Hence the required distance $= 3r = 60/7$.

68. Are the family of lines $x + 1 = \mu y = -(\mu + 1)z$ generators of the surface

$$yz + xz + xy + y + z = 0?$$

□ $x+1=\mu y=-(\mu+1)z=r \Rightarrow \left(r-1, \frac{r}{\mu}, \frac{-r}{\mu+1}\right)$ is a point on the line. Putting this point in $yz+xz+xy+y+z=0$ we see that the point satisfies the equation for all values of r. So the given family of lines generates the curve $yz+xz+xy+y+z=0$.

69. Find the condition that the plane $lx+my+nz=p$ may touch the sphere $x^2+y^2+z^2=a^2$.

□ We know the length of the perpendicular from the centre of the sphere to the tangent plane $lx+my+nz=p$ is equal to the radius of the sphere.

$$\text{i.e.,}\qquad \frac{-p}{\sqrt{l^2+m^2+n^2}}=a \Rightarrow a^2(l^2+m^2+n^2)=p^2.$$

70. Do the lines $\frac{x-1}{2}=\frac{y+1}{-3}=\frac{z+10}{8}$ and $\frac{x-4}{1}=\frac{y+3}{-4}=\frac{z+1}{7}$ intersect?

□ Obviously the lines are not parallel, also we see that lines are coplanar and hence they must be intersecting.

71. Find the image of the point $(-3, 8, 4)$ in the plane $6x-3y-2z+1=0$.

□ Equation of the line perpendicular to the plane $6x-3y-2z+1=0$ and passing through $(-3, 8, 4)$ is $\frac{x+3}{6}=\frac{y-8}{-3}=\frac{z-4}{-2}=r$.

If $(-3+6r, 8-3r, 4-2r)$ be the image of $(-3, 8, 4)$, then $\left(\frac{-3+6r-3}{2}, \frac{8-3r+8}{2}, \frac{4-2r+4}{2}\right)$ will be a point on the plane $6x-3y-2z+1=0$. This gives $r=2$. So, the image is $(9, 2, 0)$.

72. Find the equation of the surface obtained by revolving the line $y-mx=0, z=0$ about x-axis.

□ The axis of revolution is the x-axis and the equation of generatrix is $y-mx=0, z=0$. Hence, the equation of surface of revolution is

$$\sqrt{y^2+z^2}=mx$$
$$\Rightarrow \quad y^2+z^2-m^2x^2.$$

73. Find a generator of the cone $5yz-8zx+3xy=0$.

□ In this case coeff. of x^2 + coeff. of y^2 + coeff. of $z^2=0$ and the equation is a second degree homogeneous equation. So the cone has the point $(0, 0, 0)$ as vertex and has three mutually perpendicular generators. Hence all the coordinate axes are generators and one of the generator is $\frac{x}{1}=\frac{y}{0}=\frac{z}{0}$.

74. Find the equation of sphere which makes equal intercepts on the axes and has least possible radius.

☐ Let the intercepted length on the axes be k. If we take the equation of sphere as

$$(x-a)^2+(y-b)^2+(z-c)^2=r^2,$$

then it will pass through the points $(k,0,0)$; $(0,k,0)$ and $(0,0,k)$. Thus we get

$$(k-a)^2+b^2+c^2=r^2;$$
$$a^2+(k-b)^2+c^2=r^2$$
$$\text{and}\quad a^2+b^2+(k-c)^2=r^2.$$

The above equations give $a=b=c$. So

$$(k-a)^2+a^2+a^2=r^2$$

$$\text{or,}\quad 3\left[\left(a-\frac{k}{3}\right)^2+\frac{2k^2}{9}\right]=r^2$$

which gives the minimum value of $r^2=2k^2/3$ which occurs when $a=k/3$. Hence the equation of the sphere is

$$\left(x-\frac{k}{3}\right)^2+\left(y-\frac{k}{3}\right)^2+\left(z-\frac{k}{3}\right)^2=\frac{2k^2}{3}.$$

75. Find the equation of the quadric cone which passes through three coordinate axes as well as the two lines $\frac{x}{1}=\frac{y}{-2}=\frac{z}{3}$; $\frac{x}{3}=\frac{y}{-1}=\frac{z}{1}$.

☐ Equation of the cone through coordinate axes is $axy+byz+cxz=0$. If it has two generators $\frac{x}{1}=\frac{y}{-2}=\frac{z}{3}$ and $\frac{x}{3}=\frac{y}{-1}=\frac{z}{1}$ then we have $a(1)(-2)+b(-2)(3)+c(1)(3)=0$ and $a(3)(-1)+b(-1)(1)+c(3)(1)=0$. Solving we get $a=15$, $b=3$, $c=16$. So the equation of the cone is $15xy+3yz+16xz=0$.

76. Show that the line $\frac{x+2}{2}=\frac{y}{3}=\frac{z-1}{-2}$ is a generator of the cone $\frac{x^2}{4}-\frac{y^2}{9}=z$.

☐ Any point on the line $\frac{x+2}{2}=\frac{y}{3}=\frac{z-1}{-2}$ is $(2r-2, 3r, 1-2r)$. Now $\frac{x^2}{4}-\frac{y^2}{9}=z$ is satisfied by the point for all values of r. So, the given line is a generator of the surface $\frac{x^2}{4}-\frac{y^2}{9}=z$.

77. Find the equation of the plane which bisects the acute angle between the planes

$$3x-4y+12z=26;\quad -x-2y+2z+9=0.$$

☐ We write the equations of the planes with +ive constant terms as $-3x+4y-12z+26=0$; $-x-2y+2z+9=0$ and we see that $(-3)(-1)+4(-2)-12(2)=-29<0$. So the equation of the plane bisecting the acute angle between the plane is

$$\frac{-3x+4y-12z+26}{\sqrt{9+16+144}}=\frac{-x-2y+2z+9}{\sqrt{1+4+4}}$$

$$\Rightarrow\quad 3(-3x+4y-12z+26)=13(-x-2y+2z+9)$$

$$\Rightarrow\quad 4x+38y-62z=39.$$

78. Find the equation of the tangent to the hyperbolic paraboloid $5x^2 - 2y^2 = 2z$ parallel to $10x - 6y - z = 7$.

☐ Let the equation of the tangent plane is $10x - 6y - z = k$. If it touches the conicoid $5x^2 - 2y^2 = 2z$ then $\frac{10^2}{5} + \frac{(-6)^2}{-2} + \frac{2(-1)k}{1} = 0 \Rightarrow k = 1$.

79. Find the equation of the sphere which passes through the origin and touches the sphere $x^2 + y^2 + z^2 = 56$ at the point $(2, -4, 6)$.

☐ The required sphere passes through the origin i.e., centre of the given sphere and touches the sphere at the point $(2, -4, 6)$. So, $(0, 0, 0)$ and $(2, -4, 6)$ are end points of a diameter of the required sphere. Hence the equation of the sphere is

$$x(x-2) + y(y+4) + z(z-6) = 0.$$

80. Find the coordinates of a point equidistant from four points $(0, 0, 0)$, $(a, 0, 0)$, $(0, b, 0)$ and $(0, 0, c)$.

☐ We know that the equation of the sphere passing through four points $(0, 0, 0)$, $(a, 0, 0)$, $(0, b, 0)$ and $(0, 0, c)$ is $x^2 + y^2 + z^2 - ax - by - cz = 0$. So, centre of the sphere i.e., $(a/2, b/2, c/2)$ is equidistant from four points $(0, 0, 0)$, $(a, 0, 0)$, $(0, b, 0)$ and $(0, 0, c)$.

81. Verify that the plane $2x - 2y + 3z = 9$ touches the ellipsoid $x^2 + 2y^2 + 3z^2 = 9$.

☐ Here $l = 2, m = -2, n = 3$ and $a = 1/9, b = 2/9, c = 1/3$.
Now $\frac{l^2}{a} + \frac{m^2}{b} + \frac{n^2}{c} = 36 + 18 + 27 = 81 = p^2$. So the plane touches the ellipsoid.

82. Where does the plane $y - 2x = 0$ intersects the coordinate axes?

☐ The plane $y - 2x = 0$ is parallel to the z-axis as $0(-2) + 0(1) + 1(0) = 0$. So it does not meet z-axis. Again the plane passes through the origin. So it meets the x-axis and y-axis at $(0, 0, 0)$.

83. The foot of the perpendicular from the origin to the plane is $(2, 3, -1)$. Find the equation of the plane.

☐ Equation of the plane through $P(2, 3, -1)$ and perpendicular to OP, O being the origin, is $(2-0)(x-2) + (3-0)(y-3) + (-1-0)(z+1) = 0$ i.e., $2x + 3y - z = 14$.

EXERCISE

1. Find the angle of rotation of the axes for which the equation $x^2 - y^2 = a^2$ will reduce to $xy = c^2$.
2. Prove that polar of the point $(-a, 2a)$ with respect to $x^2 + y^2 - 2ax - 3a^2 = 0$ touches the parabola $y^2 = 4ax$.
3. Prove that if the polar of a point P passes through Q then the polar of Q will pass through P and if the pole of a line $L1$ lies on another line $L2$ then it's pole will lie on $L1$.

4. Find the locus of the middle points of the chords of the parabola $y^2 = 4ax$ passing through a fixed point.
5. Find the locus of the middle points of the chords of the hyperbola $\frac{x^2}{a^2} - \frac{y^2}{b^2} = 1$ passing through a fixed point.
6. The abscissas of two points A and B are the roots of the equation $x^2 + 2ax - b^2 = 0$ and the ordinates of those points are the roots of the equation $x^2 + 2px - q^2 = 0$. Find the equation of the circle describing AB as diameter.
7. Find the equation of directrix and the focus of the parabola $x^2 + 2(x + y) = 0$.

   [*Hints*: Rewrite the equation $x^2 + 2(x + y) = 0$ as $(x + 1)^2 = -2\left(y - \frac{1}{2}\right)$.]
8. (a) The straight line $\frac{x}{a} + \frac{y}{b} = 1$ cuts the axes at A and B. A line perpendicular to AB cuts the axes in P and Q. Find the locus of the point of intersection of AQ and BP.
   (b) A variable straight line through the point of intersection of $\frac{x}{a} + \frac{y}{b} = 1$ and $\frac{x}{b} + \frac{y}{a} = 1$ meets the coordinate axes at P and Q. Find the locus of the middle point of PQ.
   (c) A variable straight line AB meets x, y axes at A, B always passes through a fixed point (α, β). Find the locus of the middle point of AB.
9. The point (2, 3) is transformed to (3, 2) under a reflection about a line, find the equation of the line.
10. (a) Transform (3, 4) to the point (0, 5) by rotation.
    (b) If the expression $ax + by$ becomes $AX + BY$ under rotation, show that $a^2 + b^2$ is an invariant.
11. Find the polar equation of the straight line parallel to the initial line and at a distance 3 units from the pole.
12. Find the polar equation of the circle of radius 2 units with centre on the initial line and at a distance 2 units from the pole on the positive side.
13. Find the area of the triangle whose vertices are (1, 2, 3); (−2, 1, 5) and (3, 4, 2).
14. Are the lines $\frac{x-a}{a'} = \frac{y-b}{b'} = \frac{z-c}{c'}$ and $\frac{x-a'}{a} = \frac{y-b'}{b} = \frac{z-c'}{c}$ skew?
15. Find the shortest distance between the lines (i) $y = 0 = z$ and $y = 5, z = 2$; (ii) $x + y = 1, z = 0$ and $x = 0 = y$; (iii) $x = 0$; $y - 2z = 0$ and $x = 3, y = 0$.
16. Write down equation of the plane through the lines $x + 2 = 0 = z$ and $y = 0 = z$.
17. Show that the planes $2x + 3y - z = 0$; $3x + 3y + z = 2$ and $x - y + 2z = 5$ intersect in a point.
18. Find the equation of the plane through (1, 1, 1) and perpendicular to the straight line $3x + 2y - z - 4 = 0 = x - 2y - 2z - 5$.
19. Find the equation of the plane through the line $ax+by+cz+d = 0 = a'x+b'y+c'z+d'$ and perpendicular to x-y plane.
20. Determine the value of k such that the lines $\frac{x-1}{2} = \frac{y-4}{1} = \frac{z-2}{2}$ and $\frac{x-2}{-1} = \frac{y-8}{k} = \frac{z-8}{4}$ may intersect.

    [*Hints*: If the given lines intersect then they must be coplanar. Use the condition of coplanarity.]
21. Are the lines $\frac{x-1}{3} = \frac{y-2}{3} = \frac{z-3}{4}$ and $4x - 3y + 1 = 0 = 5x - 3z + 2$ skew?

22. Find the projection of the line joining (1, 2, 3) and (4, 2, 1) on the line

$$\frac{x-1}{3}=\frac{y-2}{4}=\frac{z-3}{5}.$$

23. Find the centre of the circle represented by $x^2+y^2+z^2=a^2$; $x+y=z$.
24. How many tangent planes can be drawn to the sphere $x^2+y^2+z^2=1$ through the line $3x-6y+5=0=2x+3z+1$?
25. Find the equation of the circle through the points (2, 0, 1); (−2, 1, 0) and (0, 3, 5).
26. Find the equation of the sphere through the circle $x^2+y^2+z^2=9$; $2x+3y+4z=5$ and the point (1, 2, 3).
27. If any tangent plane to the sphere $x^2+y^2+z^2=k^2$ intercepts lengths a, b, c on coordinate axes then show that $a^{-2}+b^{-2}+c^{-2}=k^{-2}$.
28. Find the equation of the tangent plane to the sphere $x^2+y^2+z^2-2x+6y+2z+8=0$ which passes through the line $3x-4y-8=0=y-3z+2$.
29. Find the bisector of the angle between the planes $3x-6y+2z+5=0$ and $4x-12y+3z-3=0$, which contains the origin.
30. Find the point where the line joining (2, −3, 1) and (3, −4, −5) cuts the plane $2x+y+z=7$.
31. Find the locus of the point whose distance from y-axis is 3 and identify it.
[*Hints*: In 3D distance of a point (x, y, z) from y-axis is $\sqrt{x^2+z^2}=3$ (given). So the locus of the point is $x^2+z^2=9$ and y can have any value, hence it represents a circular cylinder whose axis is y. In 2D distance of a point (x, y) from y axis is $x=3$ which is a straight line parallel to y axis.]
32. Prove that the angle between two diagonals of a cube is $\cos^{-1}(1/3)$.
33. A line in the XOY plane makes 30° with x-axis, find the d.cs of the line.

[*Hints*: The line makes 30° with x-axis and lies in XOY plane, so it makes 60° with y-axis and 90° with z-axis. So the d.cs of the line are $\sqrt{3}/2, 1/2, 0$.]
34. If a right circular cone has three mutually perpendicular generators, find its semi-vertical angle.
35. If a right circular cone has three mutually perpendicular tangent planes, find its semi vertical angle.
36. Find the condition that the planes $x=ry+qz$, $y=pz+rx$ and $z=qx+py$ pass through a line.

[*Hints*: The necessary condition that the planes pass through a line is that the coefficient determinant vanishes and other condition is automatically satisfied in this case as there is no constant term. Now calculate the value of coefficient determinant and equate it to zero to get the required condition.]
37. Find the equation of the line $8x+12y-13z-32=0=4x+4y-5z-12$ in symmetrical form.
38. Find the equation of the line which passes through (1, 1, 1) and is parallel to the line $8x+12y-13z-32=0=4x+4y-5z-12$.

[*Hints*: If the direction ratios of the line $8x+12y-13z-32=0=4x+4y-5z-12$ are a, b, c then we have $8a+12b-13c=0$ and $4a+4b-5c=0$. Solving we get $a=2$, $b=3$, $c=4$. So equation of the required line is

$$\frac{x-1}{2}=\frac{y-1}{3}=\frac{z-1}{4}.]$$

39. Find the equation of the plane through the line $8x + 12y - 13z - 32 = 0 = 4x + 4y - 5z - 12$ and passes through origin.
40. Find the area of the triangle made by the plane $\frac{x}{a} + \frac{y}{b} + \frac{z}{c} = 1$ with the coordinate axes.

    [*Hints*: The given plane intersect the coordinate axes at $(a, 0, 0)$, $(0, b, 0)$, $(0, 0, c)$. Now the area of the required triangle is given by $\Delta^2 = \Delta_x^2 + \Delta_y^2 + \Delta_z^2$, where Δx, Δy, Δz are the projections of Δ on the plane $x = 0$, $y = 0$ and $z = 0$ respectively.]
41. Find the equation of the cone whose vertex is origin and base is $x = a, z^2 + y^2 = b^2$.

    [*Hints*: Making homogeneous $z^2 + y^2 = b^2$ with the help of $x = a$ we get $z^2 + y^2 = b^2\left(\frac{x}{a}\right)^2$ i.e., $a^2(z^2 + y^2) = b^2x^2$, which is the required equation of the cone.]
42. Find the angles (in degrees) that the line $\frac{x-1}{1} = \frac{y-1}{1} = \frac{z-1}{0}$ makes with the axes.

    [*Hints*: Direction cosines of the given straight lines are $\left(\frac{1}{\sqrt{2}}, \frac{1}{\sqrt{2}}, 0\right)$ and those of co-ordinate axes are respectively $(1, 0, 0)$, $(0, 1, 0)$ and $(0, 0, 1)$ respectively. So if the line make angles α, β, γ with the coordinate axes respectively then $\cos\alpha = \frac{1}{\sqrt{2}}$, $\cos\beta = \frac{1}{\sqrt{2}}$ and $\cos\gamma = 0$ i.e., the line makes 45° with x-axis, 45° with y-axis and 90° with z-axis.]
43. Find the image of the point $(1, 3, 4)$ in the plane $2x - y + z + 3 = 0$.
44. In a triangle ABC, the coordinates of A are $(1, 2)$ and the equation of the medians through B, C are $x + y = 2$, $x = 4$ respectively. Find coordinates of B and C.
45. Find the projection of the line segment joining $(1, 2, 3)$ and $(3, -1, 5)$ on the straight line joining $(2, 1, 2)$ and $(4, 3, 1)$.

    [*Hints*: Direction cosines of the line joining $(2, 1, 2)$ and $(4, 3, 1)$ are $\frac{2}{3}, \frac{2}{3}, \frac{-1}{3}$. So the projection of the line segment joining $(1, 2, 3)$ and $(3, -1, 5)$ is $(3-1)\left(\frac{2}{3}\right) + (-1-2)\left(\frac{2}{3}\right) + (5-3)\left(-\frac{1}{3}\right) = \frac{4}{3}$ units.]

Appendix

An Important Chart

	Circle	Ellipse	Parabola	Hyperbola	Rectangular hyperbola
Standard Equation	$x^2 + y^2 = a^2$	$\frac{x^2}{a^2} + \frac{y^2}{b^2} = 1$	$y^2 = 4ax$	$\frac{x^2}{a^2} - \frac{y^2}{b^2} = 1$	$xy = c^2$
Eccentricity	$e = 0$	$e < 1$	$e = 1$	$e > 1$	$e = \sqrt{2}$
$y = mx + k$ will touch it, if	$k = \pm a\sqrt{1 + m^2}$	$k = \sqrt{a^2m^2 + b^2}$	$k = \frac{a}{m}$	$k = \sqrt{a^2m^2 - b^2}$	$k = 2c\sqrt{-m}$
Equation of tangent at (h, k)	$xh + yk = a^2$	$\frac{xh}{a^2} + \frac{yk}{b^2} = 1$	$yk = 2a(x + h)$	$\frac{xh}{a^2} - \frac{yk}{b^2} = 1$	$xk + yh = 2c^2$
Polar of (h, k) w.r.t.	Do	Do	Do	Do	Do
Chord of contact of (h, k) w.r.t.	Do	Do	Do	Do	Do
Equation of chord whose middle term is (h, k)	$xh + yk = h^2 + k^2$	$\frac{xh}{a^2} + \frac{yk}{b^2} = \frac{h^2}{a^2} + \frac{k^2}{b^2}$	$yk - 2ax = k^2 - 2ah$	$\frac{xh}{a^2} - \frac{yk}{b^2} = \frac{h^2}{a^2} - \frac{k^2}{b^2}$	$xk + yh = 2kh$
Second degree general equation in (x, y) will be, if	$a = b$, $h = 0$	$\Delta \neq 0$, $h^2 < ab$	$\Delta \neq 0$, $h^2 = ab$	$\Delta \neq 0$, $h^2 > ab$	$\Delta \neq 0$, $a + b = 0$
Parametric equation	$x = a\cos\theta$, $y = a\sin\theta$	$x = a\cos\theta$, $y = b\sin\theta$	$x = at^2$, $y = 2at$	$x = a\sec\theta$, $y = b\tan\theta$	$x = ct$, $y = c/t$